Radiation Technologies and Applications in Materials Science

This book explains various kinds of non-ionizing and high-energy radiations, their interaction with materials and chemical reactions, and conditions of various kinds of materials development technologies including applications. It covers a processing-structure-property relationship and radiations used in developing many advanced materials used in various fields. It highlights application-oriented materials synthesis and modification covering a wide variety of materials such as plastics, rubber, thermo-set, ceramics, and so forth by various radiations.

Features:

- Explains ionizing and non-ionizing radiation-assisted materials development technologies, for polymers, ceramics, metals, and carbons.
- Covers radiation-assisted synthesis, processing, and modification of all kinds of materials.
- Provides comparative studies, merits, demerits, and applications very systematically.
- Criss-crosses polymers science and technology, radiation technology, advanced materials technology, biomaterials technology, and so forth.
- Includes a section on 3D printing by LASER melting of CoCr alloys.

This book is aimed at researchers and graduate students in materials science, radiation chemistry and physics, and polymer and other materials processing.

Radiation Technologies and Applications in Materials Science

Edited by
Subhendu Ray Chowdhury

CRC Press
Taylor & Francis Group
Boca Raton London New York

CRC Press is an imprint of the
Taylor & Francis Group, an **informa** business

First edition published 2023
by CRC Press
6000 Broken Sound Parkway NW, Suite 300, Boca Raton, FL 33487-2742

and by CRC Press
4 Park Square, Milton Park, Abingdon, Oxon, OX14 4RN

CRC Press is an imprint of Taylor & Francis Group, LLC

ISBN: 978-1-032-34394-5 (hbk)
ISBN: 978-1-032-34395-2 (pbk)
ISBN: 978-1-003-32191-0 (ebk)

DOI: 10.1201/9781003321910

Dedication

I dedicate this book to the memory of my parents.

Contents

About the Editor

Dr. Subhendu Ray Chowdhury currently is a senior scientific officer in Bhabha Atomic Research Centre (BARC), Mumbai, India. He is also a visiting faculty of the Institute of Chemical Technology (ICT) [formerly UDCT], Matunga, Mumbai, India. Dr. Ray Chowdhury received his PhD in materials science from the Indian Institute of Technology (IIT), Kharagpur, India. After completion of his PhD, Dr. Ray Chowdhury worked for seven years in renown laboratories in the USA, including the Department of Materials Science and Engineering, Cornell University, and Department of Materials Science and Engineering, Pennsylvania State University. He has published around 70 journal papers, a couple of book chapters, and more than 50 conference papers in radiation-assisted materials development through processing, modification, and synthesis. He has supervised several PhD and post-graduate students. He obtained a couple of patents and transferred many technologies to Industries. His accolades and achievements include National Award for Technology Innovation 2019, by the Ministry of Chemicals and Fertilizers, Govt of India, CSIR research fellowship through NET (National Eligibility Test), and national advisory committee membership of international conferences, etc.

He was involved in a RCA/UNDP project on "Electron Beam Applications in Materials Science" in 2013, as an international expert committee member from India.

He also served as a category A, international faculty on materials science and electron beam for environmental remedy (RCA/UNDP PROJECT) in July 2017, Republic of Korea (South Korea).

He is a regular reviewer of more than 25 reputed international journals. As an experienced researcher, he delivers talks as a keynote speaker and invited speaker, and chairs many sessions at international and national conferences.

Contributors

Sayem Mohammad Abu
Institute of Power Engineering
UniversitiTenagaNasional
Kajang, Selangor, Malaysia

M.N. M. Ansari
Department of Mechanical Engineering
Universiti Tenaga + National
Kajang, Selangor, Malaysia

Pratiksha Awasthi
Department of Materials Science
    and Engineering
Indian Institute of Technology Delhi
New Delhi, India

Abhijit Bandyopadhyay
Department of Polymer Science & Technology
University of Calcutta
Kolkata, India

Shib Shankar Banerjee
Department of Materials Science and
    Engineering
Indian Institute of Technology Delhi
New Delhi, India

A. Bhattacharya
Academy of Scientific and Innovative Research,
    Council of Scientific and Industrial Research,
    Human Resource Development Centre
    Campus
Ghaziabad, Uttar Pradesh, India

Jayashree Biswal
Isotope and Radiation Application Division
Bhabha Atomic Research Centre
Trombay, Mumbai, India

Rishi Pal Chahal
Department of Physics
ChaudharyBansiLal University
Bhiwani, Haryana, India

Subhadeep Chakraborty
Department of Polymer Science & Technology
University of Calcutta
Kolkata, India

A. Wayne Cooke
Mitsubishi Chemical America,
    Performance Polymers Division
Greer-South Carolina, USA

Sumit Dokwal
Department of Biochemistry
Pt. B. D. Sharma PGIMS
Rohtak, India

Luke Dowling
Department of Mechanical and
    Manufacturing Engineering, Trinity College
The University of Dublin
Dublin, Ireland

Bharath Govind
Amity Institute of Applied Sciences
Amity University
Uttar Pradesh, Noida, India

N. L. Ishak
Radiation Processing Technology Division,
    Malaysia Nuclear Agency
Bangi, Kajang, Malaysia

Suman B. Kuhar
Department of Physics
IHL, BPSMV
Khanpur Kalan, India

Shyam Kumar
Department of Physics
Kurukshetra University
Kurukshetra, Haryana, India

Suman Mahendia
Department of Physics
Kurukshetra University
Kurukshetra, Haryana, India

S. U. Mestry
Department of Polymer & Surface Engineering
Institute of Chemical Technology (ICT)
Matunga, Mumbai, India

S. T. Mhaske
Department of Polymer & Surface Engineering
Institute of Chemical Technology (ICT)
Matunga, Mumbai, India

Safiya Nisar
Amity Institute of Applied Sciences
Amity University
Uttar Pradesh, Noida, India

Dr. Nor Azillah Fatimah Othman
Radiation Processing Technology Division
Malaysia Nuclear Agency
Bangi, Kajang, Malaysia

Tuhin Subhra Pal
Rubber Technology Centre
Indian Institute of Technology
Kharagpur, West Bengal, India

H. J. Pant
Isotope and Radiation Application Division
Bhabha Atomic Research Centre
Trombay, Mumbai, India

D. A. Patil
Department of Polymer & Surface Engineering
Institute of Chemical Technology (ICT)
Matunga, Mumbai, India

A. K. Rana
Department of Chemistry
Sri Sai University
Palampur, India

Sunita Rattan
Amity Institute of Applied Sciences
Amity University
Uttar Pradesh, Noida, India

Sudip Ray
New Zealand Institute for Minerals to Materials
    Research
Greymouth, New Zealand

Aiswarya S.
Department of Materials Science and
    Engineering
Indian Institute of Technology Delhi
New Delhi, India

J. S. Saini
Department of Mechanical Engineering
Thapar Institute of Engineering and
    Technology
Patiala, Punjab, India

Subhan Salaeh
Department of Rubber Technology and
    Polymer Science
Faculty of Science and Technology
Prince of Songkla University
Pattani Campus, Pattani, Thailand

Soumen Sardar
Department of Polymer Science & Technology
University of Calcutta
Kolkata, India

Mayank Saxena
Membrane Science and Separation Technology
    Division
Council of Scientific and Industrial Research
Central Salt and Marine Chemicals Research
    Institute (CSIR-CSMCRI)
Bhavnagar, Gujarat, India

S. Selambakkannu
Radiation Processing Technology Division
    Malaysia Nuclear Agency
Bangi, Kajang, Malaysia

Vishal Sharma
Institute of Forensic Science & Criminology
Panjab University
Chandigarh, India

V. K. Sharma
Isotope and Radiation Application Division
Bhabha Atomic Research Centre
Trombay Mumbai, India

Nischay Kodihalli Shivaprakash
Mitsubishi Chemical America
Performance Polymers Division
Greer-South Carolina, USA

Sachin Singh
Department of Mechanical Engineering
Thapar Institute of Engineering and
    Technology
Patiala, Punjab, India

A. S. Singha
Department of Chemistry
National Institute of Technology
Hamirpur, India

Nikhil K. Singha
Rubber Technology Centre
Indian Institute of Technology
Kharagpur, West Bengal, India

Prachi Singhal
Amity Institute of Applied Sciences
Amity University
Uttar Pradesh, Noida, India

Anu Surendran
International and Inter University Centre for
    Nanoscience and Nanotechnology
Mahatma Gandhi University
Kottayam, Kerala, India

Sabu Thomas
International and Inter University Centre for
    Nanoscience and Nanotechnology
Mahatma Gandhi University
Kottayam, Kerala, India

School of Chemical Sciences
Mahatma Gandhi University
Kottayam, Kerala, India

School of Energy Materials
Mahatma Gandhi University
Kottayam, Kerala, India

T. M. Ting
Radiation Processing Technology Division
Malaysia Nuclear Agency
Bangi, Kajang, Malaysia

Daniel Trimble
Department of Mechanical and Manufacturing
    Engineering, Trinity College
The University of Dublin
Dublin, Ireland

Poornima Vijayan P.
Department of Chemistry
Sree Narayana College for Women
    (affiliated to University of Kerala)
Kollam, Kerala

Foreword

I am very happy to introduce this timely publication relevant to researchers working in the field of radiation-assisted materials technology. For the last few decades, researchers around the world have been working in the field of radiation-assisted materials development using different kinds of radiations that include nonionizing electromagnetic radiation in addition to ionizing electromagnetic radiation and accelerated particle-based beams. The materials are primarily polymers extending to ceramics as well as metals. This compilation brings together myriad applications of a variety of radiations in a wide spectrum of materials highlighting the interdisciplinary nature of this subject.

This book highlights the materials technologies utilizing a wide range of radiations, namely UV, microwave, gamma, electron beam, and even laser. The list of reported materials, which have been subjected to synthesis, modification, and processing by radiation technologies includes ceramics, plastic, rubber, thermoset polymer, cellulosic material, carbon nanotube, inorganic nano-particle, metals and alloys, activated carbon, and biomass as well as coal. Radiation can form covalent bonds or break covalent bonds. The new bond formation in polymers leads to crosslinking, grafting, polymerization, and curing. On the other hand, breaking the bond of polymers can lead to degradation and help in recycling. The detail processing and mechanism of the above processes are discussed not only for a single polymeric system but also for polymeric multicomponent systems such as polymer blends and composites with wide and potentially new applications. Some recent challenges for developing materials for modern applications such as fuel cell, membrane, biomaterials, and activated carbon have been taken up by radiation technology, the details of which are discussed in this book. Another interesting aspect discussed in this book is the potential of radiations in new and emerging fields such as coal and biomass conversion into activated carbon as well as synthesis of nanoparticles for biomedical applications. Briefly, this book focuses on radiation-based material development and challenges, applications, and commercialization potential.

The information in various chapters in this book will promote and guide further progress in the field of radiation-assisted materials technology benefiting modern society. The present volume will be useful to students, researchers, and professionals associated with chemical sciences, radiation technology, materials sciences, and engineering.

Dr Ajit Kumar Mohanty
Director
Bhabha Atomic Research Centre
Trombay, Mumbai, India

Acknowledgement

It is my privilege and pleasure to acknowledge Dr. Pradeep Kumar Pujari, former director of Radio Chemistry and Isotope Group (RC & IG), Bhabha Atomic Research Centre, Mumbai, India for his continuous encouragement and support for compiling such a volume.

Editor's preface

In the last couple of decades, radiation technology has been a paragon in the processing of various materials in an expansive mode. Because of their uniqueness, non-ionizing (UV, microwave) and ionizing radiations (high-energy x-ray, gamma, electron beam, etc.) are drawing interests of researchers and are being deployed for developing a wide range of materials (polymers, ceramics, metals, advanced materials, multi-component hybrid systems, etc.) with target properties. The radiation-assisted development route includes new synthesis, processing, and modification of materials. Different radiations interact with materials following different fashions. Depending on the type of radiation and class of materials, the criteria of processing and synthesis of new materials are different. For example, the requirement of microwave radiation to interact with material for attaining the chemical reaction activation energy is polarity of the molecule. On the other hand, UV needs some extra reagent, which works as an initiator of the reaction by generating free radicals. Again, high-energy radiation (gamma, x-ray, electron beam, etc.) can ionize a wide range of non-polar and polar materials, leading to the generation of free radicals and/or ions at room temperature without an initiator. However, due to these differences, materials technologies, which deal with processing-structure properties, vary material to material and radiation to radiation. So far, there is no book available that provides comprehensive information on the diverse materials technologies by wide varieties of radiations.

Therefore, a suitable compilation, which deals with the fundamentals of various kinds of non-ionizing and high-energy radiations, different kinds of interactions with different kinds of materials and chemical reactions, conditions of various kinds of materials technologies, and then last but not the least is applications, is highly required for researchers, academicians, and industries, who are involved in this scientific research field. As the material paradigm is the understanding of the processing-structure property relationship, radiations are being used in developing many advanced materials in the field of nanotechnology, biomaterial, energy materials, and membranes and people are trying to establish a process-structure property relationship. Therefore, the aim of this book is to gather experiences and knowledge through chapters of dedicated researchers in wide varieties of radiations and intriguing materials. This book is intended to give a broad overview on the fundamentals of interactions, mechanisms, and merits and demerits of various radiations. It highlights application-oriented materials synthesis and modifications covering a wide variety of materials such as plastics, rubbers, thermo-sets, ceramics, metals, carbon nano-tubes, activated carbon, composites, blends, alloys, etc. This book also successfully highlights the opportunities and challenges for materials development and commercialization. Therefore, I hope the contents of this book will provoke new thoughts to envision more possibilities for developing more materials, which will be applied to society. I expect that the contributors' and editor's cumulative efforts will be useful for students, researchers, and industry people.

Subhendu Ray Chowdhury
Mumbai, India

1 Cross-linking of polymers by various radiations: Mechanisms and parameters

S. T. Mhaske*, S. U. Mestry, and D. A. Patil

Department of Polymer & Surface Engineering, Institute of Chemical Technology (ICT), Matunga, Mumbai, India

*Corresponding author: st.mhaske@ictmumbai.edu.in

CONTENTS

1.1   INTRODUCTION

As the name suggests, "cross-linking" in polymer terminology refers to the interlocking process of innumerable polymer chains at discrete points to emerge as a single macromolecule. The polymer molecules can be interlinked together through physical interaction (physical cross-linking) or by chemical reaction (chemical cross-linking). In the physical type of cross-linking, the secondary forces such as hydrogen bonding, steric hindrance, van der Waals forces, ionic affinity, or dative bonding play an essential role in building characteristic behaviors such as thermoplastic or thixotropic. While, in the chemical cross-linking, because of the covalent bond formation, the thermal and mechanical stability of the polymeric system increases along with the barrier properties.

DOI: 10.1201/9781003321910-1

Furthermore, chemical cross-linking restricts the reversibility of polymeric systems, and hence the products obtained are of a thermosetting type.

The degree of cross-linking segregates the polymeric types into different kinds of applications. For example, the polymer with a small number of cross-links has a suitable swelling property which can find applications in gel materials. It is very well known that the gelscan absorb and preserve a large amount of solvent. Therefore, they are applied to sanitary napkins, disposable diapers, soft contact lenses, water retention materials. On the other hand, chemically cross-linked polymers with high cross-link density hardly swell in solvents and can be used as materials with good thermal and mechanical properties and high chemical resistance. Illustrative examples of such polymers are thermosetting resins. They are cross-linked by heating and the resulting highly cross-linked network polymers are utilized as constructional, electric, and electronic materials. Highly rigid structures result from a high degree of cross-linkages. Cross-linking can also be achieved by copolymerizing functional monomers or by chemical reaction of the reactive substituent group.

Cross-linking in plastics aims to enhance thermal resistance, and in rubber, it is to create elastic properties. The cross-linked structure prevents the slippage of molecular chains in the amorphous region. Accordingly, the elongation at break decreases by cross-linking. The tensile strength of rubbers increases notably with dose but decreases after reaching the maximum value. Cross-linking makes sticky and weak rubbers transform into non-sticky and strong rubbers.

Synthetic methods for chemically cross-linked polymers are classified into cross-linking during polymerization and post-cross-linking polymer chains. In cross-linking during polymerization, chain-growth and step-growth polymerization are utilized to construct cross-linked polymers. Chain-growth polymerization of a monomer with one polymerizable group (e.g., styrene) in the presence of that with two or more polymerizable functional groups (e.g., divinylbenzene) gives the corresponding cross-linked polymer. On the other hand, cross-linked polymers in step-growth polymerization can be formed by using a monomer with three or more functional groups as a component of monomers. Post-cross-linking between polymer chains is usually carried out by the reaction between reactive groups on the polymer chains and a cross-linker having two or more reactive groups. Cross-linking reactions are typically promoted by heating or photoirradiation, though the addition of catalysts and the irradiation of radioactive rays are also used to promote cross-linking.

Furthermore, the radiation cross-linked plastic can be well understood from the example of the most common plastic, i.e., polyvinyl chloride (PVC). It is a high T_g amorphous polymer. The polyfunctional monomer is added to PVC for radiation cross-linking. The dose-dependency of tensile strength and elongation at break for PVC is the same for amorphous rubbers. The cross-linked PVC for insulation had improved hot elongation properties under elevated temperatures. The volume resistivity increases with an increased dose. Volume resistivity of polymeric insulation depends mainly on low molecular weight additives. Increased cross-linking restricts the mobility of common molecular weight additives.

Engineering plastic is defined as a plastic that exhibits higher heat resistance temperature and tensile strength. Radiation cross-linking of engineering plastics such as polybutylene terephthalate (PET) and polyamide have been commercialized. Radiation cross-linking of super engineering plastic is challenging due to aromatic rings in the main chain. High-temperature irradiation is effective for polyethersulfone, polysulfone, and polytetrafluoroethylene (PTFE). For PTFE, permanent creep decreases with increasing cross-linking density. The transparency is improved due to low crystallinity. The electrical properties such as volume resistance, dielectric loss, and dielectric constant retain are not changed by radiation cross-linking.

1.2 INSIGHTS ON THE RADIATION SOURCES

Radiant energy is one of the most effective forms of energy available to humankind. Sunlight and radioactive elements are radiation sources, potentially for various medical and industrial applications. Utilizing radiation processing should provide more advantages in comparison to chemical modification. The benefits of radiation processing include less sensitivity to moisture, higher

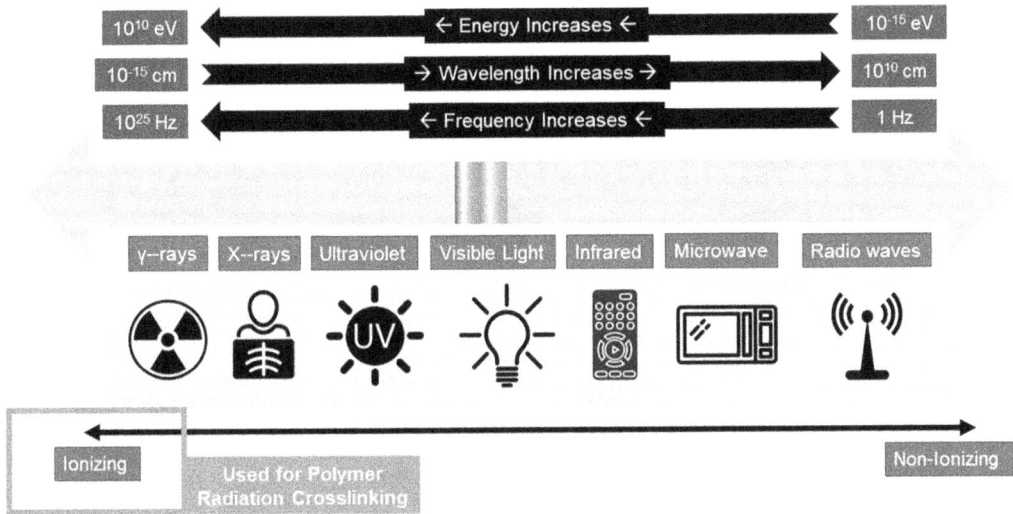

FIGURE 1.1 Electromagnetic radiations and their relation with energy, wavelength, and frequency.

throughput because of faster processing, energy-saving because processing is carried out at room temperature, high purity, and lesser toxicity because no or fewer chemicals are required, and significant reduction of volatile organic compounds as no solvent is needed. The various electromagnetic radiations and their interconnection between energy, wavelength, and frequency are sketched in Figure 1.1.

As shown in Figure 1.1, it is very well known that radiations are divided into ionizing and non-ionizing types depending upon their energies. The ionizing radiations can modify the irradiated materials' chemical, physical, and biological properties. Its recent application includes sterilization of healthcare and pharmaceutical products, which is very useful in a crisis like the pandemic of COVID-19, shelf-life extension and sprout inhibition of food and agricultural products, and material modification such as cross-linking polymers, gemstone colorization, etc.

Ionizing radiations are mainly classified into high-energy electrons (electron beams -EB), gamma rays, and X-rays. These radiations can convert monomeric and oligomeric liquids into solids and produce significant changes in the properties of solid polymers. They have a high penetration capacity compared to ultraviolet (UV) and visible radiations.

Gamma (γ) rays have been the widely explored area in the radiation processing of polymers in recent years. They are emitted from excited atomic nuclei of unstable atoms through which the nucleus rearranges itself into a state of lower energy content. Its ray is a packet of photons and has energies ranging from 10^4 to 10^7 eV. It does not impart radioactivity in a material exposed to them. They have strong penetrating power and low power throughput. The substantial operating cost and the requirement of radioisotope are the main disadvantages of using γ-rays as a radiation source. Otherwise, the equipment is easy to operate and maintain.

Further, continuous radiation requires more shielding, precautions, and persistent attenuation. The strength of gamma radiations is radioactivity, and its unit is the curie (Ci) or becquerel (Bq). The most used nuclei for the radiation processing of polymers are Cobalt-60, Cesium-137, Iridium-192, Ytterbium-169, and Thulium-170.

X-ray is another class of radiation that has been studied for the processing of polymers. They are also called Rontgen radiation from its discoverer, Wilhelm Conrad Rontgen. They have strong penetrating power, low power (throughput), and high operating costs. Electricity is a power source, and equipment is complicated to operate and maintain. It can be turned on and off, less demanding in shielding. X-rays have a wavelength in the range of 10 to 0.01 nm.

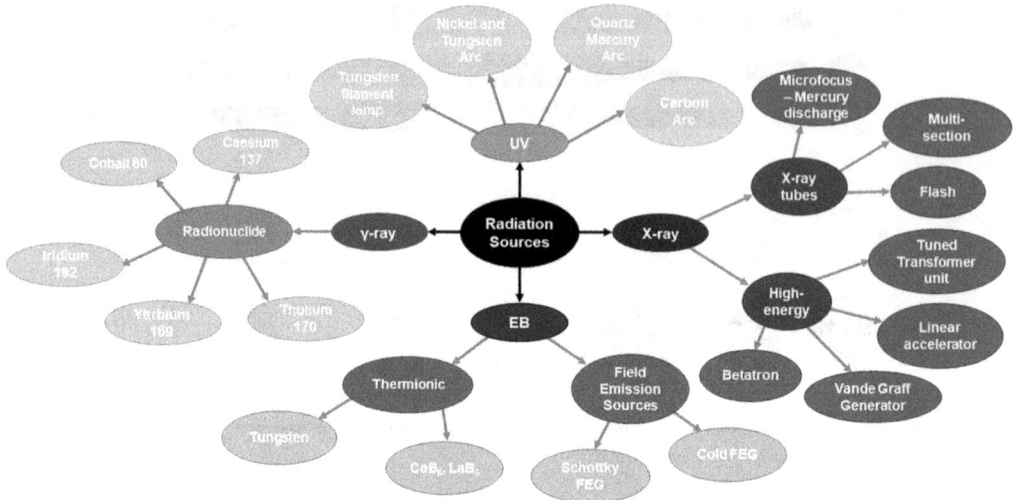

FIGURE 1.2   Commonly used radiation sources in polymer cross-linking.

Moving onto another type of radiation source is EB. In principle, a heated cathode generates fast electrons in a high vacuum. It has limited penetrating power as compared to gamma rays and X-rays. The electrons emitted from the cathode are then accelerated in an electrostatic field applied between cathode and anode. The energy gain of electrons is proportional to the accelerating voltage and expressed as electron volts (eV).

In contrast, the electrons transfer their energy to the material and undergo two interactions. The first one is the collision with electrons of the target atom resulting in material ionization and excitation. At the same time, the other is the interaction with atomic nuclei leading to the emission of X-ray photons. Figure 1.2 summarizes various radiation sources used in polymer processing.

1.3   VARIOUS MECHANISMS INVOLVED IN RADIATION CROSS-LINKING OF POLYMERS

One may define cross-linking in simple words as it is the chemical process of forming three-dimensional (3D) networks from linear polymeric chains. The radiation cross-linking uses the free radicals to create such 3D networks where the polymer may undergo various changes during the exposure time. The types of changes that polymer experiences may include cross-linking or main chain scission (Figure 1.3). These simultaneous changes compete to form a stronger or weaker polymeric network than precursor linear polymer depending upon the number of changes the precursor has undergone. One may conclude that the cross-linked structure has formed only if cross-link points are approximately two times larger than the main chain scission.

There are generally two types of cross-linked networks that are formed, i.e., H-type or T-type (Figure 1.3). The T-type of cross-linking involves the bonding through the functional groups present at the endpoint of the linear polymeric chain. At the same time, H-Type is formed through the cross-linking of the secondary radicals which are created by the breaking of side chains linked to the main chain or by the attachment of a small radical to the unsaturated (C=C) bond in the main chain. These H-type and T-type cross-linking lead to the change in end properties of the polymers; like H-type form the movable polymeric structure, i.e., chains can slide over one another and hence have flexibility (Xu et al., 2015). On the other hand, the T-type of cross-linking restricts the chain mobility and the polymers thus formed are rigid.

The cross-linking density is quantified in terms of the radiation chemical yield expressed by G values. The G value denotes the number of events per 100 eV of absorbed energy, equivalent

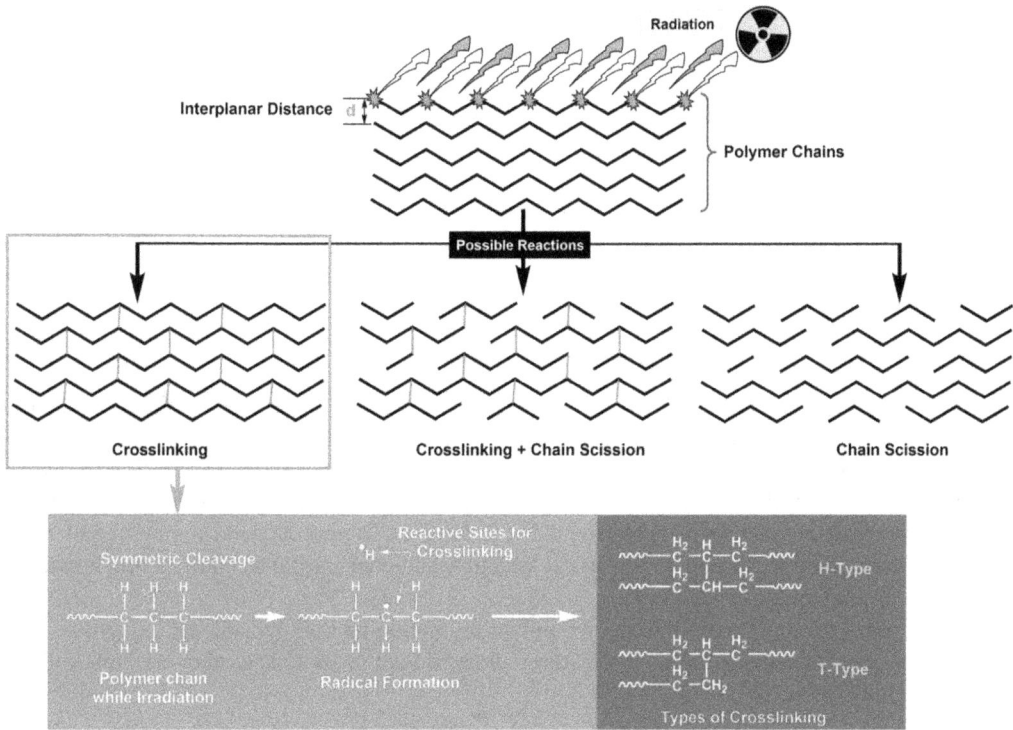

FIGURE 1.3 Possible reactions while radiation cross-linking and types of cross-linking products depending on the structure.

to 1.036×10^{-7} mol/J in SI units. G values for cross-linking G(X) and chain scission G(S) are estimated by the sol-gel analysis by applying the Charlesby–Pinner equation (Charlesby et al., 1959). The Gray (symbol: Gy) is a derived unit of ionizing radiation dose in the International System of Units (SI).

Furthermore, discussing the detection of cross-linked networks, the cross-linked networks are very tiny to observe even through scanning electron microscopy (SEM). The other characterization techniques, such as an increase in the insoluble fraction, i.e., gel content, and some mechanical tests like tensile strength or storage modulus have become evident in the existence of cross-linking. Many of the reports showed that the cross-linking points in irradiated polyethylene (PE) were observed through solid-state high-resolution ^{13}C- nuclear magnetic resonance (NMR) spectroscopy and X-ray photoelectron spectroscopy (XPS) (Fuchs & Scheler, 2000; Lappan et al., 2007; Zhong et al., 1992). The reliable evidence of cross-linking in irradiated PTFE by high-resolution solid-state ^{19}F high-speed NMR was reported by Fuchs and co-workers (Fuchs et al., 2002). There are generally two sorts of thermoplastics—in particular, semi-crystalline thermoplastics, for example, PE and polypropylene (PP), and amorphous thermoplastics, for example, poly(methyl methacrylate) (PMMA) and polystyrene (PS). In any case, there is no radiation cross-linking procedure of thermoplastics since they are not cross-linked by radiation at room temperature.

As the name suggests, the semi-crystalline polymers consist of two regions, viz., crystalline and amorphous. Polymer chains are highly ordered in the crystalline regions and possess high material density and restrictions on segmental mobility, while amorphous regions have precisely opposite properties to the crystalline areas. Tie molecules are interleaving molecular chains emerging from one crystal and connecting to another crystal through the amorphous region. These molecules contribute to the mechanical properties of semi-crystalline polymers. At the same time, it must be noted that other chemical abnormalities like unsaturation, cross-linking, or entanglements in polymeric chains are excluded in the mechanism of Tie molecules.

In the PE crystals, the length of a C-C bond is 0.154 nm, while the closest distance between C atoms of the most consolidated chains in the crystalline phase is 0.41 nm. The carbon atoms in the crystalline phase are far away, and their position is too rigidly fixed to recombine two radicals. Thus, cross-linking occurs predominantly in the amorphous regions. Although the recent advancements overview the semi-crystalline thermoplastics being cross-linked by the radiation method, the peroxide and silane method is most common in the industry. Let us take a quick run-through of the mechanisms involved in peroxide and silane cross-linking processes.

1.3.1 Overview of the cross-linking mechanism by peroxide and silane method

Figure 1.4 shows the reaction mechanism of peroxide and silane cross-linking. Generally, cross-linking by peroxide (free radical initiator) is done in the melted state where the peroxide (For Ex. AOOA) decomposes at higher temperatures and liberates free radicals (AO•). These free radicals abstract a hydrogen radical (H•) from the polymer chain (RH) to generate a polymer radical (R•). The possible reactions between these free radicals form a 3D network, and the degree of cross-linking depends upon the peroxide (chemical structure and concentration) and reaction conditions (temperature, time, and atmosphere).

In the process of silane cross-linking, the polymers containing hydrolyzable alkoxy groups react to join adjacent molecules into stable, three-dimensional cross-linked networks of siloxane

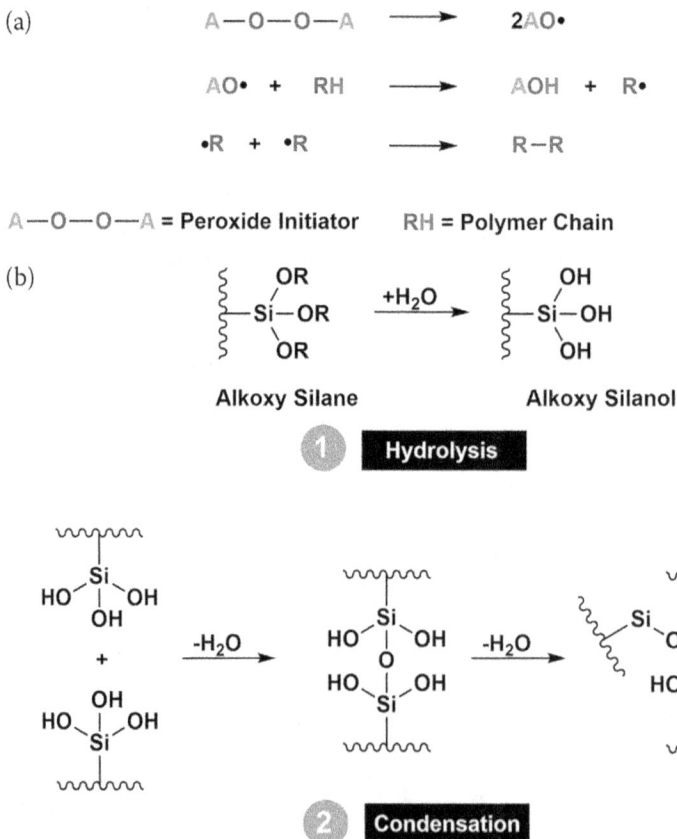

FIGURE 1.4 Mechanism of (a) peroxide cross-linking and (b) silane cross-linking of polymers. (*Content adopted from the* Marcilla et al., 2006 *and* Bengtsson & Oksman, 2006 *with the permission by Elsevier.*)

linkages in the presence of moisture. The alkoxy groups are incorporated either by grafting vinylsilanes such as vinyltrimethoxysilane (VTMS) or vinyltriethoxysilane (VTES) to polyolefin (graft copolymer) or by copolymerization of vinylsilane with an olefin (reactor copolymer). A double bond in the silane structure is a must as it leads to the direct covalent bonding between the polymeric backbone and the silane.

1.3.2 CROSS-LINKING MECHANISM OF UV IRRADIATION

The cross-linking reactions induced by UV irradiation are hugely categorized as green technology-based processes governed by low-cost electrical power inputs and low energy requirements. The other advantages include low temperature, low noise operation, and no volatile organic compounds (VOCs).

1.3.2.1 Cross-linking in polymeric solutions (water as a solvent)

The cross-linking through UV irradiation of the polymeric systems in the solution form follows water photolysis where the excited water molecule was generated $(H_2O)^*$ through the UV radiation ($\lambda < 240$ nm) absorption process (Figure 1.5). The excited water molecules are mainly found on the surface exposed to UV at the initial stage. Over time, the energy gets transferred to the molecules in bulk to produce the radicals as •H and •OH.

The excited water molecules also undergo photolysis reactions to produce oxidizing and reducing ions, i.e., H^+ and ^-OH, by intramolecular electron transfer. In the most common mechanism of water photolysis by UV irradiation, as shown in Figure 1.5, the energy transfer processes to oxygen molecules form a dioxygen radical anion (superoxide radical). In the case of complete energy transfer, relatively long-lived water molecules are produced in an excited state, as the so-called triplet state. The recombination of all the hypothetical stages gives the excited water molecule as a product (see Figure 1.5).

The limited mobility of hydroxyl radical (•OH) leads it to mainly abstract hydrogen atoms (H) to form H_2O. Furthermore, as the number of reducing and oxidizing agents (H^+ and ^-OH) increases, oxygen addition to the water solutions increases. Hence, at low concentrations of dissolved oxygen, the cross-linked polymer is obtained, and whenever there is a sufficient amount of dissolved oxygen, the photo-oxidative degradation process may be predominant. This is beneficial for water treatment and elimination of organic compounds pollutants. Still, sometimes, a part of the dissolved organic pollutants was found to precipitate, probably due to functional groups condensation (acetic

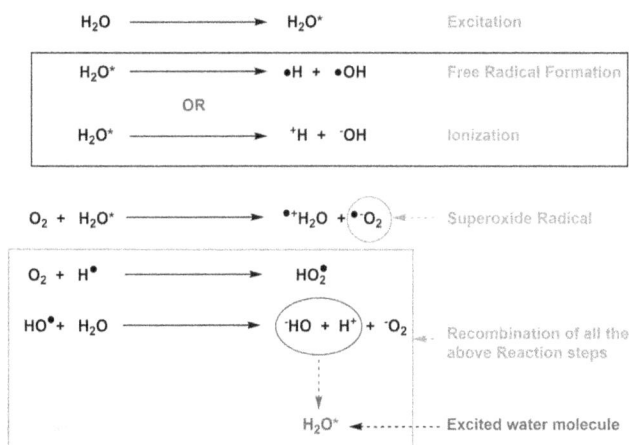

FIGURE 1.5 Mechanism of water molecule excitation and radical formations in aqueous polymeric solutions. (*Content adopted from the* Reeves & Kanai, 2017 *with the permission by Springer.*)

acid, oxalic acid, n-butanoic acid, and malonic acid), which were the most abundant degradation products detected.

The photo-oxidative degradation processes of polymers start with symmetric scission of the C–C bond to form radicals (C•). These macroradicals can recombine in the absence of oxygen (polymerization) or include the peroxy radicals (C–O–O•) (photo-oxidative degradation) (Gueven, 2004). The photons that cause degradation are mainly found in all electromagnetic radiations and can be enhanced at elevated temperatures. So, in the polymerization reaction, the concentration of oxygen should be controlled to improve the polymerization or cross-linking induced by UV irradiation.

The role of the cross-linking reaction initiation induced by UV irradiation "photo-initiator" is delayed by converting the photon energy into chemical energy in the form of a "reactive centre." But, as we have seen in the case of UV irradiation, when the water molecule absorbs photon in UV range in the absence of oxygen, it gets excited to form (•OH and •H) species with a high quantum yield (Mizera et al., 2012). The photo-initiator may exhibit an energetic wavelength and undergo rapid water photolysis to generate the initiating (•OH and •H). Subsequent polymerization occurs when reacting with a monomer molecule to start a monomer combination and build a 3D polymer network. The polymer chains are cross-linked by forming covalent bond cross chains in this process (Jacobs, 1993; Zhang et al., 2018).

1.3.2.2 Cross-linking through the radical formation

The radical cross-linking mechanism is outlined in Figure 1.6; absorbing UV leads to the formation of free radicals as a first stage. In this typical step, the energetic radicals transfer to monomer molecules that undergo unimolecular bond formation upon irradiation. The polymerization process is complete with the integration of monomer molecules. The formed radicals pass energy to another monomer molecule that keeps the polymerization or cross-linking reaction going similarly. A "reactive" photo-initiator reacts with a monomer molecule to start a polymer chain. The cross-linked density is more profound than the surface level, depending on oxygen concentration. While in the free oxygen system, the cross-linked density is expected to be higher at the surface level rather than at the deeper level due to the different yield of radicals on the side exposed to UV radiation.

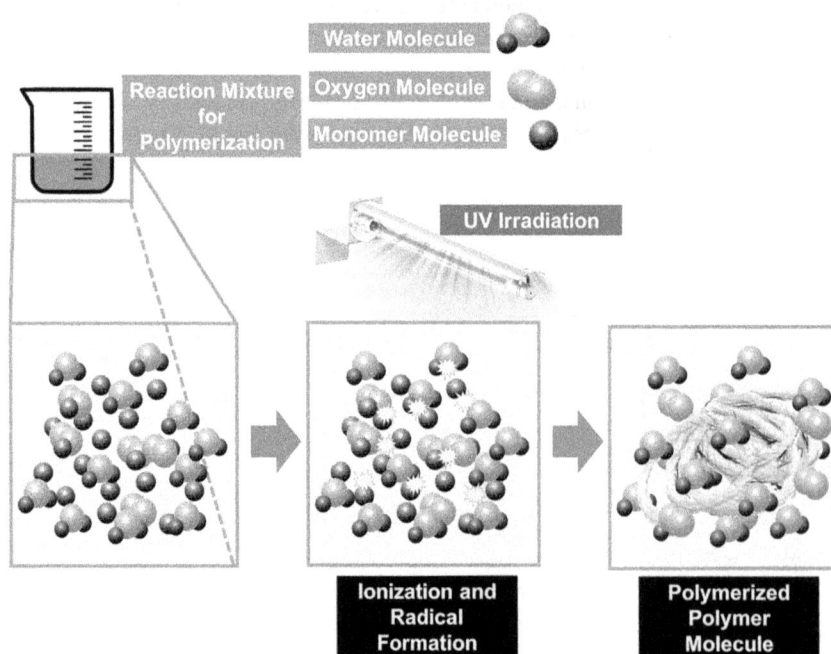

FIGURE 1.6 Mechanism of UV radiation cross-linking for aqueous polymeric solutions.

1.3.2.3 Cross-linking through anionic- and cationic-radical species

Besides photo-initiated radical cross-linking that has been outlined, there are other two photo-polymerization or cross-linking mechanisms: cationic and anionic. Additives sensitive to UV radiation were required to promote cationic and anionic polymerization. **Scheme 3** gives insights into the typical ultraviolet (UV) initiation agents used in the cationic radical type of mechanisms, namely diphenyliodonium-DPI (Ph_2I^+) and triphenylsulfonium-DPS (Ph_3S^+) (Ashcroft, 1993; J. V. Crivello & Lam, 1977; James V. Crivello & Aldersley, 2013). These compounds can generate protonic (carbonium) ions and free radicals.

The cross-linking process is also called free-radical promoted cationic cross-linking. This process undergoes rapid polymerization. For example, in the curing process of epoxy resins, the ring opens and initiates both radical and cationic polymerization mechanisms. Onium salt is sometimes called the curing agent, especially epoxy resin curing. Figure 1.7 also shows the generation of a Bronsted acid when DPI and DPS are exposed to UV radiation (Felder et al., 1978; Sasaki, 1996).

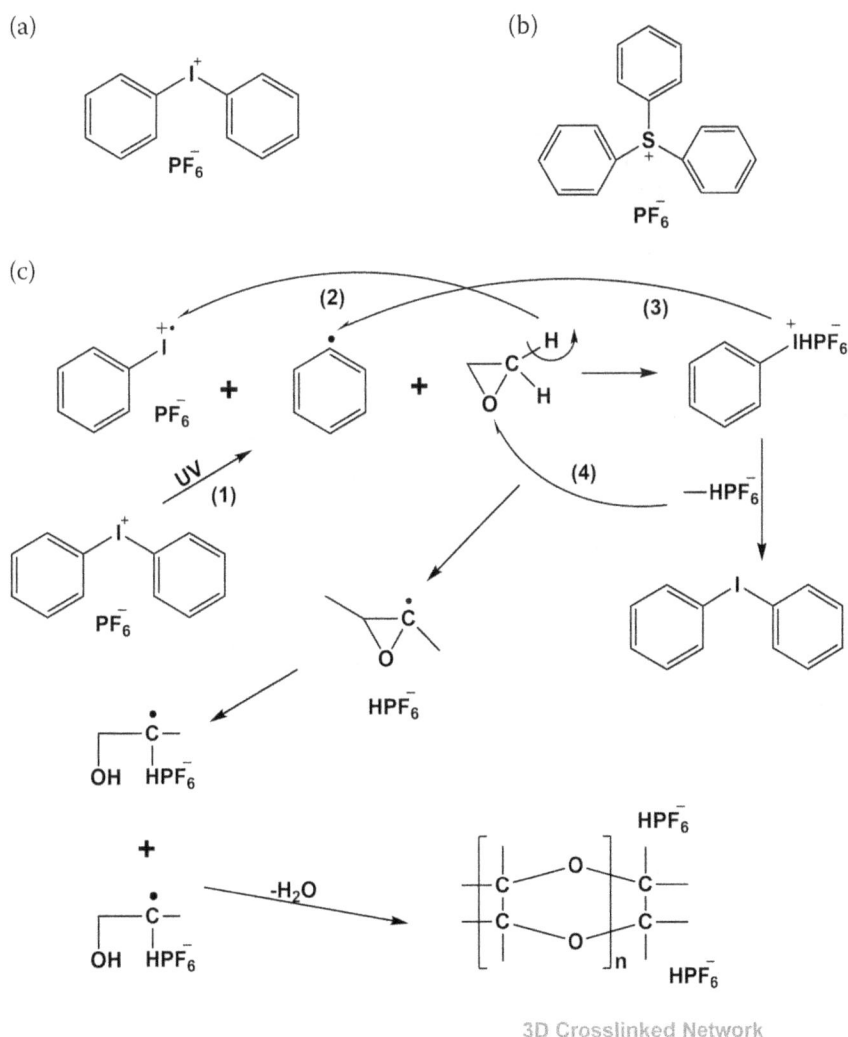

FIGURE 1.7 Structures of (a) diphenyliodonium and (b) triphenylsulfonium compounds used in the cationic-UV radiation cross-linking, (c) mechanism of cationic-UV radiation cross-linking of polymers (Ghobashy, 2018).

As depicted in Figure 1.7, an aryl iodine radical cation and aryl radicals are formed by the photolysis of DPI salt, and the following are the key points to be always considered while dealing with such mechanisms:

- Both are highly reactive species that separate protons from the monomer.
- Generating Bronsted acid $(HPF_6)^-$ and formatting diaryl iodine.
- The polymerization reaction rapidly involves the protonation of the epoxide oxygen with the potent acid.
- It is evident that the cationic polymerization of epoxies is highly quantitative and very rapid, with low activation energy.

As a conclusion of the above-claimed points, the concurrent radical and cationic reactions in hybrid systems have dramatically changed the rate of photo-polymerization (Abdelmonem & Amin, 2014; Crivello & Liu, 2000).

Kahveci and co-workers used a combination of cationic photopolymerization and anionic ring-opening polymerization mechanisms to confirm the amphiphilic graft copolymers of poly (ethylene oxide)-g-poly(isobutyl vinyl ether) (PEO-g-PIBVE) (Kahveci et al., 2014). First, poly (ethylene oxide-co-ethoxy vinyl glycidyl ether) was synthesized by anionic ring-opening co-polymerization of corresponding monomers using radiation for reaction initiation. Random anionic ring-opening copolymerization of a mixture of the gaseous ethylene oxide (EO) (boiling point = 11°C) and the respective co-monomer in an appropriate solvent can be initiated as alkalimetal alkoxide (Obermeier et al., 2011).

Table 1.1 compares the properties of various polymers obtained in different cross-linking processes.

1.3.3 CROSS-LINKING MECHANISM OF EB IRRADIATION

In this process, as the name suggests, electrons are used for the excitation of molecules and form reactive species. As we have seen, these electrons are generated in a linear beam form causing powerful and fast water radiolysis that produces hydrated electrons (e^-_{Hy}). The e^-_{Hy} is a reductive chemical species; once the solvated electron (e^-_S) hits water molecules, water becomes ionized. This depends on the time consumed for interaction and the average distance between the e^-_S and the water ions.

TABLE 1.1

Technological Comparison of Cross-Linking Methods of Polymers

Type of plastics and in general effect on the end properties	Radiation technique		
	Radiation (EB)	**Peroxide**	**Silane**
PE	Commercially available	Commercially available	Commercially available
PP		Theoretically Possible	Theoretically Possible
PVC			
Engineering plastic		Not possible	Not possible
PTFE			
Fluoropolymer		Commercially available	Theoretically Possible
Cost of compounding	Low	Medium	High
Shelf life of the compound	Long	Medium	Short
Product thickness restriction	< 10 cm	> 0.3 mm	> 0.3 mm
Rate of cross-linking	High	Low	Low
Degree of cross-linking	Medium	High	Low

The emitted electron (having energy more than 1 MeV) from the electron accelerator hits the water molecule and then scatters. The energy of an electron is lost due to bordering a pathway for secondary electrons. At the same time, the secondary electrons also have their mighty power, causing water ionization and forming copious quantities of a variety of reactive species like H_2, H_2O^+, H_2O_2, OH^-, H_3O^+, and $^•OH$, etc. (Sáenz-Galindo et al., 2018).

Typically, the chemical reactions (radiolysis) occur when the reactive species (ions and radicals) are distributed non-homogeneously in "clusters" during excitations/ionization (i.e., spurs). In contrast, homogeneously when the reactive species have diffused and thermalized. It is well-known that the ionizing and exciting species are, in fact, non-homogeneous and occur in clusters called "spurs." The electron is similar to a "stone hit" on the surface of the water-pool; by the time, the resulting "splashes" induced by "the stone hit" will expand and eventually overlap with their neighbors to result in a homogenous distribution.

Nowadays, EB radiation cross-linking has become a potential alternative to chemical cross-linking as it can significantly improve a range of thermoplastic elastomers (TPE) properties (see Figure 1.8). EB radiation is a process that is well known for its speed, simplicity, and non-thermal curing characteristics. Besides, the reprocessing characteristics and thermoplasticity of EB cross-linked TPEs can be retained considerably with a controlled irradiation dose (Rouif, 2005; Silva Aquino, 2012). Investigation of EB irradiation cross-linking on linear low-density PE (LLDPE)/polydimethylsiloxane (PDMS) blends prepared over a wide range of compositions starting from 70:30 to 30:70 (LLDPE: PDMS), by varying the radiation dose from 50 to 300 kGy, was reported by Wang and co-workers (Wang & Mao, 2012). They have also said that the number of cross-links formed was more remarkable in comparison with a chain scission up to 100 kGy irradiation for the blends, beyond which it decreased. Furthermore, after irradiation, they observed that the two-phase morphology of the blend turned into a single-phase morphology.

One of the reported studies also showed that the electrical properties of EB-irradiated LLDPE/silicone rubber (SR) could be tailored in the presence of boron nitride (BN) nanofiller at 2 vol%. The blend ratio of LLDPE/SR was fixed at 90:10. The selection of the LLDPE/SR ratio and nanofiller concentration was made based on the optimum performance achieved in the previous work (Böhler et al., 2013; Diani et al., 2006). Low irradiation doses (20, 40, 50, and 100 kGy) were employed principally to induce selective cross-linking of the SR phases.

Electron beam irradiation has been used in the wire and cable industry for more than 30 years and applied to a wide range of commodity and specialty elastomers. A survey of the types of elastomers

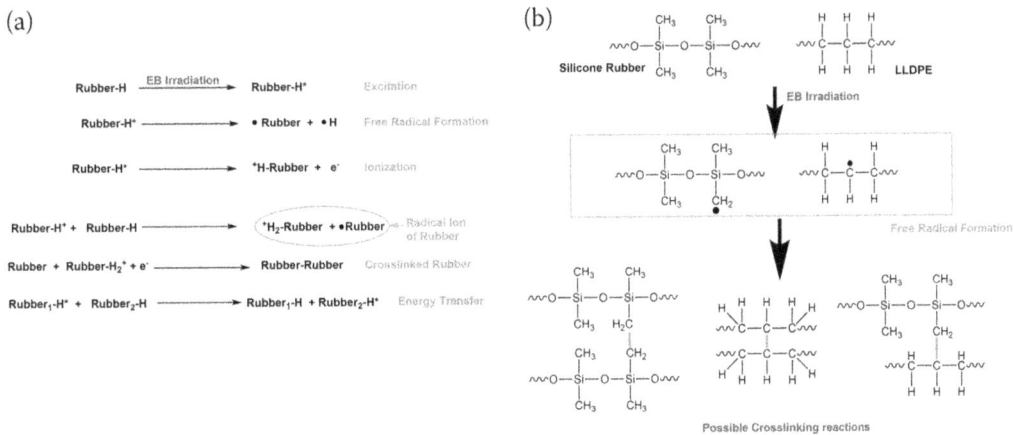

FIGURE 1.8 Mechanism of cross-linking of polymers by EB irradiation: (a) cross-linking of natural rubber, (b) crosslinking of silicone rubber (Manalia et al., 2012). (*Content adopted from the* Hill et al., 2001 *with the permission by Elsevier.*)

susceptible to radiation curing is available, as are review articles describing the electron-beam curing of commercially significant grades (Bhowmick & Vijayabaskar, 2006). Variables such as radiation dosage and the effect of polymer microstructure and chemical additives on the efficiency of electron beam cure have been studied. Some researchers also studied radiation-induced cross-linking in thermoplastic elastomers based on ethylene-propylene rubber (EPDM) and PE or polypropylene (PP) plastics (Chowdhury & Banerji, 2005; Zaharescu et al., 2000).

The reaction mechanism is similar to cross-linking with peroxides, but in this case, reaction initiation is due to EB's action and the presence of the polyfunctional monomers. Ionizing radiation produces the excitation of polymer molecules. The energies associated with the excitation are dependent on the irradiation dosage of electrons. The interaction results in free radicals formed by the dissociation of molecules in the excited state or by the interaction of molecular ions. The free radicals or molecular ions can directly connect the polymer chains or initiate grafting reactions.

1.3.4 CROSS-LINKING MECHANISM OF X-RAY IRRADIATION

Fricke (Fricke et al., 1938) had studied the radiolysis of water by X-rays in a closed vessel. Bubbles of hydrogen and carbon dioxide gas resulting from irradiation are found. X-ray photons are highly powerful, causing water radiolysis to create an ion pair e^-_{Hy}. The amount of ionization species produced is affected by the number of energetic photons absorbed (Anbar, 1961). Figure 1.9 shows that the positive and negative water molecules are unstable and can break further apart into smaller molecules. As there are the chances of formation of the free radicals and ionic species, several chemical reactions might occur like oxidation-reduction by OH^-, H^+, and e^-_{Hy} or the polymerization or cross-linking reactions by (•OH and •H and e^-_{Hy}). Also, the radiolysis products may recombine individually, forming gases such as H_2 and H_2O_2, and hence the bubbles are formed due to an evolving mixture of H_2 and O_2 gases. H_2 present in the solution or the "e^-_{Hy}" may cause reduction reactions (Mesu et al., 2007; Treutler & Ahlrichs, 1995).

1.3.5 CROSS-LINKING MECHANISM OF Y-RAY IRRADIATION

Here, two cases have been established in the polymerization induced by γ radiation: cross-linking and degradation. It is exciting to investigate the proposed mechanism of the polymerization reaction induced by γ irradiation in an aqueous system. These mechanisms present the common three

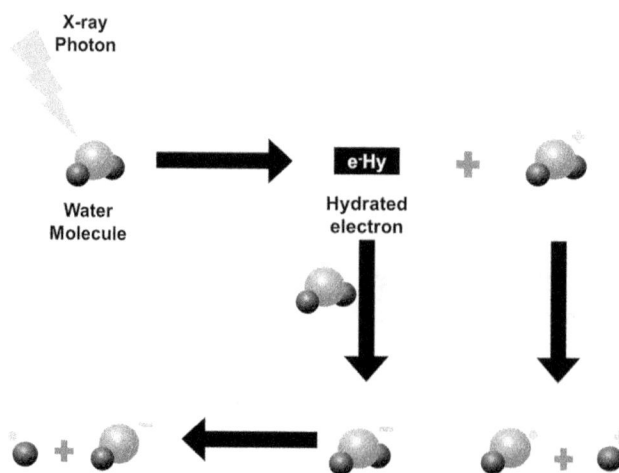

FIGURE 1.9   Mechanism of X-ray radiation cross-linking in aqueous systems (Initiation with ionization of water) (Ghobashy, 2018).

polymerization mechanisms (radical, cationic, and anionic). But, in ultra-pure irradiated water, there are no impurities that can scavenge the formed radicals and thus have lots of tracks of radiolysis to expand. In this case, the increase of radical concentrations leads to a homogeneous radical-radical combination (H_2, H_2O_2, and O_2). Also, the excess radicals can react with the reactive species formed after the cleavage of H_2, H_2O_2, and O_2. The above phenomena must be considered in polymerization or cross-linking reactions induced by γ irradiations.

The radiation chemistry of cross-linking of the water-based polymers remains the same, i.e., cleavage of photo-initiators to form reactive species like •OH, •H, and e^-_{Hy}. In addition to this, the ultra-pure water irradiated by γ rays produces 11 reactive species resulting from 20 reactions, while in acidic water, it produces 7 reactive species resulting from 10 reactions (Chatterjee et al., 1983).

Zhou (Zhou, 2012) has mentioned the cross-linking of poly(urethane-imide) network by using γ radiation in his book chapter and said the study had reported the effect of γ dosage onto the end properties of the polymeric networks. One of the studies reports the thermal properties precisely the degradation behavior of various poly(urethane-imide) systems obtained at different dosage levels of irradiation. It has shown that the degradation behavior was optimum at a certain dosage level. A further increase in irradiation dosage deteriorated the thermal properties at the initial stage of degradation. The results could relate the cross-linking with the dosage level as the cross-linking density decides the thermal degradation behavior of the polymers. As we know that good thermal degradation is attributed to the high cross-linking density and rigidity and thus the less mobility of polymeric chains at elevated temperatures. Hence, it could be inferred from the data mentioned above that in the competence of cross-linking and scission of polymeric systems, there is a decrease in the cross-linking reactions than the scission after a certain optimum level of irradiation.

Furthermore, Bhuyan and co-workers (Bhuyan et al., 2016) reported the synthesis of potato-starch and acetic acid synthesis cross-linked hydrogels using γ irradiation. They have also proposed the mechanisms involved in the hydrogels' synthesis process, as shown in Figure 1.10. They

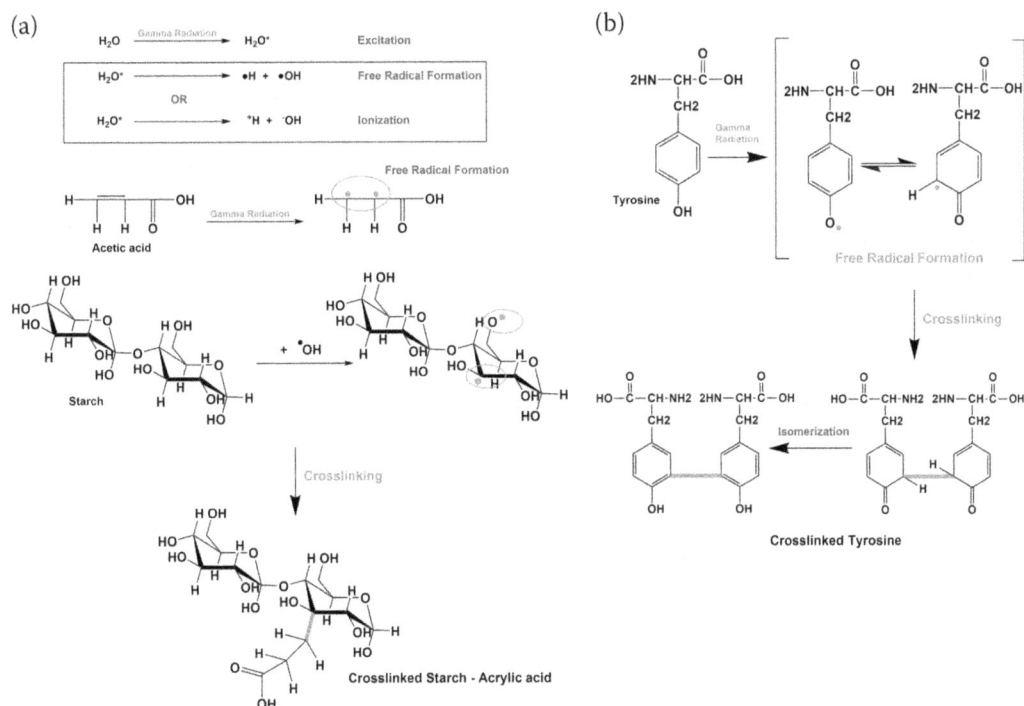

FIGURE 1.10 Mechanism of cross-linking of polymers by γ-ray irradiation: (a) cross-linking of the starch-acrylic acid copolymer, (b) cross-linking of tyrosine (Bhuyan et al., 2016).

have concluded that the 15 kGy radiation dosage level was sufficient for preparing starch/acetic acid hydrogels. Furthermore, they have also claimed that the swelling behaviors of the starch/acetic acid hydrogel can be improved by the treatment with NaOH, which could be the result of increased water affinity of the hydroxyl groups on the starch due to phenoxide ion formation.

The formations of nanoparticles (NPs) are also being carried out by the radiolysis using γ irradiation of water even in the presence of oxygen where the combination of various chemical radical scavengers (mannitol, urate, tert-butanol, isopropanol) and gases (N_2O, O_2, N_2) indicated that –OH is the main species responsible for the cross-linking of proteins (like Tyrosine). According to the studies reported by Bailey (Bailey, 1991), Varca (Varca et al., 2014), and Queiroz (Queiroz et al., 2016), the mechanism of tyrosine cross-linking by γ irradiation follows tyrosine residues attachment, generating the formation of di- and tri-tyrosines, as shown in Figure 1.10.

1.4 PARAMETERS INVOLVED IN RADIATION CROSS-LINKING OF POLYMERS

As we have seen until now, radiation cross-linking results from the recombination of polymer radicals in the amorphous region at a rubbery state. Radiation cross-linking does not occur in the crystalline and amorphous regions at a glassy state. The critical conditions for radiation cross-linking are a generation of secondary radicals in the amorphous area at a rubbery state and the mobility of polymer chains that bear secondary radicals. The response of the polymer to radiation, cross-linking, or degradation depends on the nature of the polymer.

1.4.1 GLASS TRANSITION TEMPERATURE (T_g) AND CRYSTALLINITY

As we have seen, the flexibility and rigidity of the polymeric chain are dependent on the cross-linking density of the polymeric system. Further, the molecular mobility in the polymeric chains reflects the flexibility and softness of the polymer. In the examples, we have also seen that the flexible chains are more accessible to cross-link than are rigid chains (Jiazhen, 2001). As T_g of the polymer is used as the measure of the flexibility of the polymer, one could relate that; at low temperatures below T_g, molecular motion in an amorphous region is restricted to molecular vibrations, but the chains cannot rotate or move in space. When the polymer is heated above T_g, the amorphous region becomes rubbery where segments (20–50 atoms long) of the entangled chains can move (see Table 1.2). Furthermore, differences in chemical composition create differences in T_g, and the crucial factors influencing T_g are chain stiffness, intermolecular forces, pendant groups, and stiffening groups.

The introduction of long-chain molecules in the polymeric networks decreases their T_g considerably, making them more flexible. The introduction of stiff molecules like benzene rings or

TABLE 1.2
Radiation Responses of Polyolefins and Their T_g Values

Sr. no.	Polymers	T_g (°C)	Radiation response
1	PE	−120	Crosslinking
2	LLDPE	< −25	
3	BR	−90	
4	PVC	87	Degradation
5	PVA	85	
6	PP	−10	
7	PIB	−70	
8	PS	100	Stable

rigid functional groups like C=O increases the rigidity of the polymeric network. Also, the addition of thermally stable heteroatoms such as phosphorus, silicon, boron, sulfur, etc., increases the toughness in the polymeric system, increasing the T_g value. The bulky nature of one of the ingredients used in the polymer synthesis or any such additive used in the polymer formulations harms the packing density of the polymeric system, where the voids between two polymer chains increase and thus crystallinity decreases. In this case, the effect on T_g depends upon the predominant parameter between the increased mobility of polymeric chains due to increased voids and introduced rigidity because of the bulky molecule (Figure 1.11). The grafting density also hurts the packing density of the polymeric chains and thus affects T_g and crystallinity.

Since cross-linking occurs predominantly in the amorphous regions, the radiation cross-linking efficiency of the polymer decreases with increasing crystallinity. The crystallinity depends on the structure of linear chain, polar groups, pendant groups, and degree of polymerization. The physical conditions such as the cooling rate and orientation of polymer chains also significantly influence crystallinity. The volume fraction of the amorphous region can be changed by tailoring heat treatments.

Quenching treatment (e.g., immediately immersing the polymers in icy water) after processing increases the volume fraction of the amorphous region. On the contrary, slow or isothermal crystallization or annealing decreases the volume fraction of the amorphous region. That is why quenched materials have more gel for the same radiation dose than annealed materials (Shukushima et al., 2001). One of the reports proved the effect of crystallinity on the radiation cross-linking of ethylene-propylene copolymer where the reference sample was cooled in a hot plate with water flowing, and the quenched sample was immersed immediately in icy water out of the hot-press machine. The annealed sample was prepared by annealing the reference sample at 150°C for 1 h. The effect of quenching on the crystallinity is shown by comparing the melting temperature (T_m) and the heat of fusion. The T_m of the quenched PP was 150.7°C, 10°C lower than that of the annealed PP (160.7°C).

Furthermore, the heat of fusion of the quenched PP (60.0 J/g) was smaller than that of the annealed PP (72.5 J/g). The results showed that the quenched sample generated gel more efficiently, while the annealing treatment decreased the generation of gel. The radiation cross-linking occurred in the amorphous parts of PP where the quenched polymer had a larger amorphous region compared with untreated random PP or annealed PP. Therefore, the quenched PP was easier to cross-link.

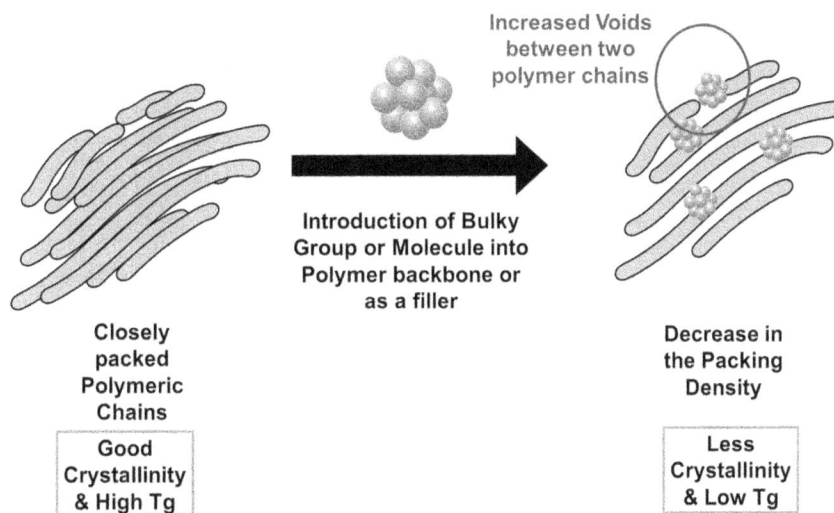

Increased Voids between two polymer chains

Introduction of Bulky Group or Molecule into Polymer backbone or as a filler

Closely packed Polymeric Chains

Good Crystallinity & High Tg

Decrease in the Packing Density

Less Crystallinity & Low Tg

FIGURE 1.11 Effect of the introduction of bulky structures on the crystallinity and Tg of polymers.

1.4.2 BOND DISSOCIATION ENERGY

As we have seen the effect of polymeric molecular structure on the mobility of polymer chains, it also regulates the yield of polymer radicals. It has been proved that the molecular structure that yields a high number of radicals often restricts molecular mobility. The yield of polymer radicals depends on the bond strength as the radicals are formed after the symmetric cleavage of the covalent bond. The secondary radicals mainly contribute to the cross-linking created by the side-chain scission, and their yields depend on the bond strength of the side chain, whereas the degradation characteristic of the polymer is governed by the bond strength of the main chain of the polymer (see Table 1.3).

The bond energy of C-H decreases in the order of primary > secondary > tertiary, whereas the tendency of bond breaking decreases in tertiary > secondary > primary. Tertiary radicals tend to proceed to β-scission, which leads to main-chain scission. The quaternary carbon atoms (β – scission) enhance the main-chain scission by irradiation, as shown in Figure 1.12(a).

1.4.3 PENDANT GROUPS AND UNSATURATION IN POLYOLEFINS

Continuing the factors related to bond dissociation energy, the placement of unsaturation present in the polymerizable compounds and the nature of the functional group present as a pendant moiety in the polymeric chain also plays a vital role in deciding the smoothness of the reaction of irradiation cross-linking. The pendant groups include the alkyl groups such as methyl ($-CH_3$), halogen atoms, and benzene rings.

TABLE 1.3

Bond Dissociation Energies of R-X(kcal/mol)

Sr. no.	R–	Example	–X		
			H	Cl	CH$_3$
1	Primary	End group of PE chain	98	80	82
2	Secondary	Methylene group in PE	94	77	75
3	Tertiary	Side chain of PP	90	75	74

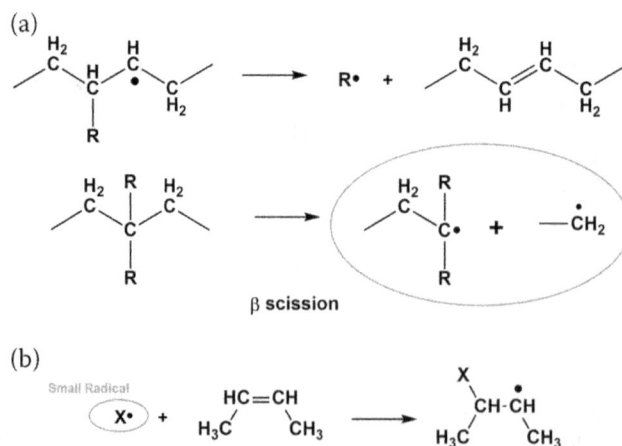

FIGURE 1.12 (a) Mechanism of scission of quaternary carbon compounds (β – scission) due to irradiation, (b) addition reaction of the small radicals to unsaturation (Sartipi et al., 2014).

The electronegativity of an atom decides the bond dissociation capacity or the leaving ability of that atom in symmetric scission reaction during the radical formation. For example, in the case of halogens, electronegativity decreases down the periodic table group, i.e., F > Cl > Br > I > At; because of the increased atomic sizes of the atoms. Also, inductive, mesomeric, and resonance effects caused due to the pendant groups lead to the change in the ability of the molecule to dissociate and thus to form a radical. Talking about the halogens, the bond dissociation energy of C-Cl is less than that of C-H because Cl is an excellent leaving group than H. Hence, the partial chlorination of a radiation degradable polymer would change it to a radiation cross-linkable polymer. For example, chlorinated PIB containing about 3.6% Cl cross-links effectively by irradiation in a vacuum (Wenwei et al., 1993). Similarly, isobutylene-isoprene rubber (IIR, butyl rubber), a copolymer of isobutylene with small isoprene (1–3 mol%), degrades by radiation, even with the presence of C=C, while chlorination of IIR to form chloro-isobutylene-isoprene rubber (CIIR) improves the cross-linking efficiency extraordinarily, as shown in Table 1.4 (Hill et al., 1995). In addition to the attachment of halogen atoms, a molecular rearrangement during dehydrohalogenation of halo-butyl rubber generates a pendant double bond that facilitates radiation cross-linking. Bromination is even more effective for enhancing cross-linking. The G(X) and G(S) of the bromo-isobutylene-isoprene rubber (BIIR) are 3.7 and 0.44, respectively. The random distribution of halogen atoms in polymer molecules is favorable for radiation cross-linking.

The presence of the methyl group suppresses the radiation cross-linking because the attachment of the methyl group reduces the bond energy of the C-C bond in the main chain and the methyl group acts as a stiffening group. For instance, Table 1.4 shows the G(X) and G(S) of PE, PP, and polyisobutylene (PIB). The substitution of the hydrogen atom in PE with a methyl group enhances the main-chain scission and reduces the cross-linking. This is due to the reduction of bond energy of the main chain C-C bond by ionization (Postolache & Matei, 2007). The cross-linking does not occur in PIB, in which two methyl groups replace two hydrogen atoms of PE. The different radiation behaviors of polyacrylate and polymethacrylateclearly show the significant effect of the quaternary carbon atom. A polyacrylate is easy to cross-link by radiation, while main-chain scission is predominant in polymethacrylate. The T_g of a polymethacrylate is higher than that of the corresponding polyacrylate. Table 1.4 compares the radiation effects of poly(methyl acrylate) (PMA) and PMMA. The methyl groups in polymethacrylate prevent the rotation of carbon atoms in the main chain, and the phenomenon is known as a steric hindrance. Thus, the flexibility of the main chain is restricted, resulting in high T_g. However, the main-chain scission of polymethacrylate is predominant even when irradiated at a higher temperature than T_g.

The dose of gelling or gel point dose (D_{gel}), G(X), and G(S) of polystyrene (PS), poly(vinyl toluene) (PVT), and poly(α-methylstyrene) (PAMS) are listed in Table 1.4 (Chernova et al., 1980). The radiation stability of PS is excellent, while main-chain scission dominates in PAMS because of the quaternary carbon atoms. The main reason is the replacement of the α-hydrogen atom in the main chain of PS by a CH$_3$ group weakens the C_α-C_β bond considerably and causes degradation of the PAMS. Furthermore, introducing a CH$_3$ group in the para-position of the benzene ring in PVT considerably reduces the resistance of the polymer to radiation. Degradation does not take place in irradiated PVT.

Generally, the phenyl group retards or inhibits the radiation reactions such as cross-linking, chain-scission, and gas evolution. This effect is the protective effect of the phenyl group and is attributed to the ability of the phenyl group to dissipate some of the absorbed energy before the bond rupture occurs.

The kind of polymerization or cross-linking C=C-X (where X is any substituent) undergoes upon the nature of the substituent, i.e., whether it has the electron-donating effect or withdrawing at the polymerization site as well as on the presence and position of unsaturation. In the presence of unsaturated C=C, radical formation increases. As shown in Figure 1.12(b), such reactions follow the addition mechanism. The typical example can be considered as radiation cross-linking of NBR consisting of acrylonitrile and butadiene segments where the butadiene part incorporates C = C to NBR, and the acrylonitrile part acts as a stiffening group (Vijayabaskar et al., 2006).

TABLE 1.4
Effects of Various Pendant-Functional Groups G(X) and G(S)

Sr. no.	Changing functional groups	Polyolefins			Polyacrylates		Rubbers		Aromatic polymers		
		PE	PP	PIB	PMA	PMMA	IIR	CIIR	PS	Poly (vinyl tolune)	Poly (a-methyl styrene)
		Increasing number of methyl groups →			Increasing number of methyl groups →		Non-Halogenated	Halogenated	Increasing number of methyl groups → (1°, 2°, 3°)		
1	G(X)	1–2	0.2–0.5	0	0.42–0.52	0.15	0	3.62 ± 0.1	0.01	0.1	0
2	G(S)	0.2–0.5	0.3–0.6	4–5	0	1.7	3.7 ± 0.1	1.71 ± 0.05	0.1	0	0.33

1.4.4 MOLECULAR WEIGHT OF POLYMERS

The molecular weight has been found to have no significant effects on the number of polymer radicals generated by irradiation, concluding that G(X) and G(S) are independent of molecular weight. However, it is very well known that the D_{gel} is inversely proportional to the weight-average molecular weight of the polymer. This means that a polymer with high molecular weight forms a gel at even a shallow irradiation dose. For a polymer with random molecular weight distribution, the D_{gel} is inversely proportional to the number-average molecular weight (M_n). The entanglement effect can support this claim in polyethylene oxide (PEO), a semi-crystalline polymer with a T_g of $-66°C$ and a T_m of 66°C. The molecular mobility of PEO is so high at the irradiation temperature around 20°C so that high molecular weight PEO would be severely entangled. The entanglement hinders the motion of molecules and impedes radical recombination. Concerning the effect of molecular weight on radiation cross-linking of PE, C = C end groups such as terminal vinyl ($CH=CH_2$) and vinylidene [$CH=C(CH_3)_2$] unsaturations enhance cross-linking efficiency of PE. The concentration of the C = C end groups per unit weight of PE decreases with increasing molecular weight. Thus, the cross-linking efficiency of PE tends to decrease with increasing molecular weight (Mitsui et al., 1975). The effect of molecular weight distribution on the efficiency of radiation cross-linking is vague. The difficulty is caused by the fact that a change in molecular weight distribution inherently affects other parameters such as vinyl content and crystallinity that affect the efficiency of radiation cross-linking. However, the high-temperature properties of radiation cross-linked PE indicate that radiation cross-linking efficiency increases with increasing molecular weight distribution (Wunsch & Dalcolmo, 1992).

1.4.5 ISOMERISM

The geometric structures define the alignment and orientation of atoms or other functional groups in the molecule in the 3D planes. The stereoisomers such as *cis-* and *trans-*polymers and stereo-regular polymers such as isotactic and syndiotactic isomers have different configurations in terms of their alignment of pendant groups in 3D. Whereas the kinds and concentrations of radicals formed by radiolysis of polymers are not directly affected by the configuration. Still, the configuration determines the mobility of polymer chains and molecular distances. Thus, the efficiency of radiation cross-linking is influenced by the isomerism indirectly.

Talking about structural isomerism, the difference between cis and trans configurations arises from the location of substituent groups on the C=C bond. *cis* refers to the configuration in which substituent groups are on the same side of a carbon-carbon double bond. In contrast, *trans* is attributed to the configuration in which the substituents are on opposite sides of the double bond. For instance, the most common example, *cis*-1,4-polyisoprene (*cis*-PI), is natural rubber, and *trans*-1,4-polyisoprene (*trans*-PI) is known as Gutta-percha. The *cis*-PI molecules are flexible at room temperature because every segment can rotate and bend. At the same time, *trans*-PI has a more compact structure than *cis*-PI and crystallizes more readily as we know that the Gutta-percha is very hard at room temperature and exhibits good tensile strength without cross-linking. As a result, *trans*-PI is more effectively cross-linked than *cis*-PI is, though the molecular mobility of *trans*-PI is lower than that of *cis*-PI (Turner, 1966). Besides, the G(X) of *cis*-PI is lower than or comparable to that of saturated PE despite higher molecular mobility and the presence of a considerable amount of unsaturation. In the case of sulfur vulcanization, the rate of cross-linking of cis-PI is generally higher than that of *trans*-PI regardless of the vulcanization system used (Boochathum & Prajudtake, 2001). The insensitiveness of *cis*-PI to radiation cross-linking may be due to the entanglement of *cis*-PI molecules. Highly flexible *cis*-PI molecules are likely to entangle with each other.

Another type of isomerism found in polymers that affects the efficiency of the radiation cross-linking is stereoregularity. In polymeric terms, isotactic is an arrangement in which all substituents are on the same side of the polymer chain. In contrast, a syndiotactic polymer chain is composed of alternating groups on both sides of the chain, and an atactic is a random combination of the groups.

Mostly, commercial PP is isotactic (i-PP), and its crystallinity is higher than syndiotactic PP (s-PP). At the same time, atactic PP (a-PP) is amorphous due to the lack of long-range order. The primary radiation effect on i-PP is main-chain scission, and the radiation cross-linking of a-PP is also tricky without a polyfunctional monomer (Schulze et al., 2008).

In contrast, s-PP gets cross-linked by irradiation with the dose of 180 KGy, and the G(X)/G(S) is 0.84. In the case of polystyrene, the G(X) and G(S) of syndiotactic polystyrene are smaller than those of atactic polystyrene, which may be attributed to the steric hindrance of the aromatic ring (Takashika et al., 1999).

1.4.6 TYPE OF POLYOLS (POLYESTERS OR POLYETHERS)

Table 1.5 shows the responses of the various polyesters to radiation. The results suggest that the methyl group prevents radiation cross-linking of polyesters as in the case of poly(hydroxybutyrate) (PHB) and poly(lactic acid) (PLA), only main-chain scission occurs by the irradiation (Mitomo et al., 1995; Nugroho et al., 2001). However, polyesters such as poly(ε-caprolactone) (PCL) and poly (butylene succinate) (PBS) with no methyl groups can be readily cross-linked by irradiation (Darwis et al., 1998; Song et al., 2001). Talking about the aromatic polyesters such as poly(ethylene terephthalate) (PET) and poly(butylene terephthalate) (PBT) are very stable against radiation because of the presence of phenyl groups in the main chain. The extent of main-chain scission of crystalline polyesters such as PET and PBT decreases with the increasing number of CH_2 groups (Bell & Pezdirtz, 1983). This explanation can be extended to the polyethers like poly(methylene oxide) (polyoxymethylene, POM, $(OCH_2)_n$), which decomposes by irradiation while PEO $[(OCH_2CH_2)_n]$ cross-links due to the increased number of CH_2 groups.

1.4.7 TYPE OF COPOLYMERS

As we have seen that PMMA degrades by irradiation because of the quaternary carbon atoms, this effect of the quaternary carbon atoms can be diluted by incorporating radiation cross-linkable components through copolymerization with an acrylate. One of the works reported the study of the effect on the G(X) and G(S) of copolymers of n-butyl acrylate (n-BA) and methyl methacrylate (MMA) by the degree of copolymerization. The results proved that the G(X) increased gradually with the content of the n-BA (radiation cross-linkable) (Turgis et al., 2003).

PE is one of the basic polymers for industrial application, and hence to increase the radiation cross-linking efficiency, various co-monomers are incorporated (Burns, 1979). Moving on to the comparison with G(X) values of commercial grades of poly(ethylene-co-vinyl acrylate) EVA and poly(ethylene-co-ethyl acrylate)(EEA) copolymers; the enhanced radiation cross-linking tendency of EVA and EEA over PE is proportional to the co-monomer content. The effect of copolymerization of vinyl monomer increases with the amorphous region and the radiation-sensitive branch structure. In EVA and EEA, the increase in the amorphous polymer content contributes to enhancing radiation cross-linking. From the structure, it is clear that EVA is more susceptible to radiation cross-linking

TABLE 1.5

Radiation Responses of Various Polyesters

Sr. no.	Polymer	Radiation responses
1	Poly(hydroxy butylate)	Degradation
2	Poly(lactic acid)	
3	Poly(caprolactone)	Cross-linking
4	Poly(butylene Succinate)	

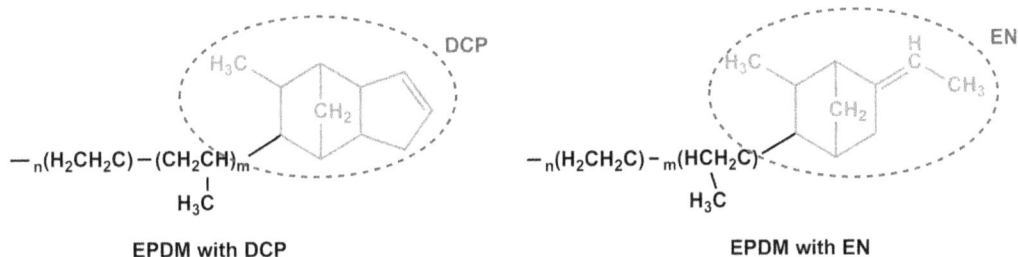

EPDM with DCP　　　　　　　　　　　**EPDM with EN**

FIGURE 1.13　Structures of the copolymers of (a) EPDM-DCP, (b) EPDM-EN.

than EEA at the same amorphous content. In EVA, the cross-linking sites are the hydrogen atom at the branch point and the methyl hydrogen atom on the acetoxy group. In EEA, cross-linking occurs primarily at the branch point on the polymer.

Poly(ethylene-co-vinyl alcohol) (EVOH) is produced by saponification of EVA, consisting of a PE part and a polyvinyl alcohol (PVA) part. The molecular mobility of EVOH increases with increasing PE content, especially when the content of ethylene is more than 50 mol% (Deng et al., 2007; Yang et al., 2010). The copolymerization can also improve the radiation degradation tendency of pure PP with ethylene. However, cross-linking of PP and ethylene-propylene copolymer (EPM) by radiation or peroxide still needs the addition of cross-linking accelerator. The efficiency of the radiation cross-linking can be further enhanced by incorporating pendant C=C groups via ter-polymerization of ethylene, propylene, and diene monomer commonly termed as EPDM. The most used diene monomers dicyclopentadiene (DCP) and 5-ethylidene-2-norbornene (EN) (see Figure 1.13) (Colomb et al., 1970; Geissler et al., 1978).

1.4.8 FLUOROPOLYMERS (FPS)

These polymers can be classified into two significant radiation reactions (cross-linking and degradation) at room temperature in an inert atmosphere. Table 1.6 depicts various radiation-degradable polymer and radiation-cross-linkable fluorine-based polymers (Forsythe & Hill, 2000).

TABLE 1.6
Radiation Responses of Various Fluoropolymers

Sr. no.	Fluoropolymers	Radiation responses
1	Polychlorotrifluoroethylene (PCTFE)	Degradation
2	Polytetrafluoroethylene (PTFE)	
3	Poly(tetrafluoroethylene-cohexafluoropropylene) (FEP)	
4	Poly[tetrafluoroethylene-coperfluoro(propyl vinyl ether)] (PFA)	
5	Poly(vinyl fluoride) (PVF)	Cross-linking
6	Poly(vinylidene fluoride) (PVDF)	
7	Poly(trifluoroethylene) (PTrFE)	
8	Poly(ethylene-co-tetrafluoroethylene) (ETFE)	
9	Poly(ethylene-co-chlorotrifluoroethylene) (ECTFE)	
10	Poly(tetrafluoroethylene-co-propylene) (TFEP)	
11	Poly(vinylidene fluoride-cotetrafluoroethylene) (VDF-TFE)	
12	Poly(vinylidene fluoridechlorotrifluoroethylene) (VDF-CTFE)	
13	ly(vinylidene fluoride-cohexafluoropropylene) (VDF-HFP)	
14	Poly[tetrafluoroethylene-coperfluoro(methyl vinyl ether)] (TFE/PMVE)	

The two main characteristic properties of radiation-degradable FPs are the absence of a C-H bond and higher T_g than room temperature. FPs with a C-H bond or no C-H bond but low T_g are prone to radiation cross-linking. One of the examples includes TFE/PVME (poly[tetrafluoroethylene-co-perfluoro(methyl vinyl ether)]), which can be cross-linked at room temperature because of its low T_g (3°C). However, it has no C-H bond.

PTFE undergoes cross-linking when it is irradiated at or above the melting point in the absence of oxygen which can be explained by the fact that the electrical repulsion force can be overcome by kinetic energy at high temperatures. Furthermore, PTFE cross-linking occurs in fluorinated ethylene polymer (FEP) and perfluoroalkoxy alkanes (PFA) irradiated above T_g. The cross-link site (tertiary CF) was detected, as shown in Figure 1.14, and T-type cross-linking (end-linking) is proposed as a cross-linking structure of PTFE (Jiazhen et al., 1993; Katoh et al., 1999; Sun et al., 1994). There are two amorphous regions in melted PTFE. One is the original amorphous region existing before melting, and the other is the new amorphous region formed from previously crystalline regions by melting. Because PTFE molecular chains are rigid and easy to crystallize, cross-linking is expected to occur mainly in the previously amorphous region. It can be assumed that the T_m of PTFE is the temperature at which the mobility of the molecular chain in the previous amorphous regions becomes vigorous to a level high enough to overcome the electrical repulsion force between PTFE radicals to recombine with each other (Lyons, 1995).

In another example, the secondary radicals of polyvinylidene fluoride (PVDF) are slightly polarized in Figure 1.14. These secondary radicals formed in PVDF can be easily combined due to the electrical attraction force between radicals. Another characteristic effect of radiation on radiation cross-linkable fluoropolymers is the generation of a large amount of hydrogen fluoride (HF). From a chemical structural point of view, the sequence of $CF_2.CF_2$ affects the radiation cross-linking efficiency. In PVDF, $CF_2.CF_2$ groups are formed by the head-to-head polymerization of vinylidene fluoride (VDF). Besides, $CF_2.CF_2$ groups are introduced by the copolymerization of tetrafluoroethylene (TFE) with VDF.

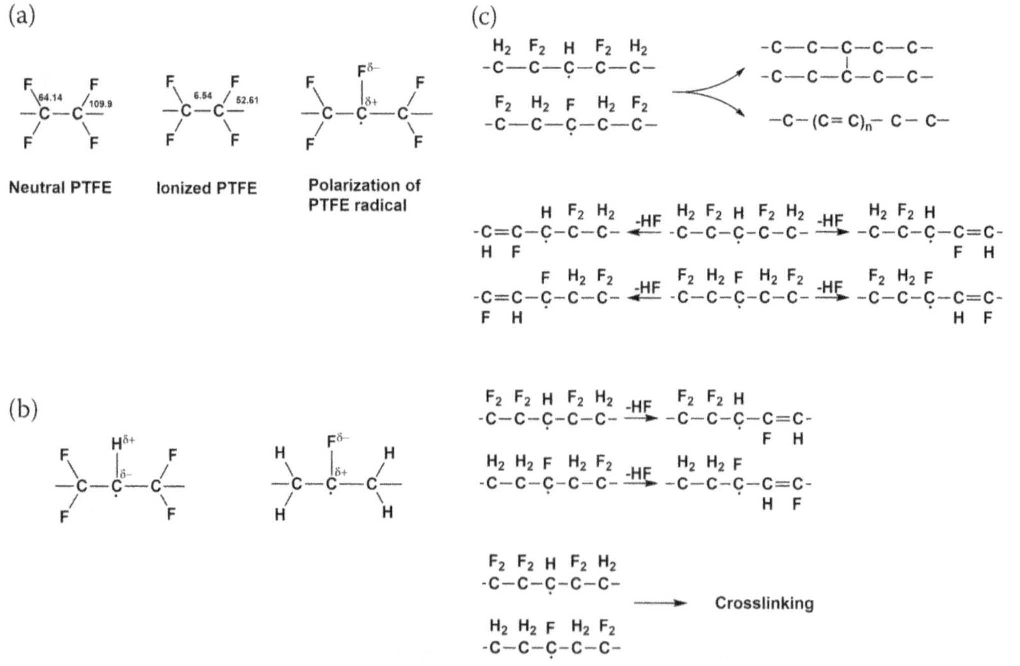

FIGURE 1.14 (a) Homolytic dissociation energies (kcal/mole) of neutral and ionized PTFE along with its polarized structure, (b) polarized structures of secondary radicals formed in PVDF, (c) radiation chemical reactions of PVDF (Schmidt et al., 2021).

The investigations have shown that the $CF_2.CF_2$ in PVDF enhances radiation cross-linking and impedes dehydrofluorination. The cross-linking and dehydrofluorination are competitive reactions of PVDF alkyl radicals (Figure 1.14). Usually, dehydrofluorination occurs on both sides of the main chain, as shown in Figure 1.14. Since CF_2CF_2 and CH_2CH_2 in PVDF would block dehydrofluorination, the probability of dehydrofluorination of alkyl radicals to form polyenyl radicals is decreased by half. Consequently, the possibility of recombination of radicals increases to form cross-links.

1.4.9 SILICON-CONTAINING POLYMERS

In the organosilicon polymers, the backbone constitutes a siloxane (Si-O) chain with high flexibility and free rotation around the Si-O axis or Si-C axis like methyl-silicone compounds. The freedom of motion increases the intermolecular distances between siloxane groups, decreasing the intermolecular forces. This results in amorphous nature leading to low modulus, low glass-transition temperature, and high gas permeability. As silicon is more electropositive than carbon, the bonds between Si and O are more ionic and have higher energies than those between C and O. The Si-O bond is about 50% ionic. Silicone rubbers undergo the radiolysis reaction, as shown in Figure 1.15.

The side chain R of general-purpose silicone rubber, MQ, is a methyl group (CH_3). Introducing a vinyl group ($CH=CH_2$) facilitates cross-linking of MQ while the phenyl group (C_6H_5) and trifluoropropyl group improve extremely low-temperature resistance and the oil resistance of MQ, respectively.

Industrially, silicone rubbers are cross-linked by organic peroxide, but the various studies also have the radiation cross-linking of silicone polymers. Alike the effect on other polymers, radiation cross-linking of silicone rubbers is also enhanced by the presence of the vinyl group while hindered by a phenyl group. The characteristic properties of silicone rubbers, like low bond energy of side-chain Si-C and high bond energy of main-chain Si-O, make them easy to cross-link by radiation. MQ, which has two methyl groups like PIB, can be readily cross-linked by radiation with the G(X) of 2.6–2.8 (Ormerod & Charlesby, 1963).

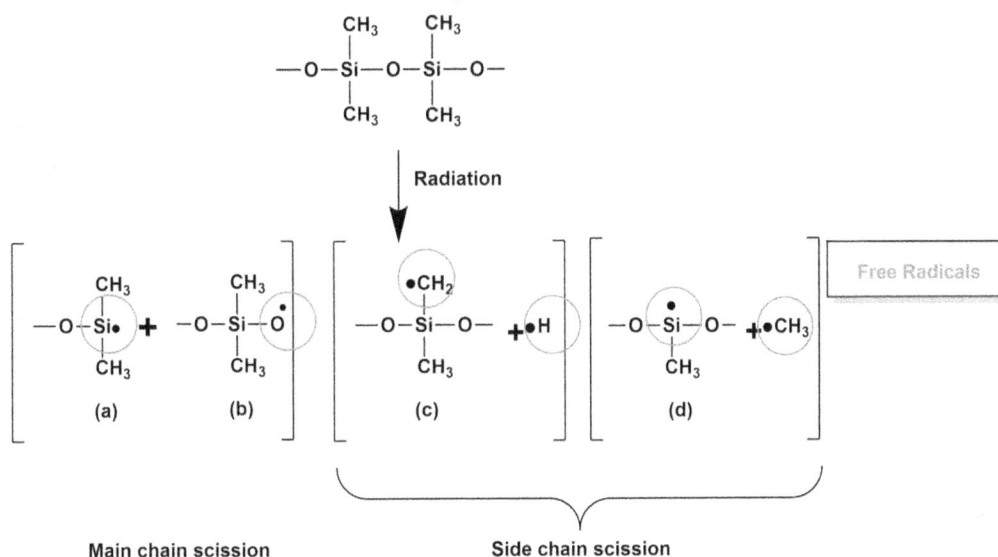

FIGURE 1.15 Radiolysis of silicon rubber. (*Content adopted from the* Wang et al., 2020 *with the permission by Elsevier.*)

TABLE 1.7

Effect of Branching Density on the Radiation Cross-Linking of PVAc

Average branching density	D_{gel} (kGy)	The average degree of polymerization	G(X)/G(S)
0.13	46.3	4500	1.39
0.27	41.6	4390	1.35
0.54	31.2	5160	1.52
0.92	24.2	5500	1.56
1.74	17.9	6490	1.42

TABLE 1.8

Comparison of the Radiation Responses of the Various Grades of PE Attributed to the Effect of Branching

Sr. no.		Grades of PE		
		LLDPE	**LDPE**	**HDPE**
1	Crystallinity (%)	50–60	45–556	65–85
2	Density (g/cc)	0.910–0.940	0.910–0.925	0.942–0.970
3	Tm (°C)	122–124	98–115	130–137
4	G(X)	2.5	1.4	1.0
5	G(S)	0.4	0.5	0.2

1.4.10 BRANCHED POLYMERS

There are two conflicting effects on the cross-linking efficiency of polymers, viz., suppressing the cross-linking due to the presence of tertiary carbon atoms and enhancing the cross-linking due to increased molecular mobility. The hydrogen atom attached to the tertiary carbon atom at the branching might be particularly susceptible to fracture by radiation. The resulting tertiary radical tends to proceed to β-bond scission resulting in the enhanced main-chain scission like in polyvinyl acetate (PVAc). Table 1.7 shows the effect of branching density on the radiation cross-linking of PVAc, where it can be seen that the G(X)/G(S) ratio is not sensitive to the branching density. However, D_{gel} and the apparent cross-linking efficiency decrease with increasing branching density in which approximately 15% of the branch points are exposed to radiation and break preferentially before the gel point (Mittelhauser & Graessley, 1969).

In the case of PE, radiation cross-linking is enhanced by introducing branching due to increased molecular mobility. Table 1.8 shows the physical properties of the G(X) and G(S) of LLDPE, LDPE, and HDPE. As we know, the LLDPE is a copolymer of ethylene and β-olefin, where but-1-ene, hex-1-ene,4-methylpent-1-ene, and oct-1-ene are used as the β-olefins. The number of short branches is 10–30 per 1000 CH_2 groups, while such short additions in PE increase the molecular mobility and radiation cross-linking efficiency (Li et al., 2002).

1.5 CONCLUSION

It can be concluded from all the compiled data and inferences drawn from the reported studies that radiation cross-linking is a very emerging type of cross-linking technique and widely studied for the processing of abundant types of polymers and very effective in terms of efficiency and selectivity.

Amongst all the kinds of radiation, being highly energetic, γ rays provide the best penetration and ionization of the target molecules, as we have seen from the example of the ionization of water molecules. But, as mentioned in all the radiation types, the foremost disadvantage of radiation cross-linking is the availability and the energy requirement for the radiation source. This disadvantage leads to the reduced cost-effectiveness of radiation cross-linking processes. Moreover, when talking about the factors affecting the radiation cross-linking of polymers, it is very much clear from the discussion described above that the chemical structure of the target polymeric molecule exceptionally decides the dosage level of the radiation energy. To address the dependency on the chemical structure, the hydrophobicity of an atom, secondary inter- and intra-molecular forces of attraction and repulsion, resonance-inductive-mesomeric effects, and the arrangement and orientation of the polymeric chains are the cruxes to be considered. These foundations decide the energy requirement and are very helpful while tailoring the end properties of the polymers during radiation processing.

REFERENCES

Abdelmonem, A. M., & Amin, R. M. (2014). Rapid green synthesis of metal nanoparticles using pomegranate polyphenols. *International Journal of Sciences: Basic and Applied Research (IJSBAR)*, *15*(1), 57–65.

Anbar, M. (1961). Photolysis of hydrogen peroxide at high light intensities in aqueous solutions enriched in O^{18}. *Transactions of the Faraday Society*, *57*, 971–982. 10.1039/TF9615700971

Ashcroft, W. R. (1993). Curing agents for epoxy resins. *Chemistry and Technology of Epoxy Resins*, Ellis Bryan, 37–71. 10.1007/978-94-011-2932-9_2, Springer Dordrecht, Germany.

Bailey, A. J. (1991). The chemistry of natural enzyme-induced cross-links of proteins. *Amino Acids*, *1*(3), 293–306. 10.1007/BF00813999

Bell, V. L., & Pezdirtz, G. F. (1983). Effects of ionizing radiation on linear aromatic polyesters. *Journal of Polymer Science: Polymer Chemistry Edition*, *21*(11), 3083–3092. 10.1002/pol.1983.170211106

Bengtsson, M., & Oksman, K. (2006). The use of silane technology in cross-linking polyethylene/wood flour composites. *Composites Part A: Applied Science and Manufacturing*, *37*(5), 752–765. 10.1016/j.compositesa.2005.06.014

Bhowmick, A. K., & Vijayabaskar, V. (2006). Electron beam curing of elastomers. *Rubber Chemistry and Technology*, *79*(3), 402–428. 10.5254/1.3547944

Bhuyan, M. M., Dafader, N. C., Hara, K., Okabe, H., Hidaka, Y., Rahman, M. M., Khan, M. M. R., & Rahman, N. (2016). Synthesis of potato starch-acrylic-acid hydrogels by gamma radiation and their application in dye adsorption. *International Journal of Polymer Science*, *2016*. 10.1155/2016/9867859

Böhler, E., Warneke, J., & Swiderek, P. (2013). Control of chemical reactions and synthesis by low-energy electrons. *Chemical Society Reviews*, *42*(24), 9219–9231. 10.1039/c3cs60180c

Boochathum, P., & Prajudtake, W. (2001). Vulcanization of cis- and trans-polyisoprene and their blends: Cure characteristics and cross-link distribution. *European Polymer Journal*, *37*(3), 417–427. 10.1016/S0014-3057(00)00137-3

Burns, N. M. (1979). The radiation cross-linking of ethylene copolymers. *Radiation Physics and Chemistry (1977)*, *14*(3–6), 797–808. 10.1016/0146-5724(79)90115-8

Charlesby, A., Pinner, S. H., & A, P. R. S. L. (1959). Analysis of the solubility behaviour of irradiated polyethylene and other polymers. *Proceedings of the Royal Society of London. Series A. Mathematical and Physical Sciences*, *249*(1258), 367–386. 10.1098/rspa.1959.0030

Chatterjee, A., Magee, J. L., & Dey, S. K. (1983). The role of homogeneous reactions in the radiolysis of water. *Radiation Research*, *96*(1), 1–19. 10.2307/3576159

Chernova, I. K., Leshchenko, S. S., Golikov, V. P., Karpov, V. L. (1980). Radiation-chemical changes in polystyrene and its methyl derivatives. *Polymer Science*, *22*(10), 2382–2394.

Chowdhury, R., & Banerji, M. S. (2005). Electron beam irradiation of ethylene-propylene terpolymer: Evaluation of trimethylol propane trimethacrylate as a cross-link promoter. *Journal of Applied Polymer Science*, *97*(3), 968–975. 10.1002/app.21795

Colomb, H. O., Trecker, D. J., & Osterholtz, F. D. (1970). Radiation-convertible polymers from norbornenyl derivatives. Crosslinking with ionizing radiation. *Journal of Applied Polymer Science*, *14*(7), 1659–1670. 10.1002/app.1970.070140701

Crivello, James V., & Aldersley, M. F. (2013). Supramolecular diaryliodonium salt-crown ether complexes as cationic photo-initiators. *Journal of Polymer Science, Part A: Polymer Chemistry*, *51*(4), 801–814. 10.1002/pola.26452

Crivello, J. V., & Lam, J. H. W. (1977). Diaryliodonium salts. A new class of photoinitiators for cationic polymerization. *Macromolecules, 10*(6), 1307–1315. 10.1021/ma60060a028

Crivello, James V., & Liu, S. (2000). Photoinitiated cationic polymerization of epoxy alcohol monomers. *Journal of Polymer Science, Part A: Polymer Chemistry, 38*(3), 389–401.

Darwis, D., Mitomo, H., Enjoji, T., Yoshii, F., & Makuuchi, K. (1998). Heat resistance of radiation cross-linked poly (ε-caprolactone). *Journal of Applied Polymer Science, 68*(4), 581–588.

Deng, P., Liu, M., Zhang, W., & Sun, J. (2007). Preparation and physical properties of enhanced radiation induced cross-linking of ethylene-vinyl alcohol copolymer (EVOH). *Nuclear Instruments and Methods in Physics Research, Section B: Beam Interactions with Materials and Atoms, 258*(2), 357–361. 10.1016/j.nimb.2007.01.300

Diani, J., Liu, Y., & Gall, K. (2006). Finite strain 3D thermoviscoelastic constitutive model for shape memory polymers. *Polymer Engineering and Science, 46*(4), 486–492. 10.1002/pen.20497

Felder, L., Kirchmayr, R., Bellus, D. (1978). Initiators for photopolymerization. US Patent 4,088,554.

Forsythe, J. S., & Hill, D. J. T. (2000). Radiation chemistry of fluoropolymers. *Progress in Polymer Science (Oxford), 25*(1), 101–136. 10.1016/S0079-6700(00)00008-3

Fricke, H., Hart, E. J., & Smith, H. P. (1938). Chemical reactions of organic compounds with X-ray activated water. *Journal of Chemical Physics, 6*(5), 229–240. 10.1063/1.1750237

Fuchs, B., Lappan, U., Lunkwitz, K., & Scheler, U. (2002). Radiochemical yields for cross-links and branches in radiation-modified poly(tetrafluoroethylene). *Macromolecules, 35*(24), 9079–9082. 10.1021/ma011529z

Fuchs, B., & Scheler, U. (2000). Branching and cross-linking in radiation-modified poly(tetrafluoroethylene): A solid-state NMR investigation. *Macromolecules, 33*(1), 120–124. 10.1021/ma9914873

Geissler, W., Zott, H., & Heusinger, H. (1978). Investigations on the mechanism of radiation induced cross-linking in ethylene-propylene-diene terpolymers. *Die Makromolekulare Chemie, 179*(3), 697–705. 10.1002/macp.1978.021790312

Ghobashy, M. M. (2018). Ionizing radiation-induced polymerization, 113-134. Djezzar, Boualem, *In Book: Ionizing Radiation Effects and Applications*, Boualem Djezzar, 113–134, Intech Open Science, London.

Gueven, O. (2004). An overview of current developments in applied radiation chemistry of polymers. *Advances in Radiation Chemistry of Polymers, IAEA-TECDOC-1420, November*, 33–39.

Hill, D. J. T., O'Donnell, J. H., Perera, M. C. S., & Pomery, P. J. (1995). High energy radiation effects on halogenated butyl rubbers. *Polymer, 36*(22), 4185–4192. 10.1016/0032-3861(95)92211-V

Hill, D. J. T., Preston, C. M. L., Salisbury, D. J., Whittaker, A. K. (2001). Molecular weight changes and scission and cross-linking in poly(dimethyl siloxane) on gamma radiolysis. *Radiation Physics and Chemistry, 62*(1), 11–17. 10.1016/S0969-806X(01)00416-9

Jacobs, P. F. (1993). Rapid prototyping & manufacturing—Fundamentals of stereolithography. *Journal of Manufacturing Systems, 12*(5), 430–433. 10.1016/0278-6125(93)90311-g

Jiazhen, S. (2001). The effect of chain flexibility and chain mobility on radiation cross-linking of polymers. *Radiation Physics and Chemistry, 60*(4–5), 445–451. 10.1016/S0969-806X(00)00419-9

Jiazhen S., Yuefang Z., Xiaoauana Z., Wanxi Z. (1993). Studies in radiation cross-linking of fluoropolymers. *Radiation Physics and Chemistry, 42*(1–3), 139–142. 10.1016/0969-806X(93)90223-H

Kahveci, M. U., Mangold, C., Frey, H., & Yagci, Y. (2014). Graft copolymers with complex polyether structures: Poly(ethylene oxide)-graft-poly(isobutyl vinyl ether) by combination of living anionic and photoinduced cationic graft polymerization. *Macromolecular Chemistry and Physics, 215*(6), 566–571. 10.1002/macp.201300794

Katoh, E., Sugisawa, H., Oshima, A., Tabata, Y., Seguchi, T., & Yamazaki, T. (1999). Evidence for radiation induced cross-linking in polytetrafluoroethylene by means of high-resolution solid-state 19F high-speed MAS NMR. *Radiation Physics and Chemistry, 54*(2), 165–171. 10.1016/S0969-806X(98)00250-3

Lappan, U., Geißler, U., & Scheler, U. (2007). The influence of the irradiation temperature on the ratio of chain scission to branching reactions in electron beam irradiated polytetrafluoroethylene (PTFE). *Macromolecular Materials and Engineering, 292*(5), 641–645. 10.1002/mame.200600495

Li, J., Peng, J., Qiao, J., Jin, D., & Wei, G. (2002). Effect of gamma irradiation on ethylene-octene copolymers. *Radiation Physics and Chemistry, 63*(3–6), 501–504. 10.1016/S0969-806X(01)00633-8

Lyons, B. J. (1995). Radiation cross-linking of fluoropolymers–A review. *Radiation Physics and Chemistry, 45*(2), 159–174. 10.1016/0969-806X(94)E0002-Z

Manalia, E., Stelescu, M. D., Craciun, G. (2012). Aspects regarding radiation cross-linking of elastomers. Boczkowska, Anna, 3-34' *In Book: Advanced Elastomers –Technology, Properties and Applications*, Intech Open Science, London.

Marcilla, A., Ruiz-Femenia, R., Hernandez, J., Garcia-Quesada, J. C. (2006). Thermal and catalytic pyrolysis of cross-linked polyethylene. *Journal of Analytical and Applied Pyrolysis*, 76(1-2), 254–259. 10.1016/j.jaap.2005.12.004

Mesu, J. G., Beale, A. M., De Groot, F. M. F., & Weckhuysen, B. M. (2007). Observing the influence of x-rays on aqueous copper solutions by in situ combined video/XAFS/UV-Vis spectroscopy. *AIP Conference Proceedings*, 882(February), 818–820. 10.1063/1.2644674

Mitomo, H., Watanabe, Y., Yoshii, F., & Makuuchi, K. (1995). Radiation effect on polyesters. *Radiation Physics and Chemistry*, 46(2), 233–238. 10.1016/0969-806X(95)00018-S

Mitsui, H., Hosoi, F., & Ushirokawa, M. (1975). Effect of double bonds on the γ-radiation-induced cross-linking of polyethylene. *Journal of Applied Polymer Science*, 19(2), 361–369. 10.1002/app.1975.070190204

Mittelhauser, H. M., & Graessley, W. W. (1969). Radiation cross-linking studies on branched polyvinyl acetate. *Polymer*, 10(1969), 439–450. 10.1016/0032-3861(69)90053-6

Mizera, A., Manas, M., Holik, Z., Manas, D., Stanek, M., Cerny, J., Bednarik, M., & Ovsik, M. (2012). Properties of selected polymers after radiation cross-linking. *International Journal of Mathematics and Computers in Simulation*, 6(6), 592–599.

Nugroho, P., Mitomo, H., Yoshii, F., & Kume, T. (2001). Degradation of poly(L-lactic acid) by γ-irradiation. *Polymer Degradation and Stability*, 72(2), 337–343. 10.1016/S0141-3910(01)00030-1

Obermeier, B., Wurm, F., Mangold, C., & Frey, H. (2011). Multifunctional poly(ethylene glycol)s. *Angewandte Chemie - International Edition*, 50(35), 7988–7997. 10.1002/anie.201100027

Ormerod, M. G., & Charlesby, A. (1963). The radiation chemistry of some polysiloxanes: An electron spin resonance study. *Polymer*, 4(1963), 459–470. 10.1016/0032-3861(63)90059-4

Postolache, C., & Matei, L. (2007). Evaluation of fundamental processes in macromolecular structures radiolysis using quantum-chemical methods. *Radiation Physics and Chemistry*, 76(8–9), 1267–1271. 10.1016/j.radphyschem.2007.02.075

Queiroz, R. G., Varca, G. H. C., Kadlubowski, S., Ulanski, P., & Lugão, A. B. (2016). Radiation-synthesized protein-based drug carriers: Size-controlled BSA nanoparticles. *International Journal of Biological Macromolecules*, 85, 82–91.

Reeves, K. G., & Kanai, Y. (2017). Electronic excitation dynamics in liquid water under proton irradiation. *Scientific Reports*, 7, 40379. 10.1038/srep40379

Rouif, S. (2005). Radiation cross-linked polymers: Recent developments and new applications. *Nuclear Instruments and Methods in Physics Research, Section B: Beam Interactions with Materials and Atoms*, 236(1–4), 68–72. 10.1016/j.nimb.2005.03.252

Sáenz-Galindo, A., López-López, L. I., Cruz-Duran, F. N. de la, Castañeda-Facio, A. O., Rámirez-Mendoza, L. A., Córdova-Cisneros, K. C., & Loera-Carrera, D. de. (2018). Badea GeorgianaIleana Radu Gabriel Lucian Applications of carboxylic acids in organic synthesis, nanotechnology and polymers. *Carboxylic Acid - Key Role in Life Sciences*. 10.5772/intechopen.74654

Sartipi, S., Makkee, M., Kaptejin, F., Gascon, J. (2014). Catalysis engineering of bifunctional solids for the one-step synthesis of liquid fuels from syngas: A review. *Catalysis Science & Technology*, 4, 893–907. 10.1039/C3CY01021J

Sasaki Y. (1996) Initiator for photopolymerization. US Patent 5,480,918.

Schmidt, M., Zahn, S., Gehlhaar, F., Prager, A., Griebel, J., Kahnt, A., Knolle, W., Konieczny, R., Schulze, A. (2021). Radiation-induced graft immobilization (RIGI): Covalent binding of non-vinyl compounds on polymer membranes. *Polumers (MDPI)*, 13, 1849. 10.3390/polym13111849

Schulze, U., Majumder, P. Sen, Heinrich, G., Stephan, M., & Gohs, U. (2008). Electron beam cross-linking of atactic poly (propylene): Development of a potential novel elastomer. *Macromolecular Materials and Engineering*, 293(8), 692–699. 10.1002/mame.200800093

Shukushima, S., Hayami, H., Ito, T., & Nishimoto, S. Ichi. (2001). Modification of radiation cross-linked polypropylene. *Radiation Physics and Chemistry*, 60(4–5), 489–493. 10.1016/S0969-806X(00)00395-9

Silva Aquino, K. A. da. (2012). Sterilization by gamma irradiation. Adrovic Feriz, *Gamma Radiation*. 10.5772/34901

Song, C. L., Yoshii, F., & Kume, T. (2001). Radiation cross-linking of biodegradable poly(butylene succinate) at high temperature. *Journal of Macromolecular Science - Pure and Applied Chemistry*, 38 A(9), 961–971. 10.1081/MA-100104947

Sun, J., Zhang, Y., Zhong, X., & Zhu, X. (1994). Modification of polytetrafluoroethylene by radiation-1. Improvement in high temperature properties and radiation stability. *Radiation Physics and Chemistry*, 44(6), 655–659. 10.1016/0969-806X(94)90226-7

Takashika, K., Oshima, A., Kuramoto, M., Seguchi, T., & Tabata, Y. (1999). Temperature effects on radiation induced phenomena in polystyrene having atactic and syndiotactic structures. *Radiation Physics and Chemistry*, *55*(4), 399–408. 10.1016/S0969-806X(99)00197-8

Treutler, O., & Ahlrichs, R. (1995). Efficient molecular numerical integration schemes. *The Journal of Chemical Physics*, *102*(1), 346–354. 10.1063/1.469408

Turgis, J. D., Vergé, C., & Coqueret, X. (2003). Composition effects on the EB-induced cross-linking of some acrylate and methacrylate copolymers. *Radiation Physics and Chemistry*, *67*(3–4), 409–413. 10.1016/S0969-806X(03)00076-8

Turner, D. T. (1966). Radiation cross-linking of a trans-1, cpolyisoprene in the liquid and solid states. *Polymer Letters*, *4*, 717–720.

Varca, G. H. C., Ferraz, C. C., Lopes, P. S., Mathor, M. Beatriz, Grasselli, M., & Lugão, A. B. (2014). Radio-synthesized protein-based nanoparticles for biomedical purposes. *Radiation Physics and Chemistry*, *94*(1), 181–185. 10.1016/j.radphyschem.2013.05.057

Vijayabaskar, V., Tikku, V. K., & Bhowmick, A. K. (2006). Electron beam modification and cross-linking: Influence of nitrile and carboxyl contents and level of unsaturation on structure and properties of nitrile rubber. *Radiation Physics and Chemistry*, *75*(7), 779–792. 10.1016/j.radphyschem.2005.12.030

Wang, J., & Mao, Q. (2012). Methodology based on the PVT behavior of polymer for injection molding. *Advances in Polymer Technology*, *32*(2013), 474–485. 10.1002/adv

Wang, P. C., Yang, N., Liu, D., Qin, Z-M., An, Y., Chen, H-B. (2020). Coupling effects of gamma irradiation and absorbed moisture on silicone foam. *Materials & Design*, *195*, 108998. 10.1016/j.matdes.2020.108998

Wenwei, Z., Xiaoguang, Z., Li, Y., Yuefang, Z., Jun, X., & Jiazhen, S. (1993). Radiation cross-linking of chlorinated polyisobutylene. *Polymer Degradation and Stability*, *41*, 5–8.

Wunsch, K., & Dalcolmo, H. (1992). Structure-reactivity relationships in radiation-induced cross-linking of polyethylenes. *International Journal of Radiation Applications and Instrumentation. Part C. Radiation Physics and Chemistry*, *39*(5), 443–448. 10.1016/1359-0197(92)90057-M

Xu, J. W., Xu, F., & Luo, Y. L. (2015). Core cross-linked H-type poly(methacrylic acid)-block-hydroxyl terminated polybutadiene-block-poly(methacrylic acid) four-armed star block copolymer micelles for intercellular drug release. *Journal of Bioactive and Compatible Polymers*, *30*(4), 349–365. 10.1177/0883911515578871

Yang, Z., Peng, H., Wang, W., & Liu, T. (2010). Crystallization behavior of poly(ε-caprolactone)/layered double hydroxide nanocomposites. *Journal of Applied Polymer Science*, *116*(5), 2658–2667. 10.1002/app

Zaharescu, T., Setnescu, R., Jipa, S., & Setnescu, T. (2000). Radiation processing of polyolefin blends. I. Crosslinking of EPDM-PP blends. *Journal of Applied Polymer Science*, *77*(5), 982–987.

Zhang, C., Zhang, M., Wu, G., Wang, Y., & Zhang, L. (2018). Radiation cross-linking and its application Adrovic Feriz, In *Radiation Technology for Advanced Materials: From Basic to Modern Applications*. NetherlandsElsevier Inc. 10.1016/B978-0-12-814017-8.00003-2

Zhong, X., Sun, J., Wang, F., & Sun, Y. (1992). XPS study of radiation cross-linked copolymer of tetra-fluoroethylene with hexafluoropropylene. *Journal of Applied Polymer Science*, *44*(4), 639–642. 10.1002/app.1992.070440410

Zhou, C., Souza Gomes Ailton De (2012). Bulk preparation of radiation cross-linking poly (urethane-Imide). *New Polymers for Special Applications*. 10.5772/48282, Intech Open Science, London.

2 Tuning desired properties by tailoring radiation grafted polymeric materials: Preparation and characterization

S. Selambakkannu*, N. A. F. Othman, T. M. Ting, and N. L. Ishak

Radiation Processing Technology Division, Malaysia Nuclear Agency, Bangi, Kajang, Malaysia

*Corresponding author: sarala@nuclearmalaysia.gov.my

CONTENTS

2.1 INTRODUCTION

Nowadays, rapid improvement in polymer chemistry in tailor-making the desirable properties of polymeric materials with specified structures for specific applications is attracting much attention (Pino-Ramos et al., 2017). Polymeric materials have been extensively used in various industrial applications based on their favorable characteristics. The performance of these materials initially depends on their properties such as biocompatibility, wettability, chemical composition, and mechanical strength. Thus, radiation processing application in surface tailoring is one of particular interest for polymeric materials in achieving desired properties.

DOI: 10.1201/9781003321910-2

Radiation processing is used worldwide, especially for sterilization, rubber vulcanization, and surface modification such as cross-linking of polymeric material, coating, and grafting. The use of radiation processing has been explored for 50 years and remains extensively continued (Cleland, Parks, & Cheng, 2003). In Malaysia, this technology has been started in the 1990s. This technology can produce new polymer chains reaction without the need for chemicals as catalysts or starters. In radiation processing, a polymeric material or product is intentionally irradiated to preserve, modify and improve its characteristics (Pino-Ramos et al., 2017). Moreover, radiation processing has several major benefits over conventional once such as being more environmentally friendly, consuming less energy than thermally or chemically processes, and cost-saving. Despite this, radiation processing has a wider range in application especially in polymeric surface tailoring to enhance the surface properties including surface coating, surface functionalization, and surface grafting (Mozetič, 2019).

Surface grafting is one of the most familiar methods that can be used to overcome the short-comings of polymers. Generally, grafting is the process of covalently attaching or developing a polymer on an existing polymer chain (Vega-Hernández et al., 2021). The main purpose of grafting is to fuse the monomer to obtain the different chemical properties of the main polymeric molecule. However, radiation-induced graft copolymerization (RIGC) is the technique for the functionalization of the surfaces of the existing polymeric materials for various types of applications (Bucio & Burillo, 2009) such as healthcare, industrial, environment, and energy sustainability. This technique is not the new trend for the surface tailoring of polymeric materials, but it is a preferred method over conventional ones that use the chemical approach (Barsbay & Güven, 2019) and also the potential to produce a variety of functions without changing physical characteristics and configuration of the materials (Setia, 2018). Generally, RIGC uses high-energy radiation which sources are electron beam (EB), gamma rays, and ultraviolet (UV) radiation.

The initial step in RIGC is to create the active site (free radicals). The samples will be placed in the zipper bag and purged with nitrogen gas (inert gas). Upon the irradiation, radicals are formed as active sites on the polymeric backbone. Subsequently, the second stage basically involves the polymeric surface modification by monomer reaction in order to form graft polymerization. In this stage, the grafting solution (monomer) is bubbled with nitrogen gas and transferred to the glass ampoule that contains the irradiated sample, and the grafting process occurs at an elevated temperature. The third stage engaged with the changes of the polymeric surface by functionalization of functional groups by chemically bonding of polar, non-polar monomers, and other derivatives. The functionalization will affect the polymer properties without impacting the backbone of the bulk polymeric material. The example of functional groups is listed in Table 2.1.

RIGC could be performed in two main routes that denominate as a pre-irradiation technique in inert or vacuum condition and per-oxidation; and simultaneous or mutual irradiation technique (Tamada, 2018). Figure 2.2 illustrates the types of radiation-induced grafting.

In the pre-irradiation method, the trunk polymer or material is irradiated either in a vacuum and inert atmospheric conditions or with the presence of oxygen. Moreover, the pre-irradiation method in vacuum or inert conditions using electron beam is commonly in practice for the preparation of RIGC polymeric materials. The mechanism of the pre-irradiation in inert conditions is as below:

$$P \xrightarrow{\text{Radiation } \gamma, \ \beta} P^{\cdot} \tag{2.1}$$

$$P^{\cdot} + M \rightarrow PM^{\cdot} \tag{2.2}$$

The polymeric substrate is irradiated in an inert atmosphere using an electron beam with the absence of dry ice in order to create the active radical sites of the polymer (Equation 2.1). Accordingly, irradiation at low temperatures will increase the degree of grafting. In the pre-irradiation method, this process is easy to control and meet for large-scale production due to separated steps of reaction and graft reaction process Tamada, 2018. After the irradiation, the

TABLE 2.1

Common Functional Groups

Functional groups	Structure	Formula
Hydroxyl	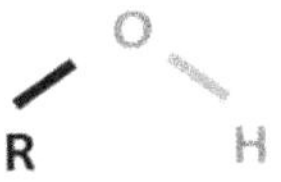	–ROH
Carbonyl	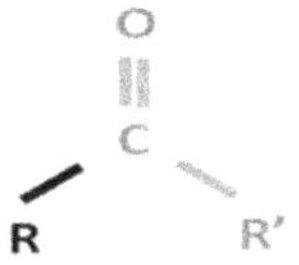	–COR
Carboxyl	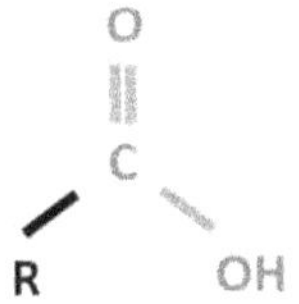	–RCOOH
Amide	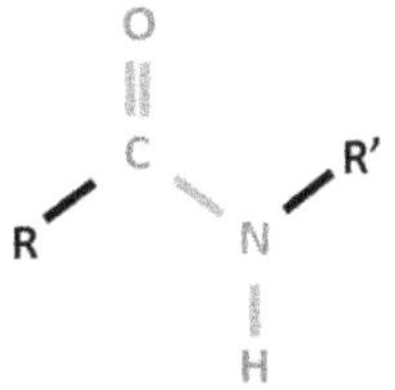	–RCONHR
Amine	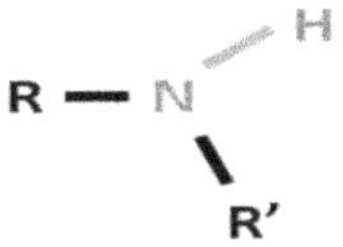	–RNHR
Sulfonate	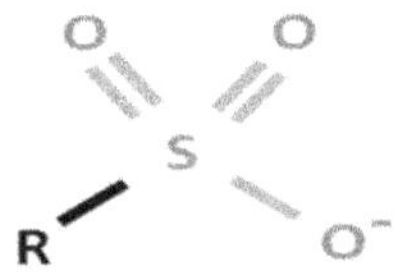	– RSO$_3$
Phosphate	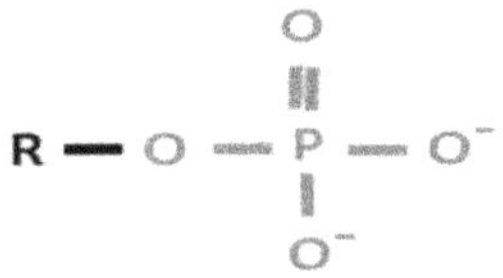	– RPO$_4$

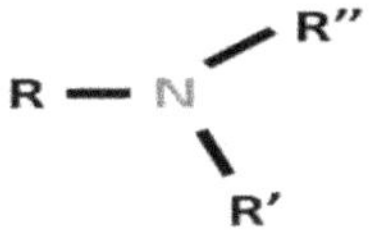

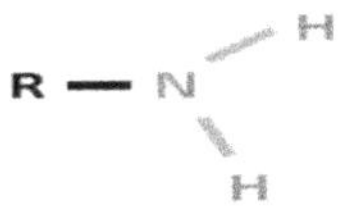

substrate should be stored at a temperature of dry ice (−78°C) in order to preserve the free radicals. The polymeric substrate is then contacted with the monomers in a vacuum or an inert condition at a certain temperature (Equation 2.2). The grafting process needs to be carried out immediately after the irradiation in order to reduce the radicals' loss. This technique provided low chances of homopolymer formation. The pre-irradiation in the presence of air or oxygen is called per-oxidation which the formation of peroxides and hydroperoxides will be formed onto the materials surface backbone (Magaña et al., 2015). The simple equation is as follows:

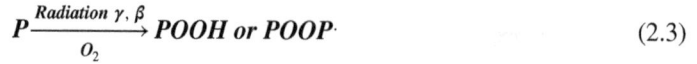

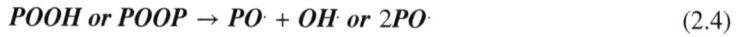

$$P \xrightarrow[O_2]{Radiation \; \gamma, \; \beta} POOH \; or \; POOP^{\cdot} \tag{2.3}$$

$$POOH \; or \; POOP \rightarrow PO^{\cdot} + OH^{\cdot} \; or \; 2PO^{\cdot} \tag{2.4}$$

$$PO^{\cdot} + OH^{\cdot} \; or \; 2PO^{\cdot} + M \rightarrow POM \tag{2.5}$$

In this way, the substrate is irradiated and forms the peroxide or hydro-peroxide products as in (Equation 2.3 and Equation 2.4), followed by high-temperature reaction with the monomers (Equation 2.5). Then the peroxides decompose to radicals for grafting initiation. The grafting process for this type of irradiation can be performed at any time due to the long periods of irradiated substrate storage Pino-Ramos et al., 2017 and the stability of peroxy products at room temperature (Kakati, Bora, & Deka, 2015). Nevertheless, the disadvantage of this method is the production of hydroxyl radical (from homolytic cleavage triggered the forming of homopolymer (Kakati et al., 2015).

The mechanism of simultaneous or mutual technique is given in the following sequence of equations:

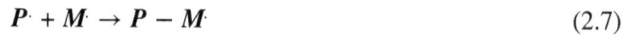

$$P + M \xrightarrow{Radiation \; \gamma, \; \beta} P^{\cdot} + M^{\cdot} \tag{2.6}$$

$$P^{\cdot} + M^{\cdot} \rightarrow P - M^{\cdot} \tag{2.7}$$

For this method, the trunk polymer will be irradiated in a monomer solution as in (Equation 2.6 and Equation 2.7) with a low dose rate in order to prevent the homopolymer (Kodama, Barsbay, & Güven, 2014; Sonnier, Otazaghine, Viretto, Apolinario, & Ienny, 2015). However, since the monomer is irradiated simultaneously to form the grafted material, thus an excessive amount of homopolymer will easy to form and reduce the efficiency of the grafting yield (Kakati et al., 2015). Therefore, the following ways are usually applied to improve the degree of grafting (Kakati et al., 2015; Zhou et al., 2015).

- Addition of inhibitors (e.g., Cu^{2+}, Fe^{2+}) to avoid unnecessary reactions.
- Lowest the irradiation dose rate in order to control the homo-polymerization.
- Regulate the homo-polymerization by the dilution of monomer so that the radicals formed in material backbone quickly take place than the formation of monomer units.
- Using swelling agents.

Upon finishing the grafting process, un-reacted monomer and homopolymers were removed by washed with the suitable solvent. Then the samples were dried until constant weight was reached. On the other hand, the degree of grating, DG (wt%) is one of the approach to quantify grafting using the equation as follows (Sutirman, Sanagi, Abd Karim, & Ibrahim, 2016):

$$\mathbf{DG}(\mathbf{wt\%}) = \frac{(\mathbf{W}_f - \mathbf{W}_0)}{\mathbf{W}_0} \times 100\% \tag{2.8}$$

This estimation can be performed based on the gravimetric method as represented in (Equation 2.8), whereby W_f and W_0 are the weights of the graft chain and trunk polymer, respectively. The degree of grafting are important in determined the grafted chain attached on the trunk polymer. There are many factors that affected the degree of grafting which will be discussed in next section. Therefore, the efficiency of the radiation-induced grafting process depends on these factors. The details about these factors will be discussed in the following section.

Myriad of approaches are presently available for polymeric materials characterization in terms of the assessment of their properties in order to provide the information upon RIGC. The comparative analysis on polymer surface for grafting confirmation can be determined using attenuated total reflection Fourier transform infrared spectroscopy (ATR-FTIR) (Mohamed, Jaafar, Ismail, Othman, & Rahman, 2017; Sakulaue, Lertvanithphol, Eiamchai, & Siriwatwechakul, 2021), X-ray photoelectron spectroscopy (XPS) (N. Othman et al., 2021), X-ray diffraction (XRD) (Kaith et al., 2015; Madani, 2011), and nuclear magnetic resonance (NMR) (Kwan & Marić, 2016). In addition, surface morphology also plays an important role in the characterization of grafted materials due to the ability to indicate the micro-structures of the grafted polymeric materials. For now, scanning electron microscopy (SEM), transmission electron microscopy (TEM), and atomic force microscopic (AFM) are the usual analysis in obtaining the evolution of structural and surface morphological of the materials (Korolkov et al., 2017). Other approaches that can be used and determined to indicate the grafted materials including thermal analysis (differential scanning calorimetry (DSC), thermogravimetric analysis (TGA) and others) (Hong, Okabe, Hidaka, & Hara, 2018; Madani, 2011), mechanical (e.g., tensile strength, dynamic mechanical analyser (DMA) (Z. Li, Tang, Dai, Wang, & Bai, 2016; Montoya-Villegas et al., 2020), rheological (Wu et al., 2016), and optical (e.g., UV-Vis, optical water contact angle) (Agrawal, Negi, Pradhan, Dash, & Samal, 2017; Madani, 2011).

In a nutshell, surface tailoring by radiation induced grafting presents the unique advantages over the other grafting methods due to its gains such as simplicity, homogeneous reaction; control over process and do not need the initiators or catalysts to initiate the polymerization (Shin, Hong, Lim, Son, & Kim, 2017). Moreover, there are other advantages of surface tailoring using radiation induced grafting such as:

- In comparison to conventional methods, radiation-induced grafting is being more economical and simple, capability in permitting the alteration of material composition within the structure (Barsbay & Güven, 2019; Nasef & Güven, 2012).
- The grafting yield can be controlled by changing the reaction conditions (e.g., irradiation dose, dose rate, monomer type, monomer concentration, solvent type) (Barsbay & Güven, 2019).
- The ability to provide polymeric material with new functions and properties improvement without chemical initiators and easy scaling (Tamada, 2018).
- Possibility of tailoring the polymeric materials in any physical forms (e.g., film, membrane, fiber, fabric, powder) at a convenient temperature (room temperature) (Ashfaq et al., 2020).

Hence, this method is an effective method in tailoring the properties of the polymeric materials.

2.2 PREPARATION OF RADIATION-GRAFTED POLYMERIC MATERIALS

The prominent intrinsic criteria involved in the preparation of radiation-grafted polymeric materials are, selection of suitable polymer substrate, radiation type, efficient monomer precursor, solvent, additives, and experimental conditions. However, the mentioned criteria ultimately, rely on the desired end application of the grafted materials. Passively, the development of multiple types of ion exchange and chelating polymeric materials with tailored physical and chemical properties by radiation-induced grafting technique favors the wide range of environmental and industrial applications such as removal of toxic metals, heavy metals, rare earth elements, organic dyes, antimicrobial films, and also as ion exchange membrane in the fuel cell (Nasef et al., 2016).

Radiation-induced graft copolymerization (RIGC) confers surface modification of polymer materials in order to cater to the specific properties required to serve the desired functions. RIGC enables the modification of almost all types of polymer surfaces, particularly in the presence of respective polar monomers that permits the desired functional properties (Nasef et al., 2021). On the other hand, surface modification of polymer materials by RIGC has proven as most versatile and useful in the creation of polymers with tailored physical and chemical properties (Pino-Ramos et al., 2017). Furthermore, polymer surface modification by RIGC is quite a simple, clean, and repeatable process. Mainly the efficiency of RIGC was measured in terms of grafting percentage (DOG). Numerous crucial factors play a role in attaining optimal DOG such as type of polymeric material, monomer, type of radiation, experimental conditions, type of solvent, and additives (Walo, 2017). These factors are tangible to control the efficiency of RIGC process and will be discussed further in the following sections.

2.2.1 TYPE OF POLYMERIC MATERIAL

Surface modification of both natural and synthetic polymers is feasible with RIGC. The most common natural polymers widely involved with RIGC are cellulose. Cellulose is known as the most abundant renewable raw materials and also a promising alternative resource, sustainable, fibrous, biodegradable, and economically feasible (Barsbay & Güven, 2019). Previously, cellulose-based fibers such as hemp (Tanasă, Zănoagă, Teacă, Nechifor, & Shahzad, 2020), flax (Tataru et al., 2020), kenaf (N. Othman, Selambakkannu, Ting, & Zulhairun, 2019), banana trunk (Selambakkannu, Othman, Bakar, Ming, et al., 2019), sisal (Wei & McDonald, 2016), cotton (Sokker, Badawy, Zayed, Eldien, & Farag, 2009), and jute (Khan, Rahaman, Al-Jubayer, & Islam, 2015) have been modified by means of RIGC to gain the required physicochemical characteristics. Meantime, modification of synthetic polymers with RIGC is also quite common which engages with polystyrene (PS), poly-propylene (PP), and polyethylene (PE), nylon (Ny) (Ting, Nasef, & Hashim, 2015), poly(ethylene terephthalate) (PET) (Chumakov et al., 2011), polyurethane (PUR) (Walo, Przybytniak, Kavaklı, & Güven, 2013), poly(tetrafluoroethylene) (PTFE) (Abd Ghani, Hamzah, Mohamed, & Salleh, 2020), and poly(vinylidene fluoride) (PVDF) (Lim & Shin, 2020). Furthermore, RIGC also enables mod-ifying the polymer surface with any size, nature, and shape such as powders, film, and fibers (Rezvova, Zhevnyk, Pak, Denisov, & Borodin, 2017). However, certain crucial factors play a role in the selection of polymeric materials for the RIGC process, such as the radiation response of natural-based polymers, glass transition, crystallinity, molecular weight, and radical yield.

Practically, most of the polymers are capable to withstand the radiation quite well. But certain polymers undergo some physical and chemical changes by exposure to radiation. Thus, radiation sensitivity onto polymers needs to be taken into consideration in selection of polymeric materials for RIGC. Investigation on the radiation response on natural-based polymers is vital as it is the common choice in the preparation of radiation-grafted polymeric materials. Literature has ac-knowledged the usage of lignocellulose fibers such as kenaf, fiber, jute, and hemp mostly for RIGC. The lignocellulose fibers are made of cellulose, hemicellulose, pectic, tannin, wax, and lignin (Sarala, Nor, Siti, & Hamdani, 2014). Lignin is one of the important components in the construction of lignocellulose fibers and serve biological functions such as water transport, me-chanical support, and resistance to various stresses. Lignin is classified as a three-dimensional polymer of aromatic compounds which consist of phenyl groups. Generally, aromatic groups are well known for their radiation resistance to organic molecules. Phenolic moieties provide greater radiation resistance to polymer materials in comparison to other aromatic groups (Ferry et al., 2013). Hence, lignocellulose fibers were subjected to pre-treatment as delignification for the partial removal of lignin. Alkaline pulping and acidic sulfite process widely used for delignification of lignocelluloses fibers. This will assist in overcoming the radiation resistance of phenolic com-pounds to attain reproducible grafting yield via RIGC. On the other hand, some polymers are classified as radiation degradable and low radiation doses are usually recommended for RIGC. Apart from that, common synthetic polymers such as polystyrene (PS), polypropylene (PP), and polyethylene (PE) responses toward radiation may vary by some functional groups that affects the amount of radio-induced radicals formed at the same dose (Ashfaq et al., 2020). Prior knowledge on the nature of polymeric materials opted for RIGC will ensure the efficiency of the process.

The glass transition point of the polymeric materials plays a crucial role in RIGC as well. Glass transition points refer to the temperature at which the carbons in the polymer chains begin to move. Moreover, the amorphous region of the polymeric materials transit from a rigid to a more flexible state at glass transition temperature and this will eventually lead to solid polymers shifting to a more rubbery form (J. Li et al., 2015). Radiation creates a latent amount of heat based on radiation type and dose applied onto the polymers. However, the radiation type and dose applied must be based on the polymers which restrict on exceeding the glass transition point that leads melting to take place (Yousif & Haddad, 2013). The crystalline/amorphous ratio also one of the aspects that affects the polymers upon radiation exposure. In the amorphous region molecules are arranged randomly and wherein the crystalline region the molecules are organized in partial patterns. Keenly, most of the radiation-induced reactions take place in amorphous region. However, a reduction in crystallinity region is also observed by proton-hopping in certain polymers. This leads to gradual diffusion of free radicals towards the amorphous region (Ashfaq et al., 2020). Therefore, highly crystalline polymeric materials will exhibit less efficiency with RIGC. Polymers with many amorphous regions are more suitable for RIGC.

The direct relation between the molecular weight of a polymer with its mechanical properties is quite significant in RIGC. The higher molecular weight of polymers indicates heavy chain entanglement. However, the excitement of the molecules by mass-energy absorption upon radiation leads to disruption in the chain arrangement (Chamas et al., 2020). Deterioration of molecular weight will take place upon loosening or breaking entangled polymers chains, which will affect the mechanical properties regarding tensile strength and elastic modulus. Moreover, an increase in temperature during the process may lead to the same consequences as well. The final performance of the grafted materials will greatly decline with enormous changes in the mechanical properties of the polymeric materials subjected to RIGC.

Radical yield refers to the total amount of radicals formed on the polymer surface by radiation. Formation of radicals upon radiation exposure due to the adsorption of the radiation energy by the substrate based on the electron density of the polymeric materials occurs spatially at random on a molecular scale (Kumar, Chaudhary, Sharma, & Verma, 2019). Deposition of energy onto polymeric materials aids the creation of reactive free radicals, which tend to easily react with molecules present in the medium. Precisely, the radicals react with appropriate functional monomers to form covalent bonds and lead to copolymerization (Venturi & D'Angelantonio, 2017). Therefore, the radical yield obtained during radiation determines the grafting percentage which owes the process efficiency. In these circumstances, the radical yield obtained by radiation is considered a quite important aspect of RIGC. The formation of radicals in polymers that are resistant such as polyimide (PI) towards radiation is quite difficult and ineffective as well. On the other hand, certain polymers such as natural polymers have proven to be incapable of radical preservation or trapping after radiation. Thus, some precautionary actions are needed such as trapping at ultra-cold temperature, $-70°C$, and storage in the absence of air.

2.2.2 MONOMER

Monomers are the chemical compounds with repetitive functional groups that are used to bond covalently with radicals formed on polymeric materials. Selection of monomers are based on the proposed application for the grafted materials. Both polar and non-polar monomers with functional moieties, such as -COOH, -OR, -OH, -SO$_3$H, -R and their derivatives are commonly used as monomers in order to impart changes in surface properties of the substrate without much influence on the bulk material (Walo, 2017). The concentration of the monomer used in RIGC is the most important aspect to ensure the process efficiency and grafting percentage (Thakur, Tan, Lin, & Lee, 2011). The moderate usage of monomers is usually adequate in order to achieve desired grafting percentage. Commonly, radiation-grafted polymeric materials required merely compatible grafting percentages around 50%–150% to function accordingly. Extremely high grafting percentage usually is due to formation of homopolymers and fusion among monomers itself. Formation of homopolymers is able to affect the

RIGC efficiency entirely. Apart from that, the homopolymers also could impart deterioration effects towards the physical properties such as tensile and robustness of the polymeric materials. Moreover, exhaustion of radicals onto the polymeric materials will lead to saturation in grafting percentage and additional monomers present in the reaction scheme merely form homopolymers (Sharif et al., 2013). Thus, the selection of adequate monomer concentration for RIGC is crucial and multiple factors such amount of radicals available for grafting, the appropriate range for grafting percentage to serve the intended purpose, and reactivity of the selected monomers needed to be considered earnestly.

The bonding of monomers with the radicals occurs by the diffusion of monomers from the bulk solution to the polymer material surface (Mazhar, Abdouss, Shariatinia, & Zargaran, 2014). The factors that hinder the diffusion ability of the monomer as a high monomer concentration and types of polymeric material surface are able to disrupt the grafting process substantially. The high monomer concentration leads to the grafting solution with greater viscosity due to the gel effect that prevents the diffusion of monomers (Suresh, Goh, Ismail, & Hilal, 2021). This will result in a reduction in grafting percentage, directly. The types of polymeric material surface in terms of hydrophilicity and hydrophobicity also could influence the diffusion of monomers as well. The nature of the polymeric surface will determine the constituent of the monomer solution that will assist the diffusion of monomers. Water can be used as a solvent for a monomer solution in a grafting scheme that uses natural polymers that are hydrophilic. However, some of the synthetic polymers are hydrophobic and only the presence of organic solvent in monomer solution can escalate the diffusion further for grafting to take place. Besides that, the radical yield onto the polymeric materials is also important in attaining a desired grafting percentage. The radical yield should be compatible with the number of monomers introduced to the grafting scheme in order to attain an optimum grafting percentage. Poor radical yield will give a very small grafting percentage, even with the presence of enormous monomers.

2.2.3   TYPE OF RADIATION

Various types of ionizing radiation sources are commonly applied in RIGC, such as gamma rays, X-rays, accelerated electrons (e-beam), and ion beams. The ionizing radiation is used in RIGC to initiate radicals on polymer surfaces and also to initiate polymerization reactions based on the type of grafting process applied. Basically, the polymers undergo excitation and ionization upon absorption of high-energy radiation. Meanwhile, those excited and ionized species will initiate the monomers towards polymerization (Charlesby, 2016). RIGC is quite famous with gamma and electron beam radiation. Usually, mutual and pre-irradiation grafting methods are preferably executed by gamma and electron beam irradiation, respectively (Walo, 2017). The gamma and electron beam radiation differs in terms of dose rate, which indicates the amount (Nasef & Hegazy, 2004). The dose rate of gamma is defined by isotope sources in the unit of kGy/h. On the other hand, the dose rate of electron beam irradiation was measured in terms of kGy/s. Thus, practically, an electron beam offers a higher radiation dose in a very short time in comparison to gamma. Obviously, this factor could significantly affect the percentage of grafting in RIGC. A greater number of radicals will be formed on the polymer surface at a higher radiation dose, which will lead to a massive grafting percentage (Nasef & Sugiarmawan, 2010). Moreover, the concentration and lifetime of radicals formed are greatly influenced by dose rate. The oxidation and duration required for termination of the graft chain after the polymerization are dependent on the dose rate, too. Low efficiency of grafting is observed at the same radiation dose with a higher radiation dose rate in a mutual grafting method. This indicates a high concentration of radical formation leads to rapid recombination of radicals, which induced termination and formation of homopolymers.

2.2.4   EXPERIMENTAL CONDITIONS

Another prominent factor that greatly influences the RIGC process are irradiation and grafting temperature. For example, the polymeric materials will be irradiated at sub-ambient temperatures

(0°C and lower) in the pre-irradiated grafting method by using dry ice to immobilize the free radicals. This will assist to restrain the combination of free radicals with each and other (Selambakkannu, Othman, Bakar, Shukor, & Karim, 2018). Moreover, radiation produced an infinite amount of heat which is capable to annihilate the free radicals formed. Thus, performing grafting at sub-ambient temperatures will help to trap the free radicals in a polymeric matrix that can be used to initiate the graft polymerization process, afterward. This will ensure the preservation of radicals and will give out a good grafting percentage (Selambakkannu, Othman, Bakar, & Karim, 2019a). As for the mutual grafting method, the radiation is supposed to be carried out above the glass transition temperature, Tg, of the polymer in order to encourage the chain segments' mobility that leads to migration of active sites towards the surface and increases the population of radicals involved in grafting (Işıkel Şanlı & Alkan Gürsel, 2011). The rigidity of the irradiated polymers below Tg readily gives access to the active centers created through irradiation on the exterior of the polymeric materials.

The thermal conditions also exert an impact on the diffusion of monomers in the solution. The monomer diffusion in the polymeric materials is mostly enhanced with the rise in temperature (Nasef & Sugiarmawan, 2010). The higher thermal energy will excite the monomers by energy transfer and increase the movement. The rise in temperature is also attributed towards an increase in the reactivity of the thermally decomposed trapped radicals, which is also known as radical deactivation. This will give a rise in grafting percentage, too.

2.2.5 TYPE OF SOLVENT AND ADDITIVES

Basically, the role of solvents in RIGC is to facilitate the swelling of polymeric materials and also to improve the monomer accessibility towards the grafting sites. Thus, the selection of suitable solvents is very crucial in achieving a good grafting yield. The solvent used in RIGC should be susceptible to radiation radiolysis and extremely capable of dissolving the desired monomers (Rath, Palai, Rao, Chandrasekhar, & Patri, 2008). On the other hand, the selection of unsuitable solvents might lead to deep distributions of polymer molecular weight. For example, non-polar solvents usually tend to pair up with ions by means of radiolysis onto the polymer backbone, which ends up with the attachment of solvent molecules, partly on the polymeric surface. On top of that, radiolytic solvents could also function as catalysts that determine the type of polymerization mechanism involved in either cationic or anionic polymerization (Walo, 2017). The grafted region that swells to the maximum in the presence of suitable solvent will offer the best and uniform grafting. This only can take place while the solubility parameter of solvent is closer to the monomer or to the polymeric material, itself. Moreover, the chain transfer constant of the solvents will also affect the grafting percentage. The higher chain transfer constant to solvent will give a low grafting percentage. Generally, water is considered the most appropriate solvent for RIGC as it has zero chain transfer constant that will minimize the wastage of monomers, as well. Solvent usually permits the migration of monomers to the radicals formed by radiation onto the polymeric material surface by controlling the viscosity of the monomers (González-Torres et al., 2011). At times, the solvent assists to reduce the viscosity of monomers in RIGC. The higher viscosity of the monomer will directly affect the grafting efficiency. Solvents also help to stabilize the monomers that exist in the emulsion state for RIGC. The stability of monomers in emulsion form is essential to attain goof grafting percentage. Separation of monomer from their emulsion state would hinder the grafting from taking place as the monomers are unable to diffuse onto the radicals in a polymer surface; in this scenario, a solvent capable of stabilizing the emulsion of monomer wholly in many grafting systems (N. A. F. Othman et al., 2020).

The presence of additives such as acids, alkali, inorganic salts, and metal ions could influence the RIGC efficiency. Certain additives possess the tendency to enhance the monomer and polymer reaction, which will boost the grafting efficiency. However, the predominant reaction between monomer and additives will lead to a drop in grafting percentage. Basically, the competition among the reaction of the monomer with polymeric material and additives leaves an impact on the grafting percentage (Bhattacharya & Misra, 2004). The acid additives plays a crucial role in the grafting system. However, the nature of acid additives used in the grafting system needs to be taken into consideration also. Acid additives such as nitric acid are able to degrade cellulose during irradiation. Meanwhile, hydrochloric acid is more likely to exert a negative effect on grafting efficiency by incorporation of chlorine with a polymer (Fauziyah & Yuwono, 2020). The radiolytic yield of H-atoms and the extent of the grafting monomer soluble in a bulk solution are the two predominant factors that are influenced by acid. Additional grafting sites will be available by the radiolytically produced H-atoms and abstracted H-atoms from the polymer base. Apart from that, the presence of acids in the grafting system tends to implicate the nature of polymeric materials and also the solvent, which could directly affect the grafting efficiency.

Mostly, the alkali treatment imposed on cellulose-based polymeric materials will lead to an increase in crystallinity, which reduces the absorption capacity of the polymer base (Sarala et al., 2014). Eventually, this will resist the diffusion of monomers onto the polymers during grafting. Grafting efficiency will be badly implicated by this as well. Moreover, the application of metal ions and salt-based additives in a grafting system preferential reduces the formation of homo-polymers, which will enhance the grafting efficiency (Goel et al., 2009). The higher efficiency of grafting is also partly due to controlled by the partitioning phenomena. However, the efficiency of grafting on cellulose reduces with with the usage of metal-based additives. The high quotient metal ions partitioned from the bulk solution to the vicinity of the cellulose-based polymeric materials reduces grafting due to depletion of metal ions in a bulk solution that reduces monomer radicals scavenging. This will eventually affect the grafting efficiency by formation of more homo-polymers. On the other hand, organic inclusion compounds such as multifunctional acrylates and urea may also increase the grafting efficiency by imposing partitioning phenomena.

2.3 CHARACTERIZATION OF THE GRAFTED POLYMERIC MATERIALS

Since grafting is exclusively a surface modification process, the characterization of graft copolymers is mostly dependent on surface characterization. The changes in polymeric materials after RIGC in terms of morphology and chemical and physical properties usually verify the efficiency of the grafting process. The number of analytical techniques that provide information on polymeric materials surface assist in comparative analyses on both grafted and non-grafted samples that prove the effects of the grafting process. However, due to the variety of combinations from polymer backbones and polymer grafts, other chemical and physical characterizations that can be grouped into direct and indirect characterization techniques also can be adapted to gather deeper understanding about the grafted materials. Direct methods are used to characterize chemical composition or structure of the grafted materials. They are frequently used to identify changes such as addition or removal of bonds between backbone polymer and grafts. Several methods, such as attenuated total reflection Fourier transform infrared (ATR-FTIR) spectroscopy, X-ray photoelectron spectroscopy (XPS), atomic forcemicroscopy (AFM), scanning electron microscopy (SEM), thermogravimetric (TGA), differential scanning calorimeter (DSC), and X-ray diffraction (XRD) measurements, help to verify the changes in polymeric materials after RIGC. The summary on polymeric materials after radiation grafting in terms of characterization techniques is shown in Figure 2.3.

2.3.1 MORPHOLOGY

The changes after RICG were prominently detected onto the surface of the polymeric materials in terms of morphology. The direct technique to observe the morphological changes after the radiation grafting process is by using a scanning electron microscope (SEM) and field emission-scanning electron microscope (FESEM). A few of our previous works reported on the morphological changes of polymeric materials after RIGC. SEM images of the polyethylene/polypropylene fibers (PEPP),

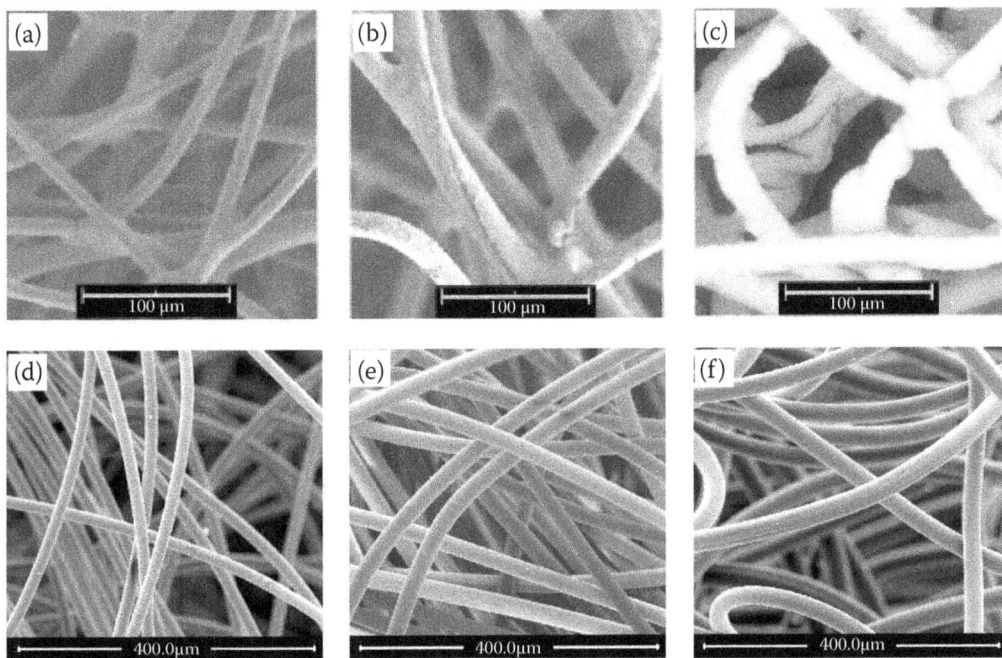

FIGURE 2.1 SEM images of (a) original nylon, (b) poly(VBC), (c) poly(GMA) grafted nylon fibers, (d) original PEPP, (e) poly(VBC), and (f) poly(GMA) grafted PEPP fibers.

poly-vinyl benzyl chloride (poly(VBC)), and poly-glycidyl methacrylate (poly(GMA)) grafted fibers are given in Figure 2.1. The diameter of the fibers increased by the grafting of poly(VBC) and poly (GMA) onto the polyethylene/polypropylene nonwoven fabrics (PEPP). It can be observed that both monomers were grafted onto the PEPP uniformly, resulting in the overall increase in the average diameter of the fibers. These suggest the confirmation on the grafting of both monomers were successful and radicals can be produced conveniently on the surface of PEPP to facilitate the graft copolymerization reaction of VBC, GMA, and other monomers. The surface of the grafted substrate with different monomers showed a similar surface morphology. Only the diameter showed a difference and this corresponds to the grafting yield.

SEM images of the original nylon fibers and grafted nylon fibers with two different monomers are presented in Figure 2.1, as well. The average diameter of the unmodified nylon fibers was increased by grafting of poly(VBC) and poly(GMA). It can be observed that both monomers were grafted uniformly onto the fibers, resulting from an almost constant diameter increased after graft polymerization. The results indicate that both monomers were compatible to be grafted onto the nylon substrate with no apparent difference can be seen on the surface of the fibers with two monomers. The relatively larger diameter of GMA-grafted nylon fibers than VBC-grafted nylon fibers was attributed to the higher percentage of grafting yield of GMA onto the nylon fibers (Pant et al., 2011). These results suggest that the grafting of VBC and GMA monomers onto nylon fibers occurred with a uniform grafting reaction across the fibers. The nylon substrate is a versatile substrate with a good compatible grafting reaction with VBC and GMA, and it is expected that other monomers can also be grafted onto the nylon.

The morphological changes on banana fibers after pre-treatment irradiation grafting and chemical functionalization had been analyzed with scanning electron microscopy (SEM). The model of SEM used was *Hitachi-SU350* at the voltage of 15.0 kV and at current of 0.2 mA. The magnification was used about 500 x. Bis Rad system was employed for gold coating of samples before imaging. The grafting of GMA onto pre-treated banana fiber was studied and the SEM images were attained as shown in Figure 2.2. Basically, banana fibers originated from the trunk of the plant. Pristine banana trunk fibers are known as hollow-centered lignocellulosic fibers and the lignin supports hold together the hollow-centered microfibrils as in bundle (Reddy & Yang, 2015). But the partial removal of lignin by acidic chlorite method is required to overcome the obstacle imposed by the phenol rings in the molecular structure of lignin upon exposure to ionizing radiation. Based on the SEM images obtained the banana fibers after the pre-treatment process was more likely loosely packed with obvious ridges among the fibrils. However, the gaps in between the fibrils clearly had diminished in banana fiber-G-GMA with the grafting yield of approximately 85.0% mainly due to the poly(GMA) coating around the fibrils that also increases the intra-chain interaction. The grafted fiber portrays thin layers of GMA co-polymer coating on the surface of pre-treated banana fibers. Moreover, the overall diameter of the fiber also clearly increases after the grafting with poly(GMA) (M Saber, Abdullah, Jamil, Choong, & Ting, 2021). This observation indicates the successful grafting of poly(GMA) onto pre-treated banana fibers.

On the other hand, the cross-section view of pre-treated banana fibers and banana fiber-G-GMA is shown in Figure 2.2, as well. This observation proves the grafting of poly(GMA) onto the banana fibers. The grafting yield of 85% is attained at irradiation dose, 30 kGy with a concentration of GMA of 3%. The pre-radiation grafting technique was used for this experiment. The grafting temperature was around 40°C and the duration was approximately 3 h. The average cell wall thickness captured was in pre-treated banana fiber and banana fiber-G-GMA (85% grafting yield) was about 1.36 µm and 2.86 µm, respectively. The increase in diameter of the microfibrils is due to the addition of grafted poly-GMA onto the pre-treated banana fibers (Barba, Peñaloza, Seko, & Madrid, 2021). The images obtained also assure the compatibility of GMA with banana fibers that uses water as a solvent. Practically, GMA exhibits partial solubility in water, and the stability of GMA in water was further enhanced by adding a small volume of surfactant. Naturally, banana fibers possess a very good absorption since the backbone of the fibers is made of cellulose. All these factors contribute toward achieving a higher grafting yield.

In one of our previous attempts, amine-functionalized fibers were successfully prepared for the adsorption of anionic dyes from an aqueous solution. The pre-treated banana was grafted with poly-GMA via the RIGC technique which was followed by chemical functionalization with trimethylamine (TMA). The fusion of TMA onto banana fiber-G-GMA takes place via the opening of the epoxy ring in GMA. The epoxy group of GMA reacts as binding reactive groups and incorporation with the amine group generates new functional groups for ion exchange or chelate formation (Selambakkannu, Othman, Bakar, Ming, et al., 2019). The amine group present in TMA functions as a precursor to covalently bond with targeted anionic dyes. Generally, the adsorption mechanism is based on the electrostatic attraction and hydrogen bonding between amine groups onto the banana fibers with anionic dye molecules. The SEM images are shown in Figure 2.2, clearly indicating the changes in the cross-section view of the banana fibers before and after the amine functionalization process. The cross-section images of fiber-G-GMA exhibit the grafting of poly(GMA) by slight coating around the fibril. But, after functionalization with TMA (TMA-fiber-G-GMA), the coating and swelling of the fibril are quite prominent and the gap in between the fibril is filled as well. This observation clearly shows the successful functionalization of TMA onto fiber-G-GMA.

2.3.2 THERMAL STABILITY

Figure 2.3 shows the thermograms of TGA and DTG analysis of the original PEPP fibers, poly (GMA), and poly(VBC) grafted PEPP fibers. PEPP fibers in their natural state revealed a pattern of single-step decomposition. The TGA analysis of original PEPP fibers showed that it was stable up to 330.5°C and decomposed at a temperature range between 330.5°C to 466.5°C with a DTG peak recorded at 427.3°C. The grafting of poly(GMA) into the PEPP fibers component leads the two-stage decomposition to take place. The first stage of thermal decomposition took place in the

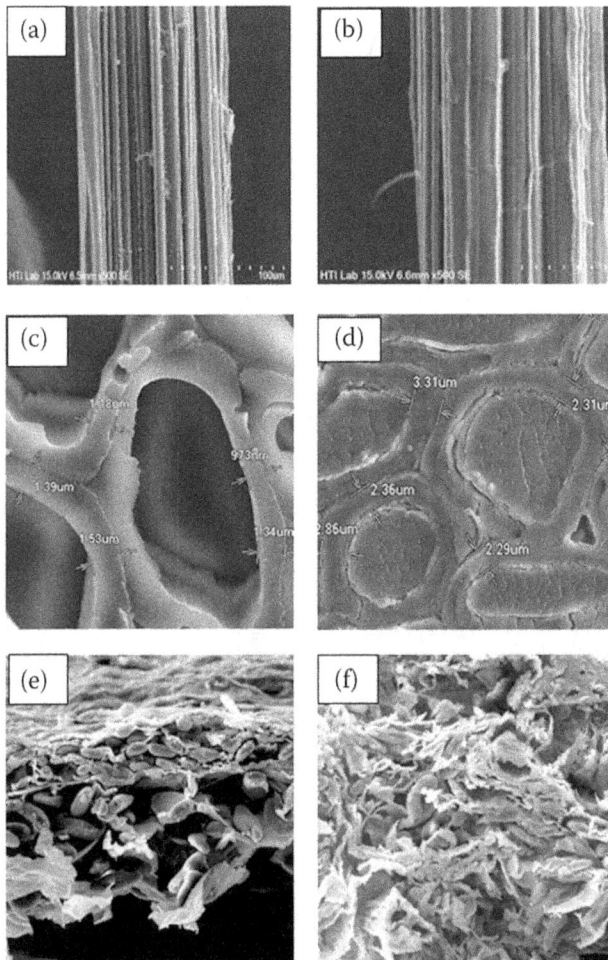

FIGURE 2.2 SEM images of (a) pre-treated banana fiber, (b) banana fiber-G-GMA (85% grafting yield), cross-section view of (c) banana fiber (pre-treated), (d) banana fiber-G-GMA, (e) banana fiber-G-GMA, and (f) banana TMA-fiber-G-GMA.

temperature range between 241.6°C to 349.9°C, with a DTG peak recorded at 283.9°C, and reflected decomposition of poly(GMA) molecular chains. Second-stage decomposition represented the PEPP molecular chains that occurred at a temperature between 349.9°C to 479.7°C, with a DTG peak observed at 441.6°C. The grafting of poly(VBC) onto the PEPP also revealed a two-stage thermal decomposition took place. The first-stage decomposition of PEPP occurred in the range of 289.7°C to 409.4°C with a DTG peak of 355.3°C, while the second-stage decomposition of poly(VBC) side chains occurred at a range of temperature in between 409.4°C to 490°C with a DTG peak at 450.3°C. The TGA result gives evidence on the addition of the new side chains of poly(GMA) and poly(VBC) onto the PEPP fibers.

Figure 2.3. also shows TGA and DTG thermograms of original nylon fibers, poly(GMA), and poly (VBC) grafted onto the nylon fibers. The thermogram of original nylon fibers demonstrates a single-step degradation pattern, with no weight loss below 350°C, after whichthermal decomposition of nylon polymer chains occurs in the temperature range of 350°C–500°C with a DTG peak at 445°C. The poly(GMA) grafted nylon fibers exhibited a three-step degradation transition, as evidenced by weight losses at temperatures of 200°C–335°C, 335°C–425°C, and 425°C–490°C with onset peaks at 300°C, 385°C, and 438°C, respectively. The poly(VBC) grafted nylon also

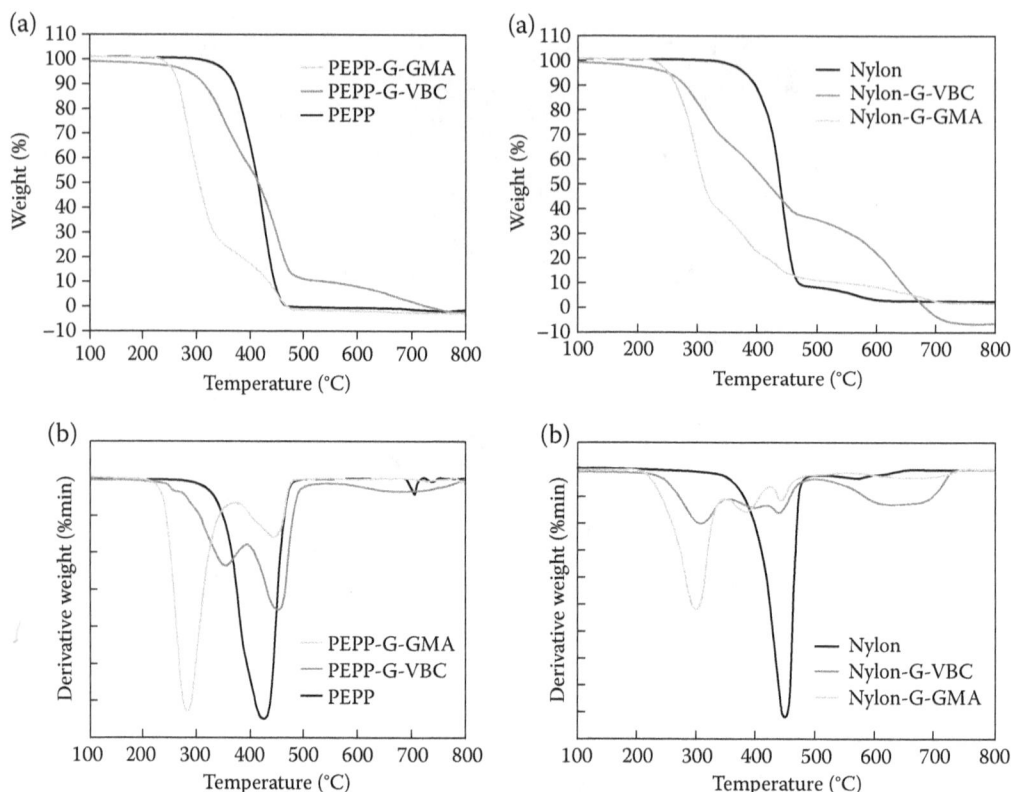

FIGURE 2.3 TGA and DTG thermograms of (a) PEPP-G-GMA, PEPP-G-VBC, PEPP, (b) Nylon, Nylon-G-VBC, Nylon-G-GMA.

exhibited a similar degradation pattern with DTG peaks at 310°C, 400°C, and 435°C for weight loss over the temperature in the range of 200°C–330°C, 330°C–420°C, and 420°C–500°C, respectively. The weight loss of poly(GMA) grafted nylon fiber was higher than poly(VBC) grafted nylon fiber, indicating that the thermal stability of poly(VBC) was higher than poly(GMA). This is attributed to the aromatic structure of poly(VBC) that provides high chemical and thermal stability than the grafted aliphatic poly(GMA) (C. Z. Liao & Tjong, 2013).

Investigation of thermal stability and upper temperature limit usage on the products that are grafted by radiation induced technique is quite essential. This technique could give insight to the thermal stability and also able to give the upper use temperature of a material, especially on a surface of modified samples. But anyhow, any temperature beyond this limit will stimulate the material to degrade. Materials lose weight in three ways which is by chemical reactions, release of adsorbed species, and decomposition. The poly(GMA) and fiber-G-GMA were analyzed for its thermal stability in inert atmosphere at heating rate of 10°C min^{-1}. The temperature range used was from a room temperature to 700°C. The nitrogen flow maintained at the rate of 50 mL min^{-1}. The grafting GMA onto kenaf fiber was performed in order to obtain GMA-grafted kenaf fiber (K.fiber-G-GMA). The thermogravimetric data for each polymer and graft copolymer are presented in Table 2.2. It is observed that the $T_{initial}$ is shifted to a higher temperature compared to cellulose, increasing from 221.8°C (treated kenaf fiber) to 230.6°C. This indicates an increase in thermal stability due to dehydration, depolymerization, and pyrolysis at a higher temperature.

According to R. K. Sharma and Zulfiqar *et al.* (Sharma, 2012), the decomposition of GMA consists of two reactions, namely, the depolymerization to a monomer, which is the major reaction, and ester decomposition. The first decomposition step is in the range of 240.0°C to 325.4°C with

TABLE 2.2
Common Functional Groups

Polymer	Number of stages	Degradation stages (°C)	Weight loss (%)	$T_{initial}$ (°C)	T_{final} (°C)
Kenaf fiber	1	30.0–112.4	7.56	211.9	449.9
	2	211.9–353.4	23.69		
	3	353.4–449.9	42.24		
Treated kenaf	1	30.0–112.6	4.73	221.8	469.9
	2	221.8–469.9	70.69		
GMA (Ndazi, Nyahumwa, & Tesha, 2008) (Khalil & Suraya, 2011)	1	240.0–325.4	N/A	240.0	413.4
	2	325.4–413.4	N/A		
GMA-grafted-kenaf fiber	1	30.0–112.6	3.79	230.6	502.8
	2	230.6–502.8	83.35		

40% weight loss is due to initiation at unsaturated chain ends, while the second decomposition step is in the temperature range of 325.4°C and 413.4°C that is associated with initiation by a random chain scission that is causing full breakage of an acrylate bond.

All samples i.e., raw kenaf fiber, treated kenaf fiber, and grafted kenaf fiber, exhibited initial weight loss from 30°C to 115°C due to evaporation of absorbed water. Weight loss was observed from 200°C to 350°C, corresponding to the volatilization of organic compound and decomposition of cellulose and hemicellulose. This is in accordance with Luo *et al.* (Luo, Strong, Wang, Ni, & Shi, 2011), where they observed the same temperature range for decomposition of water hyacinth fiber. In comparison to treated kenaf fiber, greater weight loss was found at the same temperature range for the grafted kenaf fiber. According to Choi *et al.* (Choi, Lee, Kang, & Park, 2004), the decomposition of GMA, particularly for the ester group, 2,3-epoxy-propanol group can be observed at the temperature range from 220°C to 300°C. Therefore, increasing the weight loss can be attributed to GMA grafted on the kenaf fiber backbone. A shift in the degradation temperature to a higher temperature was observed after graft copolymerization. This proved that graft copolymerization had increased the thermal stability of kenaf fiber. Final weight loss at around 500°C was due to lignin. Lignin was more stable and decomposed at a higher temperature. This thermal behavior of natural fiber has been well explained by several researchers (Kim, Yang, Kim, & Park, 2004; Zulfiqar, Zulfiqar, Nawaz, McNeill, & Gorman, 1990).

In the case of kenaf fiber, the thermogravimetric (TG) curve shows the existence of two-stage degradation. The thermal decomposition for the first weight loss is about 7.56 wt% in the temperature range of 30.0°C and 112.4°C. The initial weight loss is due to dehydration of the sample. Another reason is the presence of moisture in the sample. A similar observation has been reported by Kargarzadeh *et al.* (Kargarzadeh et al., 2012). This is because natural polymers have the tendency to absorb moisture from its surroundings. The second weight loss is 23.69% in the temperature range of 211.9°C to 353.4°C due to the loss of COO^- from the polysaccharide.

The third weight loss is at the temperature range 353.4°C to 449.9°C, which about 42.2%. The final loss of 5.37% is observed around 450.0°C to 600.0°C, representing the degradation of the remaining material into carbon residues. A number of curves are observed in differential thermal analysis (DTA) curve of kenaf fiber. This is believed to be from decomposition of lignocellulosic materials. Ndazi *et al.* (Ndazi, Nyahumwa, & Tesha, 2008) reported that cellulose decomposes thermochemically between 150°C and 500°C. Hemicellulose decomposes between 150°C and 350°C while cellulose mainly decomposes between 275°C and 350°C. Lignin is reported to decompose between 250°C and 500°C (Khalil & Suraya, 2011). A single decomposition stage is observed on the treated kenaf curve and represents only one main reaction zone. This implies that the components that decomposed at the first stage of raw kenaf fiber were not present in the treated kenaf fiber. The components were associated with removal of hemicellulose, lignin, and waxy substances during the treatment of kenaf fiber. Similar trends have been reported by other researchers who study thermal stability of cellulose after chemical treatment. Ndazi *et al.* (Ndazi et al., 2008) suggested that higher degradation temperature of rice husks treated using an alkali treatment correspond to decomposition of α-cellulose, since most of hemicellulose, lignin, and other volatile substances have been removed during the treatment.

It is observed that the initial decomposition temperature, $T_{initial}$ of 221.8°C for treated kenaf is higher than of raw kenaf fiber (211.9°C), which indicates higher thermal stability for the treated fiber. This can be explained by hemicellulose degradation in raw kenaf fiber and hemicellulose and cellulose degradation occuring in treated kenaf fiber, in which weak bonding in raw kenaf fiber has led to faster degradation. This is in a good agreement with (Khalil & Suraya, 2011) who claimed that weak bonding between the untreated kenaf bast fiber due to evolution of volatile products has caused faster degradation; hence, higher thermal stability of the kenaf fiber treated with anhydride was observed.

The influence in the crystalline region of polymeric materials upon RIGC is analyzed with differential scanning calorimeter (DSC). RIGC mainly affects the crystallinity of the polymeric materials which leads to reduction in crystalline region (Brack et al., 2004). Figure 2.4 shows the DSC thermograms of poly(GMA) and poly(VBC) grafted PEPP fibers with reference to the original PEPP fibers. There were two endothermic peaks on the thermogram of original PEPP fibers that described the melting behavior of fibers containing PE in the outer shell and PP in the fiber core (Giller, Chase, Rabolt, & Snively, 2010). The melting behaviour of the PE is represented by the first single melting peak at a temperature of 128.6°C while PP is indicated by the second double melting peaks at 152.1°C and 163.7°C. The grafting of poly(GMA) onto PEPP fibers shifted the endothermic peak of PE to 129.6°C and PP to 157.1°C and 162.4°C with a decrease in the area of the melting peak. The addition of poly(VBC) grafted onto PEPP fibers also shifted the melting peak of PE to 127.2°C and converted the double peak of PP to a single peak at a value of 156.5°C

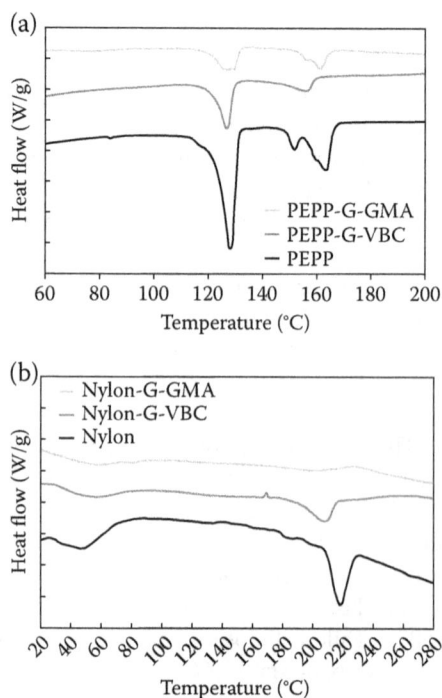

FIGURE 2.4   DSC thermograms of (a) PEPP-G-GMA, PEPP-G-VBC, PEPP, (b) Nylon, Nylon-G-VBC, Nylon-G-GMA.

with reduced intensity of the melting peaks. The DSC thermogram indicates that grafting of poly (GMA) and poly(VBC) onto PEPP fibers occurred in both the PE and PP layers of the PEPP fibers. The substantial reduction in the melting peaks of PE and PP indicates that overall crystallinity was reduced and this is attributed to the addition of amorphous composition of poly(GMA) and poly (VBC) into the fibers.

On the other hand, the DSC thermograms of original nylon and grafted fibers having different monomers are displayed in Figure 2.4 as well. The original nylon showed a single endothermic peaks, but the melting temperatures of poly(GMA) and poly(VBC) were shifted to a lower temperature with a reduced peak intensity. The thermal behavior of both grafted poly(GMA) and poly (VBC) was similar. The transition of the peak at 48°C is most likely due to the glass transition temperature (Tg) of the original nylon and shifted to a higher temperature at 60.5°C for grafted poly(GMA) and poly(VBC) fibers. The reduction in the melting temperature and peak intensity are attributed to the dilution effects resulting from the integration of the amorphous poly(GMA) and poly(VBC) side chains grafted onto the polymer backbones.

DSC thermograms of pristine PE/PP-NWF and PE/PP-NWF-G-DMAEMAare shown in Figure 2.5. Due to melting of the trunk polymer, which made of polyethylene and polypropylene, bimodal melting peaks can be seen, and the peaks are identical for both pristine and PE/PP-NWF-G-DMAEMA, respectively. The first and second melting represented by the endothermic peak of PE/PP-NWF, where the first represented the PE outer layer at 115°C with an enthalpy of 130.33 J/g. The second melting peak at 170°C with an enthalpy of 48.95 J/g is for the PP inner core. After RIGC, a slight shifting of PE melting peak from 115°C to 127°C and PP melting peaks, from 170°C to 164°C can be seen, with a reduction in area as well. The third endothermic peaks for PE/PP-NWF-G-DMAEMA are related to the degradation of DMAEMA that was grafted onto the pristine PE/PP-NWF. The summarized melting temperature (°C), melting enthalpy (ΔH_m^0), and degree of crystallinity (%) are presented in Table 2.3 and estimated by the following equation:

$$X_C = \frac{100 \times H_m}{\Delta H_m^0 \times w} \tag{2.9}$$

where X_C is the degree of crystallinity in %, ΔH_m^0 is the enthalpy of melting 100% crystallized material, H_m is the enthalpy required for melting, and w is the weight of the sample.

Crystallinity has a significant influence on the mechanical properties of a material. PE/PP is a semi-crystalline material; therefore, any modification will influence the crystallinity of the material (Bitter & Lackner, 2021). The crystallinity degree of PE/PP-NWF decreases with grafting. Thus, it can be concluded that grafting can give a nucleation effect for the PP matrix. This nucleation effect might also contribute to the improvement of Young's modulus (J. Liao, Brosse, Pizzi, & Hoppe, 2019).

2.3.3 Mechanical stability

The mechanical stability of kenaf fibers before and after RIGC with poly(GMA) was analyzed in terms of different irradiation doses. The kenaf fiber was bleached (un-grafted) for the removal of lignin in order to facilitate the grafting process via ionizing radiation for grafting to take place efficiently. However, removal of lignin with subsequent irradiation and grafting process leaves a massive effect on the mechanical strength of kenaf fiber, eventually. Therefore, tensile strength analysis onto kenaf fiber after bleaching and RIGC at different irradiation doses is very crucial to have a better understanding of the effect of the processes on the mechanical properties of cellulosic kenaf fibers. INSTRON, the universal testing machine, was used to analyze the tensile strength of the kenaf fiber. The ASTM D3379-75 Standard Test Method for Tensile Strength and Young's Modulus for High Modulus Single-Filament Materials were applied to perform this analysis (Babatunde, Yatim, Ishak, Masoud, & Meisam, 2015). The condition used for the tensile test was a

(a)

(b)

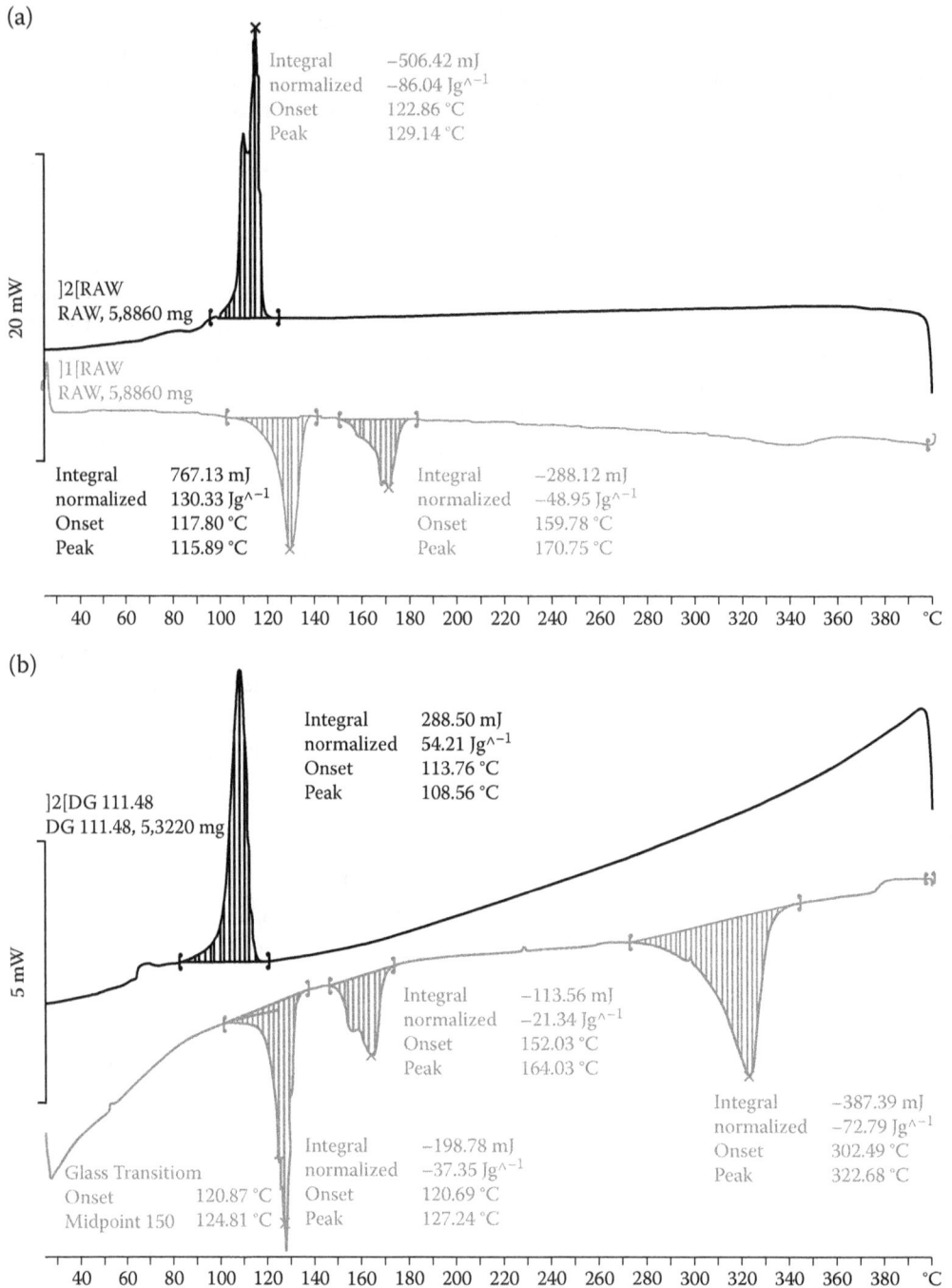

FIGURE 2.5 DSC thermograms of (a) PE/PP-NWF, (b) PP/PP-NWF-G-DMAEMA.

crosshead rate of 3 mm/min, and the gauge length was fixed at 30 mm, with a load cell capacity of 10 N. The result obtained, tensile strength, and tensile modulus are shown in Figure 2.6. The tensile strength and modulus obtained for ungrafted fibers show significant deterioration. The tensile properties reduce as the irradiation dose increases. The higher irradiation dose generates a considerable amount of heat as the reaction leads to deterioration of the kenaf fiber (Barbosa et al., 2012).

TABLE 2.3
Thermal Energy Profile

Sample	T_m	H_m	X_C
PE/PP-NWF	115.89	130.33	10.7
P-DMAEMA	124.81	37.35	3.4

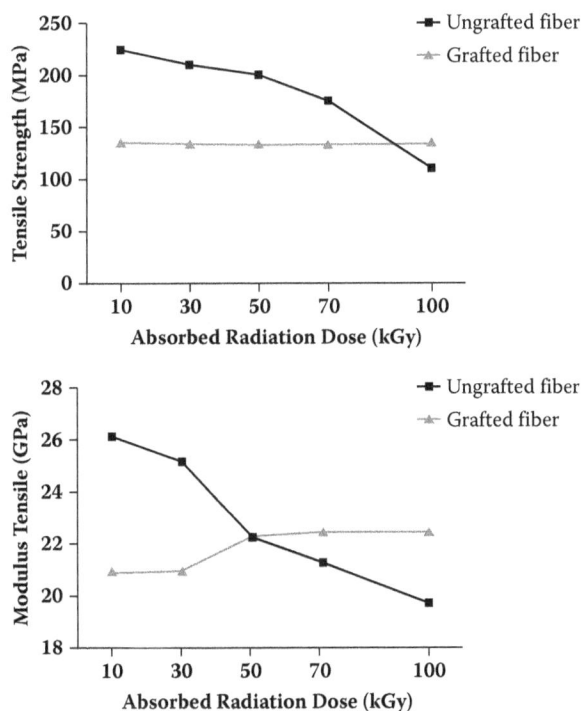

FIGURE 2.6 Tensile strength and modulus tensile for ungrafted and grafted kenaf fiber.

However, insignificant reductions in tensile strength and modulus were observed in grafted kenaf fibers as the irradiation dose increased. The thicker poly(GMA) grafted layers formed with the increase in irradiation dose. Grafting of poly(GMA) onto the fibers assists in improving chemical bonding between the GMA and fibers (Lee, 2013). Therefore, GMA onto kenaf fibers can improve the mechanical properties of the radiated fibers as the grafting yield increases.

2.3.4 CHEMICAL COMPOSITION

Figure 2.7 shows the FTIR spectra of the original PEPP fibers, poly(GMA), and poly(VBC) grafted PEPP fibers. In comparison to the original PEPP fibers, prominent peaks formed at 905, 840, and 756 cm^{-1} on a poly(GMA) grafted PEPP fibers spectrum, which corresponds to the epoxide group of GMA molecule structure. The peak of the C=O bond emerged with strong features at 1,719 cm^{-1} also attributed to poly(GMA) grafting. The grafting of poly(VBC) onto PEPP fibers leading to the appearance of the peaks at 673 to 715 cm^{-1}, indicating the C-H bending of the 1,3-disubstituted benzene ring. The presence of aromatic C-H bonds of para-substitution vibration was observed at 789 to 839 cm^{-1} on the poly(VBC) grafted PEPP fiber spectrum. A new band on the spectrum was developed at 1,264 cm^{-1} that conforms to C-Cl stretching that comes from the structure of the VBC

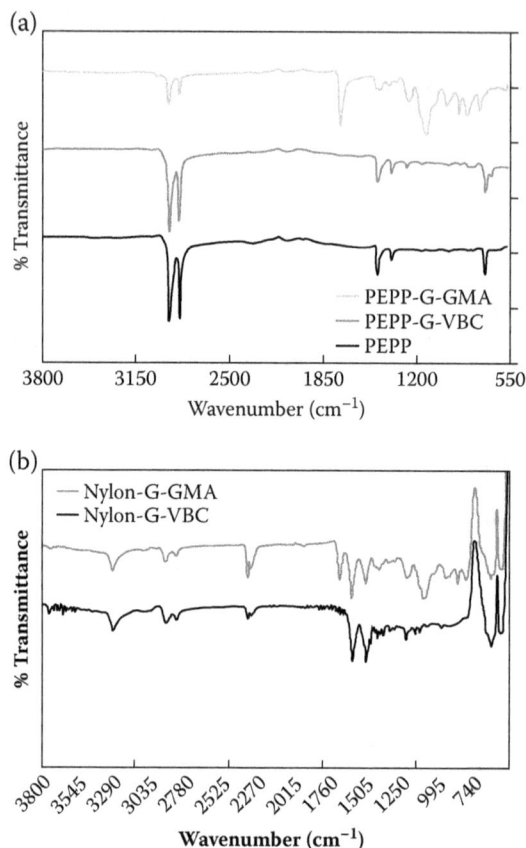

FIGURE 2.7 FTIR spectra of (a) PEPP-G-GMA, PEPP-G-VBC, PEPP, (b) Nylon, Nylon-G-VBC, Nylon-G-GMA.

molecule. The results provided evidence on the successful grafting of poly(GMA) and poly(VBC) onto the PEPP fibers.

The FTIR spectra of original nylon and grafted fibers having two different monomers are given in Figure 2.7. When nylon was grafted with GMA and VBC, the intensity of the peaks at 3,300, 3,086, 2,931, 2,859, 1,645, and 1,544 cm^{-1} were reduced. The new peak at 1,716 cm^{-1} in grafted GMA is due to C=O stretching vibration, whereas the emergence of new peaks at 1,145 and 1,127 cm^{-1} are related to the ester group's –CO stretching vibration. The peaks at 902 and 840 cm^{-1} are caused by the stretching vibration of the epoxy group in poly(GMA). The poly (VBC) grafted nylon fibers showed an increase of peak at 1,264 cm^{-1}, which corresponded to the C–Cl stretching vibration of the chloromethyl group (Wang et al., 1996; Yaacoub & Perchec, 1988). The new peak recorded at 838 cm^{-1} is mostly a result of p-disubstituted benzene (Kim et al., 2007). The peak at 3,086 cm^{-1} is attributed to the N-H bond and its intensity is reduced for both poly(GMA) and poly (VBC) grafted nylon fibers. This indicates the grafting polymerization of GMA and VBC were grafted on the nitrogen of the -CONH in nylon fibers. These findings give evidence that both GMA and VBC were successfully grafted onto nylon fibers.

FTIR also can be used to observe the chemical changes after adsorption and to predict the adsorption mechanism. After thorium adsorption, significant changes were found at peaks related to nitrogen atoms. Therefore, it is predicted that N atoms play a vital role in the adsorption of thorium since it affected all the chemical bonds related to nitrogen. In addition, the new bands at 1,326 cm^{-1} and 1,141 cm^{-1} corresponding to Th-O was observed, indicating thorium adsorption occurs at oxygen atoms as well. From this observation, adsorption mechanism can be predicted, as shown in Figure 2.8.

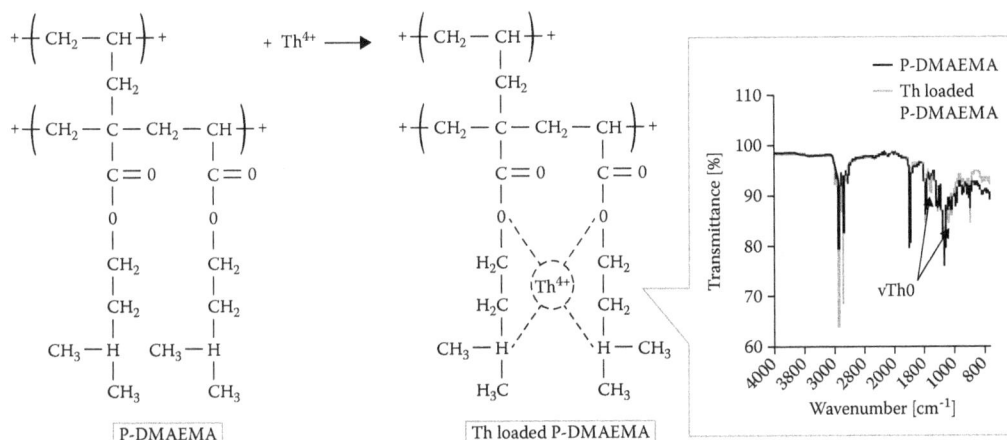

FIGURE 2.8 FTIR spectrum for P-DMAEMA and thorium loaded P-DMAEMA.

The FTIR quantitative analysis also can be performed to quantify the unknown amount of functional group present on the backbone after grafting, and therefore allowing accurate calculatation of the percentage of grafting. In our previous work on radiation grafting of GMA onto kenaf fiber, we have calculated the percentage of grafting (%) using this method. To do that, a calibration line was prepared, and the calculation needs to fit the Beer Lambert Law.

Five samples containing 20, 40, 60, 80, and 100 w/v% of GMA in an isopropanol (IPA) solvent were analyzed. Based on the relationship between the increase of GMA concentration correlates to spectral absorbance peak, a calibration line was plotted according to the Beer Lambert Law, which is a plot of absorbance peak versus the volume percent of poly(GMA). The measured absorbance of the GMA peak centered at 1,725 cm^{-1} (C=O stretching of carbonyl group, integration limits 1,729 to 1,725 cm^{-1}). A linear regression plot of peak area versus the weight per volume percent of GMA (w/v%) was obtained from the characteristic peak of 1,725 cm^{-1}, corresponding to the carbonyl group of GMA for the calibration purposes.

2.3.5 CRYSTALLINITY

X-ray diffraction is commonly used to analyze the degree of order (crystallinity) of the dedicated samples. The strong signals created from X-ray diffraction for crystalline fraction of dedicated samples to measure the crystallographic parameters such as distances between the crystalline cell units. The non-crystalline part of the analyzed samples was usually represented by broader and less clearly refined features in diffraction patterns. The operating condition for XRD machine was nickel filtered CuKα (λ = 1.542A) at 30 kV and 30 mA. The results was obtained from 2θ of 2° to 30° at a rate of 1°/min. X-ray diffractogram of original PEPP fibers, poly(GMA), and poly(VBC) grafted PEPP fibers are illustrated in Figure 2.9. The four diffraction peaks belonging to PP have 2θ at 14.26°, 17.09°, 18.74°, and 21.71° correspond to crystalline lattices of 110, 040, 130, and 041, respectively (Nishino, Matsumoto, & Nakamae, 2000). The PE comprises of two crystalline peaks at 21.71° and 24.19°, which are related to reflection planes of 110 and 200, respectively (C. Liao & Tjong, 2012). There is no new peak and shifting of the peak position of the grafted PEPP with poly(GMA) and poly(VBC). Although the poly(GMA) and poly(VBC) grafted PEPP fibers show a diffraction curve that is similar to the lattice of the original PEPP fibers, the intensity of the peaks are reduced substantially and this is attributed to the incorporation of poly(GMA) and poly (VBC), which comprise of amorphous-grafted side chains. The result provides strong evidence of successful grafting of poly(GMA) and poly(VBC) onto PEPP fibers.

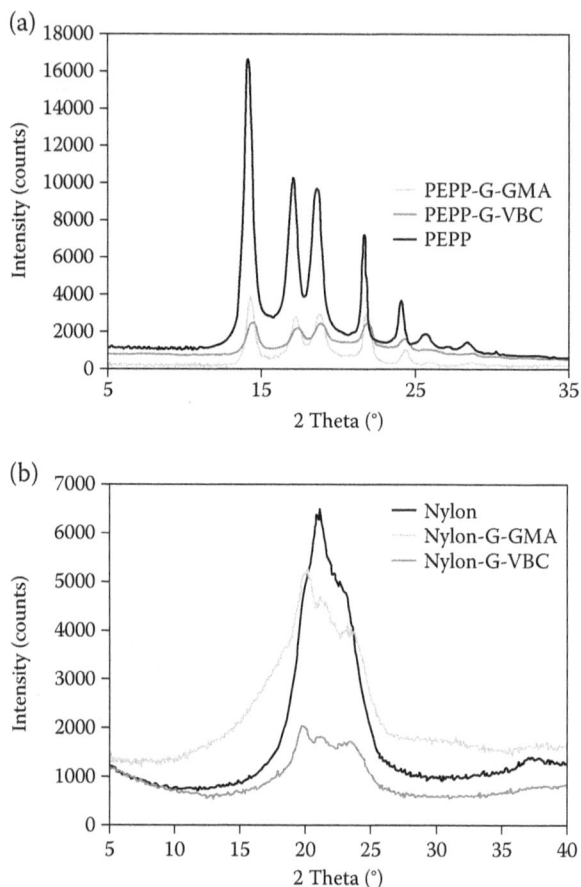

FIGURE 2.9 X-ray diffractograms of (a) PEPP-G-GMA, PEPP-G-VBC, PEPP, (b) Nylon, Nylon-G-VBC, Nylon-G-GMA.

Figure 2.9 illustrates the X-ray diffraction patterns of original nylon and grafted nylon with poly (GMA) and poly(VBC). The diffractograms of the original nylon reveals a semicrystalline polymer with a γ-form crystallization, as indicated by the peak at $2\theta = 21.2°$. This is in agreement with few studies that found $2\theta = 21.4°$ and $2\theta = 21.5°$ (Giller et al., 2010; Pant et al., 2011). The significant change in intensity of poly(GMA) and poly(VBC) grafted nylon are due to the graft polymerization of GMA and VBC onto the nylon fibers. Grafting of poly(GMA) and poly (VBC) were found to cause a reduction in the crystalline peak intensity of nylon due to the reduction in peak intensity at $2\theta = 21.2°$ followed by the appearance of new peaks at $2\theta = 19.9°$ and $2\theta = 23.5°$. The reduction of crystal structure of nylon in poly(VBC) was higher than poly(GMA). The reduction in overall crystallinity of the grafted fibers is attributed to the dilution of the crystalline structure by the amorphous poly(GMA) and poly (VBC) grafted, as well as the partial disruption of the crystallite resulting from the incorporation of poly(GMA) and poly(VBC).

2.4 FUTURE PROSPECT FOR POTENTIAL APPLICATION IN MALAYSIA

The radiation grafting process has been more attractive due in comparison to chemical grafting due to its ability of fast free radical formations without any toxic chemical intermediates like an initiator or catalyst. These chemicals often need a long time to react, and are high in toxicity. Significant reduction in chemical consumption in the process leads to a reduction of cost and, due to the simple operation, the possibility to apply this process on a large scale for mass production makes it attractive for industrial applications. Radiation-induced grafting has several advantages as shown below:

- The ability to modify the polymer surface to have very distinct properties through the choice of different monomers.
- The ease and controllable introduction of graft chain with a high density and exact localization of graft chains to the surface with the bulk properties unchanged.
- Covalent attachments of graft chain into a polymer surface assuring long-term chemical stability of introduced chains.
- Simplicity and the flexibility of initiating the reaction using various types of high-energy radiation source such as Cobalt-60, electron beam (EB), or ultra-violet.
- The graft polymerization can be initiated in wide range of temperatures, including low regions in monomers available in bulk, solution, emulsion, or even in a solid state.

Therefore, radiation grafting is an environmentally cost-effective method to develop novel material with unique properties for various potential in various applications as discussed below. However, the application of grafted material induced grafting depends on the optimum grafting yield.

- Radiation-grafted materials for separation and purification

    Among the most popular and well known usage of radiation grafted materials is to be used in separation processes. It can either be for removal of toxic and hazardous elements from aqueous media, or for recovery of valuable materials. The grafting yield around 100% usually is sufficient for separation and purification applications. In Malaysia, man-made factors due to human activities such as mining operation (Kubová, Matúš, & Bujdoš, 2005), production of metal alloys, and industrial activities (Yan-Feng et al., 2007) can contribute to excessive hazardous ionic species concentration accumulated in soil and contaminated the water system. Thus, tremendous research has been conducted, aiming to improve efficiency in removing these species from Malaysia's water system. N. A. F. Othman, Selambakkannu, & Tuan Abdullah, (2022) used imidazole functionalized grafted adsorbent to remove about 99% of aluminum with the highest adsorption capacity of 4.93 mg/g. The same group of researchers investigated the usage of radiation grafted adsorbents and the selectivity towards several ions such as Cu (N. A. F. Othman et al., 2019) and Hg (Salamun et al., 2016). Grafted adsorbent also can be used for water treatment. Water pollution from textile effluent in Malaysia requires serious attention particularly. Selambakkannu et al. confirmed that radiation-induced grafted fibers is beneficial for the removal of anionic dye (Selambakkannu, Othman, Bakar, & Karim, 2019b; Selambakkannu, Othman, Bakar, Ming, et al., 2019).
- Radiation-grafted membranes for battery and fuel cell applications

    The use of radiation-induced graft polymerization seems to have great potential in the energy sector. The use of radiation-induced grafting in energy areas also involves the preparation of ion exchange membrane for redox flow cell and fuel cell applications (Nasef, 2009; Nasef & Hegazy, 2004). This is due to the performance and economic viability of the fuel cell or battery system which required the development of membrane separator with good chemical stability, high efficiency, and low cost (Zhang, Li, Zhang, & Shi, 2015). Moreover, modification with radiation grafting enables the membrane to gain adequate water uptake and swelling ratio, without scarifying its mechanical properties, and thus is suitable for fuel cell application. On top of that, radiation grafting assists in the fabrication of the membrane on tuning the hydrophilic and hydrophobic domains to control the ionic properties, as well. Recently, there are some research vigorously conducted in Malaysia for fuel cell and battery storage applications by radiation grafting. The modification of ion exchange membranes by using radiation grafting is simple, is able to produce new polymer chains reaction without the need of chemicals initiator, reaction parameters can be controlled for combination of several functional groups, and capable of controlling the yield of grafting. In addition, a review on the radiation-grafted membrane for a fuel cell has been published by (Abdiani et al., 2019).

- Radiation grafting for health-care applications

The outbreak of COVID-19 that caused a world pandemic was an eye opener for researchers all over the world to venture into the development of material with antibacterial and antimicrobial properties. Germs such as bacteria, viruses, fungi, and protozoa can be found everywhere, especially in food, water, and air that eventually will enter our body. Therefore, the use of antimicrobial agents has become prominent to combat or eradicate these germs. Antimicrobial agents have specific inhibitory activity and mechanisms against each microorganism. Therefore, the selection of antimicrobial agents is dependent on their efficacy against a target. Radiation grafting can modify the polymer surface to have very distinct properties through the choice of antimicrobial agents (Nasef et al., 2021) have reviewed comprehensively on this subject on polymers where he summarized the progress and strategies for designing various antimicrobial polymeric surfaces using a radiation grafting technique.

Recent development on radiation grafting of polymeric materials in food industry in particular the packaging material have become one of the interesting researches that have been to explored in Malaysia. This is aligned with the government's efforts in preserving the environment by minimizing the disposable plastic packaging waste as well as maintaining the food safety and freshness. The development of active packaging using radiation grafting technique is actively conducted purposely to fulfill the safety requirements (limits the usage of preservatives/additives in food), prolong the shelf life of the food (maintain product quality and safety), and is cost effective and environmentally friendly (recycling, reduction, biodegradable). Radiation grafting was implemented in research done by (Shukri, Ghazali, Fatimah, Mohamad, & Wahit, 2014) to covalently bond low density polyethylene with an antimicrobial agent, sorbic acid. The idea is to develop an antimicrobial active packaging film, whereby the sorbic acid does not migrate into food, in the effort of improving food safety. With a similar aim, the same technique was used in the preparation of black seed oil grafted LDPE film (Shukri, Ghazali, Wahit, Hilmi, & Ibrahim, 2022).

Apart from that, the most significant requirements in consideration of bio-material for medical purposes are non-toxicity, chemical free, and feasible for the human body; for example, the fabrication of the polymeric bio-materials in the field of tissue engineering. Research that has been conducted on fabrication of the grafted film using radiation-induced grafting methods perhaps can be a potential in the biomedical devices or new bio-materials for future usage since this technique is well known as the fast and clean method and also the product produced is improved in terms of hydrophilicity and biocompatibility (Tarmizi et al., 2020). In addition, radiation-induced grafting is also a tool for improving the properties of the polymeric materials and viable as biocompatible material in implant applications (Hidzir, Radzali, Rahman, & Shamsudin, 2020; Radzali, Hidzir, Mokhtar, & Rahman, 2020).

- Radiation grafting for bio-diesel production

Radiation-induced grafting provides a number of advantages for the preparation of polymeric materials in energy applications. Most recently, the production of heterogeneous biocomposite-based catalysts (kenaf fiber) using radiation grafting is the best to ensure environmental sustainability and is more economical. In addition, the development of heterogeneous catalysts in the production of bio-diesel from palm oil–based synthesis using a radiation-induced grafting method provides new functionalization on the catalyst itself and also affords advantages in terms of being environmentally friendly, value added, cost-saving, and easy to operate (Zabaruddin et al., 2019). The other research on bio-diesel production through the use of heterogeneous catalyst, for example, is radiation-induced grafting of flax fiber for bio-diesel production from cottonseed oil (Moawia, Nasef, & Mohamed, 2016; Moawia, Nasef, Mohamed, Ripin, & Zakeri, 2019). Therefore, the development of heterogeneous catalysts using this technique demonstrate excellent feasibility efficacy in bio-diesel production with low yield of by-products as well as stablility and can be regenerated without significant loss of function.

2.5 SUMMARY

The discussion above mainly focused on the effect of radiation grafting copolymerization onto polymers in terms of structure, properties, and function. The merging of RIGC with polymer research provides extraordinary features to the polymer that widen the prospective application in multiple sectors. This book chapter covers the approach and tangible factors that influence the efficiency of the RIGC. The common approaches used to execute RIGC on polymeric materials are pre-irradiation and post-irradiation techniques in inert or vacuum conditions and per-oxidation environments. The prominent factors that directly affect the efficiency of RIGC are the type of polymeric material, monomer, type of radiation, experimental conditions, type of solvent, and additives. Apart from that, the characterization of the radiation-grafted polymeric materials in terms of physical and chemical changes was discussed in depth as well. The changes in polymeric materials upon RIGC in terms of morphology, chemical, and physical properties indicate the successful fusion of co-polymers, which provide the desired functionality. The application of radiation-grafted polymeric materials in Malaysia for separation and purification, fuel cell, healthcare, and bio-diesel production were discussed in this chapter. Large-scale commercial exploitation for the preparation of radiation-grafted polymeric materials should be developed in order to expand the electron beam curing processes to a higher level.

REFERENCES

Abd Ghani, F., Hamzah, K., Mohamed, N. H., & Salleh, W. N. W. (2020). Modification o f PTFE flat sheet film via radiation induced grafting polymerization with acrylic acid (Modifi kasi Filem Kepingan Rata PTFE melalui Kaedah Pempolimeran Cangkuk Sinaran dengan Asid Akrilik). *Sains Malaysiana*, *49*(1), 169–178.

Abdiani, M., Abouzari-Lotf, E., Ting, T. M., Nia, P. M., Sha'rani, S. S., Shockravi, A., & Ahmad, A. (2019). Novel polyolefin based alkaline polymer electrolyte membrane for vanadium redox flow batteries. *Journal of Power Sources*, *424*, 245–253.

Agrawal, G., Negi, Y. S., Pradhan, S., Dash, M., & Samal, S. (2017). Wettability and contact angle of polymeric biomaterials. In Cristina Tanzi, M. & Farè, S. (eds). *Characterization of Polymeric Biomaterials* (pp. 57–81). Elsevier.

Ashfaq, A., Clochard, M.-C., Coqueret, X., Dispenza, C., Driscoll, M. S., Ulański, P., & Al-Sheikhly, M. (2020). Polymerization reactions and modifications of polymers by ionizing radiation. *Polymers*, *12*(12), 2877.

Babatunde, O. E., Yatim, J. M., Ishak, M. Y., Masoud, R., & Meisam, R. (2015). Potentials of kenaf fibre in bio-composite production: A review. *Jurnal Teknologi*, *77*(12).

Barba, B. J. D., Peñaloza, D. P., Seko, N., & Madrid, J. F. (2021). RAFT-mediated radiation grafting on natural fibers in aqueous emulsion. *Materials Proceedings*, *7*(1), 4.

Barbosa, A., Costa, L. L., Portela, T., Moura, E., Del Mastro, N., Satyanarayana, K., & Monteiro, S. (2012). Effect of electron beam irradiation on the mechanical properties of buriti fiber. *Matéria (Rio de Janeiro)*, *17*(4), 1135–1143.

Barsbay, M., & Güven, O. (2019). Surface modification of cellulose via conventional and controlled radiation-induced grafting. *Radiation Physics and Chemistry*, *160*, 1–8.

Bhattacharya, A., & Misra, B. (2004). Grafting: A versatile means to modify polymers: Techniques, factors and applications. *Progress in Polymer Science*, *29*(8), 767–814.

Bitter, H., & Lackner, S. (2021). Fast and easy quantification of semi-crystalline microplastics in exemplary environmental matrices by differential scanning calorimetry (DSC). *Chemical Engineering Journal*, *423*, 129941.

Brack, H. P., Ruegg, D., Bührer, H., Slaski, M., Alkan, S., & Scherer, G. G. (2004). Differential scanning calorimetry and thermogravimetric analysis investigation of the thermal properties and degradation of some radiation-grafted films and membranes. *Journal of Polymer Science Part B: Polymer Physics*, *42*(13), 2612–2624.

Bucio, E., & Burillo, G. (2009). Radiation-induced grafting of sensitive polymers. *Journal of Radioanalytical and Nuclear Chemistry*, *280*(2), 239–243.

Chamas, A., Moon, H., Zheng, J., Qiu, Y., Tabassum, T., Jang, J. H., … Suh, S. (2020). Degradation rates of plastics in the environment. *ACS Sustainable Chemistry & Engineering*, *8*(9), 3494–3511.

Charlesby, A. (2016). *Atomic Radiation and Polymers: International Series of Monographs on Radiation Effects in Materials*. Elsevier.

Choi, S.-H., Lee, K.-P., Kang, H.-D., & Park, H. G. (2004). Radiolytic immobilization of lipase on poly (glycidyl methacrylate)-grafted polyethylene microbeads. *Macromolecular Research*, *12*(6), 586–592.

Chumakov, M. K., Shahamat, L., Weaver, A., LeBlanc, J., Chaychian, M., Silverman, J., … Al-Sheikhly, M. (2011). Electron beam induced grafting of N-isopropylacrylamide to a poly (ethylene-terephthalate) membrane for rapid cell sheet detachment. *Radiation Physics and Chemistry*, *80*(2), 182–189.

Cleland, M., Parks, L., & Cheng, S. (2003). Applications for radiation processing of materials. *Nuclear Instruments and Methods in Physics Research Section B: Beam Interactions with Materials and Atoms*, *208*, 66–73.

Fauziyah, B., & Yuwono, M. (2020). *The effect of acid variation on physical and chemical characteristics of cellulose isolated from Saccharum officinarum L. Bagasse*. Paper presented at the IOP Conference Series: Earth and Environmental Science. Eindhoven, The Netherlands. OIP Publishing.

Ferry, M., Bessy, E., Harris, H., Lutz, P., Ramillon, J.-M., Ngono-Ravache, Y., & Balanzat, E. (2013). Aliphatic/aromatic systems under irradiation: Influence of the irradiation temperature and of the molecular organization. *The Journal of Physical Chemistry B*, *117*(46), 14497–14508.

Giller, C. B., Chase, D. B., Rabolt, J. F., & Snively, C. M. (2010). Effect of solvent evaporation rate on the crystalline state of electrospun Nylon 6. *Polymer*, *51*(18), 4225–4230.

Goel, N., Bhardwaj, Y., Manoharan, R., Kumar, V., Dubey, K., Chaudhari, C., & Sabharwal, S. (2009). Physicochemical and electrochemical characterization of battery separator prepared by radiation induced grafting of acrylic acid onto microporous polypropylene membranes. *Express Polymer Letters*, *3*(5), 268–278.

González-Torres, M., Perez-González, A. M., González-Perez, M., Santiago-Tepantlán, C., Solís-Rosales, S. G., & Heredia-Jiménez, A. H. (2011). Effects of solvent on gamma radiation–induced graft copolymerization of acrylamide onto poly (3-hydroxybutyrate). *International Journal of Polymer Analysis and Characterization*, *16*(6), 399–415.

Hidzir, N. M., Radzali, N. A. M., Rahman, I. A., & Shamsudin, S. A. (2020). Gamma irradiation-induced grafting of 2-hydroxyethyl methacrylate (HEMA) onto ePTFE for implant applications. *Nuclear Engineering and Technology*, *52*(10), 2320–2327.

Hong, T. T., Okabe, H., Hidaka, Y., & Hara, K. (2018). Radiation synthesis and characterization of super-absorbing hydrogel from natural polymers and vinyl monomer. *Environmental Pollution*, *242*, 1458–1466.

Işıkel Şanlı, L., & Alkan Gürsel, S. (2011). Synthesis and characterization of novel graft copolymers by radiation-induced grafting. *Journal of applied polymer science*, *120*(4), 2313–2323.

Kaith, B., Sharma, R., Sharma, K., Choudhary, S., Kumar, V., & Lochab, S. (2015). Effects of O7+ and Ni9+ swift heavy ions irradiation on polyacrylamide grafted Gum acacia thin film and sorption of methylene blue. *Vacuum*, *111*, 73–82.

Kakati, D. K., Bora, M. M., & Deka, C. (2015). Cellulose graft copolymerization by gamma and electron beam irradiation. In Thakur, V. J. *Cellulose-Based Graft Copolymers* (pp. 98–107). CRC Press.

Kargarzadeh, H., Ahmad, I., Abdullah, I., Dufresne, A., Zainudin, S. Y., & Sheltami, R. M. (2012). Effects of hydrolysis conditions on the morphology, crystallinity, and thermal stability of cellulose nanocrystals extracted from kenaf bast fibers. *Cellulose*, *19*(3), 855–866.

Khalil, H. A., & Suraya, N. L. (2011). Anhydride modification of cultivated kenaf bast fibers: Morphological, spectroscopic and thermal studies. *BioResources*, *6*(2), 1122–1135.

Khan, M. A., Rahaman, M. S., Al-Jubayer, A., & Islam, J. (2015). Modification of jute fibers by radiation-induced graft copolymerization and their applications. V. K. Thakur, Ed., *Cellulose-Based Graft Copolymers: Structure and Chemistry* (pp. 209–235).Taylor & Francis.

Kim, H.-S., Yang, H.-S., Kim, H.-J., & Park, H.-J. (2004). Thermogravimetric analysis of rice husk flour filled thermoplastic polymer composites. *Journal of Thermal Analysis and Calorimetry*, *76*(2), 395–404.

Kim, S. H., Park, Y. C., Jung, G. H., & Cho, C. G. (2007). Characterization of poly(styrene-b-vinylbenzylphosphonic acid) copolymer by titration and thermal analysis. *Macromolecular Research,* 15, 587–59410.1007/bf03218835.

Kodama, Y., Barsbay, M., & Güven, O. (2014). Radiation-induced and RAFT-mediated grafting of poly (hydroxyethyl methacrylate)(PHEMA) from cellulose surfaces. *Radiation Physics and Chemistry,* 94, 98–104.

Korolkov, I. V., Güven, O., Mashentseva, A. A., Atıcı, A. B., Gorin, Y. G., Zdorovets, M. V., & Taltenov, A. A. (2017). Radiation induced deposition of copper nanoparticles inside the nanochannels of poly (acrylic acid)-grafted poly (ethylene terephthalate) track-etched membranes. *Radiation Physics and Chemistry,* 130, 480–487.

Kubová, J., Matúš, P., & Bujdoš, M. (2005). Influence of acid mining activity on release of aluminium to the environment. *Analytica Chimica Acta,* 547(1), 119–125.

Kumar, V., Chaudhary, B., Sharma, V., & Verma, K. (2019). *Radiation Effects in Polymeric Materials.* Springer.

Kwan, S., & Marić, M. (2016). Thermoresponsive polymers with tunable cloud point temperatures grafted from chitosan via nitroxide mediated polymerization. *Polymer,* 86, 69–82.

Lee, J. S. (2013). The effect of electron beam irradiation on chemical and morphological properties of Hansan ramie fibers. *Fashion & Textile Research Journal,* 15(3), 430–436.

Li, J., Zhao, J., Tao, L., Wang, J., Waknis, V., Pan, D., ... Patel, J. (2015). The effect of polymeric excipients on the physical properties and performance of amorphous dispersions: Part I, free volume and glass transition. *Pharmaceutical Research,* 32(2), 500–515.

Li, Z., Tang, M., Dai, J., Wang, T., & Bai, R. (2016). Effect of multiwalled carbon nanotube-grafted polymer brushes on the mechanical and swelling properties of polyacrylamide composite hydrogels. *Polymer,* 85, 67–76.

Liao, C., & Tjong, S. C. (2012). Mechanical and thermal behaviour of polyamide 6/silicon carbide nano-composites toughened with maleated styrene–ethylene–butylene–styrene elastomer. *Fatigue & Fracture of Engineering Materials & Structures,* 35(1), 56–63.

Liao, C. Z., & Tjong, S. C. (2013). Mechanical and thermal performance of high-density polyethylene/ alumina nanocomposites. *Journal of Macromolecular Science, Part B,* 52(6), 812–825.

Liao, J., Brosse, N., Pizzi, A., & Hoppe, S. (2019). Dynamically cross-linked tannin as a reinforcement of polypropylene and UV protection properties. *Polymers,* 11(1), 102.

Lim, S. J., & Shin, I. H. (2020). Graft copolymerization of GMA and EDMA on PVDF to hydrophilic surface modification by electron beam irradiation. *Nuclear Engineering and Technology,* 52(2), 373–380.

Luo, G. E., Strong, P. J., Wang, H., Ni, W., & Shi, W. (2011). Kinetics of the pyrolytic and hydrothermal decomposition of water hyacinth. *Bioresource Technology,* 102(13), 6990–6994.

M Saber, S. E., Abdullah, L. C., Jamil, S. N. A. M., Choong, T. S., & Ting, T. M. (2021). Trimethylamine functionalized radiation-induced grafted polyamide 6 fibers for p-nitrophenol adsorption. *Scientific Reports,* 11(1), 1–17.

Madani, M. (2011). Structure, optical and thermal decomposition characters of LDPE graft copolymers synthesized by gamma irradiation. *Current Applied Physics,* 11(1), 70–76.

Magaña, H., Palomino, K., Cornejo-Bravo, J. M., Alvarez-Lorenzo, C., Concheiro, A., & Bucio, E. (2015). Radiation-grafting of acrylamide onto silicone rubber films for diclofenac delivery. *Radiation Physics and Chemistry,* 107, 164–170.

Mazhar, M., Abdouss, M., Shariatinia, Z., & Zargaran, M. (2014). Graft copolymerization of methacrylic acid monomers onto polypropylene fibers. *Chemical Industry and Chemical Engineering Quarterly,* 20(1), 87–96.

Moawia, R. M., Nasef, M. M., & Mohamed, N. H. (2016). Radiation grafted natural fibres functionalized with alkalised amine for transesterification of cottonseed oil to biodiesel. *Jurnal Teknologi,* 78(8–3).

Moawia, R. M., Nasef, M. M., Mohamed, N. H., Ripin, A., & Zakeri, M. (2019). Biopolymer catalyst for biodiesel production by functionalisation of radiation grafted flax fibres with diethylamine under optimised conditions. *Radiation Physics and Chemistry,* 164, 108375.

Mohamed, M. A., Jaafar, J., Ismail, A., Othman, M., & Rahman, M. (2017). Fourier transform infrared (FTIR) spectroscopy. In Ismail, A., Matsuura, T., & Oatley-Radcliffe, D. (eds). *Membrane Characterization* (pp. 3–29). Elsevier.

Montoya-Villegas, K. A., Ramírez-Jiménez, A., Zizumbo-López, A., Pérez-Sicairos, S., Leal-Acevedo, B., Bucio, E., & Licea-Claverie, A. (2020). Controlled surface modification of silicone rubber by gamma-irradiation followed by RAFT grafting polymerization. *European Polymer Journal,* 134, 109817.

Mozetič, M. (2019). Surface modification to improve properties of materials (*Vol. 12,* pp. 441). Multidisciplinary Digital Publishing Institute.

Nasef, M. M. (2009). Fuel cell membranes by radiation-induced graft copolymerization: Current status, challenges, and future directions. In Javaid Zaidi, S. M. & Matsuura, T. (eds). *Polymer Membranes for Fuel Cells* (pp. 87–114). Springer.

Nasef, M. M., Gupta, B., Shameli, K., Verma, C., Ali, R. R., & Ting, T. M. (2021). Engineered bioactive polymeric surfaces by radiation induced graft copolymerization: Strategies and applications. *Polymers, 13*(18), 3102.

Nasef, M. M., & Güven, O. (2012). Radiation-grafted copolymers for separation and purification purposes: Status, challenges and future directions. *Progress in Polymer Science, 37*(12), 1597–1656.

Nasef, M. M., & Hegazy, E.-S. A. (2004). Preparation and applications of ion exchange membranes by radiation-induced graft copolymerization of polar monomers onto non-polar films. *Progress in Polymer Science, 29*(6), 499–561.

Nasef, M. M., & Sugiarmawan, I. A. (2010). Radiation induced emulsion grafting of glycidyl methacrylate onto high density polyethylene: A kinetic study. *Malaysian Journal of Fundamental and Applied Sciences, 6*(2), 93–97.

Nasef, M.M., Ting, T.M., Abbasi, A., Layeghi-moghaddam, A., Alinezhad S. Sara, Hashim K. (2016). Radiation grafted adsorbents for newly emerging environmental applications. *Radiation Physics and Chemistry, 118* , 55–60.

Ndazi, B. S., Nyahumwa, C. W., & Tesha, J. (2008). Chemical and thermal stability of rice husks against alkali treatment. *BioResources, 3*(4), 1267–1277.

Nishino, T., Matsumoto, T., & Nakamae, K. (2000). Surface structure of isotactic polypropylene by X-ray diffraction. *Polymer Engineering & Science, 40*(2), 336–343.

Othman, N., Selambakkannu, S., Azian, H., Ratnam, C., Yamanobe, T., Hoshina, H., & Seko, N. (2021). Synthesis of surface ion-imprinted polymer for specific detection of thorium under acidic conditions. *Polymer Bulletin, 78*(1), 165–183.

Othman, N., Selambakkannu, S., Ting, T., & Zulhairun, A. (2019). Modification of kenaf fibers by single step radiation functionalization of 2-hydroxyl methacrylate phosphoric acid (2-HMPA). *SN Applied Sciences, 1*(3), 1–9.

Othman, N. A. F., Selambakkannu, S., & Tuan Abdullah, T. A. (2022). *Controlled Process of Radiation-Induced Grafting by Chemical Vapour Deposition for the Synthesis of Metal Adsorbent.* Paper presented at the Key Engineering Materials.Switzerland, Scientific.net.

Othman, N. A. F., Selambakkannu, S., Tuan Abdullah, T. A., Hoshina, H., Sattayaporn, S., & Seko, N. (2019). Selectivity of copper by amine-based ion recognition polymer adsorbent with different aliphatic amines. *Polymers, 11*(12), 1994.

Othman, N. A. F., Selambakkannu, S., Yamanobe, T., Hoshina, H., Seko, N., & Abdullah, T. A. T. (2020). Radiation grafting of DMAEMA and DEAEMA-based adsorbents for thorium adsorption. *Journal of Radioanalytical and Nuclear Chemistry, 324*(1), 429–440.

Pant, H. R., Baek, W.-I., Nam, K.-T., Jeong, I.-S., Barakat, N. A., & Kim, H. Y. (2011). Effect of lactic acid on polymer crystallization chain conformation and fiber morphology in an electrospun nylon-6 mat. *Polymer, 52*(21), 4851–4856.

Pino-Ramos, V. H., Ramos-Ballesteros, A., López-Saucedo, F., López-Barriguete, J. E., Varca, G. H., & Bucio, E. (2017). Radiation grafting for the functionalization and development of smart polymeric materials. *Applications of Radiation Chemistry in the Fields of Industry, Biotechnology and Environment,* 67–94.

Radzali, N. A. M., Hidzir, N. M., Mokhtar, A. K., & Rahman, I. A. (2020). *60Co-Induced Grafting of Dual Polymer (Acrylic Acid-co-HEMA) onto Expanded Poly (tetrafluoroethylene) Membranes.* Paper presented at the AIP Conference Proceedings.Kuala Lumpur, Malaysia, AIP Publishers.

Rath, S., Palai, A., Rao, S., Chandrasekhar, L., & Patri, M. (2008). Effect of solvents in radiation-induced grafting of 4-vinyl pyridine onto fluorinated ethylene propylene copolymer. *Journal of applied polymer science, 108*(6), 4065–4071.

Reddy, N., & Yang, Y. (2015). Fibers from banana pseudo-stems. In Reddy, N. & Yang, Y. (eds). *Innovative Biofibers Ffrom Renewable Resources* (pp. 25–27). Springer.

Rezvova, M., Zhevnyk, V., Pak, V., Denisov, V. Y., & Borodin, Y. V. (2017). *Increase the strength characteristics of polymer films by radiation graft polymerization.* Paper presented at the IOP Conference Series: Materials Science and Engineering. Tomsk, Russian Federation, IOP Publishers.

Sakulaue, P., Lertvanithphol, T., Eiamchai, P., & Siriwatwechakul, W. (2021). Quantitative relation between thickness and grafting density of temperature-responsive poly (N-isopropylacrylamide-co-acrylamide) thin film using synchrotron-source ATR-FTIR and spectroscopic ellipsometry. *Surface and Interface Analysis, 53*(2), 268–276.

Salamun, N., Triwahyono, S., Jalil, A., Majid, Z., Ghazali, Z., Othman, N., & Prasetyoko, D. (2016). Surface modification of banana stem fibers via radiation induced grafting of poly (methacrylic acid) as an effective cation exchanger for Hg (II). *RSC Advances*, *6*(41), 34411–34421.

Sarala, S., Nor, A., Siti, F. M., & Hamdani, S. (2014). Delignification studies of banana fibers for radiation graft copolymerization. *Australian Journal of Basic and Applied Sciences*, *8*(15 Special), 112–118.

Selambakkannu, S., Othman, N. A. F., Bakar, K. A., & Karim, Z. A. (2019a). Adsorption studies of packed bed column for the removal of dyes using amine functionalized radiation induced grafted fiber. *SN Applied Sciences*, *1*(2), 175.

Selambakkannu, S., Othman, N. A. F., Bakar, K. A., & Karim, Z. A. (2019b). Adsorption studies of packed bed column for the removal of dyes using amine functionalized radiation induced grafted fiber. *SN Applied Sciences*, *1*(2), 1–10.

Selambakkannu, S., Othman, N. A. F., Bakar, K. A., Ming, T. T., Segar, R. D., & Karim, Z. A. (2019). Modification of radiation grafted banana trunk fibers for adsorption of anionic dyes. *Fibers and Polymers*, *20*(12), 2556–2569.

Selambakkannu, S., Othman, N. A. F., Bakar, K. A., Shukor, S. A., & Karim, Z. A. (2018). A kinetic and mechanistic study of adsorptive removal of metal ions by imidazole-functionalized polymer graft banana fiber. *Radiation Physics and Chemistry*, *153*, 58–69.

Setia, A. (2018). *Applications of Graft Copolymerization: A Revolutionary Approach Biopolymer grafting* (pp. 1–44). Elsevier.

Sharif, J., Mohamad, S. F., Othman, N. A. F., Bakaruddin, N. A., Osman, H. N., & Güven, O. (2013). Graft copolymerization of glycidyl methacrylate onto delignified kenaf fibers through pre-irradiation technique. *Radiation Physics and Chemistry*, *91*, 125–131.

Sharma, R. K. (2012). A study in thermal properties of graft copolymers of cellulose and methacrylates. *Advances in Applied Science Research*, *3*(6), 3961–3969.

Shin, I. H., Hong, S., Lim, S. J., Son, Y.-S., & Kim, T.-H. (2017). Surface modification of PVDF membrane by radiation-induced graft polymerization for novel membrane bioreactor. *Journal of Industrial and Engineering Chemistry*, *46*, 103–110.

Shukri, N. A., Ghazali, Z., Fatimah, N. A., Mohamad, S. F., & Wahit, M. U. (2014). Physical, mechanical and oxygen barrier properties of antimicrobial active packaging based on ldpe film incorporated with sorbic acid. *Advances in Environmental Biology*, 2748–2753.

Shukri, N. A., Ghazali, Z., Wahit, M. U., Hilmi, F. F., & Ibrahim, S. N. S. S. (2022). Surface analysis of grafted low density polyethylene film by FTIR and XPS spectroscopy. In Key Engineering Materials (Vol. 908), pp. 49–53. Switzerland: Trans Tech Publications Ltd.

Sokker, H. H., Badawy, S. M., Zayed, E. M., Eldien, F. A. N., & Farag, A. M. (2009). Radiation-induced grafting of glycidyl methacrylate onto cotton fabric waste and its modification for anchoring hazardous wastes from their solutions. *Journal of Hazardous Materials*, *168*(1), 137–144.

Sonnier, R., Otazaghine, B., Viretto, A., Apolinario, G., & Ienny, P. (2015). Improving the flame retardancy of flax fabrics by radiation grafting of phosphorus compounds. *European Polymer Journal*, *68*, 313–325.

Suresh, D., Goh, P. S., Ismail, A. F., & Hilal, N. (2021). Surface design of liquid separation membrane through graft polymerization: A state of the art review. *Membranes*, *11*(11), 832.

Sutirman, Z. A., Sanagi, M. M., Abd Karim, K. J., & Ibrahim, W. A. W. (2016). Preparation of methacrylamide-functionalized cross-linked chitosan by free radical polymerization for the removal of lead ions. *Carbohydrate Polymers*, *151*, 1091–1099.

Tamada, M. (2018). Radiation processing of polymers and its applications. In Kudo, H (ed). *Radiation Applications* (pp. 63–80). Springer.

Tanasă, F., Zănoagă, M., Teacă, C. A., Nechifor, M., & Shahzad, A. (2020). Modified hemp fibers intended for fiber-reinforced polymer composites used in structural applications—A review. I. Methods of modification. *Polymer Composites*, *41*(1), 5–31.

Tarmizi, Z., Ali, R., Nasef, M., Akim, A., Eshak, Z., & Noor, S. (2020). *Fabrication and Characterization of PU-g-poly (HEMA) Film for Clotting Time and Platelet Adhesion*. Paper presented at the IOP Conference Series: Materials Science and Engineering. Kuala Lumpur, IOP Publishing Ltd.

Tataru, G., Guibert, K., Labbé, M., Léger, R., Rouif, S., & Coqueret, X. (2020). Modification of flax fiber fabrics by radiation grafting: Application to epoxy thermosets and potentialities for silicone-natural fibers composites. *Radiation Physics and Chemistry*, *170*, 108663.

Thakur, V. K., Tan, E. J., Lin, M.-F., & Lee, P. S. (2011). Poly (vinylidene fluoride)-graft-poly (2-hydroxyethyl methacrylate): A novel material for high energy density capacitors. *Journal of Materials Chemistry*, *21*(11), 3751–3759.

Ting, T., Nasef, M. M., & Hashim, K. (2015). Modification of nylon-6 fibres by radiation-induced graft polymerisation of vinylbenzyl chloride. *Radiation Physics and Chemistry*, *109*, 54–62.

Vega-Hernández, M. Á., Cano-Diaz, G. S., Vivaldo-Lima, E., Rosas-Aburto, A., Hernandez-Luna, M. G., Martinez, A., ... Penlidis, A. (2021). A review on the synthesis, characterization and modeling of polymer grafting. *Processes*, *9*(2), 375.

Venturi, M., & D'Angelantonio, M. (2017). *Applications of Radiation Chemistry in the Fields of Industry, Biotechnology and Environment*. Springer.

Wang, W. F., Tan, T. L., Tan, B. L., & Ong, P. P. (1996). High-resolution FTIR spectrum and rotational structure of the V8 band of methylene chloride. *Journal of Molecular Spectroscopy*, *175*, 363–369.

Walo, M. (2017). Radiation-induced grafting. *Applications of Ionizing Radiation In Materials Processing* (*Vol. 1*, pp. 193–210): Warszawa.

Walo, M., Przybytniak, G., Kavaklı, P. A., & Güven, O. (2013). Radiation-induced graft polymerization of N-vinylpyrrolidone onto segmented polyurethane based on isophorone diisocyanate. *Radiation Physics and Chemistry*, *84*, 85–90.

Wei, L., & McDonald, A. G. (2016). A review on grafting of biofibers for biocomposites. *Materials*, *9*(4), 303.

Wu, F., Zhang, S., Chen, Z., Zhang, B., Yang, W., Liu, Z., & Yang, M. (2016). Interfacial relaxation mechanisms in polymer nanocomposites through the rheological study on polymer/grafted nanoparticles. *Polymer*, *90*, 264–275.

Yaacoub, E, & Le Perchec, P (1988). Effective catalytic activity of polymeric sulfoxides in alkylation reactions under two-phase conditions. *Reactive Polymers, Ion Exchangers, Sorbents*, 8, 285–29610.1016/0167-6989(88)90304-2.

Yan-Feng, Z., Xue-Zheng, S., Huang, B., Dong-Sheng, Y., Hong-Jie, W., Wei-Xia, S., ... Blombäck, K. (2007). Spatial distribution of heavy metals in agricultural soils of an industry-based peri-urban area in Wuxi, China. *Pedosphere*, *17*(1), 44–51.

Yousif, E., & Haddad, R. (2013). Photodegradation and photostabilization of polymers, especially polystyrene. *SpringerPlus*, *2*(1), 1–32.

Zabaruddin, N. H., Mohamed, N. H., Abdullah, L. C., Tamada, M., Ueki, Y., Seko, N., & Choong, T. S. Y. (2019). Palm oil-based biodiesel synthesis by radiation-induced kenaf catalyst packed in a continuous flow system. *Industrial Crops and Products*, *136*, 102–109.

Zhang, H., Li, X., Zhang, H., & Shi, D. (2015). Use of Porous Membrane and Composite Membrane thereof in Redox Flow Energy Storage Battery. Google Patents.

Zhou, T., Shao, R., Chen, S., He, X., Qiao, J., & Zhang, J. (2015). A review of radiation-grafted polymer electrolyte membranes for alkaline polymer electrolyte membrane fuel cells. *Journal of Power Sources*, *293*, 946–975.

Zulfiqar, S., Zulfiqar, M., Nawaz, M., McNeill, I., & Gorman, J. (1990). Thermal degradation of poly (glycidyl methacrylate). *Polymer Degradation and Stability*, *30*(2), 195–203.

3 Microwave-assisted conversion of coal and biomass to activated carbon

Sudip Ray*

New Zealand Institute for Minerals to Materials Research, Greymouth, New Zealand

*Corresponding author: sudip.ray@nzimmr.co.nz

CONTENTS

DOI: 10.1201/9781003321910-3

3.1 INTRODUCTION

Activated carbon is a ubiquitous porous sorbent material finding usage in a myriad of contemporary applications, from resolving various industrial processing issues and the household in situ remediations to mitigating complex environmental concerns. The major applications include water treatment with its ability to remove toxic compounds from the waste stream, enabling air purification by reducing greenhouse gas emissions, recovery of precious metals from the ores such as gold extraction as well as growing use in food and beverage processing, production of medical and pharmaceutical supplies for healthcare and cosmetic industries. According to the U.S. Environmental Protection Agency, adsorption by activated carbon is one of the best available drinking water treatment technologies (Web Reference, 2022a). The original use of porous carbon materials can be traced back to the prehistorical time around 3750 BCE during the Bronze Age era to remove impurities in smelted metal and for the reduction of copper, tin, and zinc ores in the manufacturing of bronze. Later on, from around 1550 BCE, its beneficial uses in treating ailments, adsorbing nuisance odors, and water purification are well documented (Dabrowski, 2001). Despite the business disruption over the last couple of years due to the COVID-19 pandemic, the demand for this material witnessed sustained growth from regular uses to a variety of emerging applications. According to a recent study by Marketsandmarkets (Web Reference, 2022b), the global activated carbon market was estimated at USD 5.7 billion in 2021 and is projected to reach USD 8.9 billion by 2026, at a CAGR of 9.3% from 2021 to 2026. The United States and Europe are exploiting this sorbent material, while an enormous demand is expected from the Asia-Pacific markets, especially in the water treatment and air purification sectors.

As the name signifies, this material mainly consists of carbon compounds. Ideally, it can be produced from a range of carbonaceous feedstock. However, high fixed-carbon containing materials from the abundant, inexpensive natural organic origin such as coal and lignocellulosic biomass, for example, woods, coconut husk, apricot kernel shell, and fruit stones, are commonly used as the precursor for large-scale production to secure techno-economic feasibility. When biomass-based feedstock is used, the material is also termed *activated charcoal* or *activated biochar*. In commercial practice, coal is the primary carbonaceous feedstock to produce activated carbon (Saleem et al., 2019).

The unique characteristic of activated carbon is its outstanding ability to remove contaminants from fluid (gas, vapor, or liquid) mediums owing to a process called adsorption through its porous structure. Adsorption is the accumulation of a chemical substance or pollutant from a fluid phase on the surface of a liquid or solid substrate, as opposed to absorption, in which the invading substance enters the substrate's bulk space. Thus, in adsorption, a mass transfer process occurs at the activated carbon, i.e., at the adsorbent surface, leading to the transfer of a pollutant molecule, i.e., adsorbate, from a bulk fluid phase to the solid surface. Such a process can occur because of nonspecific, noncovalent, intermolecular interactions termed *physisorption* through combined physical forces, e.g., the long-range London dispersion forces, a type of van der Waals forces, and the short-range intermolecular repulsion. Specific chemical interactions may also take place between the pollutant molecules and the functional groups present on the adsorbent surface, followed by chemical bond formation, known as *chemisorption* (Marsh and Rodríguez-Reinoso, 2006, Thommes, 2010). The physical forces are generally weak, and hence the adsorption process can be made reversible. The majority of the pollutants can be released from the spent adsorbent in a subsequent step through desorption, and the adsorbent can be reused. Due to the presence of abundant porous structures of the low dimensional nanoscale size, it generates a large specific surface area and pore volume where it can accommodate the pollutant molecules. Along with high hydrophobicity, distinctive molecular structure, and suitable chemical functionalities, activated carbon exhibits an extremely high affinity for many classes of organic and inorganic pollutants and excellent adsorption capacity.

The conventional slow-pyrolytic heating process to produce activated carbon from coal and biomass is time-consuming and energy-intensive. Also, there are limitations in controlling the heating process, which impact the crucial microstructure formation in the final product. Hence there is a need for alternative solutions for producing this valuable material more economically and with more precise process control to acquire superior quality for addressing advanced customized solutions. Amongst the alternative heating processes, microwave technology has been successfully used in a range of commercial and household applications such as food processing, polymer curing, and ceramic sintering. Microwave processing of materials provides an alternative approach to improve the physical properties of the materials, innovative solutions for processing materials that are hard to process, a reduction in the environmental impact of material processing, economic advantages through energy savings, less space and production time requirement, and an opportunity to produce new materials with improved microstructures that other methods cannot achieve.

This chapter reviews the use of microwave energy as a sustainable and energy-efficient process in producing high-value activated carbon from coal and biomass. The general characteristics of activated carbon, relative advantages, and limitations of coal and biomass as the feedstock are highlighted. The benefits and scopes for microwave radiation technology over the conventional pyrolysis heating process are discussed. The working principle of microwave radiation and its influence on the carbonization and activation processes are provided. Key functional properties, especially the activated carbon's adsorption capacity, underlying molecular adsorption mechanism, characterization by various analytical techniques, and the effect of process parameters are explored. Physical and chemical activation in conjunction with microwave radiation to enhance the production efficiency is assessed. The future perspective on further advancing this research area is recommended.

3.2 GENERAL ASPECTS OF ACTIVATED CARBON

Before delving into the depth of this subject matter on the production of activated carbon utilizing microwave heating technology, it is vital to address the competitive importance of this specific product against its closest counterpart, which is primarily the motivating factor for this topic of interest. Technically, char compounds such as coal char and biochar are another class of economic porous carbon adsorbents. Both activated carbon and char compounds belong to *Pyrogenic carbon*, contribute to significant environmental remediation applications, and are produced anthropogenically. In the case of activated carbon products, these char compounds are the intermediates that require additional steps to activate by physical or chemical processes. The carbon feedstock can also be converted to the activated product in a single step without going through the intermediate char formation. However, chemical reagents are still required to facilitate this process. Nevertheless, converting the carbon feedstocks to activated carbon products thus imposes an extra cost burden for introducing the activation step in its production but outperforms the char compounds due to superior adsorption capabilities. The adsorption efficiency of biochar usually fluctuates due to their limited porous structures, whereas activated carbon has a more stable and predictable efficiency (Berger, 2012) with well-developed porosity. Biochar generally takes a longer time than activated carbon to adsorb contaminants when the same amounts of these two sorbent materials are used (Oleszczuk et al., 2012). This may lead to either a need for larger amounts of biochar to have equivalent process time or a longer time to adsorb the same amount of contaminant. The former would increase costs, and the latter may pose challenges in time-constrained applications. As a result, no major commercial applications of biochar as an adsorbent exist yet (Alhashimi and Aktas, 2017), while the use of activated carbon is further expanding.

3.2.1 Production of activated carbon by conventional pyrolysis

The main thermochemical conversion routes for the conversion of coal and biomass resources into heat, electricity, fuels, chemicals, or any other value-added products are combustion or incineration (complete oxidation), gasification (partial oxidation), torrefaction (reducing water and volatiles contents under mild heating in the absence of oxygen), and pyrolysis (thermal degradation in the absence of oxygen). The pyrolysis process comprehends to be an efficient valorization technology for industrial applications to convert the carbonaceous waste stream into useful products. Pyrolysis consists of two main stages known as primary and secondary pyrolysis. Primary pyrolysis causes devolatilization of the main constitutes by dehydration, dehydrogenation, and decarboxylation processes. Secondary pyrolysis triggers the thermal or catalytic cracking of heavy compounds or char into gases such as CO, CO_2, CH_4, and H_2. There are several modes of the pyrolysis process based on the reaction rate induced by process conditions such as heating rate and residence time. They are categorized as slow, fast, and flash pyrolysis to produce the desired product (Foong et al., 2020). Fast and flash pyrolysis processes drastically intensify the degradation reactions and are more suitable for producing liquid and gaseous products. On the contrary, the slow pyrolysis process can increase the combustible characteristic of the char, reduce the carbon burn-off and gasification reactions, and increase the carbon content in the solid product in a more controlled manner compared to the fast and flash pyrolysis processes. Thus, in a typical manufacturing process, the carbon content of the feedstock material is initially increased (carbon up-gradation) by slow pyrolytic heating under an inert atmosphere to remove the moisture, and the volatile matters, followed by char formation, and this step is known as the carbonization process. In the subsequent stage, a highly porous structure could be developed in the carbonized product, which is called the activation step. Activation refers to a process that improves the adsorptive capacity of the material. It is more specifically defined as *"any process which selectively removes the hydrogen or hydrogen-rich fractions from a carbonaceous raw material in such a manner as to produce an open, porous residue"* (Lewis and Metzner, 1954). Activation of carbon progressively develops the pore structure and internal surface area of carbonaceous material. This includes enlarging existing pores, opening up pores either partially or totally filled with tars, and forming new pores. During these carbonization and activation processes, from 15% to over 70% of the weight loss may take place from the original carbon feedstock to attain the optimal properties.

Both fixed-bed and movable-bed pyrolysis reactors can be used to produce activated carbon from coal and biomass. The movable-bed reactors are pneumatic type (bubbling, spouted, circulating, or transport fluidized beds), mechanical type (rotary kiln, rake, auger, ablative, stirred), and reactors in which the charge moves under gravity (Lewandowski et al., 2019). The rotary kiln reactors are commonly used as it allows good mixing of materials with various shapes, sizes, and heating values under continuous operation and delivers good heat transfer during slow rotation of the inclined kiln, and are simple in the adjustment of the residence time of solids in the reactor (Mopoung and Dejang, 2021). Generally, at first, feedstocks are dehydrated at around 170°C and pulverized to a predetermined size before char formation and activation steps to obtain the desired properties in the final product. Then, during the initial char formation process, the conditions are set as a low process temperature (400°C–700°C), slow heating rate (0.1°C/s–1°C/s), and long solid residence time (several hours) under an inert atmosphere. The activation of the char products is typically carried out at a higher temperature (700°C–1000°C and sometimes even up to 1200°C) using physical reagents. Physical activation is principally the partial gasification of the carbonaceous material (Sajjadi et al., 2019a) in a slow and controlled manner with mild oxidizing agents such as steam and carbon dioxide (CO_2). A combination of them is also used by injecting steam into the flue gas, which contains CO_2 as one of the combustion products. Alternatively, carbonization and activation can be performed in a single step by using chemical reagents. In the chemical activation process, the carbon precursor material is first impregnated with the activating agents or soaked with the chemical solution (Sajjadi et al., 2019b), followed by providing heat energy over a relatively lower temperature range of 400°C–800°C to initiate the dehydration followed by the activation reactions for simultaneous carbonization and creation of the pores. The chemicals used for this purpose are strong mineral acids, metallic salts, or hydroxides which act as dehydrating, cross-linking, and oxidizing substances. The activated product generated by these physical and chemical reagents is then crushed and sieved to different size grades. Washing the final product is often required, especially for the chemically activated product or removing the ash impurities while making the highly pure grades. For some special grades (Llamas-Unzueta et al., 2022) to increase product strength and water stability, pulverized raw carbonaceous materials are initially kneaded with a binding agent from inorganic (clays) or organic (pitches, tars, and resins) sources before the carbonization and activation steps to form shaped products. Alternatively, activated carbon powders are extruded with a binding agent, preferably clays, to build the shaped products to harden the pellets by mild heating (400°C–600°C) to minimize the damage to the already formed porous structures.

Therefore, the production of activated carbon by the conventional slow pyrolysis method is an elaborate process and hence time-consuming, energy-intensive, and causes significant operating costs. Hence, there is a need for an alternative approach to reduce the operating costs while improving the product quality and yield. The following sections discuss the justification for choosing microwave radiation as a sustainable and energy-efficient alternative processing technology in activated carbon production.

3.2.2 TYPES, PHYSICOCHEMICAL PROPERTIES, AND APPLICATION OF ACTIVATED CARBON

Activated carbon provides the flexibility to control the porosity and chemical functionality, and thereby various types of the products can be developed with bespoke formulations for specific applications.

3.2.2.1 Types

There are primarily three different types of activated carbon produced from coal and biomass and, in some specialized cases, from polymer and fiber sources for the adsorption of a specific or a set of pollutants from different fluid phases. The majority (ca. 60%) of activated carbon products are consumed in liquid-phase adsorption processes. Prices vary depending on the raw material source and production methodologies, activated carbon's physicochemical properties, including adsorption performance, and other end-use requirements such as hydrodynamic resistance, abrasion resistance, and recyclability.

3.2.2.1.1 Powdered activated carbon (PAC)

The powdered form of activated carbon would have less than 0.18 mm fine particle size, which can pass through a U.S. 80-mesh sieve (0.18 mm). This is the most economical and major physical form of activated carbon, with a global market share of 48% (Web Reference, 2022c). PACs are mainly applied as single-use batch processes for liquid-phase adsorption. However, by adjusting the manufacturing conditions, the internal pore structures and the adsorption properties can be fine-tuned, and the product could be used to remove a variety of contaminants from fluid phases. In PAC dosing systems, the carbon particles are injected into the contaminated liquid, dispersed within the liquid for an appropriate contact time, and then removed by flocculation, sedimentation, or filtration. The fine particle size facilitates fast adsorption, but the powdered form is difficult to handle when used in fixed adsorption beds. They also cause a high-pressure drop in fixed beds, which are hard to regenerate. An activated carbon bed is frequently used as a primary treatment tool in wastewater industries, and then the spent product is removed by sedimentation with the contaminated carbon for disposal. They are also used for flue gas treatment to remove dioxins, furans, and heavy metals from waste incineration plants. They are injected into the flue gas through a distribution lance, and the exhausted carbon is then removed through an electrostatic precipitator or fabric filter.

3.2.2.1.2 Granulated activated carbon (GAC)

To overcome the limitations of PACs, activated carbons are also produced in irregular granulated form with particle sizes in the range of 0.18 to 5 mm. According to ASTM D 2867-95, a minimum of 90% particle size of GAC should be larger than 80 mesh (0.18 mm). The advantage of GAC is that it is more suitable for all the fluid phase applications and can provide a good balance of particle size, specific surface area, and head-loss characteristics. Compared to powdered form, granulated one is hard, abrasion-resistant, relatively dense to withstand operating conditions, causes low hydrodynamic resistance, is easy to handle, and lasts longer. GACs can be formulated into a module that can be removed after saturation, conveniently regenerated usually by heat treatment in steam, and can be utilized multiple times for the same application. A bigger size GAC (2 to 5 mm) is preferable for gas and vapor phase applications to control the crucial pressure drop. In contrast, a smaller size GAC (0.18 to 2 mm) is desirable for liquid phase applications to enhance the adsorption kinetics, which is more critical than the pressure drop issue. However, GACs are relatively more expensive than PACs but still have a significant global market share of 35% (Web Reference, 2022c) as it offers several operational benefits.

3.2.2.1.3 Specialized activated carbon

Activated carbon is also available in various specialized forms, accounting for the remaining 17% of the global market share. Examples of some of these specialized forms are mentioned below.

3.2.2.1.4 Extruded activated carbons (EAC)

To further enhance the practical and end-use requirements, especially improving the abrasion resistance, activated carbons are also produced in uniform cylindrical pellet form with diameters typically from 0.8 to 5 mm and 15 to 35 mm in length by an extrusion process using a suitable binder. Their high adsorption activity, specific surface area, mechanical strength, and low dust content make them ideal sorbents for gas and vapor phase applications. The high adsorptive capacity and uniformity of its shape make it particularly useful in applications where a low-pressure drop is a major consideration for removing a variety of contaminants from air and vapor streams. Pellets are also an environmentally friendly product that can be reactivated through thermal oxidation and can be used multiple times for the same application.

GAC or EAC are usually introduced as packed-bed systems, and the fluid flows through this static bed. With decreasing the contaminant concentration in the fluid, the pollutant loading on the carbon increases. It creates a concentration gradient along the column between the inlet and the outlet concentration. Eventually, it arrives at a breakthrough point when the traces of contaminants are detected in the purified fluid. The required dosing of activated carbon is estimated from this breakthrough behavior by correlating the contaminated fluid concentration and adsorber run time.

3.2.2.1.5 Activated carbon fiber

Activated carbon is also produced in fibrous form. They are available in cloth and felt formats and can be tailor-made by various activity levels, weights, and thicknesses to meet the requirements. Activated carbon fiber products are primarily made from polyacrylonitrile, Kynol (Phenolic resin), rayon, and pitch-based feedstock materials (Lee et al., 2014). It can be used for a wide range of filtration applications, including air conditioning, catalyst media, emission control, purification filters, solvent recovery, and water filtration. However, fibrous-activated carbon is relatively an expensive material.

Other specialized activated carbon such as acid-washed high purity activated carbons and chemically impregnated carbons, honeycomb, and spherical-shaped carbon are also produced as customized products. PACs, GACs, and some specialized high purity grades and impregnated particulate activated carbon are mainly produced from coal and biomass, while the activated carbon fiber, non-woven fabric, or cloth are fabricated from polymer and fiber sources. The current chapter focuses on the activated carbon produced from coal and biomass as the precursor material.

3.2.2.2 Physicochemical properties

As porosity and their structural framework are intimately related, understanding the molecular arrangements of carbon compounds in the porous material together with a description of their origins and formation mechanism are prerequisites to developing the production protocol and realizing the appropriate use of activated carbons. Activated carbon is commonly described as an amorphous form of porous carbon (Saleem et al., 2019) with a disordered structural framework, unlike other forms or allotropes of elemental carbon such as diamond, graphite, and fullerenes. Activated carbon also differs from the char compounds. The latter products contain more organic residues, are less porous, and have a lower specific surface area. The activated carbons are typical *non-graphitizable* and *isotropic* carbons. Franklin (1951) proposed a two-dimensional model of *non-graphitizable* carbon, precisely indicating the origin of porosity and representing the defective graphene layered structures in the activated carbon Figure 3.1a. The figure demonstrates the turbostratic carbon structure with randomly oriented graphitic carbon layers. These disordered packing of turbostratic carbon sheets and clusters create the pore structure consisting of interconnected nanochannels. This leads to the formation of open and closed pores; however, the open pores are most important for facilitating the adsorption process. Also, the disordered arrangement forms heterogeneity and nonuniformity of the pore space. According to Franklin's model, graphitic nanodomains are bonded by amorphous carbon (sp^3) domains, preventing graphitization and leading to micropore formation. As a result, they do not have long-range parallel stacking of graphene layers along the c-axis (third dimension). These structural phenomena differentiate activated carbon from graphites and explain the reason for their difference in hardness. The bonded, i.e., the cross-linked structure, provides the necessary structural rigidity in the highly porous structure of activated carbon, and thereby, it is *hard carbon*. Such bonding effect is weaker in graphitizing carbons, and hence graphites are *soft carbons*.

Based on electron microscopy observations, Shiraishi (1984) proposed a three-dimensional model for the microtexture of hard porous carbon, where carbon layers form the walls of irregularly shaped pores (Figure 3.1b). Yoshida et al. (1991) adopted this model to support the concept of porosity in hard porous carbon as an assembly of cage-like space of nanometer dimensions. The walls of the cages are the assembly of carbon atoms in defective micro-graphene layered structures. Figure 3.1c (Schuepfer et al. 2020) shows the current perception of the transformation of precursor molecules in the carbonization process. The major stages which occur during heat treatment causing atomic rearrangements are the following: Initially, amorphous carbon exhibits all three kinds of carbon bonds (sp^1, sp^2, and sp^3) while the layered structures begin to form at temperatures only above 700°C. On further heating, the amorphous carbon of mixed bonding transforms to sp^2-bonded amorphous carbon at around 1200°C. Nongraphitic carbon is formed in the temperature range from about 1400°C to 1700°C. Above 1700°C, the crystallites are arranged into ordered stacks of graphene sheets. Activated carbons are generally heat-treated up to 1000°C depending on the requirement of the pore structure and the activation method applied. The shaded area in Figure 3.1c represents the temperature range typically used to produce activated carbons, where the carbon compound typically belongs to the amorphous phase. Thus, with advancing the

(a)   (b)

5 nm

(c)

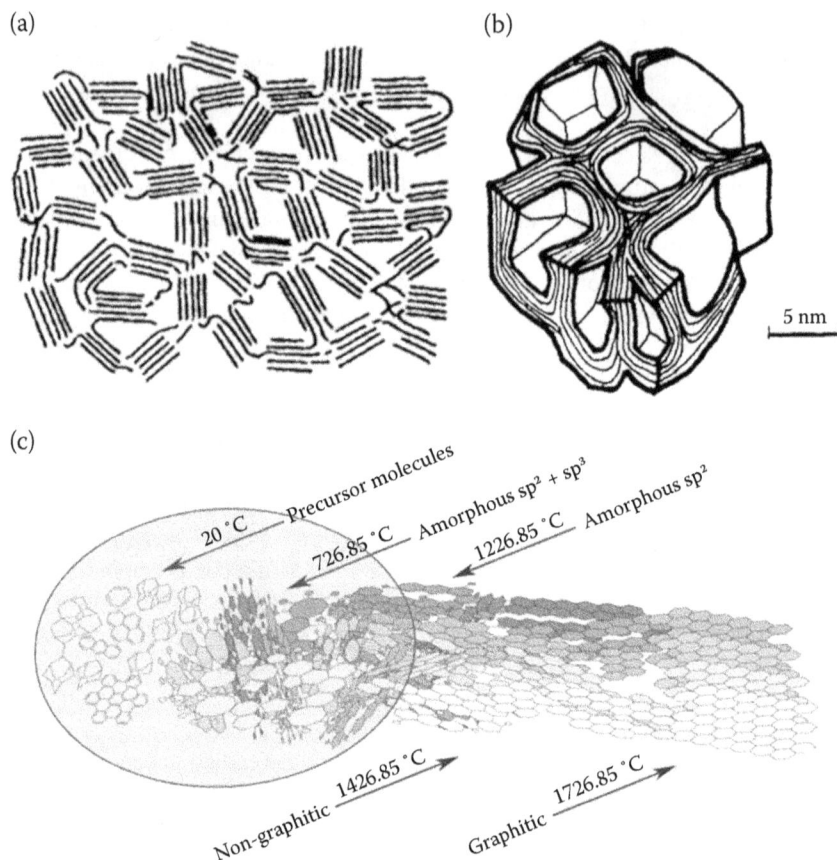

FIGURE 3.1 Schematic presentation of the possible model structures and formation of activated carbons **(a)** two-dimensional model [Reproduced with permission from reference Franklin, 1951, *Proceedings of the Royal Society of London. Series A, Mathematical and Physical Sciences* 209(1097): 196–218], **(b)** three-dimensional model [Reproduced with permission from reference Yoshida et al., 1991, *Carbon* 29(8): 1107–1111], and **(c)** transformation of precursor molecules in the carbonization process [Reproduced with permission from reference Schuepfer et al., 2020, *Carbon* 161: 359–372].

heat treatment commencing from lower ambient temperature, the molecular structure and their arrangement within the carbonized matrix increasingly resembles graphene, the structural element of graphite. However, unlike graphene, the graphitic carbon layers in the activated carbon remain highly disordered. The differences in the precursor chemical structure, molecular bonding and networking capabilities, and the processing conditions significantly influence their molecular transformation during the pyrolysis (Ray and Cooney, 2018). Accordingly, the non-graphitic and graphitic phases evolve with defective structures, which are the basis for the porosity formation and their adsorption capabilities and affect the product yield. This also manifests the importance of selecting appropriate feedstock to create the desired product.

Different non-carbon elements such as oxygen, hydrogen, sulfur, and nitrogen are commonly present in activated carbon in the form of functional groups and/or atoms chemically bonded to the carbon structure (Bansal and Goyel, 2005). The functional groups, namely carboxyl, carbonyl, phenols, lactones, and quinones present in the activated carbon significantly contribute to the uptake of pollutants from different fluid phases. These functional groups originate from precursor molecules and also during carbonization, activation, and post-chemical treatment.

3.2.2.3 Applications

Activated carbon finds a wide range of applications due to cost-effectiveness, abundance, renewability, suitable sorption properties, and environmental friendliness. Moreover, its well-developed internal pore structure, pore-volume, high specific surface area, surface functional groups, low density, good mechanical strength, chemical and thermal stability, and suitability for large-scale production, regeneration, modification of chemical functionalities for targeting specific pollutants provide an attractive choice as a commercial sorbent for air, vapor, and water treatment and developing a material system for new and advanced applications. A list of various commercial applications of activated carbon is provided in Table 3.1.

TABLE 3.1
Commercial Applications of Activated Carbon

Current Commercial Applications of Activated Carbon as Sorbent		
Fluid Phase in the Adsorption Process	Application Purpose	Example
Gas and vapor phase adsorption	General-purpose removal of pollutants in industrial, automobile, and domestic settings	Airconditioning, ventilation, and exhaust systems. This includes the removal of organic impurities by exhaust air treatment following commercial production processes often designed for removing targeted pollutants.
	Purification of process gases	Scrubbing and oil separation from compressed air.
	Purification of natural gas	Removal of H_2S and BTEX (benzene, toluene, ethylbenzene, and xylene) from natural gas streams.
	Flue gas cleaning	Removal of toxic substances such as dioxins and heavy metals, e.g., mercury at waste incineration plants.
	Odor removal	Kitchen exhaust hoods and refrigerator filters.
	Solvent recovery	Paints and coating manufacturing, rotogravure printing operations, packaging, food, and chemicals industries.
	Emission control	At filling stations during tank ventilation for motor vehicles and many industrial applications.
	Miscellaneous	Military gas masks, industrial respirators, cigarette and cigar filters to protect personnel from toxic vapor, and supporting material to prepare heterogeneous catalysts.
Liquid phase adsorption	Water treatment for removal of odor, toxic solute organic matter, non-polar substances (phenol, mineral oil, BTEX, polyaromatic hydrocarbons), heavy metals, dichlorination, and deozonation.	Wastewater, sewage and municipal water treatment, landfill seepage treatment, remediation of acid mine drainage, drinking water purification, swimming pool and aquarium water treatment, boiler feedwater treatment, condensate, and contact water treatment, ground and surface water cleanup, and various service water treatment.
	Treatment of different types of liquids in commercial production units	Sugar solutions, glucose, vegetable oils and fats, glutamate, spices, wine, beer, fruit concentrates, plant extracts, chemicals, and pharmaceuticals.
	Miscellaneous	Precious metal recoveries such as gold and silver, medical remedies for poisoning and preventing the absorption of ingested toxins in the gut, and various cosmetic applications include teeth whiteners, soaps, shampoos, and acne face masks.

3.2.3 THE KEY CONSIDERATION FOR DEVELOPING FUNCTIONAL PROPERTIES OF ACTIVATED CARBON

The practical value of activated carbon is attributed to its effectiveness in removing organic and inorganic pollutants from fluid phases through the adsorption process. The physicochemical characteristic properties of the adsorbent (pore structure and size distribution, pore volume, specific surface area, chemical functionality, density, and hardness) and the adsorbate (molecular size, chemical property, hydrophilic behavior, boiling point, and polarity), as well as operating conditions (temperature, residence time, solute adsorbate concentration, composition of the solution or gas mixture, relative humidity in the gas phase and pH value of the solution in the liquid phase), govern the adsorption process and the maximum usage of adsorption capacity. Thorough material characterization and particularly investigation of the adsorption mechanism are crucial for material development and troubleshooting to enhance the accumulation of targetted pollutants on the porous wall surfaces and within the porous cavity and to control the desorption process. According to The International Union of Pure and Applied Chemistry (IUPAC) classification (Sing et al., 1985), pores are classified into three different types as per the dimensions of the available space between the pore walls. A pore width of less than 2 nm is a micropore; when it is in the range of 2 to 50 nm is a mesopore, and when over 50 nm is called a macropore. Microporous structures are the most coveted texture to generate high surface area as they can enhance the adsorption of small-size pollutants such as in air filtration applications. Microporous structures combined with mesoporous structures are also required while dealing with both small and large size pollutants to facilitate the mass transport, adsorption process, and accommodating the pollutants within the relatively larger mesopores, such as in the wastewater treatment scenario. Macropores have less adsorption ability but contribute to adsorption kinetics by enabling the mass transport of the pollutant molecules to diffuse into the adsorption sites at the micropores and mesopores.

The mechanism of interactions between fluid molecules from the gas, vapor, or liquid phases and the carbon adsorbent is fundamentally different (Radovic et al., 2001, Bandosz, 2008). Electrostatic and dispersive (π–π dispersion, hydrogen-bonding, and the electron donor-acceptor complex formation) type relatively stronger molecular interactions occur in the liquid phase. However, the gas adsorption process is primarily based on weak van der Waals–type molecular interactions, mainly depend on the adsorbent's specific surface area, micropore volume, and pore size distribution, where electrostatic interaction would not occur. Nevertheless, in all the fluid phases, the surface chemistry of the carbon and the accessibility of the adsorptive to the inner surface of the adsorbent is crucial. The interactions between fluid molecules and pore walls are the dominant sorption feature in the micropores. The pollutant molecules from the fluid begin to fill the narrowest pores with the highest potential energy and are continued by filling larger pores and cracks afterward. This initial continuous *primary micropore filling* takes place at very low relative pressures ($P/P_o<0.01$, where P_o is the saturation pressure of the adsorptive at a given temperature) in highly energetic pores referred to as *ultramicropores* (pore width less than 0.7 nm for N_2 adsorption, i.e., of width equivalent to no more than two or three molecular diameters). In pores larger than a few molecule sizes, called *supermicropores* (pore width 0.7 to 2 nm), the interaction energy between fluid molecules and the pore walls decreases substantially. Adsorption in these pores may occur at a higher relative pressure ($P/P_o \approx 0.01$–0.2), referred to as *secondary micropore filling,* where the adsorption process is mainly driven by the collective adsorbate-adsorbate interactions (Thommes, 2010). Once these micropores are occupied, less energetic mesopores begin to fill up at relative pressures over 0.2, where the fluid-wall attraction, and attractive interactions between the fluid molecules, contribute to the sorption behavior. Thus, adsorption in mesopores initiates with the surface coverage of pore walls. This induces mesopores to behave as narrower pores of higher potential energy at the center. Consequently, multilayer adsorption takes place following the full surface coverage of walls, until the pores are completely filled. For different sizes of adsorbate molecules, the corresponding relative pressures related to the adsorption process may vary but will follow the above trend. During this pore filling through a single layer followed by multilayer formation, the adsorbate gas molecules tend to condense to a liquid-like state before the saturation pressure of the gas is reached. This phenomenon is called *capillary condensation,* which results in much higher adsorbate deposits on the adsorbent and would not be achieved through the adsorption process only (Thommes, 2010). However, multilayer adsorption from the liquid phase is unusual because of liquids' strong screening interaction forces (Moreno-Castilla, 2004).

Analyzing the shape of the adsorption isotherm can provide information on the extent of adsorption and the nature of the porous structure of the sorbents. According to their internal pore width, IUPAC (Sing et al., 1985) categorized the adsorption isotherms into six classes. The pore size is generally specified as the internal pore width (slit-like pores) or pore radius/diameter (cylindrical and spherical pores). Isotherm Type I is obtained in highly microporous materials through *physisorption* or *chemisorption* following monolayer Langmuir-type adsorption. A Type II isotherm is multilayer sigmoid type adsorption on open surfaces and nonporous or macroporous solids. A Type III isotherm appears in nonporous or highly macroporous materials. In this case, the interactions between adsorbent and adsorbate are generally weak, but strong adsorbate-adsorbate cohesive interactions exist, contributing to the adsorption process with hyperbolic isotherm nature. A Type IV isotherm occurs in mesoporous materials with the occurrence of staged adsorption, i.e., monolayer formation followed by the build-up of additional layers with *capillary condensation.* The condensed liquid follows different kinetics during the desorption process than the gas molecules directly adsorbed on the solid surface. As a result, adsorption and desorption do not follow the same isotherm curve, creating a hysteresis loop. Type V is an elongated S-type isotherm also with a hysteresis loop and observed for mesoporous adsorbents with poorly developed micropores. Type VI is unusual and can be referred to as stepwise multilayer adsorption and appears only when the sample surface contains different types of adsorption sites with energetically different characteristics. The actual pore shape and morphology are complex and diverse types that influence the hysteresis loop curve. IUPAC (Sing et al., 1985) classified the hysteresis loop into four types. In the H1 type with a narrow hysteresis loop, the distribution of pore diameter is relatively uniform, which occurs in the cylindrical or columnar-shaped pores. In the H2 type with a broad hysteresis loop, the pore types are different, and the distribution of pore diameter is relatively large, typically forming when the pore shape is neck-like, wide, or ink bottle–like. In the H3 type with a narrow hysteresis loop, the pore shape is wedged with an opening at both ends or groove pores formed by flaky particles. In the H4 type with a narrow hysteresis loop, the pore shape is slit-pore. The slit-pore structure (Figure 3.1a) is the most widely used representation of a disordered carbon micropore (Bhatia, 2017).

Characterization of functional properties of activated carbon by various analytical techniques following standard test protocols is also an important aspect of tailored product development. These include proximate (moisture, volatile matter, fixed carbon, and ash content) and ultimate (wt% of the major components carbon, hydrogen, and oxygen as well as minor components sulfur and nitrogen) analysis of feedstock, intermediate char, and final activated carbon product. Following additional experiments are also required to perform depending on the product grades and applications for scientific purposes and to meet the precise quality for commercial production and usage targets. Physicochemical tests (product yield, particle size distribution, particle density, apparent or bulk density, pressure drop, dusting attrition, mechanical strength by hardness, abrasion resistance, impact resistance and crushing strength, pH value, water and acid soluble contents, polycyclic aromatic hydrocarbon content, heavy metal content, ignition temperature), textural analysis (N_2 adsorption-desorption isotherm by BET analysis for determining pore volume, specific surface area, pore diameter and pore size distribution and adsorption mechanism), and some specific analysis such as iodine adsorption (determination of micropores), methylene blue adsorption (determination of mesopores), molasses decolorization behavior (determination of macropores and large mesopores), carbon tetrachloride or butane adsorption (determination of the pore volume), adsorption of various solvents for VOC testing (e.g., benzene, toluene, hexane and cyclohexane), adsorption of phenol for water treatment, chlorine half-length value (to determine the amount of activated carbon required to reduce the chlorine concentration in contaminated water by 50%), and phenazone adsorption (pharmaceutical purposes). Different sophisticated characterization techniques are also employed, such as spectroscopy (FTIR, Raman, XPS, EDX, XRD, XRF, ICP-MS, and NMR) for compositional and molecular analysis, microscopy (SEM, TEM, and AFM) for morphological studies and TGA, and DSC for thermal analysis.

Hence, the primary requirements for producing this sorbent material are the formation of well-developed abundant microporous and mesoporous structures and thereby high internal surface area, optimal pore size distribution, and pore volume to maximize *physisorption* process and having favorable chemical functionalities on the carbon surface for enabling interactions between pollutant molecules and the activated carbon and to assist *chemisorption* process while manufacturing at a high product yield. The prerequisites for *physisorption* and *chemisorption* conditions vary depending on the end-use requirements to target the specific pollutant(s) with their molecular size and chemical functionalities. This manifests the need for producing tailored porous carbon products. The main parameters that influence the product yield and adsorption quality are the feedstock type and their chemical composition, selection of the activating agent and their doses, reaction temperature, process heating rate, reaction atmosphere, solid residence time, and use of any catalyst to enhance the reaction kinetics. Other end-use controls such as high purity with almost no ash content, hydrodynamic resistance, abrasion resistance, and recyclability are often required for better performance of the product in specific applications. Careful selection of raw materials and pyrolysis processing method, along with optimization of the process parameters during the initial carbon-upgrading, intermediate char formation, and final activation stages of production and post-processing, is of utmost importance to customizing the product quality and achieving the maximum possible product yield. Different other adsorbents or chemical agents can also be combined with the activated carbon to boost functional use. This generates a matrix of possibilities for variation in developing different grades and formulations of specialized products to achieve the best results for targeted applications.

3.3 MICROWAVE RADIATION-INDUCED PREPARATION OF ACTIVATED CARBON

3.3.1 MICROWAVE RADIATION

Microwaves are non-ionizing electromagnetic waves that travel at the speed of light with frequencies between 300 MHz to 300 GHz, corresponding to wavelengths between 1 m to 1 mm (Huang et al., 2016) and are a part of the radiofrequency spectrum. The radiofrequency spectrum ranges between 30 kHz to 300 GHz and wavelength between 10 km to 1 mm and is divided into two types, radio waves, and microwave. Thus, microwave belongs to the radiofrequency spectrum's higher frequency and lower wavelength region. The microwave lies between infrared (frequencies between 300 GHz to 400 THz; wavelengths between 1 mm to 780 nm) and radio waves in the electromagnetic spectrum. Microwave technology is suitable for wireless transmission of signals and extensively used in radar systems, satellites, and telecommunication sectors to transmit information. The other major use is energy transmission for domestic, industrial, medical, and scientific purposes. Different international and national organizations allocate the frequencies to microwave energy users to avoid interference with communication signals. The microwave frequency band of 2.4–2.5 GHz (central frequency of 2.45 GHz at 0.122 m wavelength) and 0.902–0.928 GHz (central frequency of 0.915 GHz at 0.327 m wavelength) are customary for domestic and industrial applications (Metaxas and Meredith, 1993).

3.3.2 COMPARISON OF CONVENTIONAL HEATING AND MICROWAVE RADIATION TECHNOLOGY

Conventional slow-pyrolysis is an established process for the production of activated carbon. However, there are several limitations, especially the long processing time and relatively low heating efficiency. Microwave processing provides significant benefits over the conventional pyrolysis process in reducing the processing time and improving the heating efficiency. A comparison of these two technologies is provided in Table 3.2.

TABLE 3.2

Comparison of Conventional and Microwave Heating for Activated Carbon Preparation

Critical Parameters	Conventional Heating	Microwave Radiation Technology
Working principle and heating mechanism	The heat transfer mechanism includes convection, conduction, and thermal radiation and relies on the thermal conductivity of materials. Slow heating over a long time is required to attain the desired temperature at the core of the material. The heating process is sluggish, energy-intensive, and has high thermal inertia.	A non-contact heating method, induces fast, volumetric internal heating within the feedstock material through energy conversion. Less energy-intensive. Low thermal inertia allows immediate start-up and shut-down of the process. The heating mechanism relies on the microwave radiation absorbing capability of the material, microwave impedance matching and attenuation. The microwave generated heat is then transferred within the material system mainly by the conduction heating process.
Feedstock and activating agents	Any carbonaceous feedstock is suitable. A wide range of physical and chemical activating agents can be used.	The feedstock materials with microwave absorbing ability favor the process. The activating agents commonly used in the conventional process are also suitable for microwave processing. Some chemical activating agents such as KOH also synergistically assist in microwave absorption.
Feedstock pre-treatment	A long pre-processing time for heating and drying is necessary for the feedstock with low heating value and high moisture content grades.	Moisture and some of the inorganic impurities such as pyrites from coal have good microwave absorbing properties and assist in heating the carbon feedstock. Hence, low-grade coal and biomass with high moisture and ash content can be directly processed relatively quickly.
Reactor types	Both fixed-bed and movable-bed pyrolysis reactors can be used for batch and continuous production. The reactors are relatively less expensive but bigger in size and hence require more operating space.	Magnetron type microwave generators, rectangular waveguides for transmitting the microwave power, and single-mode- or multimode-type resonant cavity applicators are used for high heating efficiency at a lower cost.
Processing time and other influencing operating parameters	Optimization of the following key influencing operational parameters is required: reaction temperature, heating rate, solid residence time in the reactor, the flow rate of inert carrier gas and reactant gas, and the types and doses of activating agents. Slow pyrolysis necessitates a high reaction temperature (400°C–1000°C) and several hours of long processing time.	Key operating parameters and their general trend in influencing the process optimization are similar to the conventional process, as microwave radiation works synergistically with the thermal pyrolysis pathway. In addition, reactor design and types (single-mode or multimode), microwave frequency, output power, mixing intensity (stirring), microwave absorber or catalysts type, and their doses also significantly influence the product yield and quality.

(Continued)

TABLE 3.2 (Continued)

Comparison of Conventional and Microwave Heating for Activated Carbon Preparation

Critical Parameters	Conventional Heating	Microwave Radiation Technology
	Post-processing operations include washing, drying, sizing, granulation, uniform pelletization, chemical impregnation, or soaking in chemical solution as per end-user requirements.	Heating can be performed in less operation time, even in minutes (less than one hour). Capable of producing a high-quality product at a relatively lower reaction temperature (250°C–600°C). There is a possibility to minimize the number of unit operations. Similar post-processing operations are required.
Product quality and yield	It's always challenging to achieve a high yield with high microporosity and adsorption quality.	Superior solid product yield and better product quality with higher specific surface area and pore volume can be achieved from the same carbon feedstock by optimal process control. The by-product quality, such as syngas (hydrogen + CO), can be improved while the CO_2 formation can be reduced.
Product reactivation	A slow and long heating process is needed to desorb and degrade the pollutants while retaining the porous structures in the carbon product is a concern.	It can efficiently degrade and remove the pollutants from the exhausted product without much affecting the porous morphology due to the excellent microwave absorbing property of activated carbon and the volumetric heating process.
Safety and automation	Generally, a safe process and can be integrated with automation.	Microwave is regarded as a safe non-ionizing radiation technology with advanced process automation.
Technology readiness level	Matured commercial process.	Emerging technology. Processing at a large commercial scale to produce activated carbon is not yet established.

3.3.3 FUNDAMENTAL WORKING PRINCIPLES OF MICROWAVE HEATING TECHNOLOGY AND THEIR IMPLICATION IN THE PROCESSING OF CARBON FEEDSTOCKS

Based on the interaction with the electromagnetic fields, materials are classified as *conductors, insulators,* and *absorbers.* All these three types of materials are being utilized in this field of study, either in the form of microwave reactor components or as carbon-based feedstock and reactants. *Conductors* reflect the microwave, and the radiation cannot penetrate through these materials such as metal and alloys. They are used as waveguides and walls in microwave oven cavities. *Insulators* are low-loss materials and essentially transparent to microwave radiation, such as ceramics and Teflon. These materials can partially reflect and transmit most of the incident waves traveling through them and are used as supports and holders in microwave heating ovens. *Absorbers* are high-loss materials that absorb microwave radiation and convert it into energy transfer. This type of material can be heated and processed by this technology.

A schematic presentation of the working principles of microwave heating technology and their implication in the processing of carbon feedstocks are presented in Figure 3.2. Between microwave radiation's electric and magnetic field components, the electric component plays the primary role in processing solid carbonaceous materials to induce wave–material interactions. The magnetic field effect from the pristine coal and biomass impurities is negligible. In some cases, magnetic field interactions occur, for example, in the presence of metals or metal oxides in the form of impurities or when included as catalysts. The electric component of an electromagnetic field causes heating by two main mechanisms: *relaxation or polarization loss* and *ionic conduction* or *conductive loss* (Metaxas and Meredith, 1993).

3.3.3.1 Polarization loss mechanism

There are, in general, four polarization loss mechanisms: *dipolar, electronic, atomic,* and *interfacial polarisation. Electronic* and *atomic polarization* mechanisms have a negligible effect within the microwave frequency ranges.

Dipolar loss is associated with the interaction between the electric field component of microwave radiation and materials with permanent dipoles, such as water. When exposed to microwave frequencies, the dipoles of the material tend to align with the applied electric field and dissipate energy in the form of heat. This phenomenon is called the *dipolar polarization* mechanism. The ability of materials to be heated by microwaves depends on the dipole moment. Thus, at a microwave frequency of 2.45 GHz, dipolar molecules vibrate or rotate at a high-frequency rate of 2450 million times per second. The dipole field attempts to realign itself with the alternating electric field as the field oscillates. Hence, this effect is more significant in liquids with more molecular relaxation than in solid compounds. Carbonaceous feedstocks often contain a large amount of moisture trapped in their interstitial spaces between the particles. For example, as received, low-rank coal may retain 30–60 wt% moisture on a wet basis in the form of free water, bound water, and nonfreezable water (Noringa, 1998). Similarly, lignocellulosic biomass may have 50–60 wt% initial moisture (Priyanto et al., 2018). This moisture component is capable of producing a significant dipolar effect. In this process, energy is lost in the form of heat through molecular friction and collisions. Therefore with *dielectric loss,* electrical energy is converted to heat energy through kinetic energy induced by affected molecules. The amount of heat generated by this process is directly related to the freedom of molecular relaxation, i.e., the ability of the matrix to align itself with the frequency of the applied field, while the heat transfer largely depends on the intermolecular distance or packing density. Heating will not occur if the dipole does not have enough time to realign during high-frequency irradiation or reorients too quickly, especially during low-frequency irradiation with the applied field or their intermolecular distance is too far to heat transfer efficiently. Therefore, dipolar molecules such as water present in the vapor phase produced during the manufacturing process or utilized in the vapor phase reaction environment such as steam, cannot be overheated under microwave irradiation since the distance between the

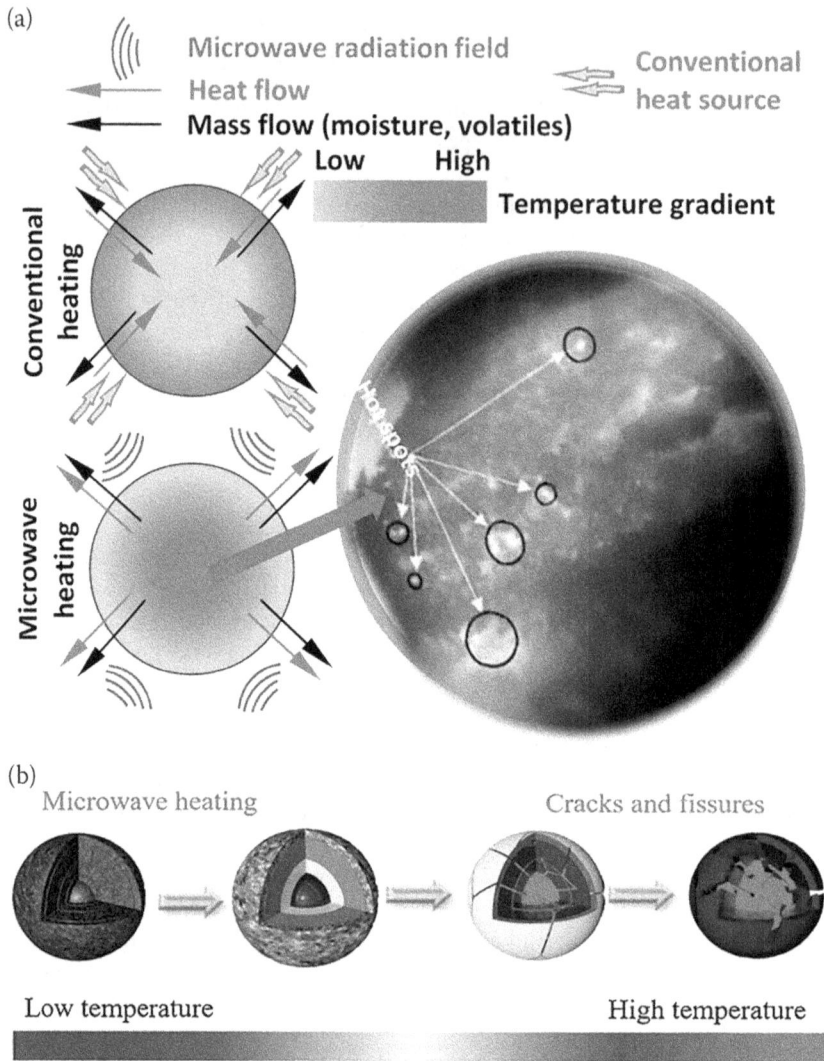

FIGURE 3.2 The working principles of microwave heating technology and their implication in the processing of carbon feedstocks: **(a)** comparison of conventional and microwave heating for activated carbon preparation. The actual hot spot observed in coal char bed during microwave heating is highlighted [Reproduced with permission from reference Liu et al., 2019, *Fuel* 256: 115899] and **(b)** schematic diagram of the evolution of cracks and fissures formation during microwave treatment of heterogeneous material containing microwave absorbing elements [Reproduced with permission from reference Lin et al., 2020, *Journal of Materials Research and Technology* 9(6): 13128–13136].

rotating water molecules would be too far to effective heat transfer. Similarly, if constrained in a solid crystal lattice, dipolar molecules will not be microwave receptive. They cannot move as freely as in the liquid state or when residing within a porous cavity.

Interfacial loss, known as the Maxwell–Wagner–Sillars (MWS) or Maxwell-Wagner polarization effect, is mainly realized in heterogeneous materials with diverse dielectric components. When the electromagnetic field is applied, polarization occurs due to charge build-up at the interfaces confined within a nonconducting medium. Differences in dielectric characteristics and electrical conductivities of different components are responsible for the redistribution of charges and subsequent *interfacial polarization.* The *interfacial polarization* also occurs on porous materials due to the different dielectric properties between materials and air medium. Thus, this mechanism influences the heterogeneous carbonaceous feedstocks, intermediate char, and the final activated carbon product.

3.3.3.2 Conductive loss

Conductive loss is the dominant loss mechanism in semiconducting carbon materials such as graphene, carbon fibers, carbon nanotubes, activated carbon, and char compounds, depending on their conductivity. In the conduction mechanism, mobile charge carriers such as electrons and dissolved ions oscillate back and forth through the material under the influence of microwave radiation's electrical component, thereby creating an electric current (Sun J et al., 2016). The electrical resistance caused by the collisions of charged species with neighboring molecules or atoms generates Joule heat. Generally, two kinds of *conductive loss* models are established. The first one is the electron migrating model, which refers to the free movement of electrons during electromagnetic wave propagation. The other one is the electron hopping model related to electron transfer between components, interfaces, and defects. The π-electrons movement in polyaromatics in carbon compounds such as char and activated carbon under the influence of electric current is a common *conductive loss* mechanism.

3.3.4 SAFETY IN MICROWAVE RADIATION TECHNOLOGY

The energy available for a specific range of electromagnetic spectrum determines the usefulness for its practical applications through wave-matter interaction. Electromagnetic radiation is an electric and magnetic disturbance traveling through space at the speed of light, 2.998×10^8 m/s (Percuoco, 2014). It contains neither mass nor charge but travels as a waveform at a constant speed in packets of radiant energy called photons, or quanta. The wave characteristics of electromagnetic radiation are expressed by Equation 3.1:

$$c = \lambda \nu \tag{3.1}$$

where c = velocity, λ = wavelength (the straight line distance of a single cycle), and ν = frequency (cycles per second, or hertz, Hz). Since the velocity is constant, any change in frequency results inversely in wavelength. The particle-like nature of electromagnetic radiation influence the interaction of photons with matter. The amount of energy (E) found in a photon is directly proportional to photon frequency (ν) and is expressed by Equation 3.2:

$$E = \nu h \tag{3.2}$$

where h = Planck's constant

The electromagnetic radiation energy is measured by an electron volt (eV), where 1 eV implies the energy gained by an electron when it is accelerated through a potential difference of 1 volt. Electromagnetic radiation accumulates energy in two forms, *excitation* and *ionization* while passing through organic and inorganic materials. *Excitation* defines the deposition of sufficient energy to raise an electron to a higher electron shell without ejection of the electron. In contrast, *ionizing radiation* such as gamma rays or X-rays can provide enough energy to eject one or more electrons from the atom (Elliott et al., 2018).

However, compared to gamma rays or X-rays in the electromagnetic spectrum, microwaves have less ability to provide higher frequencies to generate sufficient energy (>10 keV, Percuoco, 2014) to ionize an electron within an atom. Therefore microwave is generally regarded as a safe non-ionizing radiation technology for heating the suitable substances (Web Reference, 2022d). However, an excessive hot spot formation may occur during production, especially on a large scale. This would trigger a high rate of heat generation, leading to thermal runaway and subsequently causing voltage breakdown. Hence, precise temperature control during microwave processing is of utmost importance to prevent any undesired safety concerns. Microwave radiation may also produce static electricity buildup and sparks, which can cause damage to microwave hardware and reactor walls. Therefore, appropriate reactor design and implementation of microwave filters are necessary to avoid the dielectric breakdown induced by an excess of a charge buildup and to eliminate microwave radiation leakages (Beneroso et al., 2017).

3.3.5 Molecular mechanism of microwave-induced processing of coal and biomass

The mechanism underlying microwave heating of heterogeneous materials like coal and biomass depends upon the properties of the constituent energy-absorbing molecules present in them. All the common mechanisms of *dipolar polarization, interfacial polarization,* and *conduction losses* are involved in processing these carbonaceous materials and their conversion to activated carbon induced by microwave irradiation, where microwave absorbing molecules play a significant role. These absorbers are typically dielectric materials either from the pristine feedstock or develop during the heating process. They are characterized either by exhibiting a dipole movement (*dipolar polarization* in water and polar functional groups in carbon molecules) and/or possessing free charge carriers (*interfacial polarization* in heterogeneous carbonaceous feedstocks, char, and activated carbon) and/or having free electrons capable of movement under electric field propagation or able to electron transfer between components, interfaces, and defects by the electron hopping model (*conduction losses* in char and activated carbon by π-electrons). The effectivity of microwave absorption is mainly dependent on impedance matching and attenuation of microwave inside the material system. A superior impedance matching reduces direct microwave reflections. It enhances the scope for electromagnetic waves to travel inside the material system, which could be attenuated, in the form of heat by an individual dominant effect or the cumulative effect of the above mechanisms. Thus, microwave attenuation, dielectric properties, as well as the thermal conductivity of the constituent materials collectively govern the heat transfer mechanism under the influence of microwave radiation.

3.3.5.1 Dielectric properties of the feedstock and reactants

The dielectric properties of the carbonaceous materials are the most important parameter for their heating characteristic under microwave irradiation, and their heating ability is expressed by Equations 3.3 and 3.4 (Lidström et al., 2001).

$$\varepsilon = \varepsilon' - j\varepsilon'' \tag{3.3}$$

$$\tan \delta = \frac{\varepsilon''}{\varepsilon'} \tag{3.4}$$

where ε is the complex permittivity, ε' is the dielectric constant or real permittivity that represents the amount of reflected energy versus the amount of energy absorbed, ε'' is the dielectric loss factor or imaginary permittivity that measures the dissipation of electric energy into heat within the material, tan δ is the dielectric loss tangent, and $j = \sqrt{-1}$.

For a better conversion of microwave energy into heat, a moderate value of ε' combined with high values of ε'' is ideal, and consequently, it results in a high value of tan δ. In general, materials are classified as high (tan δ >0.5), medium (tan δ 0.1–0.5), and low microwave-absorbing (tan δ < 0.1) compounds (Kappe, 2004). In the case of materials transparent to microwaves, they do not have a sufficiently high loss factor in dissipating the dielectric heating. At 2.45 GHz, the tan δ value of coal was found to be 0.01 to 0.3 depending on rank and origin, which increases with increasing rank. However, for the dried and de-mineralized coals, the value diminishes to 0.008–0.044, suggesting that the organic component of the coals would behave almost like a microwave transparent material

(Binner et al., 2014). Similarly, biomass with tan δ >0.1 is low microwave-absorbing material (Li et al., 2016). Nevertheless, other material constituents such as water (tan δ 0.13; Sturm et al., 2012) and some inorganic oxides (e.g., pyrites in coal tan δ 0.225; Genn, 2013) have a significant microwave absorbing capacity. Once these coal and biomass convert to activated carbon, it becomes a high absorbing material (tan δ 0.57–0.80 or even higher; Li et al., 2016).

The interaction of microwaves with the dielectric materials generates an internal electric field (E) that induces polarization and movement of charges and is mathematically expressed by Maxwell's equation. The resistance to the induced motions due to internal, elastic, and frictional forces attenuates the electric field. These losses result in volumetric heating. The resulting thermal energy generated (Rosa et al., 2013), i.e., electromagnetic power loss per unit volume P (W/m³) by the material, is given by:

$$P = \omega \varepsilon_o \varepsilon_r'' \, |E_{rms}|^2 \tag{3.5}$$

$$\omega = 2\pi\nu \tag{3.6}$$

where ω is the angular frequency of the exciting radiation, ν is the frequency of the incident microwaves (Hz), ε_o is absolute permittivity in free space (8.85×10^{-12} C²/Nm²), ε_r'' is the imaginary part of the complex relative permittivity of the absorbing material, and E_{rms} is the root mean square of electric field strength (V/m).

3.3.5.2   Heat transfer in microwaves

Heat transfer during microwave irradiation can be modeled using Fourier's law of heat conduction (Dibben, 2001):

$$\rho C_p \frac{\partial T}{\partial t} = div\,(k\nabla T) + P \tag{3.7}$$

where ρ is the material's density (kg/m³), Cp is the specific heat capacity of the material at constant pressure (J/kg °C), $\frac{\partial T}{\partial t}$ is the change in temperature over time, k is the thermal conductivity (W/m°C), and ∇T is the temperature gradient (°C/m). Once the electric field strength is known, the thermal energy P can be calculated from Equations 3.5 and 3.6, which can be used to determine the heat transfer from Equation 3.7. Numerical methods such as the finite differential time-domain method, the finite element method, and the transmission line matrix method can be used for solving the partial differential equation.

3.3.5.3   Penetration depth of microwave field

The penetration depth, d, is one of the critical parameters to quantify the heating efficiency and uniformity of a sample by microwaves. It is defined as the distance from the surface to the place at which the magnitude of the field strength drops to e^{-1} (=0.368) of its value at the surface. Another relevant parameter, power penetration depth, Dp, is defined as the distance at which the traveling wave power density reduces to e^{-1} of its value at the surface and is half the value of the field penetration depth, d. Assuming no magnetic effects for these carbonaceous materials, the penetration depth (Peng, 2012) can be expressed as in Equation 3.8:

$$d = \frac{\sqrt{2}c}{\omega \left[\varepsilon'\left(\sqrt{1 + (tan\delta)^2} - 1\right)^{1/2}\right]} \tag{3.8}$$

where c is the velocity of light. The penetration depth is inversely related to the frequency and dielectric loss factor. Thus, penetration depth decreases with a higher frequency and high-loss materials and vice versa. The optimal sample dimension should be considered equivalent to the penetration depth to ensure an efficient and uniform microwave heating process.

In essence, when microwave field energy is transferred to the carbonaceous matrix such as coal and biomass, it primarily converts the electrical energy into kinetic or thermal energy and ultimately into heat. *Dipolar loss* acts as the dominant loss mechanism during this initial processing phase. Under the application of microwave radiation, electromagnetic waves penetrate and propagate inside these materials, and dielectric materials such as the water molecules absorb the electromagnetic radiation and convert microwave energy into thermal energy and get heated up. Subsequently, they transfer microwave-generated thermal energy to the neighboring carbon molecules mainly by the conduction heating process to initiate the carbonization process. The dipolar effect from functional groups such as C–O, C–OH, and C=O belonging to carbon molecules also contribute to the *dielectric loss*. However, the precursor carbon structure can not provide sufficient free-electron movement, and hence *conduction losses* are merely absent. Accordingly, these raw carbon precursors as heterogeneous compounds behave like weak dielectrics with a low loss material. The presence of moisture and microwave absorbing inorganic constituents is profoundly beneficial for the initial processing stage of carbonization and char formation. The actual heating effect depends on the moisture content and the amount of microwave absorbing impurities present in the feedstock. That's why microwave radiation technology provides ample scope to directly process the low-value carbon feedstocks into high-value products.

On continuing the microwave radiation, moisture and volatile compounds tend to release while the carbon feedstocks are heated up, and eventually, the primary pyrolysis process commences. The feedstock materials gradually transform to char following the physicochemical changes of the original carbon compounds. With the elimination of moisture, the *dipolar loss* effect is diminished while the carbon precursor molecules are at their initial phase of transformation and cannot significantly contribute to *interfacial polarization* and *conduction losses*. As a result, the dielectric properties of the substrate are reduced during this intermediate phase.

Nevertheless, the physicochemical changes of the carbon molecules progressively improve the dielectric properties of the carbon compounds during their transition to char and activated carbon formation (Figure 3.3) with the generation of more condensed polyaromatic compounds through the primary and secondary pyrolysis processes. Microwave irradiation then induces a flow of electrons on the surface by delocalizing π-electrons from the sp^2-hybridized carbon network, typically from these polyaromatic compounds. Accordingly, *interfacial polarization* and *conduction losses* become prominent. This flow of electrons heats up the material following Joule heating mechanisms, also known as resistive heating or ohmic heating. This *conductive loss* mechanism can produce superior heat-generating capacity in solids than the *dipolar loss* mechanism. Thus, the effect of microwave radiation is realized better in the intermediate char and activated carbon products with enhanced dielectric properties than in their carbon precursors and inevitably at the later stage of the processing of carbonaceous feedstocks when the *conductive loss* becomes the dominant loss mechanism. Incorporating a microwave absorber with coal and biomass feedstock offers a reasonable option to improve the microwave heating efficiency at the initial stage. However, the microwave penetration depth decreases with improving the dielectric properties, requiring attention and necessary adjustment to ensure uniform heating.

Polyaromatic molecules that evolved during this process consist of delocalized π-electrons, and in some instances, they acquire the freedom to move in relatively broad regions. The microwave radiation can provide sufficient kinetic energy to these free electrons and expedite them to jump out of the material, resulting in the ionization in the surrounding atmosphere, which causes plasmas and hot spot formation (Figure 3.2a). Since this plasma effect is confined to a tiny region of the space at a sub-millimeter scale and lasts for a fraction of a second, this phenomenon is called *microplasma* (Lin and Wang, 2015). At a macroscopic level, this is called *sparks* or *electric arc* formation. These

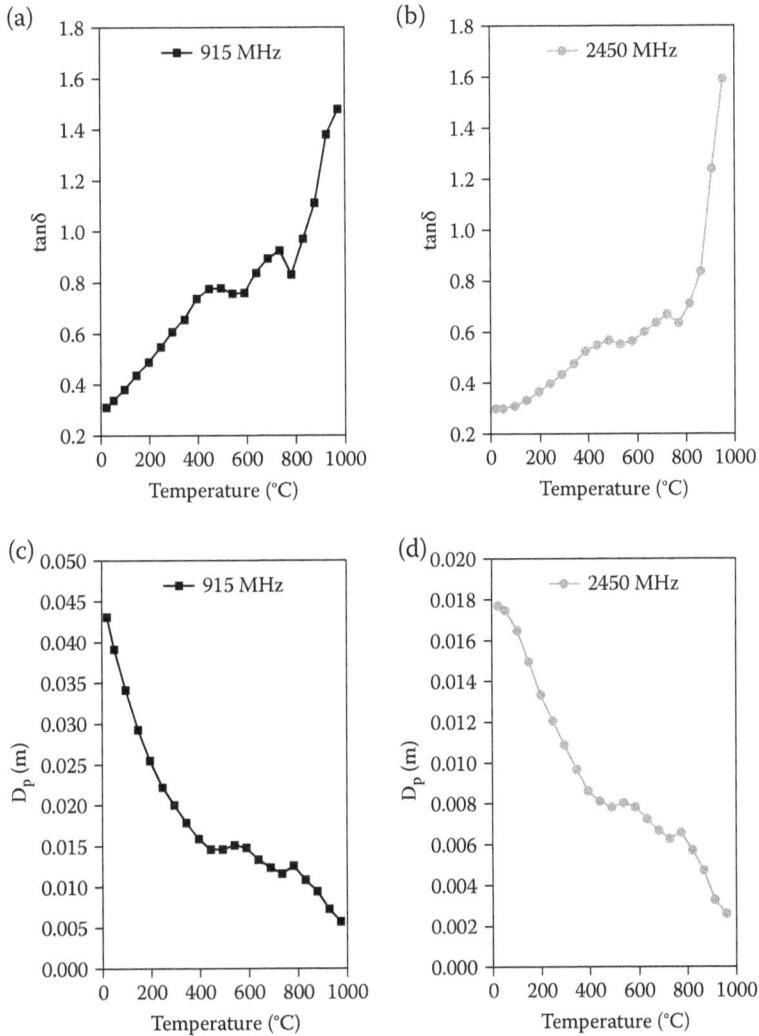

FIGURE 3.3 Influence of microwave radiation on the dielectric properties of anthracite coal. Variation of dielectric loss tangent with the temperature at (a) 915 MHz and (b) 2450 MHz. Variation of microwave penetration depth with the temperature at (c) 915 MHz and (d) 2450 MHz [Reproduced with permission from reference Peng et al., 2016, *Fuel Processing Technology* 150: 58–63].

microplasmas act as a chemical-free catalyst to stimulate the heating process but need appropriate process control to avoid any adverse effects. Therefore, the conversion of coal and biomass to activated carbon by microwave radiation treatment is a feasible process even though it follows a more complex mechanism than the processing of typical dielectric materials in a liquid phase. The molecular transformation of the carbon compounds and the associated physical and chemical changes during the microwave-induced pyrolysis process is further explained in the following relevant sections.

3.3.6 INFLUENCING FACTORS GOVERNING THE MICROWAVE PROCESSING

Activated carbons can be prepared from a wide range of carbon types of different structures. It is the porosity within activated carbons that determines their adsorption characteristics. However, the porosity develops through carbon burn-off, diminishing product yield. Hence, there is a trade-off between the porosity formation to maximize the specific surface area to ensure product quality without substantially reducing the yield to capture the product cost benefits. Judicious selection of materials based on their physicochemical properties and optimization of process parameters are essential to compromise this practical challenge of maximizing the adsorption capacity at an optimal yield. The most important variables in the microwave-assisted conversion of coal and biomass to activated carbon are the type and size of carbon feedstocks; fixed-carbon, moisture, volatile, and ash content in the feedstocks, type and doses of activating agent, operating temperature; solid residence time; microwave reactor design and type (batch or continuous, single-mode or multimode, reactor capacity); microwave frequency and output power; mixing intensity, type and flow rate of carrier gas, microwave receptor and/or catalyst type, size, and concentration.

Only a few published reports are available where the influence of different parameters under microwave processing was attempted to identify the critical issues required for process optimization. Such processes include conducting simultaneous carbonization and activation in a single-step method or carbonization and activation separately in a two-step procedure. The focal point of these studies was to explore whether microwave processing could enable achieving shorter process time, improve energy efficiency and provide better product quality and yield compared to the conventional heating process. A list of lab-scale batch trials on microwave processing of carbonaceous precursors using different activating agents is provided in Table 3.3. However, the textural properties and product yield data found in different studies provide a modest comparison, as the experiments were carried out under different conditions. Hence, selected case studies with commercially relevant feedstocks are discussed in detail. The effects of influencing parameters on the adsorption behavior, product yield, and potential applications are emphasized.

3.3.7 MICROWAVE-INDUCED PROCESSING OF COAL FOR THE PREPARATION OF ACTIVATED CARBON

Coal is a black or dark-brown colored combustible sedimentary rock derived from plant debris and comprises a complex mixture of organic substances, including volatile compounds, minerals, and moisture. The primary elemental composition of coal is carbon with variable amounts of other elements, namely hydrogen, sulfur, oxygen, and nitrogen. The organic substances present in coal are called macerals (Stach, 1968). The gradual progress of coal formation from floral or plant origin is called coalification. Based on the actual floral origin, nature, and extent of the coalification process, coal's chemical composition and textural properties vary quite substantially. Also, these variations in the chemical composition and textural properties of coal significantly influence the development of porosity and the chemical nature of activated carbon. Accordingly, it also necessitates adjusting the production methodologies to attain the desired final properties of the activated carbon. Hence, a detailed understanding of the coalification process and its influence on coal's chemical composition and physicochemical properties are the prerequisites for its effective utilization and product development.

Coal is known from most geologic periods, but 90% of all coal beds were deposited in the Carboniferous and Permian periods, i.e., around 350 to 250 million years ago, due to violent tectonic disturbances followed by earthquakes and volcanic eruptions (McGhee, 2018). Under these conditions, a massive amount of plant origin was buried in the ground or submerged in salt or freshwater. Thus, the plant-derived organic matter was preserved for millions of years, which is the primary feature of coal and the base material for activated carbon production. The following three factors determine the properties of coal:

  i. Type: variation in the original plant material and its subsequent alteration.
 ii. Rank: a difference in the degree of burial and subsequent coalification.
iii. Grade: mineral matter variation mainly originates from mud in the peat swamp. This is found in ashes after burning off the coal. High-grade coal contains less mineral matter and less moisture than its lower-grade counterparts.

TABLE 3.3
The Effect of Activating Agents on the Textural Quality and Product Yield of Activated Carbon Prepared from Coal and Biomass Feedstocks Using Microwave Technology

Microwave-assisted chemical activation

Precursor	Time (min)	MP[1] (W)	AG[2]	IR[4] (wt/wt)	Specific surface area[3] (m²/g)	Product yield (wt%)	Reference
Palm shell biochar[5]	10	1,000	Steam	5 g/min	570.8	85.3	Yek et al., 2019
Palm shell	10	1,000	Steam	5 g/min	539.8	45	Lam et al., 2020
Orange peel	15	1,000	Steam	5	305.1	31	Yek et al. 2020
Coal char	15	700	CO_2	0.755	400	34	Norman and Cha, 1995
Coconut shell	6	502	CO_2	150	625.6	37.1	Aziz et al., 2021
Orange peel	15	1,000	CO_2	5	158.5	44	Yek et al., 2020

Microwave-assisted chemical activation

Precursor	Time (min)	MP[1] (W)	AG[2]	IR[4] (wt/wt)	Specific surface area[3] (m²/g)	Product yield (wt%)	Reference
Anthracite coal	12	700	KOH	3	1770.5	–	Xiao et al., 2015
Anthracite coal[6]	10	700	KOH	1	672	–	Ge et al., 2015
Low-rank Malaysian coal	20	600	KOH	2	1100.2	63.6	Jawad et al. 2018
Sub-bituminous coal char[7]	8	726	KOH	0.5	1405	–	Uche and Okwudili (2019)
Coconut husk	6	600	KOH	1.25	1356.3	80.8	Foo and Hameed, 2012a
Bamboo biochar[8]	15	700	KOH	4	3441	39.8	Wu et al., 2015
Coconut shell	90	700	NaOH	3	2825	18.8	Cazetta et al., 2011
Jackfruit peel	7	600	NaOH	1.5	1286.7	80.8	Foo and Hameed, 2012b
Anthracite coal	12	700	H_3PO_4	3	843.6	–	Xiao et al., 2015

(Continued)

TABLE 3.3 (Continued)
The Effect of Activating Agents on the Textural Quality and Product Yield of Activated Carbon Prepared from Coal and Biomass Feedstocks Using Microwave Technology

			Microwave-assisted chemical activation				
Precursor	Time (min)	MP[1] (W)	AG[2]	IR[4] (wt/wt)	Specific surface area[3] (m²/g)	Product yield (wt%)	Reference
Bamboo	20	350	H_3PO_4	1	1432	48	Liu et al., 2010
Wood sawdust	6	600	K_2CO_3	1.25	1496	80.8	Foo and Hameed, 2012c
Orange peel	6	600	K_2CO_3	1.25	1104	81	Foo and Hameed, 2012d
Anthracite coal	12	700	$ZnCl_2$	3	934.5	–	Xiao et al., 2015
Cotton stalk	9	560	$ZnCl_2$	1.6	794.8	37.5	Deng et al., 2009

Notes

[1] MP = Microwave power
[2] AG = Activating agent
[3] Specific surface area measured by BET method
[4] Impregnation ratio (wt/wt) = Activating agent: Carbon precursor
[5] Biochar was initially produced from palm shell via microwave pyrolysis performed under vacuum at 700 W of microwave power for 25 min residence time. BET specific surface area of the biochar was 110.8 m²/g, and the yield was 40 wt%.
[6] BET specific surface area of this activated carbon was further improved from 672 to 1062 m²/g by microwave-induced post-treatment at 500 W of microwave power for 8 min residence time.
[7] Before the microwave processing, sub-bituminous coal was initially carbonized into an electric furnace and heated at the rate of 10°C/min from room temperature to 800°C for 2 h.
[8] Before the microwave processing, bamboo was initially carbonized into a muffle furnace and heated at the rate of 10°C/min from room temperature to 600°C for 2 h in an inert atmosphere.

During the geological processes of coalification under suitable temperature and pressure conditions, the buried plant materials gradually transform through a series of chemical reactions. The evolution of metamorphic grade or rank of coal during this coalification process depends on the temperature and pressure attained. Temperature is a much more critical factor involved in the catalysis of chemical changes during coalification than either pressure or geological time for the formation (Francis, 1961). The rank forms progressively according to rising temperature and pressure over time from lignite or brown coal to sub-bituminous and then bituminous, and finally anthracite, or hard or black coal. For example, bituminous coal can form at a lower temperature of 100°C–170°C, but anthracite needs a higher temperature of 170°C–275°C (Taylor et al., 1998). However, recent studies revealed that coalification could occur under mild thermal conditions of ca 113°C (Straka and Sy´korova, 2018). The tectonic pressures applied during the evolution of coal over geological time cause horizontal layering or bands, which imparts an anisotropic property to the coal compared with vertical sections. This anisotropy is more visible in the higher rank coals. Bituminous forms a dense sedimentary rock with a black or dark brown color and often appears with bands of bright and dull material. However, anthracite is harder and glossy black coal with a prominent anisotropic property.

Initially, the plant material decomposes and forms humic materials composed of polycondensed aromatic compounds with aliphatic side chains containing methoxyl, carboxyl, and carbonyl groups. At the same time, the moisture content is substantially reduced, which leads to lignite coal formation. By eliminating aliphatic side-chains containing oxygen, lignite transforms into bituminous coal. On gradual progression, with the removal of methyl and methylene groups, high volatile bituminous becomes medium volatile bituminous coal while the aromatic clusters continue to grow larger. Ring condensation and further growth of large aromatic clusters develop graphite-like structures transforming the bituminous coal into high-ranking anthracite coal (Francis, 1961). Modern peat is primarily lignin from plant tissues and also contains cellulose and hemicellulose ranging from 5% to 40%, and some minor components such as waxes and nitrogen- and sulfur-containing compounds (Andriesse, 1988). Cellulose contains about 44% carbon, 6% hydrogen, and 49% oxygen by weight, while lignin has about 54% carbon, 6% hydrogen, and 30% oxygen by weight. In contrast, the weight composition of bituminous coal is about 84% carbon, 5% hydrogen, 7% oxygen, 2% nitrogen, and 2% sulfur (William, 1973). Therefore, most of the oxygen and some hydrogen were removed during coalification while the carbon content was elevated. This is essentially a carbonization process. As carbonization in coal compounds proceeds in nature, aliphatic compounds are replaced by aromatic compounds. Subsequently, aromatic rings begin to fuse into polyaromatic compounds. These chemical changes are also accompanied by physical changes, such as decreasing average pore size. Nevertheless, coalification resembles the early phase of the carbonization of biomass in an activated carbon manufacturing process.

Coal pyrolysis can be described in terms of the stages shown in Figure 3.4 (Serio et al., 1987). During stage I, the coal may undergo some bond-breaking reactions and reduction of hydrogen bonding which may cause melting. Some light species that exist as guest molecules or are formed by breaking weak bonds are released. During stage II, further bond breaking occurs, leading to the tar and gases evolution and char formation. During stage III, the products continue to react. The char evolves secondary gases, mainly CO and H_2, while undergoing ring condensation. Representative case studies on microwave-induced coal processing for the preparation of activated carbon following physical and chemical activation are discussed below.

3.3.7.1   Microwave-induced physical activation of coal

Steam is commonly used as the physical activating agent in commercial activated carbon production by the conventional pyrolysis process. Steam gasification of coal char under microwave processing has been reported, but the focus was on carbon conversion to produce gaseous products (Liu et al., 2019). CO_2 is another frequently used physical activating agent. Norman and Cha (1995) investigated the possibility of producing activated carbon from coal char, using CO_2 as the

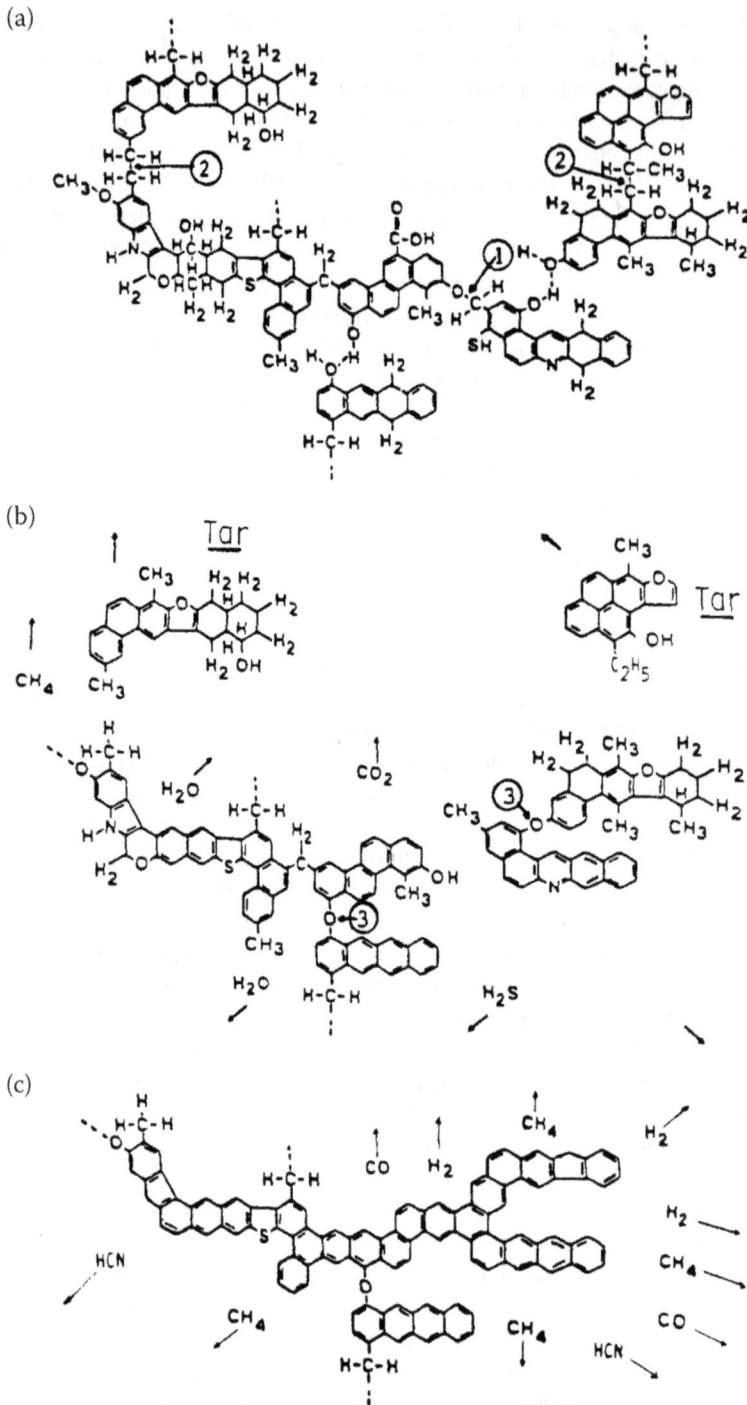

FIGURE 3.4 Hypothetical picture of coal's organic structure at successive stages of pyrolysis. The figure represents **(a)** stage I: chemical structure of raw bituminous coal, **(b)** stage II: the formation of tar and light hydrocarbons during primary pyrolysis, and **(c)** stage III: char condensation during secondary pyrolysis [Reproduced with permission from reference Serio et al., 1987, *Energy & Fuels* 1(2): 138–152].

reactant gas. To obtain the highest specific surface area of the product, a set of process parameters, namely, CO_2 inlet flow rate, input microwave energy, and char residence time, were varied. The results were correlated to the rate of carbon removal from the solid char by reacting with CO_2 under the microwave treatment. They found that the reaction rate of the coal char carbon and CO_2 reaction was mainly dependent on the microwave power input, i.e., electric field strength, and was not significantly influenced by the bed temperature. A longer solids residence time for the carbon–CO_2 reaction increased the char bed temperature, but the CO_2 conversion and the product gas flow rate remained constant. Under an optimized process condition of microwave input power of 700 W and a CO_2 inlet flow rate of 0.755 L/min, the maximum surface area of the activated carbon obtained was 400 m^2/g with the removal of 66% by weight from the original coal char feedstock. As a result of the carbon removal, the formation of a high degree of micropores in the product was also evident. Compared to the conventional process, microwave energy significantly enhanced the devolatilization of the fresh char. It reduced the solid residence time while operating at a lower temperature, thereby contributing to energy and operating cost savings. The residence time for the coal char under the direct microwave contact was 15 minutes, and the operating temperature was at about 316°C. Obtaining a similar level of specific surface area under the conventional heating process and using CO_2 as the activating agent required 3 h processing time.

3.3.7.2 Microwave-induced chemical activation of coal

KOH is the most commonly used chemical activating agent. Xiao et al. (2015) compared the microwave-assisted chemical activation of anthracite coal using KOH, H_3PO_4, and $ZnCl_2$ as the activating agent. Before the microwave processing, coal was crushed and sieved to 100 meshes, washed with distilled water, and then dried in an air oven. Initially, these three activating agents were separately impregnated with coal samples. The optimized condition for microwave processing at a frequency of 2.45 GHz under a vacuum atmosphere was identified as input microwave power of 700 W, a residence time of 12 min, and the activating agent to carbon precursor impregnation ratio (w/w) at 3:1. The activated product was washed with 10% HCl and then washed with distilled water until the filtrate became neutral, and then the product was dried and stored. Adsorption capability of KOH-treated activated product for methylene blue, and iodine was 312 and 1048.19 mg/g, respectively, which was better than the other two products activated by H_3PO_4 and $ZnCl_2$. The BET specific surface area and pore volume of the KOH-treated activated product were 1770.5 m^2/g and 0.99 cm^3/g, respectively, which was primarily microporous. Also, the activated carbon showed good adsorptive affinity towards naphthalene, phenanthrene, and pyrene and hence would be suitable for removing toxic polycyclic aromatic hydrocarbons.

In a similar study, Ge et al. (2015) converted anthracite coal to activated carbon by KOH activation in a microwave oven under an N_2 atmosphere but at a lower dose of chemical activating agent and at a less residence time. Before the microwave processing, coal was crushed and sieved to 100 meshes, washed with distilled water, and then dried in an air oven. The activating agent to carbon precursor impregnation ratio (w/w) was kept at 1:1. The activation reactions were performed at an input microwave power of 700 W and the residence time of 10 min. The activated product was then cooled to room temperature, washed with 10% HCl, and then washed with distilled water until pH 7 was reached in the residual liquid and then dried and stored. The BET specific surface area of the product was 672 m^2/g which is lower than the previous study (Xiao et al., 2015). Hence, to improve the sorption capacity and high selectivity for specific organic compound naphthalene, the above product was further modified by microwave-induced post-treatment. The input microwave power was varied at 300, 500, and 700 W settings with a frequency of 2.45 GHz and at a radiation time of 8 min. With this additional microwave treatment, the specific surface area was substantially increased, and the maximum value was obtained as 1,062 m^2/g at 500 W. This has also modified the surface chemistry of the resultant carbon material by effectively removing oxygen functionalities from the acidic groups and enhancing the surface basicity of the product within a short time. The modified product showed good potential to remove naphthalene from aqueous solutions.

3.3.8 Microwave-induced processing of biomass for the preparation of activated carbon

Recently pursuing sustainable production of activated carbon has given much attention to exploring biomass as a renewable feedstock. Biomass could be obtained from a wide range of living organisms and also from various waste streams such as municipal solid waste (e.g., kitchen waste, paper, and cardboard), agricultural and industrial waste (e.g., wood, coconut shell, sawdust, rice straw) and the effluents of wastewater treatment plants, e.g., sludge (Ethaib S, 2020). According to the United Nations Framework Convention on Climate Change (UNFCCC, 2005), biomass is defined as "*a non-fossilized and biodegradable organic material originating from plants, animals, and microorganisms. This shall also include products, by-products, residues and waste from agriculture, forestry and related industries as well as the non-fossilized and biodegradable organic fractions of industrial and municipal wastes.*" From a biochemical perspective, biomass refers to lignocellulose, starch, sugars, fats, and proteins arising from various molecules and macromolecules, including carbon compounds. Thus, biomass provides an ample scope to utilize renewable carbon sources to create high-value carbon products and support sustainability initiatives. On the contrary, the accumulation of the waste biomass often poses safety hazards and health issues. Hence, biomass valorization is no longer an alternative approach but rather an essential consideration for capturing societal benefit, developing sustainable technology, reducing waste, generating clean energy and valuable products, and realizing new horizons for a circular bio-based economy.

Activated carbon can be produced from high carbon-containing biomass, especially lignocellulosic materials from agricultural and municipal waste, as evident from a large number of research publications over the past decades. Lignocellulosic biomass mainly refers to terrestrial plants and is classified into three categories based on the chemical composition, namely hardwood, softwood, and herbaceous. Coconut shells and husks and hardwoods (e.g., *Betula spp., Carpinus spp.*) are primarily used as precursors for commercial production of regular and premium grade activated carbon. Other biomass wastes such as walnut shells, olive stones, and palm kernels have the potential for producing high-quality activated carbon, but their inconsistency in raw material supply limits the commercial potential. Activated carbon can also be produced from softwood (e.g., *Pinus spp.*) and herbaceous biomass, but they also have limited commercial uses due to their inferior adsorption properties, low material hardness, friability, and less regenerative capacity compared to coconut shell and hardwood-based products (Anderson et al., 2021). Lignocellulosic biomasses are composed of three main organic components: cellulose, hemicellulose and lignin, and some extractives and inorganic salts such as chloride, sulfate, and carbonate of Na, K, Ca, Mg, and Si. On average, the quantitative proportion of hemicellulose, cellulose, and lignin in wood biomass is observed in the range of 20%–35%, 40%–50%, and 15%–35%, respectively (Danish and Ahmad, 2018).

Pyrolysis of lignocellulosic biomass and the gradual formation of char and other products are depicted in Figure 3.5. Each constituent of lignocellulosic biomass undergoes chemical decomposition over different temperatures. Hemicelluloses generally degrade between 200°C–260°C, followed by the decomposition of cellulose in the temperature range of 240°C–350°C, while lignin degrades in the temperature range of 280°C–500°C (Seky 2013, Danish and Ahmad, 2018). Lignin is a cross-linked organic polymer that contains aromatic rings and is primarily responsible for char formation. During this pyrolytic temperature range, chain scission and mineralization of lignin and cellulosic units broke C-C and C-O bonds within the glucopyranose ring, subsequently evaporating water, CO, and CO_2 molecules, leading to the solid mass loss (Macedo, 2008). Selected case studies on microwave-induced processing of coconut shells and husks as representative biomass to prepare activated carbon following physical and chemical activation are discussed below.

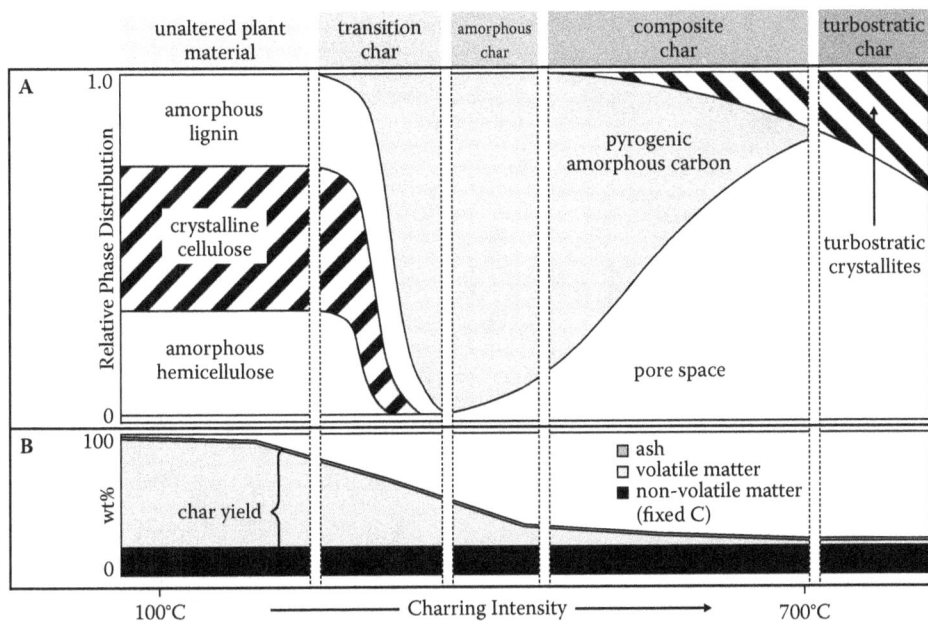

FIGURE 3.5 Dynamic molecular structure of plant biomass-derived char product across a charring gradient: (**A**) schematic representation of the four proposed char categories and (**B**) char composition [Reproduced with permission from reference Keiluweit et al., 2010, *Environmental Science & Technology* 44(4): 1247–1253].

3.3.8.1 Microwave-induced physical activation of coconut shells

In a recent study, Aziz et al. (2021) prepared activated carbon by activating the coconut shell via single-stage microwave heating under CO_2 gas flow. The adsorption performance of the product was estimated by its ability to remove toxic persistent organic pollutant dichlorodiphenyltrichloroethane (DDT). The complex connection between process variables and their responses usually follows a nonlinear relationship. They used a theoretical model based on response surface methodology to evaluate multiple variables simultaneously at a minimal number of experiments. The optimum processing conditions for preparing the activated carbon were 502 W and 6 min for radiation power and radiation time, respectively, at a CO_2 flow rate of 150 cm^3/min. This resulted in a total pore volume of 0.420 cm^3/g, BET specific surface area of 625.6 m^2/g, and an average pore diameter of 4.55 nm, corresponding to mesoporous structure and 37.1% of product yield with an ability to 84.8% of DDT removal, respectively. The 3D response surface plots for DDT removal and the product yield are presented in Figures 3.6a and 3.6b, respectively.

Under a given CO_2 flow rate, it required a specific radiation power to facilitate and augment C-CO_2, solid carbon, and reactant gas reaction to create enough functional pores to enhance the adsorption quality of the product. Thus below 490 W, even with increasing the radiation time, had caused no significant change in DDT removal. However, at radiation power above 490 W, an increase in radiation time had caused a substantial increase in the DDT removal ability of the product. On the contrary, increasing the radiation power and residence time enhanced the carbon removal process and thereby reduced the product yield. The FTIR study identified the existence of different polar groups on the activated carbon surface, while the kinetic experiments detected the role of *chemisorption* in the adsorption process. The activated carbon also showed promising outcomes with the ability to regenerate three times. Afterward, both product yield and the adsorption efficiency dropped significantly.

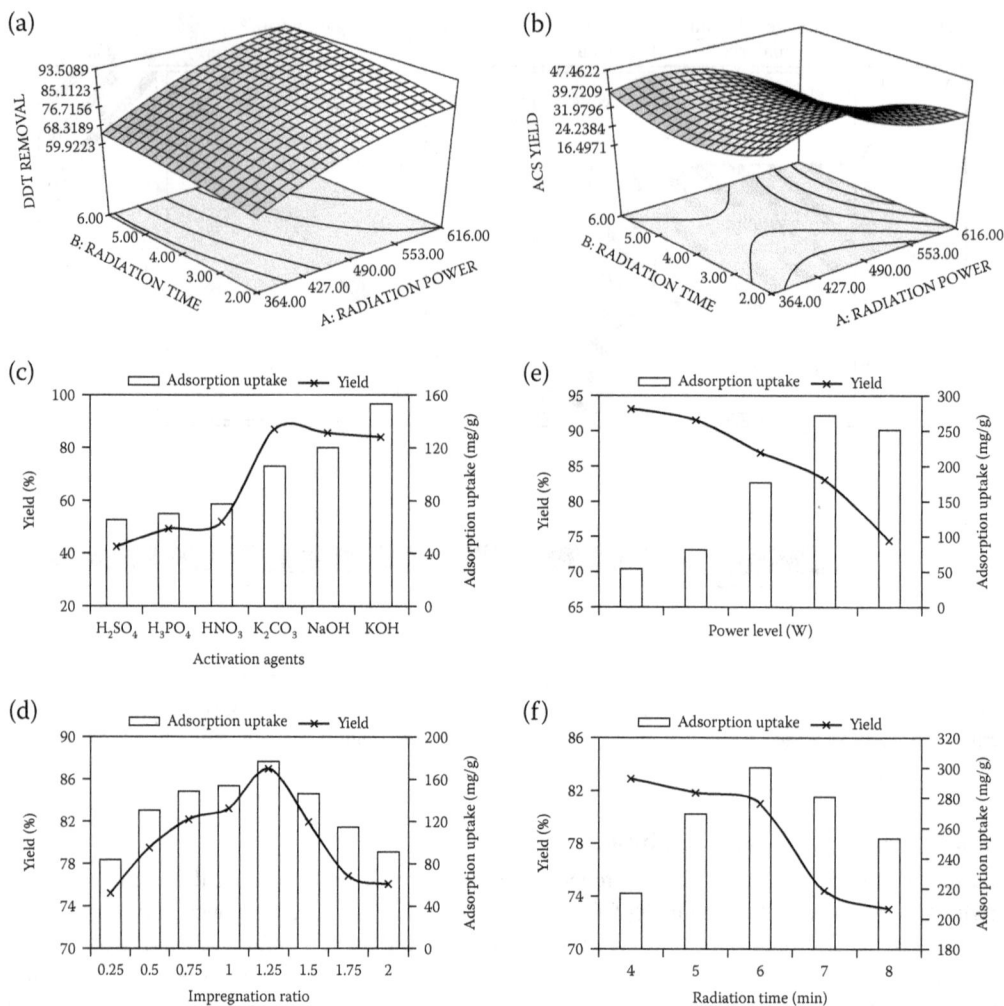

FIGURE 3.6 3D response surface plots for coconut-shell-based activated carbon produced by single-stage microwave heating under CO_2 gas flow: **(a)** DDT removal efficiency and **(b)** product yield [Reproduced with permission from reference Aziz et al., 2021, *International Journal of Chemical Engineering* 2021: 9331386]. Microwave-assisted preparation of activated carbon from coconut husk by chemical activation. Effects of **(c)** activation agents, **(d)** chemical impregnation ratio, **(e)** microwave power, and **(f)** radiation time on the carbon yield and adsorption uptake [Reproduced with permission from reference Foo and Hameed, 2012a, *Chemical Engineering Journal* 184: 57–65].

3.3.8.2 Microwave-induced chemical activation of coconut husk

Foo and Hameed (2012a) provided one of the most comprehensive studies on the microwave-assisted preparation of activated carbon from coconut husk. They analyzed the effect of activation agents and processing parameters on the product yield, textural and surface functional properties, and adsorption performance (Figures 3.6c–3.6f). The coconut husk was washed with water, dried, and ground to a 1–2 mm particle size. The initial carbonization of the feedstock in a batch size of 500 g was carried out following the conventional heat treatment in a furnace at 700°C under N_2 gas flow at 150 cm³/min. The char produced was mixed with different acids, bases, and salt type activation agents (H_2SO_4, H_3PO_4, HNO_3, K_2CO_3, NaOH, and KOH) at different impregnation ratios. The activation of the impregnated char was conducted by microwave processing at a 2.45 GHz frequency but at various microwave power levels with different exposure times under a pre-set N_2 gas flow rate of 300 cm^3/min. The resultant activated carbon was washed repeatedly with 0.1 M HCl and distilled water until pH 6–7 was reached in the residual liquid. The effects of activation agents and processing parameters on the carbon yield and adsorption uptake are presented in Figures 3.6c–3.6f. KOH was found to be the best activating agent in terms of adsorption capacity and an impressive BET specific surface area of 1,356.3 m^2/g and carbon yield of 80.8%. At a given input power of 360 W and irradiation time of 5 min, with increasing impregnation ratios of KOH to char from 0.25 to 1.25 (w/w), both the adsorption uptake and the carbon yield were improved. However, further increasing this ratio resulted in a decreasing trend. The optimized conditions found for input power and irradiation time were 600 W and 6 min, respectively.

All the above trials demonstrated the feasibility of converting coal and biomass to activated carbon by physical or chemical activating agents under microwave radiation and without the requirement of microwave absorbent. Various experiments are also attempted as a flexible implementation of microwave technology (Table 3.3). Since the dielectric properties of these carbon feedstocks are inferior to their char compounds, several researchers initially carbonized the feedstocks by conventional heating and then activated the char by microwave processing (Wu et al., 2015, Uche and Okwudili, 2019). Even pre-processing of feedstock to facilitate initial carbonization followed by steam activation using microwave has been successfully achieved (Yek et al., 2019). Also, microwave-assisted post-processing of activated carbon to improve the product quality by chemical functionalization or to reactivate exhausted products is evident (Ge et al., 2015). Thus the microwave technology performed at a laboratory scale demonstrated a feasible methodology for the production, modification, and reactivation of activated carbon originating from different carbonaceous feedstocks. However, scaled-up production trials are essential for evaluating their prospect for commercial applications, discussed in the following section.

3.3.9 MICROWAVE REACTOR DESIGN FOR PRODUCTION UP-SCALING

The reactors play a crucial role as the heating system in the microwave-assisted processing of carbonaceous feedstock and accordingly require specific design and construction through modeling and simulation. Microwave heating systems consist of three main components and some accessories for measurement and control. The main components are the microwave power source or the generator, the waveguide, and the applicator or cavity.

3.3.9.1 Generators

Microwave generators are the systems used to create microwave signals that can be tuned at a particular frequency to a specific power-level output. There are several types of microwave generators, including magnetrons, klystrons, and solid-state generators. Magnetrons are the most commonly used generators due to their high efficiency, ability to generate high microwave output power, and low cost (Meredith, 1998, Brace, 2009).

3.3.9.2 Waveguides

These are the hollow structures that direct electromagnetic waves propagation from the generator to their destination, i.e., the applicator for heating the materials. The geometry of this hollow structure or waveguide is an influencing factor in defining the wave propagation modes. Usually, rectangular-shaped hollow metal pipes are used for transmitting microwave power. The distribution pattern of electric and magnetic fields inside the rectangular waveguide varies with the frequency of electromagnetic waves. The rectangular waveguide walls reflect the electromagnetic waves from the walls and guide the electromagnetic waves to their destination in the z-direction.

3.3.9.3 Applicators

A microwave applicator is essentially a hollow cavity made of a metal structure that confines electromagnetic waves and is designed to heat a material by exposing it to a microwave field.

There are three main classes of applicators: traveling waves, near-field, and resonant applicators (Chan and Reader, 2000), where resonant applicators are the most common (Mehdizadeh, 2015). Resonant cavities can be single-mode or multimode. The multimode applicators are simple to construct and can process a wide range of workloads with different sizes and properties, but the non-uniform field strength distribution results in the formation of frequent hot and cold spots (Meredith, 1998). Single-mode cavities can provide high electric field intensity, better heating homogeneity, consistency, and predictable results. However, the sample size is limited by the half-wavelength of the applied microwave field (Metaxas and Meredith, 1993, Lidström et al., 2001).

For a large-scale production, several additional critical factors need to be considered, such as the feeding system, designing and supplying the microwave energy according to the feedstock to be processed and product output requirements, controlling the reaction temperature, heating homogeneity, adjusting the reaction environment during carbonization and activation steps, re-covering the end products and separate removal of by-products. These factors are related to microwave frequency to be used, reactor size, type (batch or continuous flow), mode (mono or multi), particle size, shape, volume and dielectric properties of the feedstock, intermediate and end pro-ducts, microwave penetration depth, heating and cooling system, and various safety measurements as well as emission control.

Waheed ul Hasan and Ani (2014) designed a pilot-scale microwave reactor to produce activated carbon suitable for steam activation, as shown in Figure 3.7. Their configuration includes a con-tinuously agitated cavity fitted with magnetron assembly and instrumentation. The size of the microwave reactor and the number of magnetrons needed to deliver the given power were cal-culated by the throughput capacity and dielectric properties of the materials to be processed. An assembly of magnetrons fitted around the reactor through waveguides provided microwave energy to the system. A gravimetric feeder system was coupled with the reactor to supply the precursor to the reactor. The temperature can be monitored using a thermocouple and controlled by a feedback temperature control loop, using microwave power as a manipulating variable. A variable speed agitator was introduced in the reactor design to ensure temperature uniformity within the reactor. N_2 can be supplied through a flow controller to create an inert environment in the reactor required for initial carbonization. The setup was suitable for steam to activate the char produced. Tars to be formed during the carbonization reaction can be drained out through the bottom of the reactor. Flue gases can be passed through a cyclone separator to recover the fine particles. The gases can then be scrubbed and vented to the atmosphere. Activated carbon products can be cooled in a solid heat exchanger and discharged on a conveyor belt for packaging.

There are a few more reports on up-scaling the production. Two selective case studies are mentioned below that demonstrate the feasibility of successfully up-scaling the activated carbon production through pilot-scale trials using microwave technology designed for batch and con-tinuous mode of operation. These trials also revealed the suitability of physical and chemical activation processes.

3.3.9.4 Case study 1: Pilot-scale production of activated carbon by chemical activation at batch scale using microwave technology

Lin et al. (2012) prepared activated carbon from sewage sludge, which is regarded as waste biomass, at a 5 kg batch size microwave reactor. Pre-dried sewage sludge was impregnated with the solutions of chemical activating agents KOH, $ZnCl_2$, and H_3PO_4, dried and crushed to the particle size of <0.5 cm, and then homogenized. About 4 kg of this dried chemically impregnated sludge was charged into the microwave heating device, and the pyrolysis was performed under a 5 L min^{-1} flow of N_2 atmosphere. The microwave heating device was composed of 12 microwave generators (magnetron and waveguide), with each magnetron operated at 2.45 GHz frequency and 800 W power coupled with a water cooling system. Microwave power was manipulated to control the reactor temperature. After pyrolysis, the products were first washed with HCl to eliminate the excess of activating agents and the fraction of soluble ash, and then the product was rinsed with

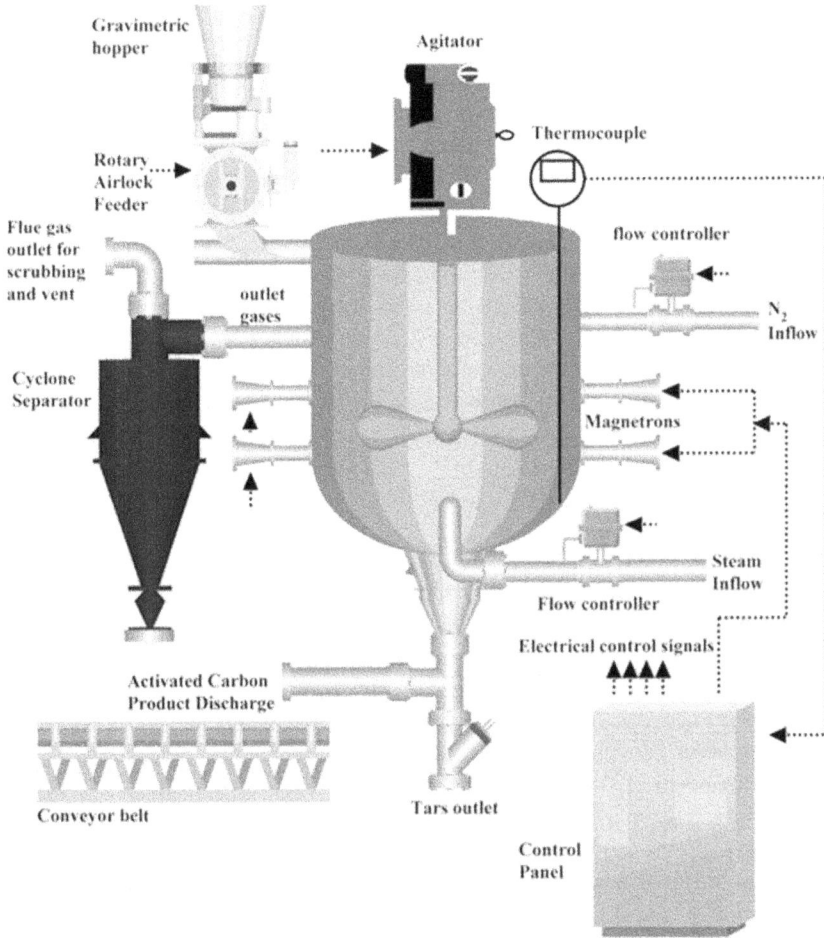

FIGURE 3.7 Continuous stirred tank microwave reactor configuration for activated carbon production by steam activation [Reproduced with permission from reference Waheed ul Hasan and Ani, 2014, *Industrial & Engineering Chemistry Research* 53(31): 12185–12191].

water, dried, and sieved to 100 mesh. The maximum product yield obtained was 72.3 wt%. KOH was found to be the most effective chemical activating agent with the highest BET specific surface area of 130.7 m^2 g^{-1} and the highest total pore volume of 0.13 mL g^{-1} prepared at a final carbonization temperature of 600°C for 20 min. A higher temperature (700°C) resulted in severe activation that had a detrimental impact on the textural quality of the product.

3.3.9.5 Case study 2: Pilot-scale production of activated carbon by physical activation in continuous mode of operation using microwave technology

Li et al. (2009) prepared activated carbon from coconut shell chars on the pilot scale using a cavity-type microwave reactor at a maximum capacity of 100 kg in the tube in a continuous operation. A water-cooled assembly of 60 magnetrons at 2.45 GHz with a power of 1 kW each was used to set up the microwave heating supply. The activated carbon was prepared in two steps, where the coconut feedstock was carbonized by conventional slow pyrolysis in an inert atmosphere followed by activation of char using steam under microwave heating. The finely ground 4.75–3.35 mm particle size pre-dried coconut shell powders were heated up to a carbonization temperature of 450°C at a heating rate of 10°C/min with conventional heating for at least 2 h under N_2 gas flow. The carbonized samples were then cooled to room temperature under N_2 flow, crushed, and sieved to obtain particles with sizes less than 2 mm. Coconut char was charged into a special ceramic tube inclining in the reactor cavity by a screw feeder and initially heated by applying microwave power 60 kW to around 860°C. The steam flow was then introduced for physical activation. The temperature was maintained at 825 ± 5°C by adjusting the microwave power and steam flow rate for 30 min. The optimum conditions were obtained as a 25 kg/h charging rate, 0.42 rev/min turning rate of ceramic tube, a steam flow rate at a pressure of 0.01 MPa, and 40 kW microwave heating power after 60 kW preactivation for 30 min. The iodine number of the microwave radiation-assisted product was 1,070 to 1,085 mg/g, which was found to be greater than the commercially available coconut shell-based activated carbons of 894 mg/g. After continuous production for 20 h, the average hourly product yield was 69.74%.

The above case studies showed the prospect of producing activated carbon at a large scale, but a thorough economic evaluation is lacking to realize their commercialization potential. Also, more improvements, especially in the continuous mode of production, are necessary, which are recommended in the following section.

3.4 FUTURE PERSPECTIVES

The studies on microwave-assisted production of activated carbon showed some unique benefits of this technology over the conventional slow-pyrolysis process; however, it also needs to overcome several limitations. There are scopes for improvement in scientific understanding of in-depth material-microwave radiation interactions at a molecular level. Selection of the suitable feedstock and their pre-treatment, advancing microwave reactor design to augment their energy and production efficiency, and cost reduction at a large scale can be performed by simulation studies. Selection of appropriate microwave frequency and executing microwave output power are also crucial factors for large-scale production. Small microwave ovens are operated at a frequency of 2.45 GHz because of precise field control but can only process materials of limited thickness due to low penetration depth. Microwave reactors used in industrial-scale operation preferably deploy a lower frequency of 0.915 GHz with a longer wavelength of 32.76 cm, allowing higher penetration depth. For example, microwave generators with high output power, such as 1,000 kW at 0.915 GHz, are available for commercial-scale continuous flow food processing (Web Reference, 2022e). The conversion of electrical energy to generate microwave energy is more efficient at the lower frequency (Haque, 1999) of 0.915 GHz (85%) than at a higher frequency of 2.45 GHz (50%). It also offers more uniform heating at a higher penetration depth and better flexibility in reactor cavity design to provide predictable field patterns in material processing. This would also enable controlling and optimizing the process conditions more effectively to maximize the product yield and enhance the adsorption qualities.

Numerical modeling and simulation studies would assist advancement in microwave reactor development with improved high power density, temperature measurement, and microwave power distribution within the reactor cavity for enhancing energy efficiency. For example, Hong et al. (2016) carried out a simulation study based on the finite element method to predict the heating behaviors of coal samples (Figure 3.8). Temperature profiles and electric field distributions suggest that in addition to microwave frequency and power, position on the electric field also affected the temperature and heating behavior of the coal sample.

Life cycle analysis and techno-economic assessments of microwave-assisted production of activated carbon need to be conducted to validate its commercial prospect on a large scale. Implementing the necessary setup toward zero-emission strategies, valuable by-products formation such as urea, and by-product recovery such as hydrogen as an energy source should be considered in the future trials, which will further boost the commercial benefit and make it an attractive business proposition.

One of the fascinating aspects of activated carbon is its potential value for futuristic applications such as energy storage, catalysis, sequestration of greenhouse gases, and microwave radiation

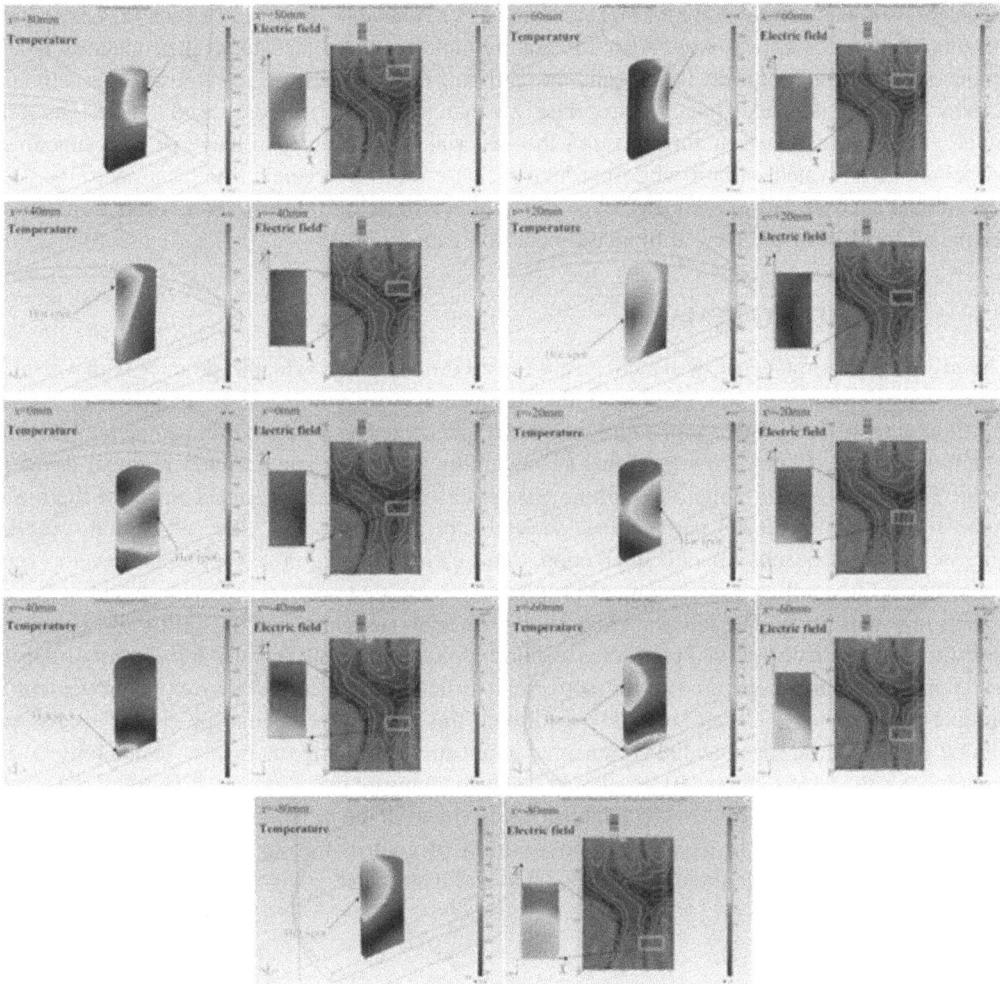

FIGURE 3.8 Evaluation of heating behavior of coal by a simulation study based on finite element method. [Reproduced with permission from reference Hong et al., 2016, *Applied Thermal Engineering* 93: 1145–1154].

absorbing purposes. Research studies revealed that carbon feedstocks from the waste origin are also a suitable precursor for creating advanced carbon products. The production of activated carbon from these low-cost, abundant coal, and biomass resources unravels the concern of their disposal issues and enables a valorization pathway toward a commercial outcome. Moreover, it can alleviate the environmental pollution caused by the waste buildup. The valorization pathway offers a higher resource recovery model than direct combustion, landfilling, composting, and other treatments. Therefore, the utilization of these carbon precursors to produce this beneficial and essential porous material should be highly encouraged.

Despite the awareness of environmental emissions and movements on climate action, coal remains the world's largest reliable source of electricity and second-largest primary energy source after oil. The task of scaling up alternative energy systems to meet rising demand while reducing or completely eliminating coal consumption is a significant practical challenge. Nevertheless, the use of coal in the energy sector and consequently the coal-derived items such as the production of activated carbon from coal precursors are expected to be continued over the coming decades. In hindsight, this has escalated exploring biomass as a sustainable feedstock, which is inexpensive and abundantly available as well. Also, integrating biomass-based feedstocks in production would render a low carbon footprint. Technically any carbonaceous biomass can be converted to activated carbon. However, the selection of appropriate raw material with high fixed carbon, their consistent supply, and the requirement for adjusting the material processing to achieve desired functional properties of activated carbon while producing cost-effectively remain crucial aspects for their commercial success and to compete in the matured coal-based product market before conquering the paradigm shift in the coal to biomass feedstock supply chain.

3.5 CONCLUDING REMARKS

The growing importance of activated carbon as a commercially valuable product with versatile uses as a high-volume sorbent material in the purification of different fluid systems is evident from this literature review. Such applications include specialized use in environmental remediation, which contributes significantly to global sustainability initiatives and enriches regional development. This chapter highlighted the current perception of the underlying mechanism of the molecular transformation from carbon feedstock to the final product, the adsorption process, microwave-carbonaceous matter interaction, and their implication for product development. There has been significant development in microwave-assisted coal and biomass-based activated carbon in recent times. The lab-scale and pilot-scale trials demonstrated the ability of microwave technology to produce the final product at a shorter process time and thereby significantly improve the energy efficiency compared to the conventional heating process. Moreover, improvement in product quality and yield can be achieved through this alternative heating process. This has exhibited great promise for potential commercial applications utilizing microwave technology. More comprehensive and systematic efforts should be taken to ensure the commercial viability of sorbent material produced by microwave, to establish the business in the progressive gas, vapor, and liquid purification market, and to explore the possibilities of contributing to high-value emerging applications, which are undoubtedly commendable and a step change to the status quo.

ACKNOWLEDGMENTS

The author would like to thank the New Zealand Ministry of Business, Innovation and Employment for their support and funding of '2017 The Regional Research Institutes Initiative' (contract number NZSMT1801), which supported this publication.

REFERENCES

Alhashimi HA, Aktas CB 2017. Life cycle environmental and economic performance of biochar compared with activated carbon: A meta-analysis. *Resources, Conservation and Recycling* 118: 13–26.

Anderson N, Gu H, Bergman R 2021. Comparison of novel biochars and steam activated carbon from mixed conifer mill residues. *Energies* 14(24): 8472.

Andriesse, JP 1988. Andriesse J. P. The main characteristics of tropical peats. In *Nature and Management of Tropical Peat Soils*, 19–43, Rome: Food and Agriculture Organization of the United Nations.

Aziz A, Nasehir Khan MN, Mohamad Yusop MF, Mohd Johan Jaya E, Tamar Jaya MA, Ahmad MA 2021. Single-stage microwave-assisted coconut-shell-based activated carbon for removal of dichlorodiphenyltri-chloroethane (DDT) from aqueous solution: Optimization and batch studies. *International Journal of Chemical Engineering*, 9331386.

Bandosz TJ 2008. Removal of inorganic gases and VOCs on activated carbons. In *Adsorption by Carbons*, ed. EJ Bottani and JMD Tascón, Oxford, UK: Elsevier Science, pp. 533–564.

Bansal RC, Goyel M 2005. Chapter 1-Activated Carbon and Its Surface Structure, In Activated Carbon Adsorption. In *Activated Carbon and Its Surface Structure* (1st Ed.), Boca Raton: CRC Press, Taylor & Francis Group, pp. 1–66.

Beneroso D, Monti T, Kostas E, Robinson J 2017. Microwave pyrolysis of biomass for bio-oil production: Scalable processing concepts. *Chemical Engineering Journal* 316: 481–498.

Berger, C 2012. *Biochar and activated carbon filters for greywater treatment – Comparison of organic matter and nutrients removal.* Uppsala, Sweden: Swedish University of Agricultural Sciences.

Bhatia SK 2017. Characterizing structural complexity in disordered carbons: from the slit pore to atomistic models. *Langmuir* 33(4): 831–847.

Binner E, Lester E, Kingman S, Dodds C, Robinson J, Wu T, Wardle P, Mathews JP 2014. A review of microwave coal processing. *Journal of Microwave Power and Electromagnetic Energy* 48(1): 35–60.

Brace CL 2009. Microwave ablation technology what every user should know. *Current Problems in Diagnostic Radiology* 38: 61–67.

Cazetta, AL, Vargas, AMM, Nogami, EM, Kunita, MH, Guilherme, MR, Martins, AC, Silva, TL, Moraes, JCG, Almeida, VC 2011. NaOH-activated carbon of high surface area produced from coconut shell: Kinetics and equilibrium studies from the methylene blue adsorption. *Chem. Eng. J.* 174: 117–125.

Chan, TVCT, Reader, HC 2000. *Understanding Microwave Heating Cavities.* Boston: Artech House Publishers.

Dabrowski A 2001. Adsorption - from theory to practice. *Advances in Colloid and Interface Science* 93: 135–224.

Danish M, Ahmad T 2018. A review on utilization of wood biomass as a sustainable precursor for activated carbon production and application. *Renewable and Sustainable Energy Reviews* 87: 1–21.

Deng, H, Yang, L, Tao, G, Dai, J 2009. Preparation and characterization of activated carbon from cotton stalk by microwave assisted chemical activation-Application in methylene blue adsorption from aqueous solution. *J. Hazard. Mater.* 166: 1514–1521.

Dibben, D 2001. Electromagnetics: Fundamental aspects and numerical modeling. In *Handbook of Microwave Technology for Food Applications*, ed. AK Datta and RC Anantheswaran, New York: Marcel Dekker, pp. 1–32.

Elliott DA, Nabavizadeh N, Seung SK, Hansen EK, Holland JM, 2018. Radiation Therapy. In *Oral, Head and Neck Oncology and Reconstructive Surgery*, ed. R Bryan Bell, Rui P Fernandes and Peter E Andersen, Saint Louis, Missouri: Elsevier, pp. 268–290.

Ethaib S, Omar R, Kamal SMM, Awang Biak DR, Zubaidi SL 2020. Microwave-assisted pyrolysis of biomass waste: A mini review. *Processes* 8(9): 1190.

Foo K, Hameed B 2012a. Coconut husk derived activated carbon via microwave induced activation: effects of activation agents, preparation parameters and adsorption performance. *Chemical Engineering Journal* 184: 57–65.

Foo KY, Hameed, BH 2012b. Potential of jackfruit peel as precursor for activated carbon prepared by microwave induced NaOH activation. *Bioresour. Technol.* 112: 143–150.

Foo KY, Hameed BH 2012c. Mesoporous activated carbon from wood sawdust by K2CO3 activation using microwave heating. *Bioresour. Technol.* 111: 425–432.

Foo KY, Hameed BH 2012d. Preparation, characterization and evaluation of adsorptive properties of orange peel based activated carbon via microwave induced K2CO3 activation. *Bioresour. Technol.* 104: 679–686.

Foong SY, Liew RK, Yang Y, Cheng YW, Yek PNY, Mahari WAW, Lee XY, Han CS, Vo D-VN, Le QV, Aghbashlo M, Tabatabaei M, Sonne C, Peng W, Lam SS, 2020. Valorization of biomass waste to engineered activated biochar by microwave pyrolysis: Progress, challenges, and future directions. *Chemical Engineering Journal* 389: 124401.

Francis W 1961. *Coal: Its Formation and Composition* (2nd Ed.). London: Edward Arnold.

Franklin RE 1951. Crystallite growth in graphitizing and non-graphitizing carbons. *Proceedings of the Royal Society of London Series A Mathematical and Physical Sciences* 209(1097): 196–218.

Ge X, Tian F, Wu Z, Yan Y, Cravotto G, Wu Z 2015. Adsorption of naphthalene from aqueous solution on coal-based activated carbon modified by microwave induction: Microwave power effects. *Chemical engineering and processing: Process intensification* 91: 67–77.

Genn GL 2013, *Novel Techniques in Ore Characterisation and Sorting.* Thesis (PhD), Australia: The University of Queensland.

Haque KE 1999. Microwave energy for mineral treatment processes – A brief review. *Int. J. Miner. Process* 57: 1–24.

Hong Y-D, Lin B-Q, Li H, Dai H-M, Zhu C-J, Yao H 2016. Three-dimensional simulation of microwave heating coal sample with varying parameters. *Applied Thermal Engineering* 93: 1145–1154.

Huang Y-F, Chiueh P-T, Kuan W-H, Lo S-L 2016. Microwave pyrolysis of lignocellulosic biomass: heating performance and reaction kinetics. *Energy*, 100: 137–144.

Jawad AH, Mehdi ZS, Ishak MAM, Ismail K 2018. Large surface area activated carbon from low-rank coal via microwave-assisted KOH activation for methylene blue adsorption. *Desalin Water Treat* 110: 239–249.

Kappe CO 2004. Controlled microwave heating in modern organic synthesis. *Angewandte Chemie International Edition* 43(46): 6250–6284.

Keiluweit M, Nico PS, Johnson MG, Kleber M 2010. Dynamic molecular structure of plant biomass-derived black carbon (Biochar). *Environmental Science & Technology* 44(4): 1247–1253.

Lam SS, Yek PNY, Ok YS, Chong CC, Liew RK, Tsang DC, Park Y-K, Liu Z, Wong CS, Peng W 2020. Engineering pyrolysis biochar via single-step microwave steam activation for hazardous landfill lea-chate treatment. *Journal of hazardous materials* 390: 121649.

Lee T, Ooi C, Othman R, Yeoh F 2014. Activated carbon fiber – the hybrid of carbon fiber and activated carbon. *Rev. Adv. Mater. Sci.* 36: 118–136.

Lewandowski WM, Januszewicz K, Kosakowski W 2019. Efficiency and proportions of waste tyre pyrolysis products depending on the reactor type—A review. *J. Anal. Appl. Pyrol.* 140: 25–53.

Lewis WK, Metzner AB 1954. Engineering, design and process development section: Activation of carbons. *Ind. Eng. Chem.* 46: 849–858.

Li L, Wang H, Jiang X, Song Z, Zhao X, Ma C 2016. Microwave-enhanced methane combined reforming by CO_2 and H_2O into syngas production on biomass-derived char. *Fuel* 185: 692–700.

Li W, Peng J, Zhang L, Yang K, Xia H, Zhang S, Guo S-H 2009. Preparation of activated carbon from coconut shell chars in pilot-scale microwave heating equipment at 60 kW. *Waste management* 29(2): 756–760.

Lidström P, Tierney J, Watheyb B, Westmana J 2001. Microwave assisted organic synthesisÐa review. *Tetrahedron* 57: 9225–9283.

Lin L, Wang Q 2015. Microplasma: A new generation of technology for functional nanomaterial synthesis. *Plasma Chem Plasma Process* 35: 925–962.

Lin Q, Cheng H, Chen G 2012. Preparation and characterization of carbonaceous adsorbents from sewage sludge using a pilot-scale microwave heating equipment. *Journal of Analytical and Applied Pyrolysis* 93: 113–119.

Lin S, Gao L, Yang Y, Chen J, Guo S, Omran M, Chen G 2020. Dielectric properties and high temperature thermochemical properties of the pyrolusite-pyrite mixture during reduction roasting. *Journal of Materials Research and Technology* 9(6): 13128–13136.

Liu Q, He H, Li H, Jia J, Huang G, Xing B, Zhang C, Cao Y 2019. Characteristics and kinetics of coal char steam gasification under microwave heating. *Fuel* 256: 115899.

Liu QS, Zheng T, Wang P, Guo L 2010. Preparation and characterization of activated carbon from bamboo by microwave-induced phosphoric acid activation. *Ind. Crops Prod.* 31: 233–238.

Llamas-Unzueta R, Montes-Mor´an MA, Ramírez-Montoya LA, Concheso A, Men´endez JA 2022. Whey as a sustainable binder for the production of extruded activated carbon. *Journal of Environmental Chemical Engineering* 10: 107590.

Macedo JS, Otubo L, Ferreira OP, Gimenez IDF, Mazali IO, Barreto LS 2008. Biomorphic activated porous carbons with complex microstructures from lignocellulosic residues. *Micro Meso Mater* 107(3):276–285.

Marsh H, Rodríguez-Reinoso F 2006. *Activated Carbon*. Oxford: Elsevier.

McGhee GR 2018. *Carboniferous Giants and Mass Extinction: The Late Paleozoic Ice Age World*. New York: Columbia University Press.

Mehdizadeh M 2015. *Microwave/RF Applicators and Probes for Material Heating, Sensing, and Plasma Generation*, (2nd Ed.), Oxford: Elsevier.

Meredith R. 1998. *Engineers' Handbook of Industrial Microwave Heating*. London: IEE.

Metaxas AC, Meredith RJ 1993. *Industrial Microwave Heating*. London: Peter Peregrinus.

Mopoung S, Dejang N 2021. Activated carbon preparation from eucalyptus wood chips using continuous carbonization–steam activation process in a batch intermittent rotary kiln. *Scientific Reports* 11(1): 1–9.

Moreno-Castilla C 2004. Adsorption of organic molecules from aqueous solutions on carbon materials. *Carbon* 42: 83–94.

Norinaga K, Kumagai H, Hayashi J-I, Chiba T 1998. Classification of Water Sorbed in Coal on the Basis of Congelation Characteristics. *Energy & Fuels* 12: 574–579.

Norman LM, Cha C 1995. Production of activated carbon from coal chars using microwave energy. *Chemical Engineering Communications* 140(1): 87–110.

Oleszczuk P, Hale SE, Lehmann J, Cornelissen G 2012. Activated carbon and biochar amendments decrease pore-water concentrations of polycyclic aromatic hydrocarbons (PAHs) in sewage sludge. *Bioresour. Technol.* 111: 84–91.

Peng Z, Hwang J-Y, Park C-L, Kim B-G, Onyedika G 2012. Numerical Analysis of Heat Transfer Characteristics in Microwave Heating of Magnetic Dielectrics. *Metallurgical and Mater. Trans. A* 43: 1070–1078.

Peng Z, Lin X, Wu X, Hwang J-Y, Kim B-G, Zhang Y, Li G, Jiang T 2016. Microwave absorption characteristics of anthracite during pyrolysis. *Fuel Processing Technology* 150: 58–63.

Percuoco R 2014. Plain radiographic imaging. In *Clinical Imaging* (3rd Ed.), ed. DM Marchiori, Saint Louis: Mosby, pp. 1–43.

Priyanto DE, Ueno S, Kasai H, Mae K 2018. Rethinking the Inherent Moisture content of biomass: Its ability for milling and upgrading, *ACS Sustainable Chem. Eng.* 6: 2905–2910.

Radovic LR, Moreno-Castilla C, Rivera-Utrilla J 2001. Carbon materials as adsorbents in aqueous solutions. In *Chemistry and Physics of Carbon* (vol. 27), ed. LR Radovic, New York: Marcel Dekker, pp. 227–405.

Ray S, Cooney RP 2018. Thermal degradation of polymer and polymer composites. In *Handbook of Environmental Degradation of Materials* (3rd Edition), Kutz, M, Ed., Amsterdam, The Netherlands: William Andrew Publishing, pp. 185–206.

Rosa R, Veronesi P, Leonelli C 2013. A review on combustion synthesis intensification by means of microwave energy. *Chem. Eng. Process. Process Intensif* 71, 2–18.

Sajjadi B, Chen W-Y, Egiebor NO 2019a. A comprehensive review on physical activation of biochar for energy and environmental applications. *Reviews in Chemical Engineering* 35(6): 735–776.

Sajjadi B, Zubatiuk T, Leszczynska D, Leszczynski J, Chen WY 2019b. Chemical activation of biochar for energy and environmental applications: a comprehensive review. *Reviews in Chemical Engineering* 35(7): 777–815.

Saleem J, Shahid UB, Hijab M, Mackey H, McKay G 2019. Production and applications of activated carbons as adsorbents from olive stones. *Biomass Conversion and Biorefinery* 9: 775–802.

Schuepfer DB, Badaczewski F, Guerra-Castro JM, Hofmann DM, Heiliger C, Smarsly B, Klar PJ 2020. Assessing the structural properties of graphitic and non-graphitic carbons by Raman spectroscopy. *Carbon* 161: 359–372.

Seky Y, Sarikanat M, Sever K, Durmuskahy AC 2013. Extraction and properties of Ferula communis (chakshir) fibers as novel reinforcement for composites materials. *Composites, B: Engineering* 44 (1): 517–523.

Serio MA, Hamblen DG, Markham JR, Solomon PR 1987. Kinetics of volatile product evolution in coal pyrolysis: Experiment and theory. *Energy & Fuels* 1(2): 138–152.

Shiraishi Y 1984. *Introduction to Carbon Materials (Kaitei Tansozairyo Nyumon in Japanese)*, Tokyo: Carbon Society of Japan, pp. 29–40.

Sing KSW, Everett DH, Haul RAW, Moscou L, Pierotti RA, Rouquerol J, Siemieniewska T 1985. Reporting physisorption data for gas/solid systems with special reference to the determination of surface area and porosity. *Pure and Applied Chemistry* 57(4): 603–619.

Stach E 1968. Basic principles of coal petrology: Macerals, microlithotypes and some effects of coalification. In *Coal and Coal-Bearing Strata*, ed. D Murchision and TS Westoll, New York: Elsevier Publishing, pp. 3–17.

Straka P, Sy´korova I,2018. Coalification and coal alteration under mild thermal conditions. *Int J Coal Sci Technol* 5(3): 358–373.

Sturm, GSJ, Verweij, MD, Van Gerven, T, Stankiewicz, AI, Stefanidis, GD 2012. On the effect of resonant microwave fields on temperature distribution in time and space. *International Journal of Heat and Mass Transfer* 55(13–14): 3800–3811.

Sun J, Wang W, Yue Q 2016. Review on microwave-matter interaction fundamentals and efficient microwave-associated heating strategies. *Materials* 9(4): 231.

Taylor GH, Teichmuller M, Davis A, Diesel CFK, Littke R, Robert P 1998. *Organic Petrology*. Berlin: Gebru¨der Borntraeger.

Thommes M 2010. Physical adsorption characterization of nanoporous materials. *Chemie Ingenieur Technik* 82 (7): 1059–1073.

Uche EC, Okwudili OJ 2019. Effect of process factors on the quality of activated carbon produced from coal by microwave-induced chemical activation. *American Journal of Engineering Research (AJER)* 8(4): 1–6.

UNFCCC 2005. Clarifications of definitions of biomass and consideration of changes in carbon pools due to a CDM project activity. *Framework Convention on Climate Change—Secretariat.* CDM-EB-20, Appendix 8.

Waheed ul Hasan S, Ani FN 2014. Review of limiting issues in industrialization and scale-up of microwave-assisted activated carbon production. *Industrial & Engineering Chemistry Research* 53(31): 12185–12191.

Web Reference 2022a. https://www.epa.gov/sciencematters/reducing-pfas-drinking-water-treatment-technologies (accessed on 20 May 2022).

Web Reference 2022b. https://www.marketsandmarkets.com/Market-Reports/activated-carbon-362.html (accessed on 20 May 2022).

Web Reference 2022c. https://www.marketsandmarketsblog.com/activated-carbon-market.html (accessed on 20 May 2022).

Web Reference 2022d. https://www.epa.gov/radtown/non-ionizing-radiation-used-microwave-ovens (accessed on 20 May 2022).

Web Reference 2022e. https://directory.newequipment.com/classified/continuous-flow-bulk-microwave-sterilization-systems-255235.html (accessed on 20 May 2022).

William R, 1973. Heat generation, transport, and storage. In *Chemical Engineers' Handbook* (5th Ed.), ed. R Perry and C Chilton.New York: McGraw-Hill.

Wu J, Xia H, Zhang L, Xia Y, Peng J, Wang S, Zheng Z, Zhang S 2015. Effect of microwave heating conditions on the preparation of high surface area activated carbon from waste bamboo. *High Temperature Materials and Processes* 34(7): 667–674.

Xiao X, Liu D, Yan Y, Wu Z, Wu Z, Cravotto G 2015. Preparation of activated carbon from Xinjiang region coal by microwave activation and its application in naphthalene, phenanthrene, and pyrene adsorption. *Journal of the Taiwan Institute of Chemical Engineers* 53: 160–167.

Yek PNY, Liew RK, Osman MS, Lee CL, Chuah JH, Park Y-K, Lam SS 2019. Microwave steam activation, an innovative pyrolysis approach to convert waste palm shell into highly microporous activated carbon. *Journal of environmental management* 236: 245–253.

Yek PNY, Peng W, Wong CC, Liew RK, Ho YL, Mahari WAW, Azwar E, Yuan TQ, Tabatabaei M, Aghbashlo M 2020. Engineered biochar via microwave CO2 and steam pyrolysis to treat carcinogenic Congo red dye. *Journal of Hazardous Materials* 395: 122636.

Yoshida A, Kaburagi Y, Hishiyama Y 1991. Microtexture and magnetoresistance of glass-like carbons. *Carbon* 29(8): 1107–1111.

4 Radiation-induced polymer modification and polymerization

Tuhin Subhra Pal and Nikhil K. Singha*

Rubber Technology Centre, Indian Institute of Technology, Kharagpur, West Bengal, India

*Corresponding author: nks8888@yahoo.com

CONTENTS

DOI: 10.1201/9781003321910-4

4.1 INTRODUCTION

Radiation energy is one of the most abundant forms of energy available to mankind. Nature provides sunlight, the type of radiation essential for many forms of life and growth. Some of the natural substances, such as radioactive elements, generate the kind of radiation that can be destructive to life, but when harnessed, it can be very useful for medical and industrial applications.

4.1.1 DIFFERENT TYPES OF RADIATION

Now the generation of radiant energy is very important. Human genius created its own devices for generating radiant energy useful on a great variety of scientific, industrial, and medical applications. Cathode-ray tubes emit impulses that activate screens of computer monitors and televisions. X-rays are used not only as a diagnostic tool in medicine but also as an analytical tool in an inspection of manufactured products such as tires and other composite structures (polymer-based composite). Microwaves are used not only in cooking or as a means of heating rubber or plastics, but also in a variety of electronic applications. Infrared radiation (IR) is used in heating, analytical chemistry, and electronics. Ultraviolet radiation (UV) has been in use for decades in medical applications, analytical chemistry, and in a variety of industrial applications. Ion beams have been widely used in commercial ion implantation in the production of semiconductor devices and also for metal surface hardening. This ion beam also has some limited applications in the processing of polymeric materials, such as treatment of polymer surfaces and polymer thin films because they have very low penetration depths. Laser beam radiation has found a wide use in medical and military applications and in several industrial applications.

Radiation-induced changes in materials depend on the origin and type of radiation and deposited energy. Energy deposition processes, in turn, depend on the origin of the radiation:

- **Particulate radiation:** Including electrons, protons, neutrons, positrons, ions.
- **Electromagnetic radiation:** Both ultraviolet (UV) and electron beam (EB) radiations are classified as electromagnetic radiations along with infrared (IR), gamma, microwave (MW), and laser beam radiations. The differences between them in frequency and wavelength are shown in Table 4.1.

Radiation is a form of energy that comes from a source and travels through some medium or through space. Radiation is emitted when an electron in an atom drops from a higher to a lower energy state. Radiation is classified into two main categories depending on its ability to ionize matter: Ionizing radiation ionizes a material with a high amount of energy and in non-ionizing radiation has a small amount of energy that cannot ionize a material.

Ionizing radiation is produced by unstable atoms possessing excess energy or mass. This type of radiation carries energy more than 10 eV that can ionize atoms and molecules and break the chemical bonds. Radioactive materials containing helium nuclei, electrons or positrons, and

TABLE 4.1
Spectrum of the Electromagnetic Radiation

Wave region	Wavelength (m)	Frequency (Hz)	Energy (eV)
Radio	>0.1	$<3 \times 10^9$	$<10^{-5}$
Microwave	$0.1–10^{-4}$	$3 \times 10^9–3 \times 10^{12}$	$10^{-5}–0.01$
Infrared	$10^{-4}–7 \times 10^{-7}$	$3 \times 10^{12}–4.3 \times 10^{14}$	$0.01–2$
Visible	$7 \times 10^{-7}–4 \times 10^{-7}$	$4.3 \times 10^{14}–7.5 \times 10^{14}$	$2–3$
Ultraviolet	$4 \times 10^{-7}–10^{-9}$	$7.5 \times 10^{14}–3 \times 10^{17}$	$3–10^3$
X-rays	$10^{-9}–10^{-13}$	$3 \times 10^{17}–3 \times 10^{21}$	$10^3–10^7$
Gamma rays	$<10^{-11}$	$>3 \times 10^{19}$	$>10^5$

photons that emit α, β, or γ rays are the common sources of ionizing radiation. Non-ionizing radiation is a type of electromagnetic radiation that cannot ionize atoms or molecules as it possesses a lower amount of energy per quantum. Out of these two, ionizing radiation is sub-categorized into direct ionizing radiation and indirect ionizing radiation.

- Direct ionizing radiation corresponds to the energy deposition in the materials by an energetic charged particle that has a Coulomb interaction with an orbital electron of the target atom.
- Indirect ionizing radiation is realized in two steps. First, fast-charged particles (electrons and protons) are released in the material due to the photon energy deposition. Second, the released, charged particles deposit their energy directly in the materials through Coulomb interactions between these particles and orbital electrons of an atom.

4.1.2 Ionizing radiation

Ionizing radiation is a radiation that travels as a particle (alpha, beta, and neutron particles) or electromagnetic wave (X and γ-rays) that carries sufficient energy to separate electrons from atoms or molecules, thereby ionizing an atom or a molecule. Ionizing radiation has become the most effective way to modify natural and synthetic polymers through cross-linking, degradation, and graft polymerization. It can modify the chemical, physical, and biological properties of the irradiated materials. In present days, the principal industrial applications of radiation are sterilization of healthcare products including pharmaceuticals, irradiation of food and agriculture products (for various end objectives, such as disinfestation, shelf-life extension, sprout inhibition, pest control, and sterilization), and materials modification (such as polymerization, polymer cross-linking, etc.).

The main three types of ionizing radiation include high-energy electrons (EB), gamma rays (γ-rays), and X-rays. These are capable of not only converting monomeric and oligomeric liquids into solids but also producing major changes in the properties of solid polymers. Due to the difference in energy, these particles and radiation have variable penetrating power, which significantly affects the living tissues. Also, in comparison to UV and visible radiations, they can penetrate considerably deeper into the material. The more energetic radiations are capable of striking electrons out of the orbits from the nucleus. The source of ionizing radiation can be both natural as well as artificial radioactive materials. There is a lot of literature (Hung, G. W. C, et al. 1974; Wolfram Schnabel et al. 1982, pp. 227; Drobny et al. 2013; Rosiak et al., 1995; Kabanov et al., 2009; Rogalski and Palmer, 2014) on the effects of ionizing radiation such as gamma rays, the high-energy electron from electron beam accelerators, and X-rays on the polymeric materials and their modification.

4.1.2.1 EB radiation

The electron beam is a flow of electrons with energy and the energy is obtained as kinetic energy when the electron moves in a high electric field. Electron beam (EB) radiation is a form of ionizing energy that is characterized by its low penetration and high dosage rates. EB irradiators produce an EB by energizing and accelerating a stream of electrons through an electromagnetic or electrostatic field. The strength of the EBs as radiation is controlled by accelerating voltage and beam current (Makuuchi and Cheng, 2011). In contrast to γ-rays, EBs generated from an accelerator are monoenergetic. EB irradiators have much higher power than γ-irradiators.

Electron beam (EB) irradiation is a useful and well-known implement for the improvement of the mechanical properties of polymers. Electron beam treatment has numerous advantages over conventional chemical or heat treatment processes. The EB process is economically feasible, faster than chemical cross-linking, and less hazardous as it eliminates the generation of undesirable by-products (like toxic and harmful peroxide residue in case of peroxide curing). Moreover, the EB radiation technique helps to maintain the purity of the processed product as it does not require any catalyst or solvent. The process commonly operates at ambient temperature, which excludes the possibility of thermal degradation for thermally sensitive components, but it can also be executed at elevated temperatures. EB irradiation does not produce scrap and waste, and the doses of radiation can be controlled easily.

4.1.2.2 Gamma rays

Gamma rays (γ-rays) is an electromagnetic radiation, which is emitted from excited atomic nuclei of an unstable atoms, the so-called radionuclides, and at the end of the process, the nucleus rearranges itself into a state of lower excitation state (i.e., lower energy content). Gamma rays are emitted by radioactive decay from radioactive isotopes and have energies in the range of 10^4 to 10^7 eV. Gamma rays penetrate matter farther than beta or alpha particles, producing ionization (electron disruption) in their path. In living cells, these disruptions result in damage to the DNA and other cellular structures ultimately causing the death of the organism. Like X-ray, γ-Rays do not create residual radioactivity in the materials exposed to them. They ionize matter by three main processes: the photoelectric effect, Compton scattering, and pair production. In the wide energy range of 100 keV to 1 MeV, Compton scattering is the main absorption mechanism, in which an incident γ-photon loses enough energy to eject an electron in an atom of the irradiated matter, and the rest of its energy is emitted as a new γ-photon with lower energy. ^{60}Co is the main source of γ-radiation.

Unlike EB or X-rays, γ-rays cannot be turned off. Once radioactive decay starts, it continues until all the atoms have reached a stable state. The most common applications of γ-rays are sterilization of single-use medical supplies, elimination of organisms from pharmaceuticals, microbial reduction in consumer products, cancer treatment, and processing of polymers (crosslinking, polymerization, degradation, etc.). It's important to note that the products that were irradiated by γ-rays do not become radioactive and thus can be handled normally.

4.1.2.3 X-rays

X-ray radiation (also known as Röntgen radiation) is a form of electromagnetic radiation. X-rays have a wavelength in the range from 10 to 0.01 nm, corresponding to frequencies ranging from 3×10^{16} to 3×10^9 and energies in the range from 120 eV to 120 keV. The great German scientist, Wilhelm Conrad Röntgen discovered it in 1895. He named it X-ray to signify an unknown type of radiation. Due to their penetrating abilities, X-rays are classified into two categories: soft X-rays (0.12 to 12 keV) and hard X-rays (12 to 120 keV). As X-rays are a form of ionizing radiation, they can be dangerous to a living organism.

There are two different atomic processes that can produce X-ray photons. One process produces bremsstrahlung (from German, meaning "breaking radiation") and the other produces K-shell or characteristic emission. Both processes involve a change in the energy state of electrons. X-rays are generated when an electron is accelerated and then made to rapidly decelerate usually due to interaction with other atomic particles. In an X-ray system, a large amount of electric current is passed through a tungsten filament, which heats the filament to several thousand degrees centigrade to create a source of free electrons. A large potential has been established between the filament (the cathode) and the target (the anode). The two electrodes are in a vacuum. The electric potential between the cathode and the anode pulls electrons from the cathode and accelerates them, as they are attracted towards the anode, which is usually made of tungsten. X-rays are generated when the electrons are given up the energy as they interact with the orbital electron or nucleus of an atom. The interaction of the electrons in the target results in the emission of a continuous radiation spectrum and also characteristic X-rays from the target material. Thus, the difference between γ-rays and X-rays is that the γ-rays originate in the nucleus and X-rays originate in the electrons outside the nucleus (surrounding it) or are produced in an X-ray generator.

4.1.2.4 UV radiation

UV radiation causes photooxidative degradation which results in the breaking of the polymer chains, produces free radicals and reduces the molecular weight, causing deterioration of mechanical properties. The energies associated with near-ultraviolet radiation is 72–97 kcal/mol. Common covalent bonds encountered in polymers have bond dissociation energies that are either lower or within this energy range. The ultraviolet radiation is absorbed by the polymer and results in the formation of photoexcited singlet (S) and triplet (T) species to transfer the absorbed energy to cause photochemical reactions. Light-induced damage to the polymer can also take place in the presence of UV radiation. In most systems, a variety of different wavelengths competing for photophysical processes, such as phosphorescence (from ($T_1 \rightarrow S_0$) transition) or fluorescence (from ($S_1 \rightarrow S_0$) transition), may prevent the chemical reaction. When photochemical reactions of polymers do take place, they tend to involve the triplet excited states of molecules rather than the ground or singlet stage species because of the relatively long lifetime of the former. The lowest excited triplet state, T_1, is formed by radiation less intersystem crossing from the lowest excited singlet state, S_1.

In the presence of UV radiation, photopolymer or light-activated resin changes its properties. These changes are often revealed structurally, for example, hardening of the material occurs as a result of cross-linking when exposed to light and this process is called curing. Typically, a photopolymer consists of a mixture of multifunctional monomers and oligomers, which can be polymerized in the presence of light (UV radiation) either through internal or external initiation.

Most commonly, photopolymers undergo a process called curing, where oligomers are cross-linked upon exposure to light, forming what is known as a network polymer. The result of photocuring is the formation of a thermoset network of polymers. UV radiation is a relatively soft radiation technique that is widely used in paints and coating technology.

4.1.3 LASER BEAM RADIATION

The term "LASER" stands for light amplification by the stimulated emission of radiation. One basic type of laser consists of a sealed tube, containing a pair of mirrors, and a laser medium that is excited by some form of energy to produce visible light, or invisible ultraviolet or infrared radiation (Figure 4.1). An "active medium", which can be a gas, solid, liquid, or microchip, is stimulated by an external "energy source" so that it produces lots of little photons of the same wavelength. These photons bounce around inside the cavity in all directions. However, some are reflected off the two mirrors which causes them to go through the laser material again and again. These photons can knock into atoms inside the materials causing them to generate yet more photons. This is the stimulated emission and amplification bit of a laser.

A laser is a device that emits light (electromagnetic radiation) through a process of optical amplification based on the stimulated emission of photons. Like this, some laser emits radiation in the form of light. Others emit radiation that is invisible to the eye, such as ultraviolet or infrared radiation. In general, laser radiation is not in itself harmful and behaves much like ordinary light in its interaction with the body. Laser radiation should not be confused with radio waves, microwaves, or the ionizing x-rays or radiation from radioactive substances such as radium.

Besides a wide use in medical and military applications, lasers are used in industrial applications for cutting, welding, material heat treatment, marking parts, and noncontact measurement of parts.

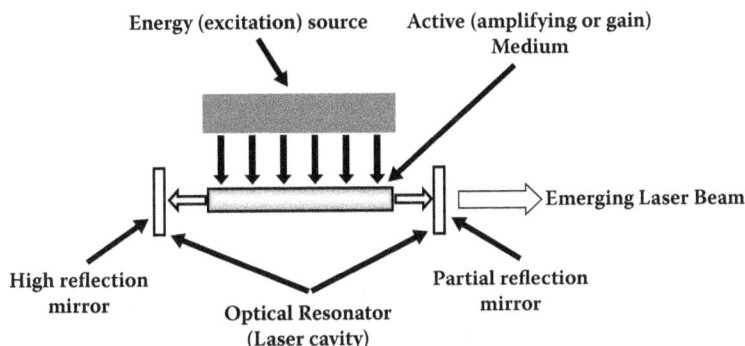

FIGURE 4.1 Formation of laser.

4.1.4 Ion beam radiation

An ion beam is a type of particle beam consisting of ions. High-energy ion beams are produced like EBs by particle acceleration, with the difference that a cyclotron, another particle accelerator is used. Due to the difference in the linear energy transfer (LET), the average energy deposited in the material used by a projectile particle along its path, the irradiation effects of an ion beam are different from ionizing radiation. Generally, the LET of an ion beam is larger than that of EB, depending on the particle mass and energy.

Ion beam radiation has been used extensively in commercial applications for ion implantation in the production of semiconductor devices and surface hardening of metals. In the semiconductor industry, a system with an acceleration voltage rating from 10 kV to 100 kV is used for ion implantation. But, over the past decade, ion beams have been adapted for commercial use for the processes involving polymers.

Initially, their applications had included treatment of polymer surfaces and thin films because of their very low depth penetration (Clough, 2001), but recent reports in the literature about their use in modifications of mechanical properties of carbon-fiber-reinforced plastics (Kudoh, Sasuga and Seguchi, 1996; Seguchi et al., 1999), in surface modification of polytetrafluoroethylene (PTFE) (Choi, Kim and Noh, 2007), and the production of fuel cells (Yamaki et al., 2019).

4.1.5 Microwave radiation

Microwaves are a form of "electromagnetic" radiation; that is, they are waves of electrical and magnetic energy moving together through space. Microwaves are used to detect speeding cars and to send telephone and television communications. Industry uses microwaves to dry and cure plywood, to cure rubber and resins, to raise bread and doughnuts, and to cook potato chips. But the most common consumer use of microwave energy is in microwave ovens. Recently, microwave radiation is being used for polymerization (Bai and Mu, 2019) of several vinyl monomers.

Microwave irradiation systems used for chemical reactions have two cavity types: multiple- and single-mode cavities. Multiple-mode cavities are usually used for domestic microwave ovens. Conversely, single-mode cavities are suited to small-scale systems. These cavities are designed to generate only one mode of microwave energy. Microwaves generate a center of high electromagnetic field intensity, and the energy distribution in the reaction vessel is homogenous.

Single mode microwaves are used mostly in chemistry and research labs. In these, the cavity is tuned to the frequency of the magnetron – 2.45 GHz. This allows for a uniform field. There is no interference, and therefore no hot or cold spots. The microwave field is completely homogeneous. Because of this, there is no reflected energy, and no need for an isolator. These characteristics allow single-mode microwaves to be much smaller than multimode, and usually of a much lower power as there is a 100% transfer of energy into the sample.

Multimode microwaves also require an isolator to protect the magnetron from reflected energy. This do not let any microwaves bounce back and hit the magnetron. It absorbs the reflected energy and turns it into heat. It's important that all microwave energy must be absorbed in a multimode cavity. The rest of the part what is not absorbed by the food will be absorbed by the isolator.

4.1.6 Basics of radiation effects on polymers

Irradiation of polymers with high-energy radiation (gamma rays, X-rays, electron beams, ion beams) leads to the formation of very reactive intermediates in the forms of excited states, ions, and free radicals. These intermediates are almost immediately gone through in several reaction pathways that results in the arrangement or formation of a new bond structure. The final effects of these reactions are the formation of oxidized products, grafts, cross-linking, and scissioning of main chains or side chains, which is also called degradation. The degree of these transformations depends on the nature of the polymers and the conditions of treatment before, during, and after irradiation and close control of these factors make the modification of polymers possible by radiation processing.

4.1.7 ADVANTAGES AND DISADVANTAGES OF RADIATION IN POLYMERS

Polymerization and modification of polymers using radiation processes do not require any metal catalyst or initiator that are often toxic and need to be purified or removed from the final product. Importantly, by the radiation process, the polymerization and the polymer modification can be done without using any solvent. So, the radiation process is considered one of the most useful, clean processes to prepare different polymeric materials. The advantages of polymerization or polymer modification by radiation process are the following:

- There is no need to add any catalyst or any initiator to start the polymerization reaction.
- Significant reduction of volatile organic compounds because no solvents is used.
- Higher purity and lower toxicity because no or less toxic chemicals are needed.
- They can be applied to polymerize at a wide range of monomers.
- The active initiation sites are homogeneously distributed in the bulk material.
- They can be used to graft off polymers of any nature (like fibers, films, powders, etc.).
- The higher rate of production because of faster processing.

There are also some disadvantages of radiation processing. Industrially, the operational cost of radiation processing depends heavily on the volume, so it can be significantly higher than chemical modification when the volume is not high enough. Sometimes the properties achieved by radiation processing are still poorer to those that can be achieved by chemical modification.

4.2 POLYMER MODIFICATION USING RADIATION

"High-energy" radiations, e.g., X-rays and gamma rays or accelerated electrons, are today extensively and economically used in the polymer industry for the production of new or modified polymers exhibiting interesting properties. The properties of the polymer, such as strength, smart-responsiveness, therapeutic effect, conductivity, and so on are not only determined by the choice of monomers but the result of the modification. Also, the post-modification is required to make hydrophobic polymers to hydrophilic polymers for the application in an aqueous environment. Furthermore, the architecture of the polymer can also be designed through modification. In drug delivery applications, many therapeutic agents can be conjugated to the polymer via modification.

Radiation processing was used early on for polymer modification. The irradiation of polymeric materials with ionizing radiation (gamma rays, X-rays, accelerated electrons, ion beams) leads to the formation of very reactive intermediates, free radicals, ions, and excited states. These intermediates can follow several reaction paths that result in disproportion, hydrogen abstraction, arrangements, and/or the formation of new bonds. Nowadays, the modification of polymers covers radiation cross-linking, radiation-induced polymerization (graft polymerization and curing) and the degradation of polymers. Figure 4.2 shows a typical technique of radiation processing for polymers. Radiation curing is used on a large scale for the high-speed production of improved coatings. Radiation cross-linking is an established technology in the wire and cable industry. It imparts to the modified polymer improved resistance to solvents and high temperatures. Radiation grafting is a method for modifying more intensely the properties of a polymer. The method can be used for several applications, such as introducing polar groups into non-polar polymers, increasing or reducing the wettability of a surface, imparting to a polymeric surface better compatibility with a specific coating and many others.

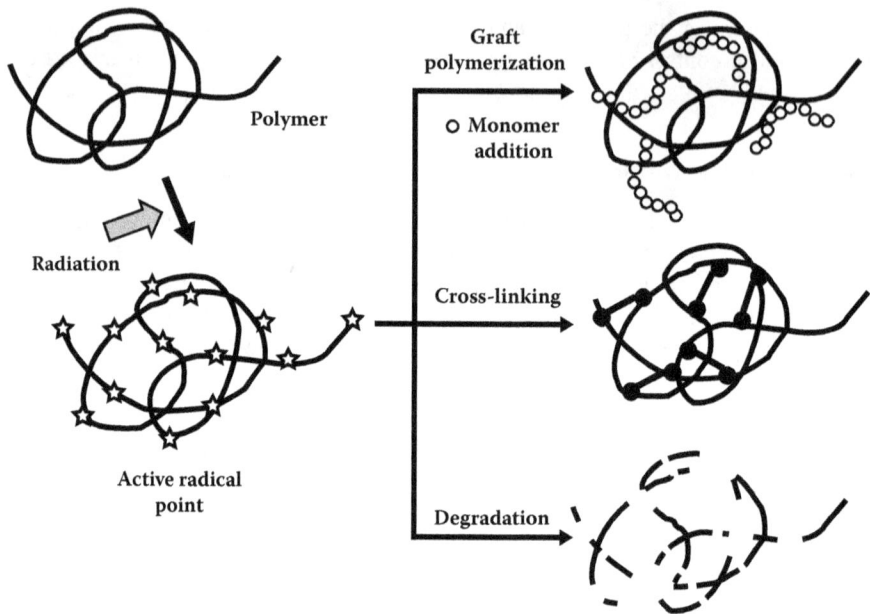

FIGURE 4.2 Typical technique to process polymer with radiation. (Adaptable from H. Kudo (ed.), *Radiation Applications, An Advanced Course in Nuclear Engineering 7*, 2018). (Kudo, 2018).

4.2.1 RADIATION CROSS-LINKING

Radiation-induced cross-linking of the polymer has been playing the leading role of radiation processing. Radiation cross-linking is the relatively mature technology and commercially most important area of radiation chemistry of polymers. The advantages of radiation cross-linking are rapid, easiness of control, and cost-effectiveness. Cross-linking is recombination between the formed radicals. When radiation is used, the solubility and the heat resistance can be improved without a chemical cross-linking agent. The main chemical issue related to the radiation cross-linking is that of enhancing the cross-linking and reducing the oxidative degradation during irradiation. This is the one reason to use electron beams for cross-linking to reduce oxidative degradation.

4.2.2 RADIATION DEGRADATION

The opposite of cross-linking, degradation (polymer chain scission) is the basis of radiation treatments aimed at enhancing processing characteristics of polymers. Scission and cross-linking can happen simultaneously and in general, which one dominates is determined by the chemical structure. Polyethylene and polyvinyl chloride are classified as cross-linking type, whereas polypropylene and polystyrene are classified as a degradation type. The yields of cross-linking and scission can be evaluated using the Charlesby–Pinner equation. The atmosphere during irradiation can affect the result of irradiation; for example, polyethylene (PE), which is a cross-linking type, can change to a degradation type under an oxygen atmosphere. Polytetrafluoroethylene (PTFE) is known as a degradation type as it reduces particle size and molecular weight, under irradiation, but the polymer can be cross-linked by irradiation just above its melting temperature.

4.2.3 RADIATION CURING

The original meaning of curing is a simple change from a liquid state to a solid-state. The radiation curing means a radiation-induced polymerization of a prepolymer-monomer mixture to form a three-dimensional network (cross-linking) involving a radical polymerization between double bonds in the prepolymer and monomer. The radiation-curable prepolymers generally are low to medium molecular weight, having mono- or multi-functional unsaturated materials such as unsaturated polyesters and acrylateprepolymers. Generally, electron beams are used for the radiation curing. The main features of radiation curing are summarized as follows: (1) Pollutants and solvent-free process, (2) rapid process, (3) application versatility, (4) low-energy requirements, (5) low-temperature operation, (6) enhanced product durability, (7) small space requirements, (8) single-component coating. But the main disadvantage of radiation curing is initial investment and running cost. Therefore, the more economical curing method using UV is more popular in the world.

4.2.4 RADIATION GRAFTING

Graft polymerization using radiation can be useful for the modification of polymers. There are three radiation grafting techniques available for polymer modification, which are (1) The pre-irradiation method, (2) simultaneous method, and (3) the peroxidation method (will be discussing in details later). From a practical viewpoint, the pre-irradiation method is preferable because of less homopolymers formation and easier control of the degree of polymerization. Graft polymerization is usually based on free radical reactions. In this case, the monomer reacts with the active free radical sites available on the surface of the polymer. This will lead to attachment of selective grafts to the surface of the original polymer. In graft coupling reactions, a polymer with functional end groups can be grafted using the reactive sites on the surfaces of another polymer.

4.2.4.1 Surface modification using radiation

Radiation treatment is being widely investigated as a means of altering surface properties of polymeric materials, including films, powders, fibers, and molded objects. Plenty of work has involved the irradiation of non-polar polyolefins (PE, PS, PP, fluoro polymers, etc.) as a means of inducing polar groups at surfaces, in order to enhance such properties as wettability, printability, adhesion with other materials or with biological components, or further chemical modification (Choi, Koh and Jung, 1996; Choi et al., 1999; Zhang et al., 2000). Many times, such irradiations take place in the presence of reactive gasses such as O_2 or NH_3, in order to form reactive polar groups at the surfaces (carboxylic acids, alcohols, amines, etc.).

Although ions, electrons, and gamma rays have all been utilized in surface treatment schemes, ions and low-energy electrons are particularly well suited for such applications due to their inherently limited penetration depth. For the surface modification, the following radiation types like gamma, UV, laser, plasma, and microwaves have been frequently utilized for improving biocompatibility of polymers.

4.2.4.1.1 γ-irradiation

High-energy photons (gamma rays) can be used for the surface treatment of polymers via free radical formation mechanism. These photons generated free radicals on the polymer surface and free molecules which can lead to propagation or termination reactions such as recombination. When the treatment is carried out in an oxygen atmosphere the free radicals present on the polymer surface, react with oxygen and resulting in the formation of oxygenated species on the polymer surface. These oxygenated species improve the mechanical properties and the adhesion tendency of the polymer.

The reactions of produced free radicals are responsible for the radiation-induced modifications to the polymer surface. Nechifor et al. studied the effect of gamma radiation on porous polymeric membranes that were obtained via the alloying of poly (hydroxy urethane) (PHU) and poly (vinyl alcohol) (PVA) in different concentrations. The results demonstrated that an increase in the dosage of gamma radiation enhanced the porosity and hydrophilic properties of the membranes (Matter, 2009). Madrid et al. reported the introduction of glycidyl methacrylate (GMA) via gamma irradiation to obtain the surface modification of microcrystalline cellulose (MCC). The grafting of GMA onto the polymer surface was to impart additional functional groups to help tailor the properties of the polymer surface (Madrid and Abad, 2015). Since there is a lack of studies involving gamma radiation as a tool for surface treatment, it is difficult to analyze the advantages of gamma radiation over a variety of polymers.

4.2.4.1.2   UV irradiation

Among the various strategies employing for the radiation and discharges, UV-assisted surface modification seems to be an easily applicable and economical method (Decker, 1998; Weibel et al., 2009). The principle of UV treatment involves the activation of chemical reactions that occur in the presence of an initiator or a sensitizer that absorbs UV irradiation and is excited to a stable triplet state from a singlet state. When UV radiation is exposed on the polymer surface, the surfaces undergo photo cross-linking, photo-oxidation in air, or photochemical reactions in a reactive atmosphere. As a result of UV radiation, it generates reactive grafting site by removing of hydrogen present on the polymer surface. UV-induced surface modification has been studied extensively for various industrial and biological applications. It has also applications to disinfect packaging materials, disinfection of surfaces, and curing and activation of surfaces.

Recently, UV-radiation-induced surface modification has received much attention. Literature dealing with UV-induced surface changes of various polymers has been reviewed to bring out the importance of this methodology. Ramanathan (Rajajeyaganthan et al., 2011) studied UV surface modification of polystyrene (PS), polyurethane, polysulfone (PSU), and polypropylene (PP) in the presence of acrylic vapor, and further polyurethane and polysulfone samples alone were treated in presence of trimethoxy propyl silane (TMPSi). These polymers presented a permanent hydrophilic surface even after 65 days of the treatment process in the presence of acrylic vapor, whereas the polyurethane and polysulfone samples became hydrophobic when treated in the presence of (TMPSi). In a similar study, hydrophobic recovery of biomaterials was investigated by Connell et al. They showed that hydrophobic recovery proceeds at a different rate for each polymer, it is generally a two-phase process and that surfaces are still more hydrophilic after 28 days than the original untreated state (O'Connell et al., 2009). Toan et al. studied the UV surface modification of fluorocarbon polymer. It has been found that there is a slight decrease in the fluorine content with the formation of carbonyl groups resulting in the increase of hydrophilicity of polymer fragments (Toan Le et al., 2013). Pieracci et al. (Pieracci et al., 2000) had reported the use of a UV-assisted graft polymerization approach to increase the antifouling properties of the membrane surfaces by increasing their hydrophilicity. Subedi et al. showed that UV of shorter wavelength (254 nm) was efficient in improving the wettability of the polycarbonate compared to longer wavelength (375 nm). Further, they found that with the increasing time interval of UV irradiation with shorter wavelength resulted in the increasing wettability of polycarbonate (Subedi, Tyata and Rimal, 2009). In summary, from the outcome of the above-mentioned researches, we can conclude that UV treatment improves the wettability and hydrophilicity. Specifically, when the thermoplastic polymers like polysulfone and PP are treated with UV in the presence acrylic acid vapors, their wettability is found to increase significantly. For synthetic aromatic polymer like PS, the UV treatment increase the hydrophobicity resulting in the formation of super-hydrophobic surfaces.

4.2.4.1.3   Laser treatment

Surface modification of polymers using a laser is an emerging technology and offers several benefits when compared to other surface modification methods. This method enables precise surface modification with little surface damage. They are rather simple techniques, easily controlled, and environmentally clean and safe processes. But, the investment and operational costs of laser surface treatments are still high, compared with other surface modification methods, such as flame and corona treatments, which are the most commonly used methods in the industry. Laser surface modifications are mainly used in areas such as electronics, optoelectronics, aerospace, materials processing, and automotive industries where a high degree of precision is required. When we compared with the other light sources, laser light is coherent and highly focused. In the presence of high voltage, the atoms present in the ground state are excited to a high energy state, and as a result, laser light is produced. There are two basic mechanisms involved in achieving the surface modification using laser irradiation: (a) thermal process and (b) photochemical process (Ozdemir and Sadikoglu, 1998; Fabbri and Messori, 2017).

In recent days, laser treatment plays a vital role in polymer surface treatment to improve biocompatibility. Khorasani et al. studied the morphological change of PDMS surface using CO_2-pulsed laser (Khorasani, Mirzadeh and Kermani, 2005). They found decreased surface wettability, and the result of in vitro analyses indicated that the platelet adhesion was reduced on laser-treated PDMS. Hence, they finally concluded that laser irradiation on silicone rubber is a multipurpose technique to produce an anti-thrombogenic surface for biomaterial applications. Breuer et al. reported the surface modification of polypropylene (PP) using photochemical laser action to enhance its adhesion properties (Breuer, Metev and Sepold, 1995). Wang et al. studied the effects of PMMA exposed to femtosecond laser pulses at various laser fluencies and focus distances. It resulted in the controlled modification of surface wettability of PMMA. This change in the wettability was supposed to be caused dominantly by laser-induced chemical structure modification and not by a change in surface roughness (Wang et al., 2015). Suggs et al. used Kr-F excimer laser (248 nm) for surface modification of polymers, namely polymethylmethacrylate (PMMA), glycol-modified polyethyleneterephthalate (PETG), and polytetrafluoroethylene (PTFE). They observed increased surface roughness of treated PMMA and PTFE, compared to untreated (Suggs, 2002). Dadsetan et al. studied the effect of CO_2-pulsed laser when it exposed onto the polyethyleneterephthalate (PET) surface. They found that the crystallinity in the surface region decreased due to laser irradiation and also decreased in contact angle with increasing laser pulse radiation (Dadsetan, Mirzadeh and Sharifi, 1999). Mirzadeh et al. used CO_2-pulsed laser on ethylene-propylene rubber (EPR), which was surface grafted with acrylamide (AAm) and 2-hydroxyethylmethacrylate (HEMA). They observed that in grafted EPR, macrophage adhesion and cellular damage decreased after laser irradiation (Mirzadeh, Ekbatani and Katbab, 1996). The above summary covers a few research studies carried out on polymers that come under the silicone, polyester, and polyacrylate family. From the results and observations, we can conclude that laser treatment increases the surface roughness and surface energy of polyacrylate families, which will improve their cell compatibility. Furthermore, for silicone polymers, it is also reported that laser treatment produces super-hydrophobic surfaces.

4.2.4.1.4 Plasma treatment

Plasma treatment is probably the most versatile method for the surface modification of many materials and of growing interest in biomedical engineering, especially for polymers to increase biocompatibility. It is relatively simple, easy to implement, and cost-effective. The plasma used to modify the surface of the polymers with non-polymerizable gases such as argon, oxygen, nitrogen, or fluorine in a vacuum system. Plasmas are capable of exerting four major effects through surface cleaning, surface etching, surface cross-linking, and surface modification. The interaction of plasma with the polymer surface can cause hydrogen separation from polymeric chains and free radical creation. Plasma forms the free radicals only on the surface. Therefore, the grafting is limited to the near-surface layer only.

Currently, plasma treatment has received much attention. Siegel et al. studied the properties of polyethylene LDPE, polytetrafluoroethylene, PS, and polyethyleneterephtalate (PET) modified by Ar plasma. They found that under the plasma discharge, the polymers were taken away and their surface morphology and roughness were changed dramatically and eventually leading to an increase in the wettability of the polymer surface (Siegel et al., 2008). Rezinckova et al. studied surface modification of polyethylene (PE), PTFE, PS, PET, and PP treated by Ar plasma. They noticed that the dramatic change in surface morphology and roughness (Řezníčková et al., 2011). Melnig et al. treated the polyurethane surface by helium plasma at atmospheric pressure and implant Ar^+ ion to increase biocompatibility as well as promoted cell adhesion using the lactate segment, i.e., poly(lactaturethane) (Melnig et al., 2005). Arefi et al. treated the surface of polypropylene (PP) with nitrogen plasma, which resulted in increased wettability and altered the surface conductivity and the adhesive properties of the polymer (Osada, 1988). Recently, Wieland and his team studied a systematic and comprehensive plasma activation on various polymeric surfaces by optimizing different parameters, including power, time, substrate temperature, and gas composition. They achieved the highest protein immobilization efficiency on polymers. Slepicka et al. treated the polypropylene (PP) surface with argon plasma. As a result, initially, oxygen concentration on the PP surface was high and the contact angle was low but with ageing oxygen concentration decreased and contact angle increased (Jaganathan et al., 2015). Aidun et al. studied the surface modification of PU-CNT electrospun modification using O_2-plasma. They used polyvinyl alcohol (PVA) and 3-glycidoxypropyl-trimethoxysilane (GPTMS) for the modification to improve physicochemical and in vitro properties for efficient bone reconstruction. Slepicka et al. determined the surface properties of PET, HDPE, PTFE, and PLLA after plasma treatment (Slepička et al., 2013). Junkar et al. used radiofrequency (RF) oxygen and nitrogen plasma on the PET surface. Their results showed that by oxygen and nitrogen plasma treatment, the surface chemistry, wettability, and morphology are altered (Junkar, Cvelbar and Lehocky, 2011).

4.2.4.1.5 *Microwave treatment*

Microwaves are the electromagnetic (EM) waves of wavelengths ranging between 100 μm and 1 m. Generally, microwaves are used in medical applications like computed tomography (CT), microwave surgery, but recently it has been proved that microwaves can also be used successfully for surface modification of polymers and fabrics. Keshel et al. exposed polyurethane surface to microwave plasma with oxygen and argon gases. As a result, they found improvement in surface roughness, and adhesion properties when the polyurethane is modified with oxygen plasma in comparison with argon plasma (Keshel et al., 2011). Rabiei et al. used microwaves on polyetheretherketone (PEEK) deposited with hydroxyapatite (HA). They concluded that microwaves can be used to increase the biocompatibility of polyetheretherketone (PEEK) (Rabiei and Sandukas, 2013). Ginn et al. used an unmodified ''kitchen microwave oven'' as the source of microwaves to modify the surface of PDMS. The results illustrated a profound increase in the hydrophilicity as well as cell adhesion properties of the PDMS surface (Ginn, Steinbock and Distillers, 2003). Badey et al. modified the surface of PTFE by microwave plasma treatment. The results showed an improvement in the wettability and hydrophilicity of surface (Badey et al., 1994). Mutel et al. treated the PP surface with low-pressure microwave cold nitrogen plasma. The microwave treatment improved the adhesion properties of epoxy resin on PP surface and increased the wettability and hydrophilicity of the polymer surface (Mutel et al., 1988). Microwave treatment utilized for various polymers that come under different polymer families has been summarized in this section.

4.2.5 RADIATION-INDUCED GRAFT POLYMERIZATION

For many decades, radiation-induced grafting has been a method to functionalize the surfaces of the polymer, so that they can be used in a variety of applications, such as biomedical, environmental and industrial uses (Nasef and Güven, 2012). Radiation grafting changes the surface of the polymeric materials by modifying with a polar or non-polar monomer having functional groups, such as –COOH, –OR, –OH, –NH_2, –SO_3H, –R, and their derivatives, which affect surface properties without the influence on the bulk material.

Electron beam (EB), gamma rays, and ultraviolet (UV) radiation can be used to generate active sites (free radicals) on a polymeric surface that can then react with vinyl monomers to form a graft copolymer. A graft copolymer can be defined as branched copolymer composed of a main chain of a polymer backbone onto which side chain grafts (branches) are covalently attached. The polymer backbone may be a homopolymer or copolymer and differs in chemical structure and composition from the graft material. The beauty of grafting is that it is possible to construct a material whose bulk is consist of one type of polymer but whose surface is made of different type of polymer. Using this approach, one can initiate with a bulk material (chosen based on good mechanical strength and low cost) and then "grows" a second polymer type on the material's surface.

The modification of polymer surfaces can be achieved by conventional grafting or by reversible addition-fragmentation chain transfer (RAFT)–mediated grafting methods. But with these conventional grafting methods, the molecular weight and polydispersity of grafted chains cannot be controlled. As a result, it shows a broad chain length distribution, which means the surface is covered with grafted chains of different lengths (Figure 4.3a). To get the controlled molecular weight of graft copolymers and narrow chain length distribution, a novel method using a controlled radical polymerization (CRP) of reversible addition-fragmentation chain transfer polymerization grafting has been used (Moad et al., 2005; Moad, Rizzardo and Thang, 2008), as shown in Figure 4.3b.

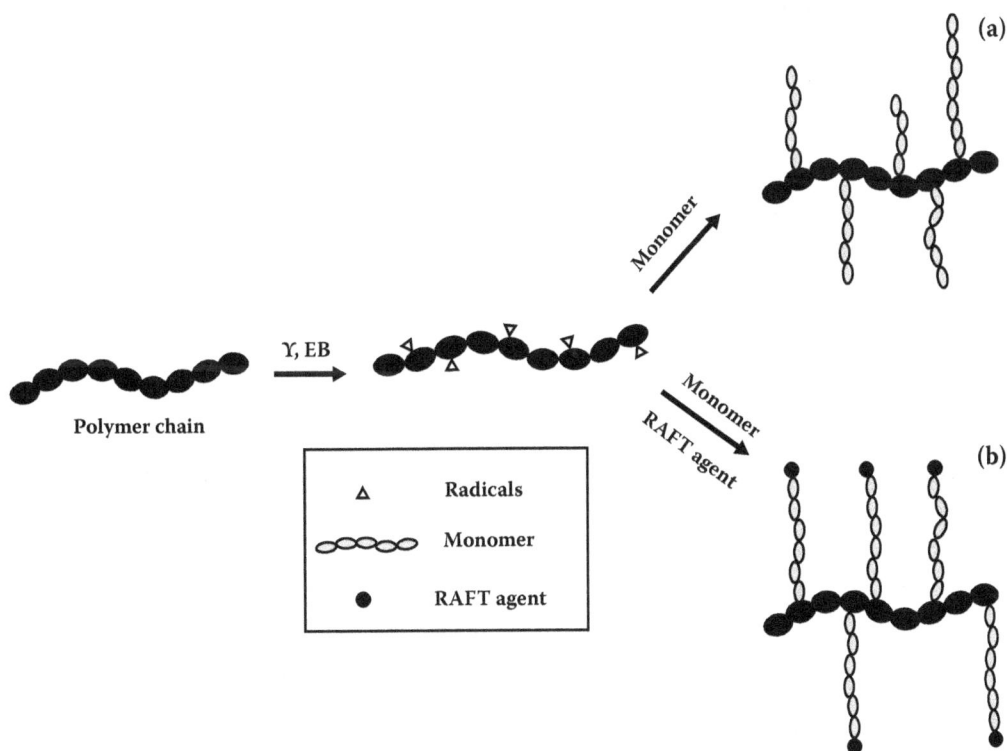

FIGURE 4.3 Radiation grafting of polymer using (a) conventional and (b) RAFT-mediated polymerization initiated by ionizing radiation.

4.2.5.1 Free-radical grafting

Macromolecules under irradiation undergo homolytic fission and this forms free radicals on the polymer. In the radiation technique, the medium (vacuum or air) is more important than the presence of initiator in the system. The lifetime of free radical depends upon the nature of the backbone polymer. There are three radiation grafting techniques are available for polymer modification: (1) the pre-irradiation method, (2) simultaneous method, and (3) the peroxidation method (Figure 4.4) (Garnett, 1979). Out of them first two methods are the major applicable method of radiation grafting for the modification of polymer (Nasef and Hegazy, 2004). The selection of either method depends on the polymer to be modified, the reactivity of monomers, and the radiation source.

In the pre-irradiation method (Uflyand et al., 1992; Bhattacharya, Das and De, 1998; Chen et al., 2003; Marmey, Porté and Baquey, 2003; Yamaki et al., 2003), the polymer backbone is first irradiated, usually in a vacuum or inert gas to produce relatively stable free radicals, which are then reacted with a monomer, usually at elevated temperatures (Dworjanyn et al., 1993). An important advantage of the pre-irradiation method is that less homopolymer is formed and easy to control the degree of polymerization. In the simultaneous method, the polymeric material is immersed in the monomer solution (or in pure monomer) and exposed to ionizing radiation (Bhattacharyya, Maldas and Pandey, 1986; Kaur et al., 1993, 1994, 1996; Basu et al., 1994; Aich, Bhattacharyya and Basu, 1997). Irradiation can be performed in air or inert atmosphere (e.g., nitrogen) usually using gamma sources. This grafting process can occur via free radical or ionic mechanism (Huglin and Johnson, no date; Zatikyan and Grushevskaya, 1988). This is the simplest and most commonly used method chosen for polymeric material surface modification and is suitable for substrates that are sensitive to radiation. The peroxidation method, which is the least often used of all the irradiation techniques, involves irradiation of the substrate in the presence of air or oxygen. This produces stable peroxide and hydroperoxides on the substrate,

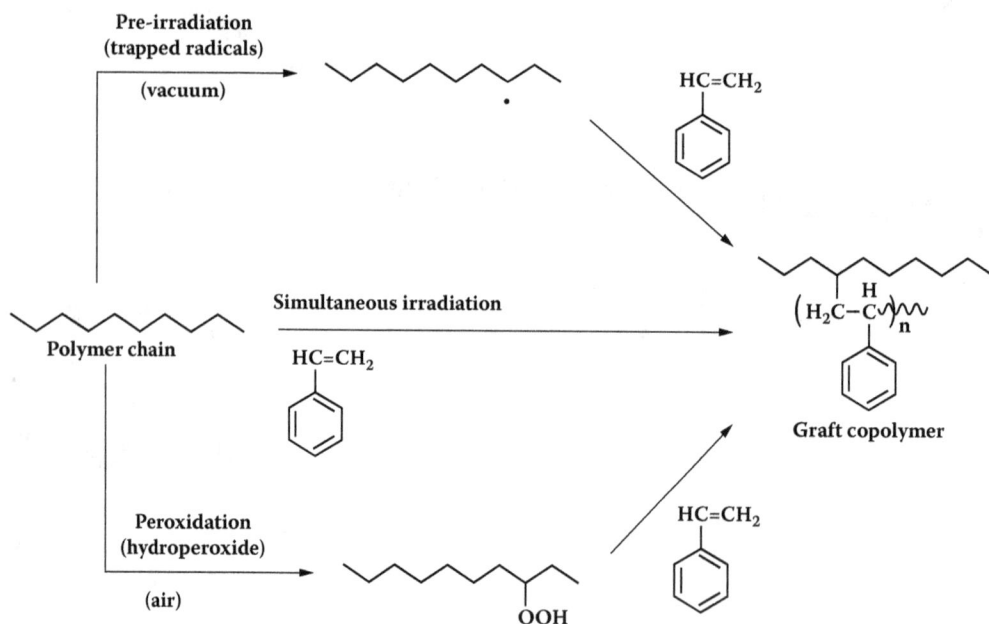

FIGURE 4.4 Schematic diagram of different radiation grafting techniques for polymer modification.

and the substrate can be stored until the combination with the monomer is possible. A monomer, with or without solvent, is then reacted with the activated peroxy branch polymer in the air or under vacuum at elevated temperatures to form the graft copolymer. The advantage of this method is the relatively long shelf life of the intermediate peroxy branch polymers before the final grafting step (Dworjanyn et al., 1993).

4.2.5.2   Ionic grafting

Radiation grafting can also proceed through an ionic mode, in which the ions form through high-energy irradiation. Ionic grafting can be of two different types: cationic or anionic. The polymer is irradiated to form the polymeric ion and then reacted with the monomer to form the grafted copolymer. The potential advantage of ionic grafting is a high reaction rate. Hence, small radiation doses are sufficient to bring about the required grafting. The cationic grafting initiated from the backbone is shown in Figure 4.5 (Path I). An alternate cationic grafting mechanism can proceed through monomer radical cation, which subsequently forms a dimer. Charge localization in the dimer occurs in such a way that the dimer radical cation then reacts with the radical produced by the irradiation of the polymer, e.g., see Figure 4.5 (Path II). Kitamura et al. explored that; MeV proton beams (Figure 4.6) have been successfully applied as ionizing radiation to induce graft polymerization of acrylonitrile to prepare amidoxime-type adsorbents on polyethylene film substrates (Kitamura et al., 2004). Mazzei et al. also reported that polypropylene (PP) films were preirradiated in the air using a 25 MeV proton beam and grafted with methyl methacrylate (MMA), styrene and acrylic acid (AAc) (Mazzei et al., 2003). When the ion beams irradiated polymer films, the H-molecule liberated and formed a chemically activated trail in the substrate. A variety of carbon radicals, C=C bonds, C-C bonds, and cross-linking form in the chemically activated trail.

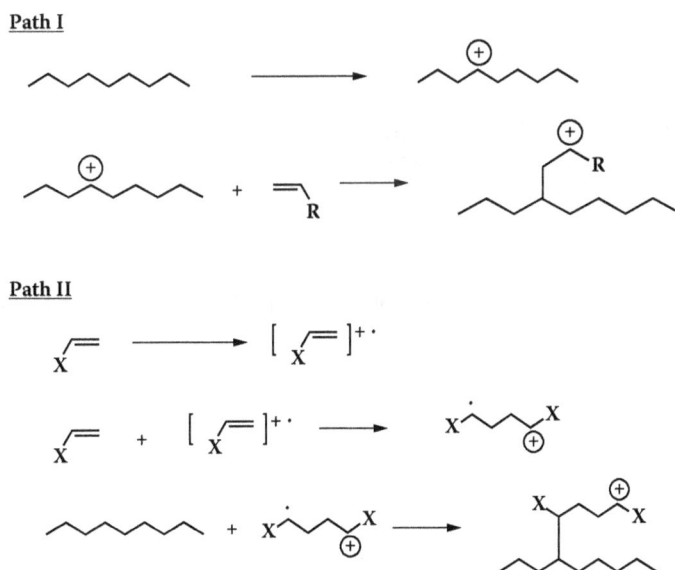

FIGURE 4.5   Path I. Reaction mechanism of cationic grafting initiated from the backbone. Path II. Reaction mechanism of cationic grafting initiated through monomer. (Bhattacharyya and Maldas, 1984). (Adaptable from *Prog. Polym. Sci.* 29 (2004) 767–814).

Radiation-induced grafting differs from chemical initiation in many aspects. In a mechanistic way, as in a radiation technique, the initiator is not required; free radical formation is on the backbone polymer/monomer, whereas in a chemical method, a free radical form first onto the initiator and then is transferred to the monomer/polymer backbone. Chemical initiation frequently brings about problems arising from local heating of the initiator, an effect that is absent in the formation of free-radical sites by radiation, which is only dependent upon the absorption of high-energy radiation. Because of the large penetration power of higher-energy radiation, methods using radiation initiation provide the opportunity to perform grafting at different depths of the base polymer matrix. Furthermore, the molecular weight of the products can be better controlled in a radiation technique and these are also capable of initiation in solid substrates.

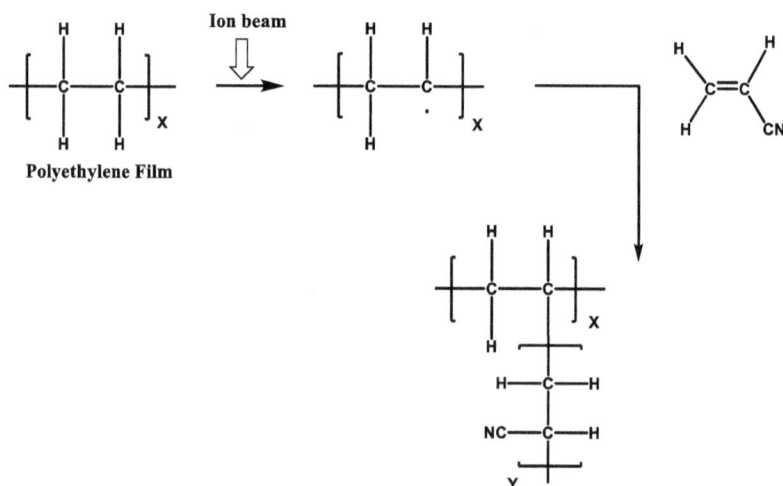

FIGURE 4.6   Reaction scheme of grafting of acrylonitrile onto polyethylene film employing MeV proton beams. (Kitamura et al., 2004). (Adaptable from A. Kitamura et al. *Radiation Physics and Chemistry* 69 (2004) 171–178).

4.2.5.3   Photochemical grafting

When a chromophore on a macromolecule absorbs light, it goes to an excited state, which may dissociate into reactive free radicals, from where the grafting process is initiated. If the absorption of light does not lead to the formation of free-radical sites through bond rupture, this process can be promoted by the addition of photosensitizers, e.g., benzoin ethyl ether, dye; such as anthraquinone sulphonate and aromatic ketones (such as benzophenone, xanthone). That means the grafting process by a photochemical technique can proceed in two ways: with or without a sensitizer (Bellobono et al., 1981; Kubota et al., 2001; Peng and Cheng, 2001). The mechanism without a sensitizer involves the generation of free radicals on the backbone, which react with the monomer free radical to form the grafted copolymer. On the other hand, in the mechanism "with sensitizer", the sensitizer forms free radicals, which can undergo diffusion so that they abstract hydrogen atoms from the base polymer, producing the radical sites required for grafting (Figure 4.7). Uchida et al. (Uchida, Uyama and Ikada, 1990) reported the graft polymerization of acrylamide (AM) on the surface of poly (ethylene terephthalate) (PET) film, with simultaneous UV irradiation without a photosensitizer. In this process, the addition of NaIO$_4$ plays a crucial role in the removal of any oxygen present in the oxygenate monomer solution, as oxygen is the strong inhibitor of radical polymerization.

(I) Without Sensitizer

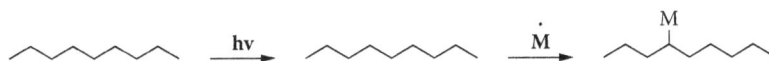

(II) With Sensitizer

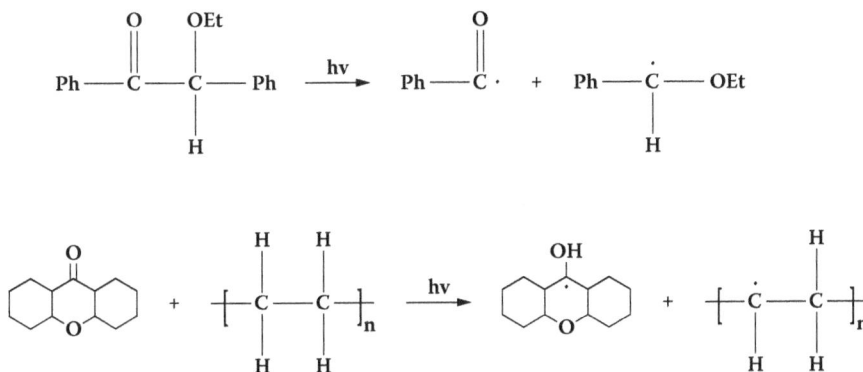

FIGURE 4.7 Mechanism for photochemical grafting method. (Bhattacharya and Misra, 2004). (Adaptable from *Prog. Polym. Sci.* 29 (2004) 767–814).

4.2.5.4 Plasma radiation grafting

Plasma surface activation and grafting is a two-step process of plasma surface activation followed by the incorporation of functional groups and reactive sites to the surface activated polymer. Plasma surface activation is used to be done by exposing the polymer to a noble gas to form the free radicals on the polymer surface which can initiate the grafting process as well as the incorporation of a new functional group. Free radicals' density on the plasma-activated polymer surface plays a crucial role in determining the degree of grafting. The low density of free radicals results in incomplete grafting, whereas high free radical density results in loss of reactivity due to recombination/termination reactions as the radicals are in close proximity and easily accessible to each other.

The exposure of the plasma-activated polymer surface for subsequent grafting reaction can be achieved through the following processes: (1) immersing the plasma-activated surface directly into a solution with a reactive agent; (2) direct contact of the plasma-activated surface with monomers from gas phase; or (3) by exposure to oxygen, etc. Shourgashti et al. used oxygen plasma to activate the surface of polyurethane film and exposed the plasma-induced polyurethane for subsequent grafting reaction to the solution of vinyl siloxy terminated polydimethylsiloxane polymer (Shourgashti, Khorasani and Khosroshahi, 2010). Wang et al. reported during a plasma-induced vapor phase grafting polymerization (PIVPGP) of acrylic acid on cotton fabrics to enhance silver nanoparticle loading on the fabric, which exhibited excellent laundry durability (Wang et al., 2017).

4.2.5.5 Enzymatic grafting

The enzymatic grafting technique is pretty new. The principle involved is that an enzyme initiates the chemical/electrochemical grafting reaction (Chen et al., 2000). Cosnier et al. (Cosnier et al., 2000) reported the enzymatic grafting on poly(dicarbazole-N-hydroxysuccinimide) film, thionine and toluidine blue have been irreversibly bound to the poly(dicarbazole) backbone, and the grafting of polyphenol oxidase (PPO) on polydicarbazole has been reported.

4.3 RADIATION-INDUCED POLYMERIZATION

Polymerization and modification of polymers using radiation processes do not require any metal catalyst or initiator, which are often toxic and needed to be purified or removed from the final product. Importantly, by the radiation process, the polymerization and the polymer modification can be done without using any solvent or in benign solvent (like water, etc.). So, the radiation process is considered one of the most useful green processes and clean technologies to prepare different polymeric materials (Lawrence, 2003).

There are different types of radiation processes. The gamma (γ) radiation technique is one of the most widely used methods for grafting and modifying the polymers (Lawrence, 2003). This γ radiation was found to induce polymerization in simple monomers in the 1930s. Electron beam radiation (EBR) process is also a very useful technique in which the polymerization or polymer modifications are initiated electrical discharges. The major research works on radiation polymerization or radiation modification started in the 1950s. UV radiation is a relatively soft radiation technique that is widely used in paints and coating technology.

In all the above processes, active radicals and/or ions are generated, which leads to initiated reactions, such as polymerization, cross-linking, or polymer degradation under different conditions. In this context, we should know about the two terms: ionization and excitation. If the radiation energy (particulate or EM) transfers energy to an orbiting electron which is equal to or greater than that electron's binding energy, then the electron is ejected from the atom. The positively charged atom and the ejected electron are called an ion pair. Radiation with energy greater than 13.6 eV is considered ionizing. If the radiation energy is less than the binding energy of the electron, then the electron got excited with the external radiation energy and raised to a higher energy state (to more outer orbital) but is not ejected. In both excitation and ionization, an electron shell is left with a "hole" that must be filled in order to return the atom to lower energy. The filling of these holes comprises an important source of secondary radiation called characteristic radiation.

4.3.1 GAMMA RADIATION

The gamma rays (γ-rays) are generated by the radioactive decay when the secondary electron interacts with the irradiated materials. The energy of γ-rays is much higher than UV light and slightly higher than X-rays. Gamma-rays ionize matter by three main processes: (1) the photoelectric effect, (2) compton scattering, and (3) pair production. A gamma ray is a packet of electromagnetic energy photons. These photons are the most energetic ones in the electromagnetic spectrum and have energies in the range from 10^4 to 10^7 eV. Gamma rays produce ionization (electron disruption) in their path. The most widely used radioactive isotopes used in medical and industrial applications are cobalt 60 (^{60}Co), cesium 137 (^{137}Cs), and iridium 192 (^{192}Ir).

The most common applications of γ-rays are sterilization of single-use medical supplies, elimination of organisms from pharmaceuticals, microbial reduction in and on consumer products, cancer treatment, and processing of polymers (cross-linking, polymerization, degradation, etc.). Hasanain et al. described the use of gamma irradiation for the sterilization of pharmaceutics formulations and medical devices, such as syringes, needles, and cannulas (Hasanain et al., 2014). Jeong et al. prepared polypyrrole/polyvinylpyrrolidone (PPy/PVP) hydrogel with different concentrations and content through polymerization and cross-linking induced by gamma-ray irradiation at 25 kGy to optimize the mechanical properties of the resulting PPy/PVP hydrogel (Jeong et al., 2020). Nechifor et al studied the property modification of a porous polymer membrane, obtained from poly(hydroxy-urethane) (PHU) and poly(vinyl alcohol) (PVA) induced by gamma radiations (Matter, 2009). Also, when the dose increases, the gamma radiations induce changes in the chemical structure followed by degradation. Xu et al. studied fabrication of polystyrene (PS) microspheres coated with β-cyclodextrin (β-CD) via gamma-ray induced emulsion polymerization in a ternary system of styrene/β-CD/water (St/β-CD/water) (Xu et al., 2012). Wang et al. reported the synthesis of snowman-like magnetic/nonmagnetic nanocomposite asymmetric particles (SMNAPs) via seeded emulsion polymerization initiated by γ-ray radiation (Wang et al., 2012). Basak and co-author (Basak et al., 2010) have investigated the adhesion between vulcanized EPDM and unmodified NR through a co-curing technique. Here, surface modification of EPDM vulcanizate has been carried out using gamma radiation in the presence and absence of trimethylolpropane triacrylate (TMPTA) as a sensitizer. Seko et al. developed a fibrous polymer adsorbent for the selective removal of radioactive Cs from contaminated waters by introducing ammonium 12-molybdophosphate (AMP) onto polyethylene nonwoven fabric through a radiation-induced emulsion graft polymerization technique (Seko et al., 2018).

4.3.2 ELECTRON BEAM (EB) RADIATION

The ionizing EB radiation produced by the electron beam accelerator produces extremely reactive species like free radicals and ions which modify the molecular structure of polymeric material leading to cross-linking or degradation. Typically, both of these processes (cross-linking or degradation) occur simultaneously and which will dominate mainly depends on the chemical structure of the polymer and applied irradiation conditions like irradiation dose, environment, and temperature. The strength of the EBs as radiation is controlled by accelerating voltage and beam current.

EB curing of the polymer has been carried out in a wide range of fields, for example in electrical wire and cable application, coating and packaging industries, etc. Work on electron beam radiation of different polymer and their blends over the last few decades has revealed that radiation curing has got the potential to play an important role in the modification of polymer blend systems. Recently, Dutta et al. investigated the effects of electron beam radiation (EBR) on the blends of ethylene vinylacetate/thermoplastic polyurethane (EVA/TPU) at two different blend ratios prepared via a melt blending technique. They studied that the influences of electron beam irradiation on morphology, mechanical, thermal, and dynamic mechanical properties of the blend system in details and to develop an economically feasible useful blend system (Dutta et al., 2015). Murray et al. studied the influence of electron beam irradiation on surface properties of medical-grade polyurethane through a diverse array of characterization techniques such as FTIR, XRD, SEM, DSC, and dynamic frequency sweep (Murray et al., 2013). Majumder et al. described the effect of electron beam–induced surface treatment in the presence of a polyfunctional monomer on the permeability behavior of EPDM rubber (Sen Majumder et al., 2000). Mandal et al. studied the EB-modification of chlorinated polyethylene (CPE) and how it was affected by radiation dose as well as the sensitizers. The modified CPE was characterized by using Fourier transform infrared (FT-IR), differential scanning calorimetry (DSC), thermogravimetric analysis (TGA), and elemental analysis (Mandal, Mongal and Singha, 2011). Datta et al. investigated the influence of electron beam radiation at two radiation doses (25 and 50 kGy) on the properties of a high vinyl (~50%) styrene-butadiene-styrene block copolymer filled at different concentrations of a nano-silica as the reinforcing agent. The role of a silane coupling agent in the presence of nano-silica to achieve better properties was also investigated (Datta et al., 2011). Halder et al. investigated successful grafting of methyl methacrylate (MMA) and butyl acrylate (BA) on butyl rubber (IIR) by using EB radiation (Haldar and Singha, 2006).

4.3.3 X-RAY IRRADIATION

X-rays are essentially produced in the form of bremsstrahlung radiation (photons) from a metallic target, only when an electron beam is allowed to impose on a metallic target. The yield is determined by the atomic number, the thickness of the target, and the current of the incident EB: the higher the atomic number of the target, the higher the X-ray intensity. The bremsstrahlung photons are not monoenergetic but have a distribution over a range of energies. The penetration of X-rays is similar to that of γ-rays, but dose distribution may be more complicated (Makuuchi and Cheng, 2011).

X-rays are primarily used for diagnostic radiography in medicine and crystallography. Other distinguished uses are X-ray microscopic analysis, X-ray fluorescence as an analytical method, and industrial radiography for the inspection of tires and welds (European Environment Agency (EEA), 2019). Recent reports and patents cover the use of X-rays in the processing of variety of parts made from polymers, particularly in the form of advanced fiber-reinforced composites (Berejka and Eberle, 2002; Wang et al., 2008).

4.4 CONCLUSION

This chapter presents a very detailed introduction of different types of radiation. The radiation process is considered one of the most useful green processes to prepare different polymeric materials. Here, we mainly focused on the polymer modification using radiation. In recent days, radiation-induced modification of polymer surfaces has gained widespread acceptance. So, in this chapter, we have reviewed a few important works carried out on polymers utilizing different types of radiation. Almost all polymers modified using radiation have been summarized to exhibit similar changes like change in adhesion on the surface, surface roughness, hydrophobicity, wettability, surface energy, and surface reactivity, etc. Also, we have sketched different mechanistic approaches for grafting by different radiation techniques. Hence, the above-discussed radiation-treated polymers can be utilized for medical and industrial applications.

ACKNOWLEDGMENT

TSP is thankful to the Indian Institute of Technology, Kharagpur, for the Institute fellowship. BARC, Mumbai is gratefully acknowledged for funding several projects related to radiation curing and processing of polymers or elastomers to which NKS was involved.

REFERENCES

Aich, S., Bhattacharyya, A. and Basu, S. (1997) 'Fluorescence polarization of N-vinyl carbazole grafted on cellulose acetate film and its electron transfer reaction with 1,4-dicyanobenzene', *Radiation Physics and Chemistry*, 50(4), pp. 347–354. doi: 10.1016/S0969-806X(97)00046-7.

Badey, J. P. *et al.* (1994) 'Surface modification of polytetrafluoroethylene by microwave plasma downstream treatment', *Polymer*, 35(12), pp. 2472–2479. doi: 10.1016/0032-3861(94)90365-4.

Bai, X. and Mu, Z. (2019) 'Research and Development of Microware Irradiation Technology in Polymer Synthesis', *IOP Conference Series: Materials Science and Engineering*, 562(1). doi: 10.1088/1757-899X/562/1/012120.

Basak, G. C. *et al.* (2010) 'Characterization of EPDM vulcanizates modified with gamma irradiation and trichloroisocyanuric acid and their adhesion behavior with natural rubber', *Journal of Adhesion*, 86(3), pp. 306–334. doi: 10.1080/00218460903479305.

Basu, S. *et al.* (1994) 'Spectroscopic evidences for grafting of N-vinyl carbazole on cellulose-acetate film', *Journal of Polymer Science Part A: Polymer Chemistry*, 32(12), pp. 2251–2255. doi: 10.1002/pola.1994.080321206.

Bellobono, I. R. *et al.* (1981) 'Photochemical grafting of acrylated azo dyes onto polymeric surfaces, IV', Die Angewandte Makromolekulare Chemie, 100(1532), pp. 135–146.

Berejka, A. J. and Eberle, C. (2002) 'Electron beam curing of composites in North America', *Radiation Physics and Chemistry*, 63(3–6), pp. 551–556. doi: 10.1016/S0969-806X(01)00553-9.

Bhattacharya, A., Das, A. and De, A. (1998) 'Structural influence on grafting of acrylamide based monomers on cellulose acetate', *Indian Journal of Chemical Technology*, 5(3), pp. 135–138.

Bhattacharya, A. and Misra, B. N. (2004) 'Grafting: A versatile means to modify polymers: Techniques, factors and applications', *Progress in Polymer Science (Oxford)*, 29(8), pp. 767–814. doi: 10.1016/j.progpolymsci.2004.05.002.

Bhattacharyya, S. N. and Maldas, D. (1984) 'Graft copolymerization onto cellulosics', *Progress in Polymer Science*, 10(2–3), pp. 171–270. doi: 10.1016/0079-6700(84)90002-9.

Bhattacharyya, S. N., Maldas, D. and Pandey, V. K. (1986) 'Radiation-induced graft copolymerization of N-vinyl carbazole and methyl methacrylate onto cellulose acetate film', *Journal of Polymer Science Part A: Polymer Chemistry*, 24(10), pp. 2507–2515. doi: 10.1002/pola.1986.080241011.

Breuer, J., Metev, S. and Sepold, G. (1995) 'Photolytical pretreatment of polymers with UV-laser radiation', *Materials and Manufacturing Processes*, 10(2), pp. 229–239. doi: 10.1080/10426919508935018.

Chen, J. *et al.* (2003) 'Grafting of polyethylene by γ -radiation grafting onto conductive carbon black and application as novel gas and solute sensors', *Radiation Physics and Chemistry*, 67(3–4), pp. 397–401. doi: 10.1016/S0969-806X(03)00074-4.

Chen, T. *et al.* (2000) 'Enzymatic grafting of hexyloxyphenol onto chitosan to alter surface and rheological properties', *Biotechnology and Bioengineering*, 70(5), pp. 564–573.

Choi, S. C. *et al.* (1999) 'Hydrophilic group formation on hydrocarbon polypropylene and polystyrene by ion-assisted reaction in an O2 environment', *Nuclear Instruments and Methods in Physics Research, Section B: Beam Interactions with Materials and Atoms*, 152(2), pp. 291–300. doi: 10.1016/S0168-583X(99)00120-2.

Choi, W., Koh, S. and Jung, H. (1996) 'Surface chemical reaction between polycarbonate and kilo-electron-volt energy Ar + ion in oxygen environment', *Journal of Vacuum Science & Technology A: Vacuum, Surfaces, and Films*, 14(4), pp. 2366–2371. doi: 10.1116/1.580024.

Choi, Y. J., Kim, M. S. and Noh, I. (2007) 'Surface modification of a polytetrafluoroethylene film with cyclotron ion beams and its evaluation', *Surface and Coatings Technology*, 201(9-11 SPEC. ISS.), pp. 5724–5728. doi: 10.1016/j.surfcoat.2006.07.063.

Clough, R. L. (2001) 'High-energy radiation and polymers: A review of commercial processes and emerging applications', *Nuclear Instruments and Methods in Physics Research, Section B: Beam Interactions with Materials and Atoms*, 185(1–4), pp. 8–33. doi: 10.1016/S0168-583X(01)00966-1.

Cosnier, S. *et al.* (2000) 'Poly(dicarbazole-N-hydroxysuccinimide) film: A new polymer for the reagentless grafting of enzymes and redox mediators', *Electrochemistry Communications*, 2(12), pp. 827–831. doi: 10.1016/S1388-2481(00)00131-4.

Dadsetan, M., Mirzadeh, H. and Sharifi, N. (1999) 'Effect of CO2 laser radiation on the surface properties of polyethylene terephthalate', *Radiation Physics and Chemistry*, 56(5–6), pp. 597–604. doi: 10.1016/S0969-806X(99)00293-5.

Datta, S. *et al.* (2011) 'Influence of silica and electron beam radiation on the properties of a high vinyl styrene-butadiene-styrene block copolymer', *Journal of Polymer Research*, 18(5), pp. 1185–1196. doi: 10.1007/s10965-010-9522-1.

Decker, C. (1998) 'The use of UV irradiation in polymerization', *Polymer International*, 45(2), pp. 133–141.

Drobny, J.G. *et al.* (2013) *Ionizing Radiation and Polymers*, Elsevier Inc., 83-99. doi: 10.1016/b978-1-4557-7881-2.00011-0.

Dutta, J. *et al.* (2015) 'Exploring the influence of electron beam irradiation on the morphology, physico-mechanical, thermal behaviour and performance properties of EVA and TPU blends', *RSC Advances*, 5(52), pp. 41563–41575. doi: 10.1039/c5ra03381k.

Dworjanyn, P. A. *et al.* (1993) 'Novel additives for accelerating radiation grafting and curing reactions', *Radiation Physics and Chemistry*, 42(1–3), pp. 31–40. doi: 10.1016/0969-806X(93)90198-4.

El-saftawy, A. A. (2016) 'REGULATING THE PERFORMANCE PARAMETERS OF ACCELERATED PARTICLES. PhD thesis. In Department of Physics Faculty of Science Zagazig University', (January 2013).

European Environment Agency (EEA), Air quality in Europe-2019 report, (2019) 済無*No Title No Title*. doi: 10.1017/CBO9781107415324.004.

Fabbri, P. and Messori, M. (2017) *Surface Modification of Polymers: Chemical, Physical, and Biological Routes, Modification of Polymer Properties*. Elsevier Inc. doi: 10.1016/B978-0-323-44353-1.00005-1.

Garnett, J. L. (1979) 'Grafting', *Radiation Physics and Chemistry*, 14(1–2), pp. 79–99. doi: 10.1016/0146-5724(79)90014-1.

Ginn, B. T., Steinbock, O. and Distillers, F., (2003) Polymer Surface Modification Using Microwave-Oven-Generated Plasma, 2003, 19, pp. 8117–8118.

Haldar, S. K. and Singha, N. K. (2006) 'Grafting of butyl acrylate and methyl methacrylate on butyl rubber using electron beam radiation', *Journal of Applied Polymer Science*, 101(3), pp. 1340–1346. doi: 10.1002/app.23005.

Hasanain, F. *et al.* (2014) 'Gamma sterilization of pharmaceuticals – A review of the irradiation of excipients, active pharmaceutical ingredients, and final drug product formulations', *PDA Journal of Pharmaceutical Science and Technology*, 68(2), pp. 113–137. doi: 10.5731/pdajpst.2014.00955.

Huglin, B. and Johnson, B. L. Role of Cations in Radiation Grafting and Homopolymerization, *Journal of Polymer Science*, 7, pp. 1379–1384.

Hung, G. W. C. (1974) 'The radiation chemistry of macromolecules. Vol. 2', *Microchemical Journal*, 19(1), pp. 100–102. doi: 10.1016/0026-265x(74)90106-4.

Jaganathan, S. K. *et al.* (2015) 'Review: Radiation-induced surface modification of polymers for biomaterial application', *Journal of Materials Science*, 50(5), pp. 2007–2018. doi: 10.1007/s10853-014-8718-x.

Jeong, J. *et al.* (2020) Gamma Ray-Induced Polymerization and Cross-linking for Optimization of PPy/ PVP Hydrogel as Biomaterial, Polymers 2022, 12, 111.10.3390/polym12010111.

Junkar, I., Cvelbar, U. and Lehocky, M. (2011) 'Plasma treatment of biomedical materials', *Materiali in Tehnologije*, 45(3), pp. 221–226.

Kabanov, V. Y. *et al.* (2009) 'Radiation chemistry of polymers', *High Energy Chemistry*, 43(1), pp. 1–18. doi: 10.1134/S0018143909010019.

Kaur, I. *et al.* (1993) 'Viscometric studies of starch-g-polyacrylamide composites', *Journal of Applied Polymer Science*, 47(7), pp. 1165–1174. doi: 10.1002/app.1993.070470704.

Kaur, I. *et al.* (1994) 'Graft copolymerization of acrylonitrile and methacrylonitrile onto gelatin by mutual irradiation method', *Journal of Applied Polymer Science*, 54(8), pp. 1131–1139. doi: 10.1002/app.1994. 070540817.

Kaur, I. *et al.* (1996) 'Viscometric, conductometric, and ultrasonic studies of gelatin-g-polyacrylamide composite', *Journal of Applied Polymer Science*, 59(3), pp. 389–397. doi: https://onlinelibrary.wiley.com/ doi/10.1002/(SICI)1097-4628(19960118)59:3%3C389::AID-APP2%3E3.0.CO;2-L.

Keshel, S. H. *et al.* (2011) 'The relationship between cellular adhesion and surface roughness for polyurethane modified by microwave plasma radiation', *International Journal of Nanomedicine*, 6, pp. 641–647. doi: 10.2147/ijn.s17180.

Khorasani, M. T., Mirzadeh, H. and Kermani, Z. (2005) 'Wettability of porous polydimethylsiloxane surface: Morphology study', *Applied Surface Science*, 242(3–4), pp. 339–345. doi: 10.1016/j.apsusc.2004.08.035.

Kitamura, A. *et al.* (2004) 'Application of proton beams to radiation-induced graft polymerization for making amidoxime-type adsorbents', *Radiation Physics and Chemistry*, 69(2), pp. 171–178. doi: 10.1016/ S0969-806X(03)00439-0.

Kubota, H. *et al.* (2001) 'Introduction of stimuli-responsive polymers into regenerated cellulose film by means of photografting', *European Polymer Journal*, 37(7), pp. 1367–1372. doi: 10.1016/S0014-3057(00)00257-3.

Kudo, H. (ed.). (2018) Radiation Applications. An Advanced Course in Nuclear Engineering Book 7. 10.1 007/978-981-10-7350-2, Springer.

Kudoh, H., Sasuga, T. and Seguchi, T. (1996) 'High energy ion irradiation effects on mechanical properties of polymeric materials', *Radiation Physics and Chemistry*, 48(5), pp. 545–548. doi: 10.1016/0969-806X (96)00077-1.

Lawrence, C. A. (2003) *Spun Yarn Technology*. Library of Congress Cataloging-in-Publication Data, 10.12 01/9780203009581.

Madrid, J. F. and Abad, L. V. (2015) 'Modification of microcrystalline cellulose by gamma radiation-induced grafting', *Radiation Physics and Chemistry*, 115, pp. 143–147. doi: 10.1016/j.radphyschem.2015.06.025.

Makuuchi, K. and Cheng, S. (2011) *Radiation Processing of Polymer Materials and Its Industrial Applications*, John Wiley & Sons. doi: 10.1002/9781118162798.

Mandal, K., Mongal, N. R. and Singha, N. K. (2011) 'Electron beam modification of chlorinated polyethylene', *Rubber Chemistry and Technology*, 84(2), pp. 137–146. doi: 10.5254/1.3560021.

Marmey, P., Porté, M. C. and Baquey, C. (2003) 'PVDF multifilament yarns grafted with polystyrene induced by γ-irradition: Influence of the grafting parameters on the mechanical properties', *Nuclear Instruments and Methods in Physics Research, Section B: Beam Interactions with Materials and Atoms*, 208(1–4), pp. 429–433. doi: 10.1016/S0168-583X(03)00887-5.

Nechifor, C.D. (2009) 'The Influence of gamma radiations on physico-chemical', *Rom. Journ. Phys.*, 54, pp. 349–359.

Mazzei, R. *et al.* (2003) 'Radiation grafting of different monomers onto PP foils irradiated with a 25 MeV proton beam', *Nuclear Instruments and Methods in Physics Research, Section B: Beam Interactions with Materials and Atoms*, 208(1–4), pp. 411–415. doi: 10.1016/S0168-583X(03)00621-9.

Melnig, V. *et al.* (2005) 'Improvement of polyurethane surface biocompatibility by plasma and ion beam techniques', *Journal of Optoelectronics and Advanced Materials*, 7(5), pp. 2521–2528.

Mirzadeh, H., Ekbatani, A. R. and Katbab, A. A. (1996) 'Surface modification of ethylene-propylene rubber by laser grafting of acrylic acid', *Iranian Polymer Journal (English Edition)*, 5(4), pp. 225–230.

Moad, G. *et al.* (2005) 'Advances in RAFT polymerization: The synthesis of polymers with defined end-groups', *Polymer*, 46(19 SPEC. ISS.), pp. 8458–8468. doi: 10.1016/j.polymer.2004.12.061.

Moad, G., Rizzardo, E. and Thang, S. H. (2008) 'Radical addition-fragmentation chemistry in polymer synthesis', *Polymer*, 49(5), pp. 1079–1131. doi: 10.1016/j.polymer.2007.11.020.

Murray, K. A. *et al.* (2013) 'The influence of electron beam irradiation conducted in air on the thermal, chemical, structural and surface properties of medical grade polyurethane', *European Polymer Journal*, 49(7), pp. 1782–1795. doi: 10.1016/j.eurpolymj.2013.03.034.

Mutel, B. *et al.* (1988) 'Treatment of polymer surfaces by low pressure microwave plasmas', *Revue de Physique Appliquée*, 23(7), pp. 1253–1255. doi: 10.1051/rphysap:019880023070125300.

Nasef, M. M. and Güven, O. (2012) 'Radiation-grafted copolymers for separation and purification purposes: Status, challenges and future directions', *Progress in Polymer Science*, 37(12), pp. 1597–1656. doi: 10.1016/j.progpolymsci.2012.07.004.

Nasef, M. M. and Hegazy, E. S. A. (2004) 'Preparation and applications of ion exchange membranes by radiation-induced graft copolymerization of polar monomers onto non-polar films', *Progress in Polymer Science (Oxford)*, 29(6), pp. 499–561. doi: 10.1016/j.progpolymsci.2004.01.003.

O'Connell, C. *et al.* (2009) 'Investigation of the hydrophobic recovery of various polymeric biomaterials after 172 nm UV treatment using contact angle, surface free energy and XPS measurements', *Applied Surface Science*, 255(8), pp. 4405–4413. doi: 10.1016/j.apsusc.2008.11.034.

Osada, Y. (1988) 'Plasma polymerization and plasma treatment of polymers. Review', *Polymer Science U.S.S.R.*, 30(9), pp. 1922–1941. doi: 10.1016/0032-3950(88)90041-X.

Ozdemir, M. and Sadikoglu, H. (1998) 'A new and emerging technology: Laser-induced surface modification of polymers', *Trends in Food Science and Technology*, 9(4), pp. 159–167. doi: 10.1016/S0924-2244 (98)00035-1.

Peng, T. and Cheng, Y. L. (2001) 'PNIPAAm and PMAA co-grafted porous PE membranes: Living radical co-grafting mechanism and multi-stimuli responsive permeability', *Polymer*, 42(5), pp. 2091–2100. doi: 10.1016/S0032-3861(00)00369-4.

Pieracci, J. *et al.* (2000) 'UV-assisted graft polymerization of N-vinyl-2-pyrrolidinone onto poly(ether sulfone) ultrafiltration membranes: Comparison of dip versus immersion modification techniques', *Chemistry of Materials*, 12(8), pp. 2123–2133. doi: 10.1021/cm9907864.

Rabiei, A. and Sandukas, S. (2013) 'Processing and evaluation of bioactive coatings on polymeric implants', *Journal of Biomedical Materials Research - Part A*, 101 A(9), pp. 2621–2629. doi: 10.1002/jbm.a.34557.

Rajajeyaganthan, R. *et al.* (2011) 'Surface modification of synthetic polymers using UV photochemistry in the presence of reactive vapors', *Macromolecular Symposia*, 299–300(1), pp. 175–182. doi: 10.1002/masy.200900128.

Řezníčková, A. *et al.* (2011) 'Comparison of glow argon plasma-induced surface changes of thermoplastic polymers', *Nuclear Instruments and Methods in Physics Research, Section B: Beam Interactions with Materials and Atoms*, 269(2), pp. 83–88. doi: 10.1016/j.nimb.2010.11.018.

Rogalski, M. S. and Palmer, S. B. (2014) Charlesby A., 'Atomic radiation', Pergamon Press Ltd., *Advanced University Physics*, pp. 745–761. doi: 10.1201/b16000-42

Rosiak, J. M. *et al.* (1995) 'Radiation formation of hydrogels for biomedical purposes. Some remarks and comments', *Radiation Physics and Chemistry*, 46(2), pp. 161–168. doi: 10.1016/0969-806X(95)00007-K.

Schnabel, W. et al . (1982). *Polymer Degradation: Principles and Practical Applications*. Oxford University Press. 10.1002/pol.1982.130200907

Seguchi, T. *et al.* (1999) 'Ion beam irradiation effect on polymers. LET dependence on the chemical reactions and change of mechanical properties', *Nuclear Instruments and Methods in Physics Research, Section B: Beam Interactions with Materials and Atoms*, 151(1–4), pp. 154–160. doi: 10.1016/S0168-583X(99) 00132-9.

Seko, N. *et al.* (2018) 'Development of a water purifier for radioactive cesium removal from contaminated natural water by radiation-induced graft polymerization', *Radiation Physics and Chemistry*, 143(April 2017), pp. 33–37. doi: 10.1016/j.radphyschem.2017.09.007.

Sen Majumder, P. *et al.* (2000) 'Barrier properties of electron-beam-modified EPDM rubber', *Journal of Applied Polymer Science*, 75(6), pp. 784–795. doi: https://onlinelibrary.wiley.com/doi/10.1002/(SICI) 1097-4628(20000207)75:6%3C784::AID-APP7%3E3.0.CO;2-9.

Shourgashti, Z., Khorasani, M. T. and Khosroshahi, S. M. E. (2010) 'Plasma-induced grafting of poly-dimethylsiloxane onto polyurethane surface: Characterization and in vitro assay', *Radiation Physics and Chemistry*, 79(9), pp. 947–952. doi: 10.1016/j.radphyschem.2010.04.007.

Siegel, J. *et al.* (2008) 'Ablation and water etching of plasma-treated polymers', *Radiation Effects and Defects in Solids*, 163(9), pp. 779–788. doi: 10.1080/10420150801969654.

Singh, A. and Silverman, J. (Hanser Publishers, Oxford University Press, New York1993) Radiation Processing of Polymer, Journal of Polymer Science: Part A: Polymer Chemistry, 3499–1993. doi: 10.1002/pola.1993. 080311341.

Slepička, P. *et al.* (2013) 'Surface characterization of plasma treated polymers for applications as biocompatible carriers', *Express Polymer Letters*, 7(6), pp. 535–545. doi: 10.3144/expresspolymlett.2013.50.

Subedi, D., Tyata, R. and Rimal, D. (2009) 'Effect of uv-treatment on the wettability of polycarbonate', *Kathmandu University Journal of Science, Engineering and* Technology, 5(II), pp. 37–41. Available at: http://www.ku. edu.np/kuset/vol5_no2/4_dec_DPsubedi_EffectofUVtreatment_07_Jan_2009_edited_original.pdf.

Suggs, A. (2002) 'Kr-F laser surface treatment of poly (methyl methacrylate, glycol-modified poly (ethylene terephthalate), and polytetrafluoroethylene for enhanced adhesion of escherichia coli K-12', Master thesis. Available at: http://hdl.handle.net/10919/35182, *Stress: The International Journal on the Biology of Stress*.

Toan Le, Q. *et al.* (2013) 'Mechanism of modification of fluorocarbon polymer by ultraviolet irradiation in oxygen atmosphere', *ECS Journal of Solid State Science and Technology*, 2(5), pp. N93–N98. doi: 10.1149/2.003305jss.

Uchida, E., Uyama, Y. and Ikada, Y. (1990) 'A novel method for graft polymerization onto poly(ethylene terephthalate) film surface by UV irradiation without degassing', *Journal of Applied Polymer Science*, 41(3–4), pp. 677–687. doi: 10.1002/app.1990.070410317.

Uflyand, I. E. *et al.* (1992) 'Polymers containing metal chelate units. VI. Post-graft polymerization of metal chelate monomers based on 1-phenyl-4-methylpent-4-en-1,3-dione', *Reactive Polymers*, 17(3), pp. 289–296. doi: 10.1016/0923-1137(92)90275-7.

Wang, B. *et al.* (2015) 'Surface wettability modification of cyclic olefin polymer by direct femtosecond laser irradiation', *Nanomaterials*, 5(3), pp. 1442–1453. doi: 10.3390/nano5031442.

Wang, C. H. *et al.* (2008) 'Optimizing the size and surface properties of polyethylene glycol (PEG)-gold nanoparticles by intense x-ray irradiation', *Journal of Physics D: Applied Physics*, 41(19). doi: 10. 1088/0022-3727/41/19/195301.

Wang, C. X. *et al.* (2017) 'In situ synthesis of silver nanoparticles on the cotton fabrics modified by plasma induced vapor phase graft polymerization of acrylic acid for durable multifunction', *Applied Surface Science*, 396, pp. 1840–1848. doi: 10.1016/j.apsusc.2016.11.173.

Wang, F. W. *et al.* (2012) 'Synthesis of snowman-like magnetic/nonmagnetic nanocomposite asymmetric particles via seeded emulsion polymerization initiated by γ-ray radiation', *Journal of Polymer Science, Part A: Polymer Chemistry*, 50(22), pp. 4599–4611. doi: 10.1002/pola.26274.

Weibel, D. E. *et al.* (2009) 'Ultraviolet-induced surface modification of polyurethane films in the presence of oxygen or acrylic acid vapors', *Thin Solid Films*, 517(18), pp. 5489–5495. doi: 10.1016/j.tsf.2009.03.204.

Xu, D. *et al.* (2012) 'Synthesis and characterization of β-CD-coated polystyrene microspheres by γ-ray radiation emulsion polymerization', *Macromolecular Rapid Communications*, 33(22), pp. 1945–1951. doi: 10.1002/marc.201200437.

Yamaki, T. *et al.* (2003) 'Radiation grafting of styrene into cross-linked PTEE films and subsequent sulfonation for fuel cell applications', *Radiation Physics and Chemistry*, 67(3–4), pp. 403–407. doi: 10.1016/ S0969-806X(03)00075-6.

Yamaki, T. *et al.* (2019) 'Nano-Structure Controlled Polymer Electrolyte Membranes for Fuel Cell Applications Prepared by Ion Beam Irradiation', *ECS Transactions*, 3(1), pp. 103–112. doi: 10.1149/1. 2356129.

Zatikyan, L. L. and Grushevskaya, L. N. (1988) Modification of polyolefins and polyvinylchloride by the radiation-induced graft polymerization, Int J Radiat Appl Instrum Part C. Radiation Physics and Chemistry, 31, pp. 579–585.

Zhang, Y. *et al.* (2000) 'Surface modification of poly(tetrafluoroethylene) films by low energy Ar+ ion-beam activation and UV-induced graft copolymerization', *Nuclear Instruments and Methods in Physics Research, Section B: Beam Interactions with Materials and Atoms*, 168(1), pp. 29–39. doi: 10.1016/ S0168-583X(99)00719-3.

5 Radiation-induced graft copolymerization – A facile technology for polymer surface modification and applications

Subhadeep Chakraborty, Soumen Sardar, and Abhijit Bandyopadhyay[*]

Department of Polymer Science & Technology, University of Calcutta, Kolkata, India

*Correspondence author: abpoly@caluniv.ac.in; abhijitbandyopadhyay@yahoo.co.in

CONTENTS

DOI: 10.1201/9781003321910-5

5.1 INTRODUCTION

The most suitable way for producing functionalized polymeric material is graft copolymerization, which has been popularized amongst researchers over the years. It is by this reaction, that side chains of monomers are tangled from the polymeric backbone, imparting interesting properties as follows (D. Kumar, Pandey, Raj, & Kumar, 2017) (Figure 5.1):

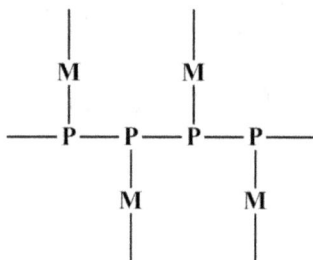

FIGURE 5.1   Graft copolymer; P – polysaccharide backbone, M – monomer.

Grafting yield or degree of grafting is defined as the increase of mass in the side chain of the graft polymer calculated in percentage (%). There are several methods available for the process of graft copolymerization listed as below:

    i. Plasma-initiated graft copolymerization
    ii. UV-light induced photo-initiation graft copolymerization
    iii. Microwave mediated graft copolymerization
    iv. Chemical initiation grafting
    v. Thermo-mechanical initiation (melt grafting)
    vi. Radiation-induced graft copolymerization (RIGC) with ionizing radiation

There are two different routes for reaching the structure of graft copolymers A_n and B_m, shown as follows (Setia, 2017) (Figures 5.2 and 5.3):

    a. By cross-linking of two unlike polymers A_n and B_m
    b. By initiating radicals on the surfaces of polymer A_n and grafting via monomer B

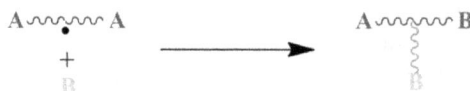

FIGURE 5.2   Cross-linking of two unlike polymers.

FIGURE 5.3   Radical initiation on the surface of the polymer.

Both the routes have one thing in common; it is that both reaction pathways require activation of the polymer molecules. It can be achieved via a variety of physical and chemical processes but high-energy radiation is particularly suited for this purpose.

The surfaces of the polysaccharide can be modified by the conventional grafting method, but it has some major drawbacks. The molecular weight and polydispersity of the grafted chain cannot be controlled in the conventional grafting method. It is due to this reason that the surface is covered with varying lengths of grafted chains (Manaila & Craciun, 2019).

Overall, radiation-induced graft copolymerization has major advantages, including modification of the surface up to the bulk of the backbone polymers, which is not possible in the case of photo- or plasma-initiated copolymerization that imparts the modification only on the surfaces of the polymer. Moreover, the simplicity and flexibility of reaction initiation with commercially available ionizing radiation sources make the RGIC more advantageous for the preparation and modification of polymeric surfaces. Commercial radiation sources such as electromagnetic radiation of γ-rays (Co-60) and particulate radiation such as electron beam (EB) are used (Barsbay & Güven, 2019; Ma, Peng, & Zhai, 2018; Sun & Chmielewski, 2017).

The so-called 'hero' RIGC proceeds, following the pathway of the free radical mechanism. When high-energy radiation falls on the surface of the virgin polymer, free radical sites are generated on its surface. These free radicals start the process of polymerization of vinyl monomers, resulting in the formation of graft copolymers (Ishihara, Asai, & Saito, 2020) (Figure 5.4).

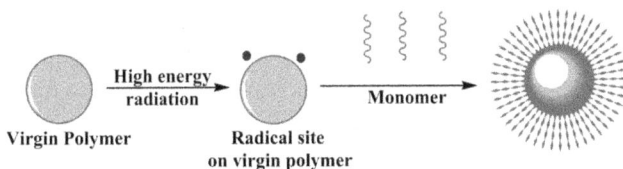

FIGURE 5.4 Graft copolymerization using high-energy radiation.

When the ionizing radiation interacts with the polymeric backbone, it generates uniformly distributed carbon-centered radicals that initiate the process of graft copolymerization up to a thickness of a few mm (Salamun et al., 2016). Ion beams with high linear energy transfer and short stopping range can be used to irradiate the polymer, but if the radiation has spatial variation in the order of nm, it appears as a spot and it gets distinctly separated from the area which are not irradiated. Those spots can be further grafted via polymerization directly or after being etched chemically (Zubair, Nasef, Ting, Abdullah, & Ahmad, 2020). However, if the number of ions is not sufficient enough to have the desired grafting yield, the polymeric backbone is again irradiated with high-energy radiation to generate more amounts of radicals (D. K. Kakati, Bora, & Deka, 2015).

The degree of grafting can also be controlled or rather adjusted by selecting irradiation and parameters of the reaction to develop the required copolymers for the specific purpose. Radiation-induced graft copolymerization can be performed over a wide range of temperatures, including the temperatures for monomers in bulk, solution, or emulsion (Ke, Drache, Gohs, Kunz, & Beuermann, 2018).

The modification of the surfaces of a virgin polymer by grafting using this process has some advantages that are listed below:

- The process is relatively simple, clean, and also repeatable.
- Polymers which may be in the form of film, fiber, membrane, fabric, or powder can be surface modified and the monomers which can be polymerized by free radicals such as styrene, vinyl amide, vinyl chloride, vinyl acetates, methacrylates, etc. can be grafted easily by this process.
- The graft copolymer thus obtained contains relatively no impurities.
- The adjustment in the degree of grafting can be done by controlling the reaction conditions such as concentration of monomer, the temperature of the reaction, the atmosphere under which the reaction is performed, type of solvent used, the addition of homopolymerization suppressor, the addition of acid, and also the radiation exposure parameters such as radiation type, dosage rate of radiation, time of irradiation, and dose.
- Radiation-induced grafting does not require the initiation of grafting by heating unlike chemical methods and hence the structure of the polymer substrate is not distorted and hence monomers that are sensitive to temperature can be grafted safely.

- It can achieve narrow molecular weight distribution by controlling the molecular weights of the grafted chains (pendant groups).
- A wide range of functional monomers can be used for grafting e.g., methacrylates, acrylates, acrylamides, styrene, dienes, acrylonitriles, and vinyl monomers.
- Polymers with well-defined topologies and architecture can be prepared e.g., gradient, block, star, comb, or hyperbranched copolymers.
- Formations of block copolymers are also possible by macromolecule chain extension by the addition of other monomers (Benoit, Hawker, Huang, Lin, & Russell, 2000; Chapiro, 1977; N. Kakati, Assanvo, & Kalita, 2019).

From the application point of view, graft copolymers prepared by radiation-induced grafting have numerous uses, such as:

- The hydrophilic and/or hydrophobic characteristics of the polymer can be tuned (improved or reduced (Cheon & Jeun, 2019).
- The blood compatibility of the medical devices can be modified (B. Singh & Singh, 2019).
- In tissue engineering, cell adhesion and growth on scaffolds can be influenced by grafting (B. Singh & Kumar, 2018).
- Lubricity of implants can be improved (Pino-Ramos, Flores-Rojas, Alvarez-Lorenzo, Concheiro, & Bucio, 2018).
- Membranes that are used in batteries, fuel cells, chromatography can be designed by this method (Pathania, Sharma, & Sethi, 2017).
- Metal ion adsorbents can also be prepared (Saleh, Ibrahim, Elsharma, Metwally, & Siyam, 2018).
- The graft copolymers can also be used as efficient flocculants for wastewater treatment (Mittal, Ray, & Okamoto, 2016).

5.2 CLASSIFICATION OF RADIATION-GRAFTED COPOLYMERS

It is necessary to identify the different physical forms (morphologies) of these materials along with their functionality to have a better understanding of the role of the radiation-induced functionalized graft copolymers in the field of separation and purification (Barsbay & Güven, 2019) (Figure 5.5).

The classification of the grafted copolymers can be done under various categories, as follows:

1. Origin of the material
2. Chemical nature of the grafted copolymers
3. Mechanism of separation and
4. Functions

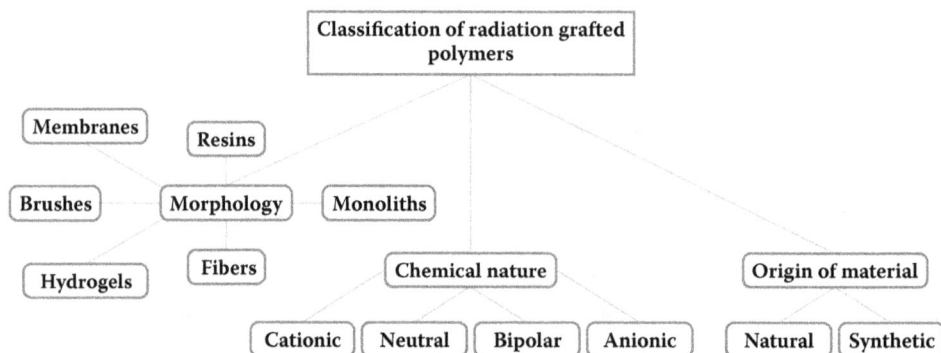

FIGURE 5.5 Classification of radiation-grafted copolymers.

According to the morphological viewpoint, grafted copolymers can be available in the form of a bead, gel, fiber, fabric, and membranes, along with varying physical and chemical characteristics.

According to the chemical point of view, radiation-grafted copolymers can be available in the ionic form (cationic, anionic, or bipolar) and neutral based on the type of monomers used for grafting or the reaction performed to impart ionic properties (Le Moigne, Sonnier, El Hage, & Rouif, 2017).

Based on the origin of the material, most of the radiation-induced grafted copolymers are synthetic e.g., polymers having modified backbone such as PE (polyethylene), PP (polypropylene), PVDF (polyvinylidene fluoride), PTFE (poly(tetrafluoroethylene)), and PETFE (poly(ethylene-co-tetrafluoroethylene)). Moreover, by modifying the backbone of the natural polymers (polysaccharides) such as cellulose, alginate, guar gum, xanthan gum, starch, agar-agar, chitosan, etc., grafted copolymers can be formed (Madrid, Cabalar, & Abad, 2018).

Based on function or mechanism of separation, radiation-induced grafted copolymers can be classified into several types such as ion-exchangers, polymer-ligand exchangers, chelating copolymers, hydrogels, affinity grafted copolymers, and polymeric electrolytes (Nasef & Güven, 2012).

5.3 CLASSIFICATION OF THE RADIATION-INDUCED GRAFTING METHODS

Radiation-induced graft copolymerization can be performed by two main methods:

1. Simultaneous irradiation (mutual or direct)
2. Pre-irradiation method

In the first method i.e., simultaneous irradiation, the polymeric backbone is irradiated keeping it immersed in the pure monomer or solution of the monomer. The side reactions i.e., homopolymerization may be suppressed by giving low irradiation dosage or adding inhibitors. Figure 5.6 illustrates the various procedures of radiation-induced grafting.

Whereas in the pre-irradiation method, the polymeric backbone is irradiated in a vacuum or inert atmosphere to form free radicals and subsequently brought in contact with monomers under controlled conditions (Omichi, Ueki, Seko, & Maekawa, 2019).

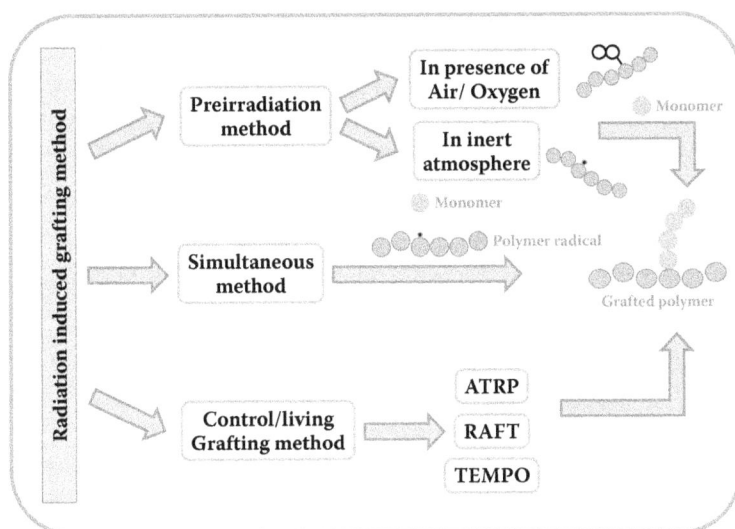

FIGURE 5.6 Classification of radiation-induced grafting.

5.3.1 Mutual or simultaneous irradiation method

The polymeric material in this method is immersed in the solution of monomer (or in the pure polymer itself) and it is exposed to the radiation source causing ionization. The process of irradiation can be done in an inert atmosphere (in presence of argon or nitrogen) using gamma rays as a source of radiation. It is the most common and simplest method chosen for modifying the surfaces of polymeric materials and is also suitable for polymeric materials that are sensitive to radiation (Saito, Fujiwara, & Sugo, 2018).

The general mechanism for the polymerization process is as follows:

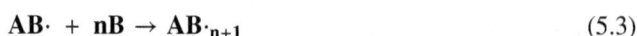

$$AH \rightarrow A\cdot + H\cdot \tag{5.1}$$

$$A\cdot + B \rightarrow AB\cdot \tag{5.2}$$

$$AB\cdot + nB \rightarrow AB\cdot_{n+1} \tag{5.3}$$

When ionizing radiation falls on the surface of the polymeric materials, active sites are generated on the surface (Equation 5.1). Then the radical, A, reacts with a molecule of monomer B, initiating the process of graft copolymerization (Equation 5.2). After the process of initiation, the propagation of the polymeric chain occurs and this occurs by successive addition of the monomer to the polymeric radical (Equation 5.3).

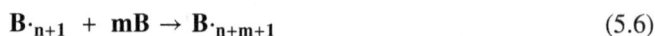

$$B \rightarrow B\cdot \tag{5.4}$$

$$B\cdot + nB \rightarrow B\cdot_{n+1} \tag{5.5}$$

$$B\cdot_{n+1} + mB \rightarrow B\cdot_{n+m+1} \tag{5.6}$$

However, it is due to the exposure of the grafting mixture to the radiation causing ionization, active sites can be generated in the polymeric backbone, monomer, and to a lesser extent in the solvent. It is for this reason; side reactions take place which diminishes the degree of grafting due to less involvement of monomer in the process of grafting. Rather, homopolymerization occurs simultaneously, leading to the formation of a homopolymer, which is not at all desirable and it occurs as shown above (Madrid, Abad, Yamanobe, & Seko, 2017).

Practically, to suppress the degree of homopolymerization and increase the grafting efficiency, reaction conditions should be optimized, which can be done by adding inhibitors such as salts of Cu or Fe or inorganic acids in small amounts, selecting solvents, the lower dose of radiation, etc.

This method of grafting is usually preferred when:

1. The polymeric backbone is highly sensitive to radiation.
2. When fast grafting is required (Moawia, Nasef, Mohamed, & Ripin, 2016).

5.3.2 Pre-irradiation method

Grafting via the pre-irradiation method involves the following steps:

1. The polymeric backbone is irradiated with ionizing radiation to generate free radical species acting as active species for grafting in the presence of air or an inert atmosphere.
2. The reaction of the monomer is initiated with the irradiated polymeric backbone.
3. Heating will then support the propagation reaction with peroxides.

$$AH \rightarrow A\cdot + H\cdot \tag{5.7}$$

$$A\cdot + O_2 \rightarrow AOO\cdot \tag{5.8}$$

$$AOO\cdot + AH \rightarrow AOOH + A\cdot \tag{5.9}$$

$$AOOH \rightarrow AO\cdot + OH\cdot \tag{5.10}$$

$$AO\cdot + B \rightarrow AOB\cdot \tag{5.11}$$

$$AOB\cdot + nB \rightarrow AOB\cdot_{n+1} \tag{5.12}$$

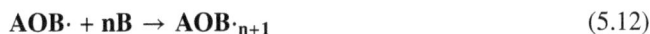

Peroxide radicals are generated due to oxidation of alkyl radical when the polymeric substrate is exposed to ionizing radiation. This again when coming in contact with the polymeric backbone will lead to the formation of hydrogen peroxides. The hydrogen peroxides then get decomposed to alkoxy radicals on heating. The generated radical will then undergo grafting with the monomers. It is due to the stability of the hydrogen peroxides, that the grafting reactions can be performed sometime later after the irradiation process (Sharif et al., 2013).

Another method of doing pre-irradiation is by using species that are non-oxidizable upon irradiation. However, in that case, for achieving a high concentration of radical, the polymer should be irradiated using a high dose and the reaction must be performed under an inert atmosphere or in a vacuum. The grafting reaction should be done immediately after the irradiation of the polymeric backbone. An advantage of grafting via pre-irradiation is the formation of less amount of homopolymer; however, the hydroperoxide species can undergo homopolymerization upon thermal decomposition.

In this method of grafting, only the polymeric backbone is exposed to the ionizing radiation and not the monomer and hence the formation of homopolymer is somewhat suppressed.

* **Differences between mutual and pre-irradiation methods:** Table 5.1

TABLE 5.1
Difference Between Mutual and Pre-Irradiation Method

Parameters	Mutual/Simultaneous	Pre-Irradiation
Type of radiation	Gamma, EB	EB, gamma
Absorbed dose	Low (10 kGy or less)	High (100 kGy and more)
Dose rate	Low [kGy/h]	High [kGy/s]
Irradiation time	Long [h]	Short [min] or [s]
Atmosphere	Inert gas	Air/inert gas/vacuum
Side reactions: homopolymerization	High	Low
Temperature	Ambient	Irradiation: ambient/low temperature, graft polymerization: high temperature (peroxide decomposition)

5.4 RAFT-MEDIATED GRAFTING

The field of graft copolymerization by controlled radical polymerization has been explored widely by researchers e.g., atom transfer radical polymerization (ATRP), nitroxide mediated polymerization (NMP), and reversible addition-fragmentation chain transfer polymerization (RAFT).

The polymers with narrow molecular weight distribution can be successfully obtained from this method. But among all these techniques of grafting, only RAFT-mediated graft copolymerization can be done with gamma or electron beam radiation. The radicals formed due to radiation on the polymeric backbone react with monomers in presence of RAFT agent and graft copolymer results. Unlike RAFT, all other methods require either heat or photoinitiator for the formation of active sites on the polymeric backbone (Figures 5.7 and 5.8).

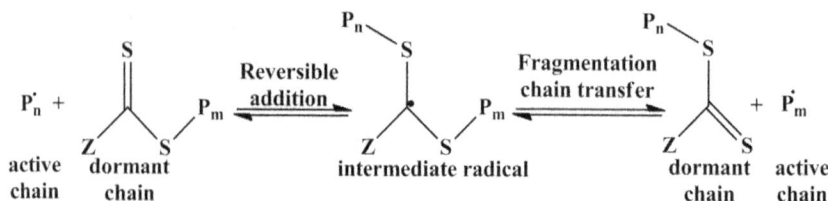

FIGURE 5.7   Radiation-mediated grafting using RAFT polymerization.

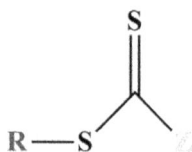

FIGURE 5.8   RAFT agent.

RAFT-mediated graft polymerization by radiation can be performed by both a mutual or simultaneous method and pre-irradiation method. The only difference occurs due to the presence of the RAFT agent. The Z group determines the rate of addition and fragmentation rates in polymerization, whereas the R group gets cleaved to form R, which can reinitiate polymerization.

The vital step of RAFT polymerization is the chain equilibrium reaction between the radical and the propagating polymeric chain (Kodama, Barsbay, & Güven, 2014).

5.5   FACTORS CONTROLLING THE RADIATION-INDUCED GRAFTING

The radiation-induced graft copolymerization can be controlled by varying the parameters of the reaction and the environment. The effect of various factors in radiation-induced graft copolymerization has been discussed in detail below.

5.5.1   Nature of monomer

In the mutual grafting method, if the G-radical for the polymeric backbone is greater than the G-radical for monomer, then radiation-induced grafting is more favored. Monomers that tend to swell the component will probably improve bulk grafting. The smaller-sized monomers tend to migrate through the surface and promote grafting. It's necessary to know whether a monomer is polymerizing in bulk or solution homogeneously or heterogeneously. Control of the grafting is very difficult in the case of heterogeneously polymerizing monomers. Grafting, which is a chemical reaction, depends on the experimental parameters such as substrate's composition, monomer form and concentration, dose and dose rate, additive presence, temperature, storage conditions, and atmosphere. GMA, a typical reactive monomer, is also known as a monomer of precursors. GMA is one of the most flexible and convenient functionalized vinyl monomers. The ester group in GMA's molecular structure is resistant to hydrolysis at ambient temperature within a wide range of pHs (Pal, Majumder, & Bandyopadhyay, 2016).

5.5.1.1 Monomers for radiation-grafted membrane

TABLE 5.2
Different Types of Monomers

Monomer	Abbreviation
Acrylic monomer	
Acrylic acid	AA
Methyl acrylate	MA
Acrylamide	AAm
Acrylonitrile	AN
Methyl methacrylate	MAA
Methacrylate	MAA
Glycidyl methacrylate	GMA
Vinyl acetate	VAc
Vinyl monomer	
Styrene	St
4-Vinylpyridine	4-VP
2-Vinylpyridine	2-VP
N-Vinylpyrrolidone	N-VP

The following hydrophilic monomers were commonly used on membrane preparation: acrylic acid, 4-vinyl pyridine, N-vinylpyrrolidone. Such monomers polymerize much faster than styrene, and experimental conditions in which the grafting reaction is not impeded by excessive homopolymerization have to be sought in each case. With a surfactant, the water-insoluble monomer is dispersed in water to form stable emulsions. Emulsion stability is dependent on the ratio of monomer to surfactant. The stability of the emulsion during the greasing process should be preserved. The grafting of emulsions improves the yield of grafting. It is an environment-friendly alternative to bio-solvent grafting (Elsharma, Saleh, Abou-Elmagd, Metwally, & Siyam, 2019).

5.5.2 NATURE OF POLYMERS

The surface modification of all the types of synthetic as well as natural polymers can be done by the employment of the radiation-induced grafting method. The degree of grafting and the properties of the final grafted product depends mainly on the structure and morphology of the polymer. Hence the polymer needs to meet those requirements. The polymers that degrade when high irradiation falls on their surface should not be treated with a high dose of irradiation. Hence, preirradiation is generally not suitable for that kind of polymer. Moreover, chemical and radiation-induced grafting can be done for radiation-resistant materials (Awang et al., 2017).

The amount of radicals formed in the polymeric substrate strongly determines the degree of grafting in the radiation-grafted polymers. The radicals formed as a result of irradiation and the subsequent conversion of radicals (paramagnetic species formed as a result of irradiation) can be efficiently tracked by the use of electron paramagnetic resonance (EPR) spectroscopy. This method is very useful for comparing the number of radicals formed in the polymeric substrate and also to predict the behavior of the grafting e.g., when different polymers are irradiated at the same dosage amount, keeping all the conditions the same, the amount of radicals formed in the polymeric substrates varied from one polymer to the other. The same is the case for the degree of grafting of the radiation-induced polymers using the mutual method of grafting (Thayyath S. Anirudhan & Senan, 2011).

The crystallinity of the polymer also affects the degree of grafting. When natural polymers are used as a backbone, then the generation of the radicals should be done more carefully as the radiation may cause degradation of the polymer. Moreover, when the polymeric waste is used as grafting material, then proper cleaning should be done.

5.5.3 NATURE OF SOLVENT

The degree of grafting for the same monomer, dosage, atmospheric reaction, temperature, etc. can differ considerably depending on the solvent, its properties, and behavior during the grafting process. The form of solvent influences not only the efficiency of the grafting but also the homogeneity of the grafted chains, which can be obtained with good swelling solvents. Hydrophilic monomers are commonly used for grafting water and alcohol. However, the solvent should be selected experimentally for each grafting mechanism, in particular for the grafting processes mediated by RAFT (Aminabhavi & Deshmukh, 2016).

5.5.4 NATURE OF RADIATION

Grafting caused by radiation may often be done with gamma radiation (^{60}Co, ^{137}Cs) and EB radiation. Compared to the pre-irradiation process, where the chosen energy is an electron beam, the generation of free radicals is normally achieved using a gamma source. The key distinction between these two forms of radiation is the amount of exposure i.e., described as the dose administered within a given period. The dose limit for isotope sources is comparatively small (kGy/h), whereas it is high (kGy/s) for electron beams. With a gamma source, however, the irradiation period is significantly longer (h) than by utilizing a high-energy electron beam (min or s). This aspect influences the degree of grafting considerably. The higher the dosage, the more radicals are produced in the polymeric material which has a direct effect on the grafting degree. The dosage intensity influences radical concentration and lifetime, degradation, and period after that the growth graft chains are terminated. In mutual or simultaneous grafting, an improvement in the dosage intensity results in poorer grafting performance with the same dosage, as the high radical concentration improves their recombination, contributing to a faster termination cycle and further homopolymerization (Bhunia, Goswami, Chattopadhyay, & Bandyopadhyay, 2011).

5.5.5 TEMPERATURE

The temperature has a direct effect on the degree of grafting in the grafted copolymers. The degree of grafting can be calculated as:

$$\text{Degree of grafting} = \frac{Weight\ of\ the\ graft\ copolymer - weight\ of\ the\ polysaccharide\ taken}{Weight\ of\ polysaccharide\ taken} \times 100\%$$

Various types of species, such as allyl, alkyl, and peroxy radicals, are formed upon irradiation by electron beam onto the surface of the polymeric backbone. The stability of each kind of radical depends on the temperature.

5.5.6 TEMPERATURE FOR IRRADIATION

In the case of pre-irradiation grafting, the irradiation of the polymeric backbone should be done at a lower temperature to resist the recombination of the radical itself during the process of irradiation. Then, the radicals are stored and, after that, the graft polymerization will be carried out with those stored radicals.

This is not the case for mutual grafting. In the case of mutual grafting, the temperature at which the polymeric backbone is irradiated with the electron beam affects the degree of grafting. If the polymer on which grafting is to be performed, is irradiated with a temperature greater than that of its glass transition temperature, T_g, then the mobility of the polymeric chains increases heavily due to which they migrate towards the sites where grafting will occur, thus decreasing the degree of grafting. On the other hand, if the irradiation is done at a temperature below the glass transition temperature, then the polymeric chains remain rigid and hence the monomer should approach the polymer to an effective contact. Therefore, the grafting should be performed at an optimum temperature, and hence its selection is necessary to get a higher grafting efficiency (Ghobashy & Khafaga, 2017).

5.5.7 TEMPERATURE OF GRAFTING

Generally, with increasing the temperature of the reaction, the rate of reaction increases. Thus, for graft copolymerization reaction, with increasing the temperature, the rate of grafting increases. Several reactions occur simultaneously, which can be accelerated by increasing the temperature and leads to their termination. Moreover, when the temperature is increased:

i. The process of grafting increases due to the change in the kinetics of grafting.
ii. Decomposition of the peroxide radicals and generation of the active site for polymerization and, hence, chain propagation (Golshaei & Güven, 2017).

5.6 RESISTING THE FORMATION OF HOMO-POLYMERS

In radiation-mediated grafting, the role of the homopolymerization suppressor is very important. When ionizing radiation falls on the surface of the reactants, active radicals are supposed to be formed on the surfaces of both the polymeric backbone and the monomer. The homopolymer formation reaction will be favorable if the concentration of monomer is high as well as high reactivity. But, this is not desirable since in that case surface modification of the polymeric backbone will not be achieved. So, the suppression of homopolymerization can be achieved by adding a very small amount of suppressor. Inorganic metal salts such as iron (II) chloride ($FeCl_2$), copper (II) chloride ($CuCl_2$), copper (II) sulfate($CuSO_4$), and ammonium iron (II) sulfate (Fe$(NH_4)_2(SO_4)_2(6H_2O)$) (Mohr's salt) can be used as additives for suppressing homopolymer formation. These additives are used in aqueous solutions of monomers and after dissolving they become hydroxyl radical scavengers, thus reducing homopolymerization. There is an optimum concentration of these salts to be used as an additive. Moreover, if a larger concentration of additives will be added, it will affect the degree of grafting. Some studies also showed that, with increasing the amount of copper and iron salts as additives, the degree of grafting increases (Hassan & Zohdy, 2018).

5.7 ROLE OF ADDITIVES

The degree of grafting can be increased by the addition of mineral acids such as hydrochloric acid, sulfuric acid, etc. to the system of the monomer solution. Under the condition of low pH, water radiolysis generates hydrated electrons that get trapped by protons to convert them into hydrogen atoms. Those species have a greater ability to abstract hydrogen from the polymeric backbone than their precursor electrons. As a result, the yield of radicals formed in polymeric chains increases, leading to the enhancement of active sites that initiate grafting.

Also, the addition of poly-functional monomers can accelerate the reaction of polymerization.

5.8   CHARACTERIZATION OF THE GRAFT COPOLYMERS

The grafted copolymers should be extracted using suitable solvents before performing the characterization after successful completion of the graft copolymerization process to remove the unreacted monomers, free polymers, and residual additives. After washing, the product is dried to a constant weight. The degree of grafting (DG) is determined by using the following equation:

$$\text{Degree of Grafting } (\%) = \frac{W - W_0}{W_0} \times 100\%$$

where
   W = weight of the grafted polymer
   W_0 = weight of non-grafted polymer

There are numerous analytical techniques available to gather information regarding the surface of a polymer, which can be in the range of micron to the nanometer.

The comparative analysis of the grafted and non-grafted samples can confirm the effects of grafting: attenuated total reflection Fourier transforms infrared (ATR-FTIR) spectroscopy, Raman spectroscopy, X-ray photoelectron spectroscopy (XPS), atomic force microscopy (AFM), atomic scanning electron microscopy (SEM), and contact angle (CA) measurements.

The analysis of structural and morphological changes can be done by differential scanning calorimetric (DSC), thermo-gravimetric analysis (TGA), dynamic mechanical analysis (DMA), X-ray diffraction (XRD), and others (Oraby, Senna, Elsayed, & Gobara, 2016).

5.9   CLASSIFICATION OF RADIATION-INDUCED GRAFT COPOLYMERS

It's crucial to grasp the many physical forms (morphologies) of radiation-grafted functional copolymers, as well as their functionality, to comprehend their significance in purification and separation. Grafted copolymers may be categorized into distinct groups based on their material origin, chemical nature, separation process, and functionality. Grafted copolymers come in a variety of morphological forms including beads, gels, fibers, fabrics, and membranes, each with its own set of physical and chemical properties. Radiation-grafted copolymers can be ionic or neutral chemically, depending on the kind of grafted monomer and the chemical treatment used to bestow the ionic properties.

The bulk of radiation-grafted copolymers are synthetic; for example, modified synthetic backbone polymers like polyethylene, polypropylene (PP), poly(vinylidene fluoride) (PVDF), poly (tetrafluoroethylene) (PTFE), and poly(ethylene-co-tetrafluoroethylene) (ETFE). Graft copolymers with modified natural backbone polymers such as cellulose, starch, alginate, and chitosan are also used. Grafted copolymers can be divided into numerous categories based on their function or method of separation, such as ion exchangers, polymer-ligand exchangers, chelating copolymers, hydrogels, affinity grafts copolymers, and polymer electrolytes.

5.9.1   Grafted synthetic adsorbents

Graft adsorbents are chelating polymers with modified synthetic polymer matrices including immobilized functional groups. These adsorbents may be made from a wide range of functional group-containing polymers (e.g., polyamines, polyacrylonitrile, polyacrylamides, and PAA) as well as backbone polymers (e.g., PP, PE, and PTFE) as carriers for functional groups with high affinity for certain metals. This provides a lot of possibilities for custom selective separation applications. The development of complexes between the functional groups on the graft adsorbents and the metal ions to be removed is attributable to the graft adsorbents' adsorption processes (Yantasee et al., 2004).

Graft adsorbents come in a variety of shapes and sizes, including beads, fibers, textiles, and membranes. RIGC of active monomers such as AAc or AN on polymer films or fibers (PE and PP) is frequently used to make composite metal complex adsorbents, which are then treated with a solution of metal ions.

An alternate technique for preparing the chelating adsorbent is to graft a nonpolar monomer, such as GMA, followed by epoxy ring-opening by chemical treatment with disodium iminodiacetate and washing in a suitable metal solution. The presence of carboxylate groups in the grafted layers causes metal acrylates complexes to develop, which vary in valence depending on the metal ion. RIGC of carboxylate monomers complexes different metal ions, including Fe^{3+}, Co^{2+}, Ni^{2+}, Cr^{3+}, Mn^{2+}, Ag^{+}, Cu^{2+}, Cd^{2+}, and Rh^{2+}, onto various backbone polymers, followed by chemical treatments with metal solutions (Choi & Nho, 2000; Turmanova & Atanassov, 2007; Yang, Peng, Wang, & Liu, 2010).

5.9.2 GRAFTED BIOADSORBENTS

Grafted bioadsorbents are chelating copolymers with natural polymer (polysaccharide) backbones that have been chemically changed with ionic groups, such as chitosan, alginate, starch, and cellulose. The instability of physical and chemical characteristics of their native forms including some functional groups (e.g., acetamido groups of chitin, hydroxyls in polysaccharides, and primarily carboxyls and sulfates in polysaccharides of marine algae) resulted in low selectivity of these bioadsorbents (J. P. Chen & Yang, 2005).

Because bioadsorbents are less expensive than synthetic ion exchange resins, interest in chemically modifying natural polymers to make them is developing. Several graft chemical groups were applied to polysaccharides to provide ionic properties, including phosphate, sulfonate, carboxyl, amido, amino, and hydroxyl groups, with cross-linking enhancing the bioadsorbents' durability.

Chitosan-based bioadsorbents are effective resins used in a variety of environmental remediation applications, including the selective removal of heavy metals, ions, and dyes. Because of their flexibility, bioadsorbents can be employed in a variety of forms including flakes, powder, gels, beads, and fibers. Due to their limited surface area and lack of porosity, flake and powder forms of chitosan-based adsorbents are less appropriate for use in industrial-scale columns, resulting in a high-pressure drop. Modified beads made from cross-linked chitosan, on the other hand, have high porosity, wide surface area, and good stability in acidic environments, making them the best choice for industrial applications (Varma, Deshpande, & Kennedy, 2004).

Grafted chitosan bioadsorbents were also made by grafting a variety of functional groups onto the chitosan backbone to increase the density of ionic sites and improve sorption selectivity for different metal ions. Ethylenimine, epichlorohydrin, AA, MA, carboxylic anhydrides, methylpyridine, crown ether, and ethylenediaminetetra acetic acid were among the newly grafted functional groups (Guibal, 2004).

5.9.3 HYDROGELS

Hydrogels are insoluble in water, and are very hydrophilic polymers with three-dimensional network topologies. Hydrogels absorb a large quantity of water, in the range of a few hundred to thousands of percent of their dry weight due to their super absorption power that allows for unrestricted water exchange among external polluted water preparations. Depending on the nature of the network, hydrogels may be split into two categories: (1) covalently cross-linked networks (chemical or permanent hydrogels) and (2) physical or reversible hydrogels generated by secondary interactions.

The cross-linking in permanent hydrogels may be done using UV light or ionizing radiation, which may be applied at room temperature on a wide range of hydrophilic polymer types, without the inclusion of initiators or other chemicals that leave undesirable residues. On the other hand, reversible hydrogels feature network architectures that are kept together by molecular entanglements and/or secondary forces such as ionic bonding, hydrogen bonding, or hydrophobic forces.

Due to the existence of clusters of molecular entanglements and hydrophobic or ionic related domains, these hydrogels are not homogenous. They also have faults in their transit networks due to the occurrence of free chain ends and loops (Hoffman, 2012).

Cross-linking always improves the stability of hydrogels. In this context, for the manufacture of hydrogels based on polysaccharides, a new cross-linking process using ionizing radiation (gamma-rays or EB) in a paste-like condition of the polymer was described (Kume, Nagasawa, & Yoshii, 2002; K. Singh, Ohlan, Saini, & Dhawan, 2008; Wach, Mitomo, Nagasawa, & Yoshii, 2003; Yoshii et al., 2003; L. Zhao, Mitomo, Nagasawa, Yoshii, & Kume, 2003).

For hydrogels, ionizing radiation is used. Cross-linking has several benefits over traditional approaches, including the ability to activate the reaction without the need for cross-linking chemicals (or initiators) and the ability to commence the reaction at room temperature. Varying the irradiation dosage may regulate the degree of cross-linking, which has a big impact on the characteristics of the hydrogels (Giammona, Pitarresi, Cavallaro, & Spadaro, 1999).

5.10   SEPARATION SYSTEMS

The use of radiation-grafted functional copolymers for separation or purification necessitates the selection of a suitable system/process to fulfill the productivity and selectivity requirements. A combination of factors relating to the separating polymer, process parameters, and treated stream properties determine the performance and efficiency of a separation or purification process. This comprises the separating polymer's chemical composition and shape, as well as process operating variables including flow modes, flow rates, and the physical and chemical characteristics of the treated stream. The separating polymer, for example, is the system's brain, controlling the separation process's selection and mode of operation.

Radiation-grafted functional copolymers come in four fundamental morphologies: (i) membranes, (ii) resins/beads, (iii) hydrogels, and (iv) fibers and textiles, as mentioned before. Applying such a wide range of morphologies for separation necessitates the development of engineering systems capable of controlling operating parameters and overall system performance to achieve high productivity and selectivity. The functional copolymers have been housed in a variety of separation system topologies.

Three major factors influence the process of selecting separating materials, system configuration, and operating mode: (1) the characteristics of the treated stream, such as the species present and their concentrations relative to the targeted species and pH, (2) the system operating parameters, particularly temperature and pressure, required to achieve the desired purity in the product, and (3) economic viability or the cost of operation. As a general guideline, build the separation system to fulfill the goal stream criteria while using a separating polymer that allows for optimal efficiency and cost-effectiveness. Because radiation-grafted functional copolymers may be made to order, there's a good chance that more than one technically successful material can be found to fulfill all separation goals.

5.11   APPLICATION OF RADIATION-INDUCED GRAFT COPOLYMERS

The field of separation and purification has emerged globally in the scientific world using the graft copolymers due to its bio-degradable nature and cheap materials needed for its preparation and further modifications (if needed) including its ease of preparation (Rajaram & Das, 2008) and low production cost (Halder & Islam, 2015). It is due to this reason, various types of copolymers have been developed and still being developed, including hydrogels, resins, fiber, membranes, and fabrics. The following separations and purification can be done with these copolymers:

a. Purification of solvent
b. Purification of solution
c. Ions extraction

Radiation-grafted copolymers can be applied to the separation of gaseous and non-aqueous mixtures (Ochoa-Segundo et al., 2020).

The first type of separation in aqueous media is done when dissolved ions are to be removed as impurities and the solvent is required in the pure state. In most cases, ions are considered to be impurities. Moreover, the solvent purification method does not require the use of an ion-selective membrane (González-Torres et al., 2016).

The second type of purification, the undesirable solutes that may be toxic, are to be removed from the product of a solution of a certain composition. The most common example is the removal of toxic metal ions or impurities from the industrial effluent. Another example is the removal of toxic elements from the blood by hemodialysis. The third type of purification is needed when a dissolved compound is to be extracted from the solution. An example of this kind of purification is the extraction of gold from leaching solutions and uranium and vanadium from seawater (Takács et al., 2005).

The different types of functional groups present in the graft copolymers can do different types of separation and purification as per the requirement (Vajihinejad, Gumfekar, Bazoubandi, Rostami Najafabadi, & Soares, 2019).

This involves desalination of brackish water and seawater, softening of rough water, processing caustic soda, acid extraction, water electrolysis, oxidation of solvents, isolation of solvent mixtures, elimination of heavy metal ions and pigments from wastewater, isolation of toxic acids from agricultural drainage, disposal of contaminants, separation of proteins and peptides, and purification of blood (hemodialysis) (Salehizadeh, Yan, & Farnood, 2018).

This chapter describes the application of radiation-grafted copolymers in the treatment of wastewater.

5.12 TREATMENT OF WASTEWATER WITH BIO-BASED FLOCCULATING AGENT

Because of the depleting natural supplies of fresh water and the scarcity in many distant regions, interest in water treatment is growing across the world. It is commonly known that the majority of the water on the planet (96%) is in the form of salty water found in the world's seas, and is thus unfit for human consumption. Furthermore, another 3% of water is trapped in glaciers and ice, leaving only 1% available for human use. A wide range of polymeric materials in various physical forms have been produced and used in a wide range of water treatment applications.

In water purification systems, radiation-grafted copolymers in the form of membranes and adsorbents have been utilized to remove ions, particles, organic compounds, and microorganisms. Membranes are made up of a wide range of copolymers with a wide range of properties such as pore size, molecular weight cutoff, and ion rejection. For water treatment applications, many separation techniques based on selective and nonselective membranes, such as electrodialysis, ion exchange, reverse osmosis, and ultrafiltration, have been developed.

A column containing synthetic resin beads is used in an ion-exchange technique for water treatment. Depending on the ion sites in the resin, the resin selectively adsorbs cations or anions from the treated water. The ion exchange process continues until all accessible exchange sites are filled, resulting in resin depletion and subsequent regeneration using appropriate chemicals.

Because reverse osmosis (RO) is the most cost-effective technology for removing 95%–99% of all impurities from water, RO membranes can effectively reject any particles, bacteria, and organics with a molecular weight greater than 300 Daltons (including pyrogens). There are already a great number of RO membranes available to suit a wide range of rejection needs.

The development of society and hence the so-called urbanization is mainly based on the development of industries and thus industrialization has been increased enormously. It is due to this, the environment gets polluted either directly or indirectly (Kadooka et al., 2017). The worst thing happens when the freshwater bodies get polluted and due to this the marine life is endangered (Lee, Robinson, & Chong, 2014). The discharged water when mixed with nearby water bodies has the full potential to destroy the aquatic life (Braşoveanu, Koleva, Nemţanu, Koleva, & Paneva, 2018).

Thus, it is a pertaining topic to treat wastewater before its discharge to make it free from toxic elements. Toxic elements may be present in the form of cations, anions, or suspended organic matters (B. Singh & Singh, 2020). Various processes are available for the treatment of wastewater including sedimentation, filtration, coagulation, membrane separation, microfiltration, etc. But most of these methods are either time-consuming or less effective or require the usage of the non-biodegradable counterpart (Bazoubandi & Soares, 2020). Flocculation in this case plays an important role (Mate, Mishra, & Srivastava, 2020). It is defined as the aggregation of particles either by charge neutralization or particle bridging and settling down of the agglomerate due to the force of gravity as the molecular mass increases (Kadooka et al., 2017). Polysaccharides in this case play a vital role due to their long-chain structures and more porosity, they can accommodate a large number of particles (Nandi, Changder, & Ghosh, 2019). But, the disadvantage of using polysaccharides as flocculating agents is their lower shelf life and low efficiency. For this reason, the polysaccharides are being modified using synthetic vinyl monomers which increases their radius of gyration as well as molecular weight (Pal et al., 2016). Thus, the grafting reaction is performed by using ionizing radiation and the advantage of grafting by this method is that it is eco-friendly and also requires much less time than chemical modification (Molatlhegi & Alagha, 2017).

Hegazy et al. (Hegazy, El-Rehim, & Shawky, 2000) had done the selective removal and recovery of metals from industrial wastewater. It is an issue for the environment as well as for the economy. There are a variety of heavy metals that should be removed from waste solutions before they are released into the environment. As a result, research on the fabrication of hydrophilic membranes with both anionic and cationic exchangers has been conducted. To achieve these properties in the required membranes, a trial was conducted using radiation graft copolymerization of binary monomers containing anionic and cationic exchangers such as acrylic acid/2- and 4-vinyl pyridine (AAc/2-VP) (AAc/4-VP) onto commercially available polymeric substrates such as low-density polyethylene (LDPE). The circumstances under which the grafting process will progress uniformly are specified. The produced grafted membranes were characterized and certain chosen features were evaluated, and the prospect of their practical usage in wastewater treatment from heavy and hazardous metals such as Pb, Zn, Cd, Fe, and so on was investigated. Using the atomic absorption approach, the metal uptake by such produced membranes was determined. The efficacy and durability of the membrane were examined. The LDPE-g-P (AAc/2VP) membranes had a greater absorption for a given metal than the LDPE-g-P (AAc/4VP) membranes. By treating the membrane with 0.1 N HCl for 2 hours at room temperature, the chelated metal ions were readily desorbed. The selectivity of the membrane towards different metals was determined using a combination of two or three metals in the same feed solution.

Anirudhan et al. (T. S. Anirudhan & Rejeena, 2012) conducted the graft copolymerization process of glycidyl methacrylate onto nanocellulose (NC) in the presence of ethyleneglycoldimethacrylate as a cross-linker followed by immobilization of poly(acrylic acid)-modified poly (glycidyl methacrylate)-grafted nanocellulose (PAPGNC) (acrylic acid). Thermogravimetric (TG) analysis, X-ray diffraction (XRD), scanning electron microscopy (SEM), and Fourier transform infrared (FTIR) investigations were used to describe the hydrogel. PAPGNC was tested for its ability to adsorb chicken egg white lysozyme (LYZ) from aqueous solutions. At pH 6.0, the maximum adsorption was discovered, and the adsorption capacity reached saturation in less than 2 hours. The kinetic data was discovered to follow a pseudo-second-order chemisorption model. The equilibrium data agrees well with the Langmuir and Jovanovich isotherm models, indicating that LYZ covers the PAPGNC surface monolayer. At 30°C, the maximum adsorption capacity was reported to be 148.42 mg/g using the Langmuir isotherm model. An endothermic adsorption process was discovered by thermodynamic analysis. With 0.1 M NaSCN, the spent adsorbent was successfully degraded. PAPGNC is a promising material for recovering LYZ from aqueous solutions, according to the results of this study.

Anirudhan et al. (Thayyath Sreenivasan Anirudhan & Rejeena, 2014) synthesized poly(acrylic acid-co-acrylamide-co22acrylamido-2-methyl-1-propane sulfonic acid)-grafted nanocellulose/poly (vinyl alcohol) composite. P(AA-co-AAm-co-AMPS)-g-NC/PVA is a new cellulose-based SAPC that might be used as a drug delivery vehicle. The antibiotic amoxicillin was chosen as a model medicine since it is used to treat Helicobacter pylori-induced peptic and duodenal ulcers. The graft copolymerization procedure was used to make P(AA-co-AAm-co-AMPS)-g-NC/PVA, which was characterized using FTIR, XRD, SEM, and DLS. Equilibrium swelling investigations were carried out to examine the SAPC's stimuli-response behavior, and it was discovered that equilibrium swelling was affected by pH, contact duration, temperature, ionic strength, cross-linker concentration, and PVA content. Using varying amounts of amoxicillin, the maximum drug encapsulation efficiency was discovered. Drug release tests were conducted in the simulated stomach and intestinal fluids, with the release peaking in the intestinal fluids after 4 hours. At a pH of 7.4, the drug release kinetics were studied using Peppas' potential equation and shown to follow a non-Fickian mechanism. As a result of the drug release tests, P(AA-co-AAm-co-AMPS)-g-NC/PVA appears to be a promising vehicle for in vitro amoxicillin delivery into the gastrointestinal tract.

Salmeiri et al. (Salmieri, Khan, Safrany, & Lacroix, 2015) studied, using gamma radiation, the effect of 2-hydroxyethyl methacrylate (HEMA) grafting on the structure and physicochemical properties of methylcellulose (MC)-based films. In an MC-based aqueous formulation containing 1% MC, 0.5% vegetable oil, 0.25% glycerol, and 0.025% Tween®80, HEMA (0.1%–1%, w/w) was integrated. Casting was used to make the films, which were subsequently subjected to radiation (5–25 kGy). The films containing 1% HEMA had the greatest puncture strength (PS) values (282 N/mm) at 10 kGy, according to the results. Due to a putative grafted complex MC-HEMA, measurements of water vapor permeability (WVP) revealed that films subjected to high radiation doses (25 KGy) developed the greatest barrier characteristics (5.5 g mm/m^{2}day KPa). After grafting, Fourier transform infrared (FTIR) examination revealed a drop in the vinyl vibration band from HEMA at 1,636 cm^{-1}, implying that graft polymerization took place via a HEMA reaction from vinylidene into methylene group, accompanied with HEMA homopolymerization. Furthermore, FTIR peaks showed that greater radiation doses (0–100 KGy) enhanced MC-g-HEMA copolymer grafting. Scanning electron microscopy (SEM) analysis of MC-g-HEMA films revealed a smoother look. As a result of these findings, it was possible to confirm that gamma radiation was used to successfully graft poly HEMA onto a MC polymer.

Anirudhan et al. (T. S. Anirudhan, Deepa, & Binusreejayan, 2015) studied a new technique of creating an adsorbent, using cellulose derived from cotton, poly(itaconic acid)-poly(methacrylic acid)-grafted-nanocellulose/nano bentonite composite (P(IA/MAA)-g-NC/NB, for the efficient removal of Uranium (VI), [(U(VI)] from aqueous solutions has been developed. The absorbent was made by graft copolymerizing methacrylic acid (MAA) and itaconic acid (IA) on a nanocellulose/nano bentonite (NC/NB) composite using ethylene glycol dimethacrylate (EGDMA) as a cross-linker and potassium peroxydisulphate (KPS) as an initiator on a nanocellulose/nano bentonite (NC/NB) composite. FTIR, XRD, SEM-EDS, TG, and potentiometric titrations were used to characterize the adsorbent. Two monomers' carboxylic groups boost the efficient adsorptive elimination of U(VI). The adsorbent's ability was investigated and improved. The ideal pH was discovered to be 5.5. For the elimination of U(VI) from 100 mg/L, the adsorbent dosage was determined to be 2.0 g/L. The pseudo-second-order model was used to analyze the kinetic data. The equilibrium was reached after 120 minutes, and the Sips isotherm model was shown to be accurate. 0.1N HCl is used to desorb the adsorbed U(VI). The recurrent use of the adsorbent for the extraction of U(VI) from aqueous solutions was demonstrated in six cycles of adsorption-desorption experiments. The adsorbent's practical efficacy and effectiveness were examined using simulated nuclear industrial effluent, and 0.45 g/L adsorbent was shown to be sufficient for U(VI) removal.

Mesbah et al. (Mesbah et al., 2021) for the first time created a new cellulose acetate-g-poly (2-acrylamido-2-methyl propane sulfonic acid-co-methyl methacrylate) copolymer via free radical polymerization. FT-IR, ^{1}H NMR, and EDX were used to confirm the chemical structure of the graft copolymer. The thermal changes were studied using the TGA and DSC. Various grafting characteristic measures such as grafting efficiency (percent), grafting yield (percent), and add-on value (percent) were found after factors impacting the grafting process were investigated. Simple solution casting was used to create flexible membranes with various graft copolymer compositions. Ion exchange capability (IEC), water uptake (WU), and proton conductivity (r) were investigated as physicochemical parameters. The IEC, WU, and conductivity of these membranes were all greater than the clean CA. The maximal proton conductivity of the CA-g-poly (2-acrylamido-2-methylpropane sulfonic acid-comethyl methacrylate) copolymer membrane (68%; add-on percent) was 6.44×10^{-3} S/cm, compared to 0.035×10^{-3} S/cm for the pure CA. As a result, choosing the right graft copolymer composition allows for fine-tuning of physical properties, which might lead to applications like polyelectrolyte fuel cell membranes or biodiesel generation.

Pakhira et al. (Pakhira, Chatterjee, Mallick, Ghosh, & Nandi, 2021) synthesized PVDF-g-P (tBAEMAran-OEGMA), PVBO] from a poly(vinylidene fluoride) graft random copolymer of t-butyl aminoethyl methacrylate (tBAEMA) and oligo(ethylene glycol) methyl ether methacrylate (OEGMA, Mn = 300). [PVDF-g-P(tBAEMAran-OEGMA), PVBO]PVBO-1 in water forms huge multimicellar aggregates (MMcA); however, at pH 12, the bigger aggregates crumble into much smaller micelles, creating non-conjugated polymer dots (NCPDs), as seen by transmission electron microscopy and dynamic light scattering experiments. The reversible fluorescence on/off behavior is also observed when the temperature is reduced or increased. According to a theoretical study, most amino groups become neutral at high pH and have a strong tendency to form aggregates as a result of crowding of a large number of carbonyl and amine groups, minimizing the HOMO-LUMO gap, showing an absorption peak in the visible region, and causing aggregation-induced emission.

Zhu et al. (Zhu, Zheng, Shi, & Wang, 2021) synthesized a new generation of xanthan gum compounds that are less harmful to the environment, XG-AA/AM/AMPS, by grafting acrylic acid (AA), acrylamide (AM), and 2-acrylamido-2-methylpropane sulfonic acid (AMPS) onto XG and CGA drilling fluids. XG-AA/AM/AMPS were used as a foam stabilizer and effectively developed drilling fluids with a temperature resistance of 180°C. The XG-AA/AM/AMPS-based CGA drilling fluids were shown to have appropriate rheology qualities within 180°C, including considerable shear thinning behavior, greater low shear rate viscosity and apparent viscosity, and suitable rheological parameters, all of which were favorable to cuttings transporting. The filtration capacity of produced CGA fluid may be adjusted within 13.5 mL at 140°C–180°C. Furthermore, multiple methods (PSD, XRD, SEM, and polarizing microscope) were used to investigate the high-temperature resistance mechanism of XG-AA/AM/AMPS, which may be ascribed to the dual impacts of XG-AA/AM/AMPS on improving clay particle dispersion and aphron stability.

Zhao et al. (C. L. Zhao, Deng, Gao, & Wu, 2021) synthesized the novel amphiphilic acylated dextran-g-polytetrahydrofuran (AcyDex-g-PTHF) graft copolymers were successfully synthesized using a combination of living cationic ring-opening polymerization of tetrahydrofuran (THF) to prepare to live PTHF chains with different molecular weights (Mn, PTHF) of 800–2,800 g/mol and nuclei. The incompatibility of the hard dextran backbone and soft PTHF branches, as well as the restricted crystallization of the backbone, cause microphase separation in the graft copolymer. This copolymer has excellent red blood cell hemocompatibility, good HeLa cell biocompatibility, and great resistance to bovine serum albumin adsorption. pH-sensitive controlled drug release behavior is seen in microspheres (1 m) of graft copolymers loaded with the medication ibuprofen. Furthermore, the AcyDex-g-PTHF/Ag nanocomposites are effective against *E. coli* and *S. aureus*. This new AcyDex-g-PTHF graft copolymer is hemocompatible, biocompatible, and antifouling, with potential applications in biological and medical domains.

Zhao et al. (X. Zhao, Wang, Song, & Lou, 2020) used microwave-assisted free radical copolymerization; a binary flocculant (sodium alginate-dimethyl diallyl ammonium chloride, SAD) was effectively produced. The synthetic process was optimized based on the flocculation properties of yellow 7GL dye, with an amount of initiator of 0.8 wt% (equal molar ratio of ammonium peroxydisulphate and sodium bisulfite as complex initiator), sodium alginate: dimethyl diallyl ammonium chloride=1:1 (molar ratio), and a microwave irradiation time of 18 minutes at 280 W. The color removal ratio for the 100 mg/L yellow 7GL simulated wastewater was 73.5% at a SAD dose of 425 mg/L, according to the data. Under a wide range of flocculant doses and ambient pH, the SAD maintained outstanding decolorization ratios. It's possible that the flocculation process is a mix of charge neutralization and bridging action. The proposed SAD flocculant has a broad application promise in the treatment of dye wastewater due to its easy synthesis procedure, eco-friendliness, and excellent decolorization ratio.

Zhou et al. (Zhou, Zhou, & Yang, 2018) used a $KMnO_4/HIO_4$ initiator method, and a bio-copolymer was created by grafting starch with acrylamide (AM) and dimethyl diallyl ammonium chloride (DMDAAC). The response surface methodology was used to improve the initiation and grafting reactions. The optimal conditions for the start reaction were 0.28 mmol $KMnO_4$, 0.25 mmol HIO_4, and 67.64°C, with the greatest copolymer yield of 13.43 g produced from 4 g of raw starch. The best conditions for the grafting process were a temperature of 68.71°C, a molar ratio of (AM + DMDAAC)/starch anhydroglucose units of 2, and an AM/(AM+DMDAAC) molar ratio of 0.34, and a copolymer with a maximum cationic degree of 1.54 meq/g. The improved grafting method showed a high raw material utilization and a high grafting efficiency of 97.120% ± 14%. The improved graft copolymer's flocculation ability was also examined. The graft starch was shown to be efficient in removing reactive colors and dispersing colors from wastewater. Graft starch had a dye removal efficacy that was about 10% greater than polyacrylamide. As a result of the improved initiation and grafting reactions, graft starch was shown to be a potential ecologically friendly flocculant.

Lohar et al. (Shelar-Lohar & Joshi, 2019) synthesized the chelating amidoximated sodium alginate graft copolymer (Alg-g-An) and modified utilizing the acrylonitrile grafted sodium alginate (Alg-g-An) copolymer. FTIR, FESEM, and TGA were used to characterize the graft copolymer. On Alg-g-AO, the dyes methylene blue (MB) and safranin O (SO) were adsorbed. The adsorption kinetic and isothermal studies revealed that the kinetic model for both dyes was in good agreement with the second pseudo model and Langmuir isotherm. The desorption and regeneration experiments describe the adsorbent Alg-g-potential AO's capacity and reusability. These findings demonstrated that synthesized Alg-g-AO is a competent and viable adsorbent for use in wastewater treatment.

Sharma et al. (Sharma, Naushad, Pathania, Mittal, & El-desoky, 2015) synthesized a binary graft copolymer by graft copolymerization using the vinyl monomer acrylic acid (AAc) and a binary combination of AAc and acrylamide changed hibiscus cannabinus fiber (AAm). To maximize grafting yield, the various reaction parameters were tuned. At a molar concentration of 0.35M for AAc and 0.4M for AAm, the optimal percent grafting for AAc and binary mixture (AAc + AAm) was determined to be 93.6% and 74.6%, respectively. FTIR and scanning electron microscopy were used to examine raw AAc grafted H. cannabinus fiber (Hcf-g-polyAAc) and AAc + AAm grafted H. cannabinus fiber (Hcf-g-poly-AAc + AAm). The modified H. cannabinus fibers were tested as a possible dye removal option in an aqueous environment.

Sarkar et al. (Sarkar, Pal, Ghorai, Mandre, & Pal, 2014) synthesized for the removal of hazardous malachite green dye (MG) from aqueous solution, a biodegradable adsorbent based on amylopectin and poly (acrylic acid) (AP-g-PAA). The graft copolymer was produced and evaluated using FTIR spectroscopy, GPC analysis, SEM analysis, and XRD analysis, among other methods. According to biodegradation research, the copolymer is biodegradable. The adsorbent has a high capability for removing MG from an aqueous solution (Qmax, 352.11 mg g-1; 99.05% of MG was removed in 30 minutes). The graft copolymer's point to zero charges (pzc) has been found to play a major influence in adsorption effectiveness. The pseudo-second-order and Langmuir isotherm models are used to describe the adsorption kinetics and isotherm, respectively. Thermodynamic characteristics indicate that dye absorption is a spontaneous process. Finally, the desorption investigation reveals that the adsorbent has remarkable regeneration efficiency.

Kumar et al. (N. Kumar, Banerjee, & Jagadevan, 2020) synthesized Dextrin-g-PMETAC, a cationic graft copolymer made by grafting 2-(methacryloyloxy) ethyl trimethyl ammonium chloride (METAC) onto the dextrin polymer. The biomass harvesting efficiency of three indigenous microalgal strains, Scenedesmus sp. CBIIT(ISM), Micractinium sp. NCS2, and Chlorella sp. NCQ was investigated using the flocculation efficiency of this graft cationic copolymer. The results showed that the produced copolymer was particularly efficient at dewatering microalgae. At optimal flocculant dosages of 25 ppm, 35 ppm, and 45 ppm, the biomass harvesting efficiency of the dextrin-g-PMETAC for Scenedesmus sp. CBIIT(ISM), Micractinium sp. NCS2, and Chlorella sp. NCQ was 98.87% ± 0.17%, 98.19% ± 0.22%, and 97.56% ± 0.28%, respectively. In terms of proteins, carbohydrates, lipids, and residual ash content, a comparison of biomass composition of polymer induced flocculated biomass and centrifuged biomass revealed that biomass harvesting using dextrin-g-PMETAC did not affect microalgal biomass features. According to the findings, dextrin-g-PMETAC might be used as an effective polymeric flocculant for microalgae harvesting.

Chen et al. (L. Chen et al., 2019), synthesized maleoyl chitosan-graft-poly(acrylamide–acryloxyethyl trimethyl ammonium chloride) [MHCS-g-P(AM-DAC)] by UV radiation to increase the solubility and flocculation effectiveness of chitosan. The qualities of flocculants for algae removal were researched in depth. Over the synthesis process of MHCS-g-P(AM-DAC), the impacts of factors such as monomer concentration, MHCS content, illumination time, photoinitiator concentration, cationic degree, pH, and grafting efficiency were investigated. Different analytic approaches were used to characterize the MHCS-g-P(AM-DAC) that was obtained. The effects of dose, pH, and G value on the flocculating characteristics of MHCS-g-P(AM-DAC) on algae removal were also explored, demonstrating that it was more successful than organic flocculant (CPAM) and inorganic coagulant (PFS and PAC). The acquired zeta values were used to determine the flocculation processes of MHCS-g-P(AM-DAC).

Pal et al. produced hydrolyzed graft copolymer (Hyd. PVP-g-PAM) and used it as a flocculant in the destabilization of aqueous nanoparticles suspension (multi-walled carbon nanotubes or MWCNTs). Partially alkaline hydrolysis of a premade graft copolymer of polyvinyl pyrrolidone yielded a new product (Hyd. PVP-g-PAM) (PVP-g-PAM). Due to a substantially wider radius of gyration as a result of straightening and uncoiling of the grafted chains, this unique product exhibits much better flocculation effectiveness than the parent graft copolymer (PVP-g-PAM) as well as PVP. Using a typical jar test protocol, the product's flocculation competency was studied and compared to that of PVP-g-PAM and the commercial flocculant polyacrylamide (PAM) in coal-fine and aqueous solutions of MWCNTs. The results show that the synthetic flocculant is more effective at flocculation than PAM. In MWCNT suspension, the ideal dose as a flocculant of the synthesized product was 1 ppm, whereas in coal-fine it was 0.75 ppm.

Tudu et al. (Tudu, Pal, & Mandre, 2018) synthesized a graft copolymer (AP-g-PAA) using different polymers such as starch amylopectin (AP), polyacrylic acid (PAA). The acquired findings were evaluated, and they showed that employing AP-g-PAA, the iron ore grade was increased from 58.49%–67.52%, with a 95.08% recovery. In addition, employing AP, 64.45% Fe was produced with an 88.79% recovery. Similarly, PAA upgraded the grade to 63.46% Fe with an 82.10% recovery. Characterizing concentrates with X-ray diffraction (XRD) and electron probe microanalysis (EPMA) methods further supports the findings.

Zhou et al. (Zhou et al., 2018) created a bio-copolymer by grafting starch with acrylamide (AM) and dimethyl diallyl ammonium chloride (DMDAAC) using a $KMnO_4/HIO_4$ initiator method to produce an efficient and ecologically acceptable flocculant. The response surface methodology was used to improve the initiation and grafting reactions. The optimal conditions for the start reaction were 0.28 mmol $KMnO_4$, 0.25 mmol HIO_4, and 67.64°C, with the greatest copolymer yield of 13.43 g produced from 4 g raw starch. The best conditions for the grafting process were a temperature of 68.71°C, a molar ratio of (AM + DMDAAC)/starch anhydroglucose units of 2, and an AM / (AM+DMDAAC) molar ratio of 0.34, and a copolymer with a maximum cationic degree of 1.54 meq/g. The improved grafting method showed a high raw material utilization and a high grafting efficiency of 97.120% ± 14%. The improved graft copolymer's flocculation ability was also examined. The graft starch was shown to be efficient in removing reactive colors and dispersing colors from wastewater. Graft starch had a dye removal efficacy that was about 10% greater than polyacrylamide. As a result of the improved initiation and grafting reactions, graft starch was shown to be a potential ecologically friendly flocculant.

REFERENCES

Aminabhavi, T. M., & Deshmukh, A. S. (2016). Polymeric hydrogels as smart biomaterials. In Kalia, S. (ed). *Polymer and Composite Materials*. pp. 45–71. Springer International Publishing. 10.1007/978-3-319-25322-0

Anirudhan, T. S., Deepa, J. R., & Binusreejayan. (2015). Synthesis and characterization of multi-carboxyl-functionalized nanocellulose/nano bentonite composite for the adsorption of uranium(VI) from aqueous solutions: Kinetic and equilibrium profiles. *Chemical Engineering Journal*, *273*(VI), 390–400. 10.1016/j.cej.2015.03.007

Anirudhan, T. S., & Rejeena, S. R. (2012). Poly(acrylic acid)-modified poly(glycidyl methacrylate)-grafted nanocellulose as matrices for the adsorption of lysozyme from aqueous solutions. *Chemical Engineering Journal*, *187*, 150–159. 10.1016/j.cej.2012.01.113

Anirudhan, Thayyath S., & Senan, P. (2011). Adsorption of phosphate ions from water using a novel cellulose-based adsorbent. *Chemistry and Ecology*, *27*(2), 147–164. 10.1080/02757540.2010.547487

Anirudhan, Thayyath Sreenivasan, & Rejeena, S. R. (2014). Poly(acrylic acid-co-acrylamide-co-2-acrylamido-2-methyl-1-propane sulfonic acid)-grafted nanocellulose/poly(vinyl alcohol) composite for the in vitro gastrointestinal release of amoxicillin. *Journal of Applied Polymer Science*, *131*(17), 8657–8668. 10.1002/app.40699

Awang, N. A., Salleh, W. N. W., Hasbullah, H., Yusof, N., Aziz, F., Jaafar, J., & Ismail, A. F. (2017). Graft copolymerization of acrylonitrile onto recycled newspapers cellulose pulp. *AIP Conference Proceedings*, *1885*. 10.1063/1.5002438

Barsbay, M., & Güven, O. (2019). Surface modification of cellulose via conventional and controlled radiation-induced grafting. *Radiation Physics and Chemistry*, *160*(February), 1–8. 10.1016/j.radphyschem.2019.03.002

Bazoubandi, B., & Soares, J. B. P. (2020). Amylopectin-graft-polyacrylamide for the flocculation and dewatering of oil sands tailings. *Minerals Engineering*, *148*(January), 106196. 10.1016/j.mineng.2020.106196

Benoit, D., Hawker, C. J., Huang, E. E., Lin, Z., & Russell, T. P. (2000). One-step formation of functionalized block copolymers. *Macromolecules*, *33*(5), 1505–1507. 10.1021/ma991721p

Bhunia, T., Goswami, L., Chattopadhyay, D., & Bandyopadhyay, A. (2011). Sustained transdermal release of diltiazem hydrochloride through electron beam irradiated different PVA hydrogel membranes. *Nuclear Instruments and Methods in Physics Research, Section B: Beam Interactions with Materials and Atoms*, *269*(16), 1822–1828. 10.1016/j.nimb.2011.05.011

Braşoveanu, M., Koleva, L. S., Nemţanu, M. R., Koleva, E. G., & Paneva, T. P. (2018). Graphical user interface for investigation and optimization of electron beam induced grafting of starch. *Journal of Physics: Conference Series*, *1089*(1), 1–7. 10.1088/1742-6596/1089/1/012017

Chapiro, A. (1977). Radiation induced grafting. *Radiation Physics and Chemistry*, *9*(1–3), 55–67. 10.1016/0146-5724(77)90072-3

Chen, J. P., & Yang, L. (2005). Chemical modification of Sargassum sp. for prevention of organic leaching and enhancement of uptake during metal biosorption. *Industrial and Engineering Chemistry Research*, *44*(26), 9931–9942. 10.1021/ie050678t

Chen, L., Sun, Y., Sun, W., Shah, K. J., Xu, Y., & Zheng, H. (2019). Efficient cationic flocculant MHCS-g-P (AM-DAC) synthesized by UV-induced polymerization for algae removal. *Separation and Purification Technology*, *210*(July 2018), 10–19. 10.1016/j.seppur.2018.07.090

Cheon, J., & Jeun, J. (2019). Preparation of acrylic acid grafted polypropylene by electron beam irradiation and heavy metal ion adsorption property. *Composites Research*, *32*(6), 335–341.

Choi, S. H., & Nho, Y. C. (2000). Adsorption of UO22+ by polyethylene adsorbents with amidoxime, carboxyl, and amidoxime/carboxyl group. *Radiation Physics and Chemistry*, *57*(2), 187–193. 10.1016/S0969-806X(99)00348-5

Elsharma, E. M., Saleh, A. S., Abou-Elmagd, W. S. I., Metwally, E., & Siyam, T. (2019). Gamma radiation induced preparation of polyampholyte nanocomposite polymers for removal of Co(II). *International Journal of Biological Macromolecules*, *136*, 1273–1281. 10.1016/j.ijbiomac.2019.06.081

Ghobashy, M. M., & Khafaga, M. R. (2017). Radiation synthesis and magnetic property investigations of the graft copolymer poly(Ethylene-g-Acrylic Acid)/Fe3O4 Film. *Journal of Superconductivity and Novel Magnetism*, *30*(2), 401–406. 10.1007/s10948-016-3718-5

Giammona, G., Pitarresi, G., Cavallaro, G., & Spadaro, G. (1999). New biodegradable hydrogels based on an acryloylated polyaspartamide cross-linked by gamma irradiation. *Journal of Biomaterials Science, Polymer Edition*, *10*(9), 969–987. 10.1163/156856299X00568

Golshaei, P., & Güven, O. (2017). Chemical modification of PET surface and subsequent graft copolymerization with poly(N-isopropylacrylamide). *Reactive and Functional Polymers*, *118*(February), 26–34. 10.1016/j.reactfunctpolym.2017.06.015

González-Torres, M., Vargas-Muñoz, S., Del Real, A., De Jesús Ruíz-Baltazar, Á., Reyes-Cervantes, E., Del Pilar Carreón-Castro, M., & Rodríguez-Talavera, R. (2016). Surface modification of poly(3-hydroxybutyrate-co-3-hydroxyvalerate) by direct plasma-radiation-induced graft polymerization of N-hydroxyethyl-acrylamide. *Materials Letters*, *175*, 252–257. 10.1016/j.matlet.2016.04.005

Guibal, E. (2004). Interactions of metal ions with chitosan-based sorbents: A review. *Separation and Purification Technology*, *38*(1), 43–74. 10.1016/j.seppur.2003.10.004

Halder, J., & Islam, N. (2015). Water Pollution and its Impact on the Human Health. *Journal of Environment and Human*, *2*(1), 36–46. 10.15764/eh.2015.01005

Hassan, M. S., & Zohdy, M. H. (2018). Adsorption kinetics of toxic heavy metal ions from aqueous solutions onto grafted jute fibers with acrylic acid by gamma irradiation. *Journal of Natural Fibers*, *15*(4), 506–516. 10.1080/15440478.2017.1330721

Hegazy, E. S. A., El-Rehim, H. A. A., & Shawky, H. A. (2000). Investigations and characterization of radiation grafted copolymers for possible practical use in waste water treatment. *Radiation Physics and Chemistry*, *57*(1), 85–95. 10.1016/S0969-806X(99)00312-6

Hoffman, A. S. (2012). Hydrogels for biomedical applications. *Advanced Drug Delivery Reviews*, *64*(SUPPL.), 18–23. 10.1016/j.addr.2012.09.010

Ishihara, R., Asai, S., & Saito, K. (2020). Recent progress in charged polymer chains grafted by radiation-induced graft polymerization; Adsorption of proteins and immobilization of inorganic precipitates. *Quantum Beam Science*, *4*(2), 20. 10.3390/qubs4020020

Kadooka, H., Kiso, Y., Goto, S., Tanaka, T., Jami, M. S., & Iwata, M. (2017). Flocculation behavior of colloidal suspension by use of inorganic and polymer flocculants in powder form. *Journal of Water Process Engineering*, *18*(February), 169–175. 10.1016/j.jwpe.2017.05.011

Kakati, D. K., Bora, M. M., & Deka, C. (2015). Cellulose graft copolymerization by gamma and electron beam irradiation. *Cellulose-Based Graft Copolymers: Structure and Chemistry*, 79–88. 10.1201/b18390

Kakati, N., Assanvo, E. F., & Kalita, D. (2019). Alkalinization and graft copolymerization of pineapple leaf fiber cellulose and evaluation of physic-chemical properties. *Polymer Composites*, *40*(4), 1395–1403. 10.1002/pc.24873

Ke, X., Drache, M., Gohs, U., Kunz, U., & Beuermann, S. (2018). Preparation of polymer electrolyte membranes via radiation-induced graft copolymerization on poly(Ethylene-alt-tetrafluoroethylene) (ETFE) using the cross-linker N,N′-methylenebis(acrylamide). *Membranes*, *8*(4), 1–16. 10.3390/membranes8040102

Kodama, Y., Barsbay, M., & Güven, O. (2014). Radiation-induced and RAFT-mediated grafting of poly(hydroxyethyl methacrylate) (PHEMA) from cellulose surfaces. *Radiation Physics and Chemistry*, *94*(1), 98–104. 10.1016/j.radphyschem.2013.07.016

Kumar, D., Pandey, J., Raj, V., & Kumar, P. (2017). A review on the modification of polysaccharide through graft copolymerization for various potential applications. *The Open Medicinal Chemistry Journal*, *11*(1), 109–126. 10.2174/1874104501711010109

Kumar, N., Banerjee, C., & Jagadevan, S. (2020). Cationically functionalized dextrin polymer as an efficient flocculant for harvesting microalgae. *Energy Reports*, *6*, 2803–2815. 10.1016/j.egyr.2020.09.040

Kume, T., Nagasawa, N., & Yoshii, F. (2002). Utilization of carbohydrates by radiation processing. *Radiation Physics and Chemistry*, *63*(3–6), 625–627. 10.1016/S0969-806X(01)00558-8

Le Moigne, N., Sonnier, R., El Hage, R., & Rouif, S. (2017). Radiation-induced modifications in natural fibres and their biocomposites: Opportunities for controlled physico-chemical modification pathways? *Industrial Crops and Products*, *109*(February), 199–213. 10.1016/j.indcrop.2017.08.027

Lee, C. S., Robinson, J., & Chong, M. F. (2014). A review on application of flocculants in wastewater treatment. *Process Safety and Environmental Protection*, *92*(6), 489–508. 10.1016/j.psep.2014.04.010

Ma, J., Peng, J., & Zhai, M. (2018). Radiation-grafted membranes for applications in renewable energy technology. In Wu, G., Zhai, M., & Wang, M. *Radiation Technology for Advanced Materials: From Basic to Modern Applications*. pp. 207–247. Elsevier. 10.1016/B978-0-12-814017-8.00007-X

Madrid, J. F., Abad, L. V., Yamanobe, T., & Seko, N. (2017). Effects of chain transfer agent on the electron beam-induced graft polymerization of glycidyl methacrylate in emulsion phase. *Colloid and Polymer Science*, *295*(6), 1007–1016. 10.1007/s00396-017-4088-7

Madrid, J. F., Cabalar, P. J. E., & Abad, L. V. (2018). Radiation-induced graft polymerization of acrylic acid and glycidyl methacrylate onto abaca/polyester nonwoven fabric. *Journal of Natural Fibers*, *15*(5), 625–638. 10.1080/15440478.2017.1349713

Manaila, E., & Craciun, G. (2019). Biodegradable hydrogels based on acrylamide, acrylic acid and sodium alginate synthesized by electron beam irradiation. *Acta Physica Polonica A, 135*(5), 1063–1064. 10. 12693/APhysPolA.135.1063

Mate, C. J., Mishra, S., & Srivastava, P. K. (2020). Design of low-cost Jhingan gum-based flocculant for remediation of wastewater: Flocculation and biodegradation studies. *International Journal of Environmental Science and Technology, 17*(5), 2545–2562. 10.1007/s13762-019-02587-x

Mesbah, F., El Gayar, D., Farag, H., Tamer, T. M., Omer, A. M., Mohy-Eldin, M. S., & Khalifa, R. E. (2021). Development of highly ionic conductive cellulose acetate-g-poly (2-acrylamido-2-methylpropane sulfonic acid-co-methyl methacrylate) graft copolymer membranes. *Journal of Saudi Chemical Society, 25*(9), 101318. 10.1016/j.jscs.2021.101318

Mittal, H., Ray, S. S., & Okamoto, M. (2016). Recent progress on the design and applications of polysaccharide-based graft copolymer hydrogels as adsorbents for wastewater purification. *Macromolecular Materials and Engineering, 301*(5), 496–522. 10.1002/mame.201500399

Moawia, R. M., Nasef, M. M., Mohamed, N. H., & Ripin, A. (2016). Modification of flax fibres by radiation induced emulsion graft copolymerization of glycidyl methacrylate. *Radiation Physics and Chemistry, 122*, 35–42. 10.1016/j.radphyschem.2016.01.008

Molatlhegi, O., & Alagha, L. (2017). Adsorption characteristics of chitosan grafted copolymer on kaolin. *Applied Clay Science, 150*(September), 342–353. 10.1016/j.clay.2017.09.032

Nandi, G., Changder, A., & Ghosh, L. K. (2019). Graft-copolymer of polyacrylamide-tamarind seed gum: Synthesis, characterization and evaluation of flocculating potential in peroral paracetamol suspension. *Carbohydrate Polymers, 215*(January), 213–225. 10.1016/j.carbpol.2019.03.088

Nasef, M. M., & Güven, O. (2012). Radiation-grafted copolymers for separation and purification purposes: Status, challenges and future directions. *Progress in Polymer Science, 37*(12), 1597–1656. 10.1016/j.progpolymsci.2012.07.004

Ochoa-Segundo, E. I., González-Torres, M., Cabrera-Wrooman, A., Sánchez-Sánchez, R., Huerta-Martínez, B. M., Melgarejo-Ramírez, Y., … Rodríguez-Talavera, R. (2020). Gamma radiation-induced grafting of n-hydroxyethyl acrylamide onto poly(3-hydroxybutyrate): A companion study on its polyurethane scaffolds meant for potential skin tissue engineering applications. *Materials Science and Engineering C, 116*, 111176. 10.1016/j.msec.2020.111176

Omichi, M., Ueki, Y., Seko, N., & Maekawa, Y. (2019). Development of a simplified radiation-induced emulsion graft polymerization method and its application to the fabrication of a heavy metal adsorbent. *Polymers, 11*(8), 1–11. 10.3390/polym11081373

Oraby, H., Senna, M., Elsayed, M., & Gobara, M. (2016). Radiation-induced graft copolymerization of vinyl monomer onto polyolefin polymeric substrate: Preparation and characterization. *The International Conference on Chemical and Environmental Engineering, 8*(13), 43–61. 10.21608/iccee.2016.35032

Pakhira, M., Chatterjee, D. P., Mallick, D., Ghosh, R., & Nandi, A. K. (2021). Reversible stimuli-dependent aggregation-induced emission from a "nonfluorescent" amphiphilic PVDF graft copolymer. *Langmuir, 37*(16), 4953–4963. 10.1021/acs.langmuir.1c00310

Pal, A., Majumder, K., & Bandyopadhyay, A. (2016). Surfactant mediated synthesis of poly(acrylic acid) grafted xanthan gum and its efficient role in adsorption of soluble inorganic mercury from water. *Carbohydrate Polymers, 152*, 41–50. 10.1016/j.carbpol.2016.06.064

Pathania, D., Sharma, A., & Sethi, V. (2017). Microwave induced graft copolymerization of binary monomers onto luffa cylindrica fiber: Removal of congo red. *Procedia Engineering, 200*, 408–415. 10.1016/j.proeng.2017.07.057

Pino-Ramos, V. H., Flores-Rojas, G. G., Alvarez-Lorenzo, C., Concheiro, A., & Bucio, E. (2018). Graft copolymerization by ionization radiation, characterization, and enzymatic activity of temperature-responsive SR-g-PNVCL loaded with lysozyme. *Reactive and Functional Polymers, 126*, 74–82. 10. 1016/j.reactfunctpolym.2018.03.002

Rajaram, T., & Das, A. (2008). Water pollution by industrial effluents in India: Discharge scenarios and case for participatory ecosystem specific local regulation. *Futures, 40*(1), 56–69. 10.1016/j.futures.2007.06.002

Saito, K., Fujiwara, K., & Sugo, T. (Eds). (2018). Innovative polymeric adsorbents. In *Innovative Polymeric Adsorbents*. Springer Nature Singapore Pte Ltd. 10.1007/978-981-10-8563-5

Salamun, N., Triwahyono, S., Jalil, A. A., Majid, Z. A., Ghazali, Z., Othman, N. A. F., & Prasetyoko, D. (2016). Surface modification of banana stem fibers via radiation induced grafting of poly(methacrylic acid) as an effective cation exchanger for Hg(II). *RSC Advances, 6*(41), 34411–34421. 10.1039/c6ra03741k

Saleh, A. S., Ibrahim, A. G., Elsharma, E. M., Metwally, E., & Siyam, T. (2018). Radiation grafting of acrylamide and maleic acid on chitosan and effective application for removal of Co(II) from aqueous solutions. *Radiation Physics and Chemistry, 144*(II), 116–124. 10.1016/j.radphyschem.2017.11.018

Salehizadeh, H., Yan, N., & Farnood, R. (2018). Recent advances in polysaccharide bio-based flocculants. *Biotechnology Advances*, *36*(1), 92–119. 10.1016/j.biotechadv.2017.10.002

Salmieri, S., Khan, R. A., Safrany, A., & Lacroix, M. (2015). Gamma rays-induced 2-hydroxyethyl methacrylate graft copolymerization on methylcellulose-based films: Structure analysis and physicochemical properties. *Industrial Crops and Products*, *70*, 64–71. 10.1016/j.indcrop.2015.02.056

Sarkar, A. K., Pal, A., Ghorai, S., Mandre, N. R., & Pal, S. (2014). Efficient removal of malachite green dye using biodegradable graft copolymer derived from amylopectin and poly(acrylic acid). *Carbohydrate Polymers*, *111*, 108–115. 10.1016/j.carbpol.2014.04.042

Setia, A. (2017). Applications of graft copolymerization: A revolutionary approach. In Thakur, V. K. *Biopolymer Grafting: Applications*. pp.1–44. Elsevier. 10.1016/B978-0-12-810462-0.00001-6

Sharif, J., Mohamad, S. F., Fatimah Othman, N. A., Bakaruddin, N. A., Osman, H. N., & Güven, O. (2013). Graft copolymerization of glycidyl methacrylate onto delignified kenaf fibers through pre-irradiation technique. *Radiation Physics and Chemistry*, *91*, 125–131. 10.1016/j.radphyschem.2013.05.035

Sharma, G., Naushad, M., Pathania, D., Mittal, A., & El-desoky, G. E. (2015). Modification of Hibiscus cannabinus fiber by graft copolymerization: Application for dye removal. *Desalination and Water Treatment*, *54*(11), 3114–3121. 10.1080/19443994.2014.904822

Shelar-Lohar, G., & Joshi, S. (2019). Amidoximated functionalized sodium alginate graft copolymer: An effective adsorbent for rapid removal of cationic dyes. *Materials Today: Proceedings*, *26*(2), 3357–3362. 10.1016/j.matpr.2019.10.130

Singh, B., & Kumar, A. (2018). Radiation-induced graft copolymerization of N-vinyl imidazole onto moringa gum polysaccharide for making hydrogels for biomedical applications. *International Journal of Biological Macromolecules*, *120*, 1369–1378. 10.1016/j.ijbiomac.2018.09.148

Singh, B., & Singh, B. (2019). Developing a drug delivery carrier from natural polysaccharide exudate gum by graft-copolymerization reaction using high energy radiations. *International Journal of Biological Macromolecules*, *127*, 450–459. 10.1016/j.ijbiomac.2019.01.075

Singh, B., & Singh, B. (2020). Radiation induced graft copolymerization of graphene oxide and carbopol onto sterculia gum polysaccharide to develop hydrogels for biomedical applications. *FlatChem*, *19*, 100151. 10.1016/j.flatc.2019.100151

Singh, K., Ohlan, A., Saini, P., & Dhawan, S. K. (2008). Composite – Super paramagnetic behavior and variable range hopping 1D conduction mechanism – Synthesis and characterization. *Polymers for Advanced Technologies*, *November 2007*, 229–236. 10.1002/pat

Şolpan, D., Torun, M., & Güven, O. (2010). Comparison of pre-irradiation and mutual grafting of 2-chloroacrylonitrile on cellulose by gamma-irradiation. *Radiation Physics and Chemistry*, *79*(3), 250–254. 10.1016/j.radphyschem.2009.09.008

Sun, Y., & Chmielewski, A. G. (2017). *Applications of Ionizing Radiation in Materials Processing*. Institute of Nuclear Chemistry and Technology. Retrieved from http://www.ichtj.waw.pl/ichtj/publ/monogr/m2017_1.htm

Takács, E., Wojnárovits, L., Borsa, J., Papp, J., Hargittai, P., & Korecz, L. (2005). Modification of cotton-cellulose by preirradiation grafting. *Nuclear Instruments and Methods in Physics Research, Section B: Beam Interactions with Materials and Atoms*, *236*(1–4), 259–265. 10.1016/j.nimb.2005.03.248

Tudu, K., Pal, S., & Mandre, N. R. (2018). Comparison of selective flocculation of low grade goethitic iron ore fines using natural and synthetic polymers and a graft copolymer. *International Journal of Minerals, Metallurgy and Materials*, *25*(5), 498–504. 10.1007/s12613-018-1596-5

Turmanova, S., & Atanassov, A. (2007). Forming complexes of carboxyl-containing copolymers with metal ions – Electrochemical and physicomechanical properties. *Journal of the University of Chemical Technology and Metallurgy*, *42*(1), 35–40.

Vajihinejad, V., Gumfekar, S. P., Bazoubandi, B., Rostami Najafabadi, Z., & Soares, J. B. P. (2019). Water soluble polymer flocculants: Synthesis, characterization, and performance assessment. *Macromolecular Materials and Engineering*, *304*(2), 1–43. 10.1002/mame.201800526

Varma, A. J., Deshpande, S. V., & Kennedy, J. F. (2004). Metal complexation by chitosan and its derivatives: A review. *Carbohydrate Polymers*, *55*(1), 77–93. 10.1016/j.carbpol.2003.08.005

Wach, R. A., Mitomo, H., Nagasawa, N., & Yoshii, F. (2003). Radiation cross-linking of carboxymethylcellulose of various degree of substitution at high concentration in aqueous solutions of natural pH. *Radiation Physics and Chemistry*, *68*(5), 771–779. 10.1016/S0969-806X(03)00403-1

Yang, Z., Peng, H., Wang, W., & Liu, T. (2010). Crystallization behavior of poly(ε-caprolactone)/layered double hydroxide nanocomposites. *Journal of Applied Polymer Science*, *116*(5), 2658–2667. 10.1002/app

Yantasee, W., Lin, Y., Fryxell, G. E., Alford, K. L., Busche, B. J., & Johnson, C. D. (2004). Selective removal of copper(II) from aqueous solutions using fine-grained activated carbon functionalized with amine. *Industrial and Engineering Chemistry Research, 43*(11), 2759–2764. 10.1021/ie030182g

Yoshii, F., Zhao, L., Wach, R. A., Nagasawa, N., Mitomo, H., & Kume, T. (2003). Hydrogels of polysaccharide derivatives cross-linked with irradiation at paste-like condition. *Nuclear Instruments and Methods in Physics Research, Section B: Beam Interactions with Materials and Atoms, 208*(1–4), 320–324. 10.1016/S0168-583X(03)00624-4

Zhao, C. L., Deng, J. R., Gao, Y. Z., & Wu, Y. X. (2021). Hemocompatible, biocompatible and antifouling Acylated dextran-g-polytetrahydrofuran graft copolymer with silver nanoparticles: Synthesis, characterization and properties. *Materials Science and Engineering C, 123*(May 2020), 111998. 10.1016/j.msec.2021.111998

Zhao, L., Mitomo, H., Nagasawa, N., Yoshii, F., & Kume, T. (2003). Radiation synthesis and characteristic of the hydrogels based on carboxymethylated chitin derivatives. *Carbohydrate Polymers, 51*(2), 169–175. 10.1016/S0144-8617(02)00210-2

Zhao, X., Wang, X., Song, G., & Lou, T. (2020). Microwave assisted copolymerization of sodium alginate and dimethyl diallyl ammonium chloride as flocculant for dye removal. *International Journal of Biological Macromolecules, 156*, 585–590. 10.1016/j.ijbiomac.2020.04.054

Zhou, H., Zhou, L., & Yang, X. (2018). Optimization of preparing a high yield and high cationic degree starch graft copolymer as environmentally friendly flocculant: Through response surface methodology. *International Journal of Biological Macromolecules, 118*, 1431–1437. 10.1016/j.ijbiomac.2018.06.155

Zhu, W., Zheng, X., Shi, J., & Wang, Y. (2021). A high-temperature resistant colloid gas aphron drilling fluid system prepared by using a novel graft copolymer xanthan gum-AA/AM/AMPS. *Journal of Petroleum Science and Engineering, 205*(April), 108821. 10.1016/j.petrol.2021.108821

Zubair, N. A., Nasef, M. M., Ting, T. M., Abdullah, E. C., & Ahmad, A. (2020). Radiation induced graft copolymerization of amine-containing monomer onto polyethylene coated propylene for CO2 adsorption. *IOP Conference Series: Materials Science and Engineering, 808*(1), 1–8. 10.1088/1757-899X/808/1/012028

6 Irradiation-induced effect on polymer: From mechanism to biomedical applications

Sumit Dokwal*

Department of Biochemistry, Pt. B. D. Sharma PGIMS, Rohtak & Kalpana Chawla Medical College, Karnal, India

Suman Mahendia*

Department of Physics, Kurukshetra University, Kurukshetra, India

Rishi Pal Chahal

Department of Physics, Chaudhary Bansi Lal University, Bhiwani, India

Vishal Sharma

Institute of Forensic Science & Criminology, Panjab University, Chandigarh, India

Suman B. Kuhar

Department of Physics, IHL, BPSMV, Khanpur Kalan, India

Shyam Kumar

Department of Physics, Kurukshetra University, Kurukshetra, India

*Corresponding authors: drsumitdokwal80@gmail.com; smahendia@kuk.ac.in

CONTENTS

DOI: 10.1201/9781003321910-6

6.1 INTRODUCTION

Polymers are the macromolecules that contain the number of monomers united with some covalent bonds [Gedde, Mikael, and Hedenqvist, 2019]. The process through which the monomers are connected together is termed *polymerization* and they play very crucial roles in everyday life as plastic materials [Brun et al., 2001; Thompson, 2001; Liu, Guo, and Yang, 2002; Fink, 2004; Laskarakis, Gravalidis, and Logothetidis, 2004; Fahmy, 2007; Lee, Y. -M. Kim, and Y. C. Kim, 2007]. Polymers are one of the largest and most promising classes of materials being extensively used on a large scale in biomedical applications [Muppalaneni and Omidian, 2013; Jiang, Liu, and Feng, 2011; Qiu and Park, 2012]. The underlying reason of wide diversity in polymer-based applications is that the polymers can be easily molded into any shape and design, which results in abundant types of structures with desired physical, chemical, and bio-mimetic properties [Eid and Hegazy, 2012; Croisier and Jerome, 2013; Li et al., 2017; Mitrousis, Fokina, and Shoichet, 2018]. This certifies their extensive use in various medical applications. Moreover, many biopolymers produced by living organisms are fundamental to biological functions. Therefore, the field of biomaterials and related technology dealing with tissue engineering, their re-pairmen, and restoration of their function must be propelling. Ceramics, metals, natural or synthetic polymers, and composites are widely used as biomaterials. Further, radiation processing has also found its place in the widespread polymer industry for various applications like sterilization, irradiated packaging for food, polymer modification, etc. [Sharma et al., 2013; Nguyen and Liu, 2013; Singh R., Singh D., and Singh A., 2016; Shintani, 2017; Tamada, 2018]. Therefore, a systematic study of irradiation in polymer is very helpful to evaluate the impact on their performances, such as mechanical strength, molecular weights, physical and chemical properties, glass transition, and melting point. However, it is expected that after irradiation, properties of polymers may undergo some physical and chemical changes like change in color, odor, stiffness, softness, and chemical resistance or melting temperature [El-Ahdal, 2001; Kaczmarek and Podgorski, 2007; Nouh, 2012; Abramowska et al., 2015; Cieśla et al., 2017]. Irradiation-induced effects can be used for tailoring of polymers with relative ease and low production cost, which favors the successful production of polymer-based materials with a wide range of properties and functionality [Banoriya, Purohit, and Dwivedi, 2017; Bernard et al., 2018; Constantin et al., 2019]. This could be the reason of promoting research on polymeric-based biomaterials and related technologies, especially the study of irradiation effects on them. Thus, radiation-induced studies are keen to generate the convergence of concept, innovation, and research at the interface between material science, polymer chemistry, biomaterial engineering, and radiation technology [Mishra et al., 2002; Salmawi, 2007; Gupta, 2011; Gupta et al., 2015; Abramowska, 2015; Cao et al., 2020].

The interaction of radiation with polymers primarily results in chain scission and cross-linking of produced monomers. This completely alters the initial polymer structure by ionization, free radical formation, cross-linking, carbonization, free-radical production, displacing atoms, sputtering, bond scissor, etc. which finally result in disintegration of monomers and production of saturated and unsaturated groups. These alterations in structure of polymers are named defects, which are the reason for the change in their properties [Chahal et al., 2011; Brenner and Hall, 2007, Ahlbom and Feychting, 2003].

The irradiation of the polymer can be broadly divided into two categories: electromagnetic irradiation and ion beam irradiation. The electromagnetic radiation can be further divided into two categories: non-ionizing radiation and ionizing radiation. The radiations with frequency up to the frequency of visible light are generally considered as non-ionizing radiation. These radiations can interact with polymers in a variety of ways. So, in this chapter, we put our emphasis on structure and properties of polymers, types of radiation, and radiation-induced effects on polymer properties. Later, the subject of this chapter will be oriented on a few radiation-induced effects on polymers for biomedical applications.

6.2   TYPES AND SOURCES OF RADIATION WITH THEIR BIOLOGICAL IMPACTS

Radiation is a form of energy that travels from its source in the form of energy/particle waves. Before understanding the effect of radiation exposure on polymers, we should have adequate knowledge of types and sources of radiation. Natural radiation sources like cosmic rays, radioactive substance (rocks, soil, plant) is everywhere, they are known as background radiations. Depending upon their wavelength, they range from longer wavelength (radio waves) to shorter wavelength (gamma rays), as shown in Figure 6.1. This figure shows each portion of the EM spectrum with relation to our day-to-day life from lowest energy/longest wavelength (at the top) to highest energy/shortest wavelength (at the bottom). Radio waves: Our radio captures radio waves emitted by radio stations, bringing your favorite tunes. Radio waves are also emitted by stars and gases in space. Microwave: Microwave radiation will cook your popcorn in just a few minutes, but is also used by astronomers to learn about the structure of nearby galaxies. Infrared: Night vision goggles pick up the infrared light emitted by our skin and objects with heat. In space, infrared light helps us map the dust between stars. Visible: Our eyes detect visible light. Fireflies, light bulbs, and stars all emit visible light.

Ultraviolet: Ultraviolet radiations are emitted by the sun and are the reason skin tans and burns. "Hot" objects in space emit UV radiation as well. X-ray: A dentist uses X-rays to image your teeth, and airport security uses them to see through your bag. Hot gases in the universe also emit X-rays. Gamma ray: Doctors use gamma-ray imaging to see inside your body. The biggest gamma-ray generator of all is the universe.

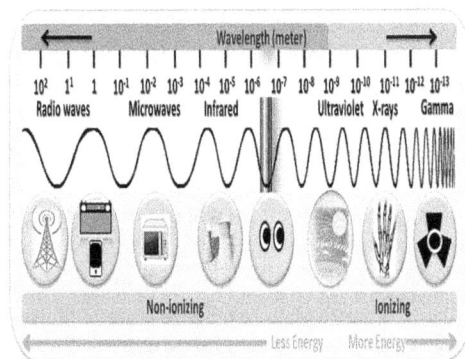

FIGURE 6.1 Types of radiation covering longer-wavelength to shorter-wavelength regions.

Further, in terms of ability to ionize matter, two categories of radiation are (a) non-ionizing and (b) ionizing radiation. Generally, radiations with more energy i.e., shorter wavelength are known as ionizing radiations and vice versa.

6.2.1   NON-IONIZING RADIATION

Non-ionizing radiation may be emitted by natural as well as man-made sources; they are essentially classified into optical and electromagnetic (EM) field radiation. Choosing energy of radiation as a parameter, the optical radiations are classified as ultraviolet (UV), visible, and infrared (IR) radiations; while the EM field is sub-grouped into a RF (radio frequency) region that consists of microwave and radio waves with very high and low frequency [Ahlbom and Feychting, 2003;

Brenner and Hall, 2007]. As non-ionizing radiation comprises the long wavelength with low energy (<12.4 eV), they only be sensed by human in the form of heat in their intensity.

6.2.1.1   Ultraviolet (UV) radiation

The human eye is not sensitive to UV radiation i.e., human eye can't see UV radiation. The most common UV radiation we are familiar with is sunlight, which has majorly three kinds: UVC, UVB, and UVA. Radiation in wavelength range: 200-280 nm known as UVC rays, mostly absorbed by the earth's ozone layer and thus not reaching earth's surface. UVA (longer wavelength: 315-400 nm) and a small amount of UVB (mid-wavelength: 280-315 nm) rays are transmitted through the atmosphere. Upon human exposure, it's UVB radiation that reaches only to the epidermis i.e outer layer of our skin while UVA penetrates through the dermis i.e., middle layer of skin. UVC is among the highest-energy portions with low penetration depth, therefore prolonged exposure to it causes low risk to skin cancer, cataracts, or permanent vision loss. Any skin burns and eye injuries caused from UVC exposure usually resolve within a week with no known long-term damage. As such, UVC is blocked by our atmosphere, but artificial lamps or lasers are used that may cause human exposure. As there is no sharp boundary in UVC and UVB wavelengths, therefore, some UVB radiations may also be emitted with UVC lamps [Ng, 2003]. Additionally, some UVC lamps generate ozone whose high-level exposure may also cause chronic respiratory diseases like asthma or increase vulnerability to respiratory infection.

However, both mid-range wavelength UV radiation i.e., UVB rays and long-wavelength UV radiation i.e., UVA rays, can cause damage to skin, promote cell degradation, damage to soft tissue, skin damage, and eye inflammatory reactions. Sunburn is a sign of high-radiation dose received over a short time and the prolonged UV exposure results in premature aging and skin cancer [Gajšek et al., 2013; Nations U, Programme E, 2016]. Besides disadvantages, UVB exposure also has some benefits to our skin as it helps in producing vitamin D3, which, along with calcium, is very important for our bone and muscle health [Hameed and Akhtar, 2019]. However, its exposure dose has been decided over vitamin D level, skin color, sunscreen, latitude, and altitude, etc.

6.2.1.2   Optical and infrared (IR) radiation

The visible light covers the wavelength range from 380 to 760 nm (roughly 400–800 nm). The arc lamps, spotlights, gas and vapor discharge tubes, etc. are a few examples of intense broad-spectrum man-made sources of visible light. The visible light is not harmful to living organisms in most circumstances; but a sudden exposure may cause an aversion response like eye blinking. This response is very quick and in generally occurs within a quarter of a second. Further, the sun is a major source of IR radiation, also known as thermal radiation, and experienced in the form of heat. Not only sun emits IR radiation but many man-made appliances like home electrical appliances, electrical bulbs and heaters, etc. are also good sources of IR radiations. However, for human beings, in general, IR radiations are not considered hazardous, but when received in excessive quantity they are considered potentially hazardous. IR radiations are mainly categorized into the following three regions: near-infrared (nearest to visible spectrum) IR-A: ranging from 0.76 to 1.4 micrometer; mid-infrared (IR-B) with wavelength 1.4 to 3 micrometer and far-infrared (IR-C) with wavelength 3 micrometer to 1 mm. The wavelength of IR radiation is the major parameter on which injury threshold and damage mechanism depends. Mostly, IR and optical radiation cause skin burn and skin photo-aging with their continuous exposure [Holzer and Elmets, 2010; ICNIRP, 2013]. For eyes, different types of damages can occur, and they are as follows:

- 1.4 micrometer to 1 mm IR radiation may cause thermal damage of the cornea.
- Thermal damage of the iris by the radiation with wavelength range ~ 0.38–1.4 micron.
- Thermal damage of crystalline lens by near IR radiations of wavelength approximately 800–3,000 nm.
- Retina may be thermally damage by IR radiation in wavelength range 0.38–1.4 microns
- Retina may be damaged photochemically by "blue-light" principally in wavelength range 0.38–0.55 micron (0.3–0.55 micron for the aphakic eye)

6.2.1.3 Radio frequency (RF) radiation

The RF radiation (radio and microwave radiations ranging from 3 kHz–300 GHz) does not have sufficient energy to remove electrons from an atom, and thus are non-ionizing radiation. But, if received in large amounts by human body, heat is produced that may cause burns and body tissue damage [Singh et al., 2018]. Very low levels of frequency of RF radiations are useful for heating food items like in a microwave oven. Microwaves absorbed in food causes its water molecules to vibrate thereby producing heat and they do not made food radioactive. Further, from safety purposes, the outer wall of microwave ovens are shielded and it is ensure that microwave radiations are coming only when oven is ON with door completely closed and are contained within it only. However, by any means or by damage to the oven, if radiation is leaked out, they could cause burning. However, RF radiations are not strong enough to cause damage to DNA directly. Another well-known area of RF wave exposure includes mobile phones and some recent epidemiological studies have shown that heavy mobile exposure can cause effects on brain tissue or sometimes cause brain cancer [Miller et al., 2019]. Other risk factors are related to neurological disorder, learning and memory defects, immunity imbalance, and anomalies in reproductive systems are also of great concern. Therefore, limitation has been put to the use of mobile phones and other sources of radio waves. Sources of RF radiation can be natural (sun, outer space, lightning striking in the sky, etc.) as well as man-made. Many man-made sources for radio waves include transmission of signals for radio and TV broadcasts, cordless phones, cell phones and cell phone towers, satellite, radar, WiFi, Bluetooth devices, and body scanners used for security screening (smart meter/millimeter wave scanners).

6.2.2 IONIZING RADIATION

As the name suggests, these radiations are capable of ionizing the materials. Such radiations can be in form of particles as well as waves. The particle forms of ionizing radiation include alpha, beta, neutron, and ionic (low energy and swift heavy ion) while wave forms include (X-rays and gamma rays). These radiations are hazards to us and have a special symbolic representation, shown in Figure 6.2.

FIGURE 6.2 Hazardous radiation sign.

Further, penetration power of these ionizing radiations, which leaves an impact on living tissues, depends on its energy. More is the energy; more will be the penetration as the large energetic radiations will be able to eject electrons out from an orbit of an atom. Further, from a readability point of view, we subcategorize ionizing radiation as of nuclear and atomic origin.

6.2.2.1 Nuclear origin

This subcategory includes the radiation that originated either inside or outside the nucleolus of an atom. Typical ionizing subatomic particles (alpha particles, beta particles, and neutrons) and high-energy gamma radiation are found from the nucleus in radioactive decay while X-rays (high-energy waves) are coming out from the electron cloud of an atom. Figure 6.3 schematically represents the origin and penetration power of various ionizing radiations that are of atomic origin.

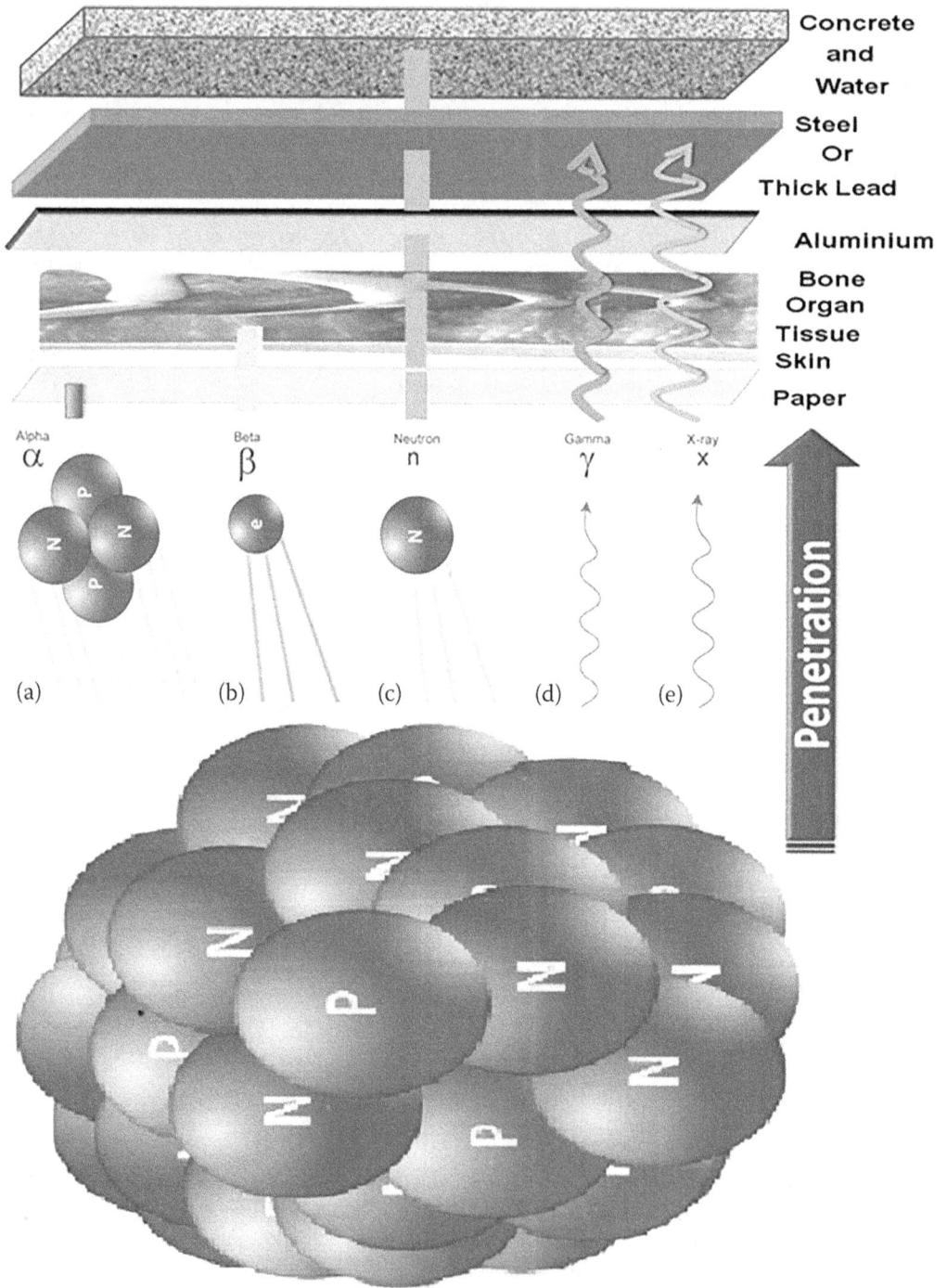

FIGURE 6.3 Schematic representation for the emission of (a) an alpha (α) particle; (b) a beta particle (β); (c) a neutron (n); (d) gamma (γ) rays from the nucleus of an atom; and (e) X-rays from the electron cloud of an atom with their corresponding penetration.

When a radioactive decay occurs by changing the parent nuclei to a daughter one with an atomic number less than two units of magnitude and atomic weight less than four units in magnitude, alpha radiation is produced. Thus, an alpha particle is a bare He^4 nucleus with two protons and two neutrons [Bodner and Rhea, 1978]. Due to their large ionizing power, owing to its charge and mass, the interaction of alpha particles is strong with matter with a penetration depth of only a few centimeters in air. Whether it is their large ionizing power or small penetration depth, alpha particles aren't able to cross the paper and outer layer of dead skin cells, but are capable if an alpha emitting substance is ingested in food or air, thereby causing serious cell damage. However, beta radiation occurs during the radioactive process and is in the form of either an electron or a positron [UNSCEAR, 2008]. Due to a smaller mass than alpha particles, its penetration depth is large, up to few meters in air and can be stopped by a thick piece of plastic, or even a stack of paper, or thin metal plate like aluminum (Al). Its penetration depth in skin is nearly a few centimeters, thereby causing somewhat of a potential health hazard. However, the main threat is still primarily from internal emission from ingested material.

The next type of nuclear origin radiation consists of spontaneous or induced nuclear fission. The penetration depth of neutron radiation is a few thousands of meters in air, making them most penetrating. For effectively stopping them, a hydrogen-rich material, such as concrete or water, is required. Because of no charge, they cannot directly ionize an atom but if they are absorbed into a stable atom they can ionize it indirectly by making it unstable and more likely to emit off ionizing radiation of another type [Bodner and Rhea, 1978; UNSCEAR, 2008]. In fact, neutrons are the only type of radiation that is capable of turning the other materials radioactive. Another ionizing radiation that is of wave nature is called gamma radiation, which is also emitted from an unstable nucleus during a radioactive process [Martin, 2013, Ionizing Radiation, Part 1, Vol 75]. It consists of photons of high energy and carries no mass or charge. Because of these, they have more penetration power and travel very far in air; they lose (on average) half their energy for every 500 feet. For stopping gamma rays, a thick or dense layer of material with a high atomic number such as lead or depleted uranium is required.

6.2.2.2 Atomic origin

Likewise, another ionizing radiation is an X-ray, which has a wave nature. But the primary difference is originating from the electron cloud, generally released from de-excitation of a tightly bound electron. They are usually longer wavelength and lower energy than gamma radiation i.e., wavelength of X-rays are ranging from 0.01 to 10 nm and energies in the range 124 eV to 124 k eV. This energy is enough to eject out an electron from atoms and dislocate molecular bonds and is harmful to living tissue and produces a risk of radiation-induced cancer [Bodner and Rhea, 1978; Martin, 2013; Radiation IARC Monographs, Volume 100D]. Radiation sickeness is caused with a very high radiation dose of X-rays when exposed over a short period of time. This ionizing radiation finds its application in killing malignant cells for the treatment of cancer.

An ion beam radiation is a type of charged particle beam consisting of ions ranging from few eV to MeV to GeV. The energy of GeV is mainly used in particle physics, for material science engineering we are concentrating up to a few MeV energy only. On basis of energy, the role of ions in material modification can be divided in three regions: (i) very low energies, up to a few keV, suitable for synthesis of thin films by sputtering; (ii) energies from tens of keV to a few MeV for implantation related processes; and (iii) the swift heavy ions (SHI), energies >1 MeV/ nucleon, where electronic energy loss is dominant. Energetic ion beams emerged as a powerful technique in material modification, which is capable of generating different macro/nanostructures in remarkable ways on the surface or in the near-surface region of a solid. Figure 6.4a represents the schematic of energy deposition by ion projectiles when interacting with solids.

At a biomedical interface, ion beams produce novel and interesting research [Podgorsak, 2005; Grandfield and Engqvist, 2012]. In addition to exciting possibilities emerging in material science

(a)

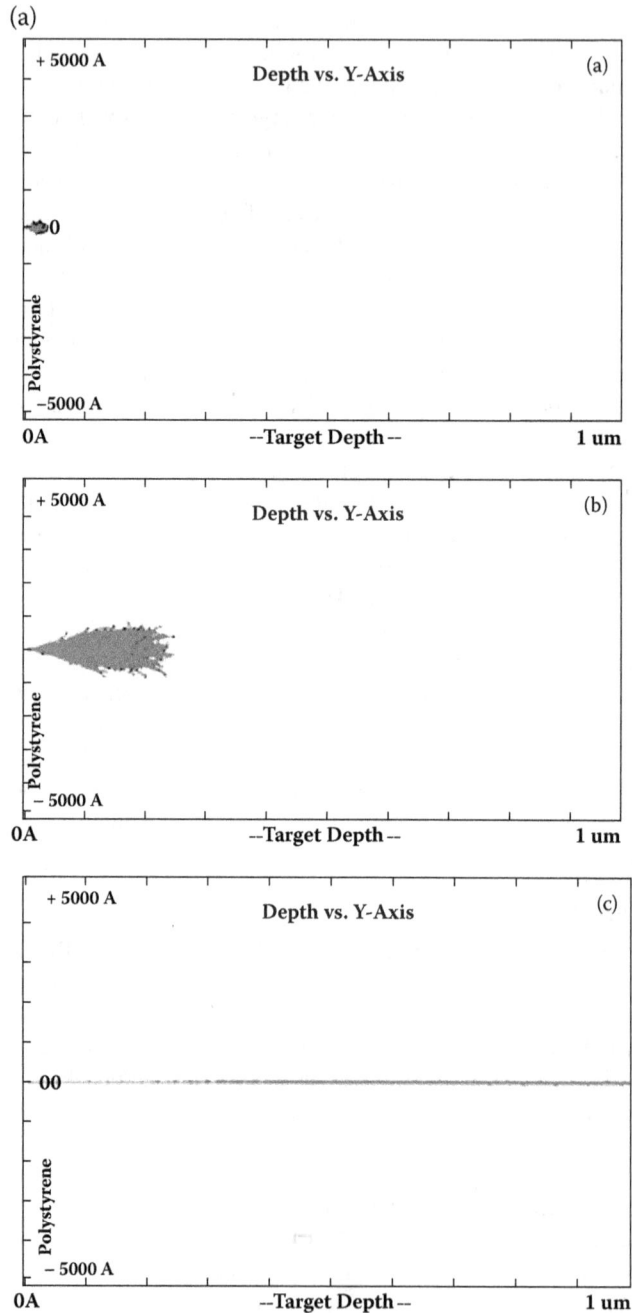

FIGURE 6.4A Scheme of energy deposition when ion projectiles (Ar+) interact with solids: (a) very slow charged ions of 10 keV kinetic energy, (b) slow 100 keV charged ions, and (c) swift ions of 1,000 MeV kinetic energy (Based on SRIM 2008 calculations).

engineering, they have diverse applications in biotechnology, biomedical engineering, medicine, etc.; therefore, we put our focus on the thrust of ion beam irradiation in the creation of functional materials, devices, and systems through control of matter on the nanometer length scale and exploitation of novel phenomena and properties (physical, chemical, biological) at that length scale.

Electron-beam processing is employed for the following three kinds of product modifications:

- To improve mechanical, thermal, chemical, and other properties of polymer-based products by cross-linking.
- Recycling of materials by material degradation.
- Medical and pharmaceutical goods sterilization by radiation processing.

As the name indicates, the electron-beam irradiation involves electrons usually of large energy, to treat an object for a variety of purposes. Electron beam energies can be controlled and may vary from few keV to several MeV, depending on the required penetration depth. In order to estimate or analyze the extent of modification by passage of incident energetic ion into material medium, it is essential to estimate the energy loss amount.

GENERAL PROCEDURE TO EVALUATE -dE/dx (ELECTRONIC & NUCLEAR)

Let the instantaneous energy of the incident charged particle, which traverses a target medium with N scattering centers, be E and $d\sigma(E,Q)$ is the differential cross-section for the transfer of energy Q by the incident ion of energy E. The number of interactions N_0 taking place in a thickness dx of the target medium in which the energy Q is transferred to the scattering centers can be given as

$$N_0 = N \ d\sigma(E, Q) \ dx \tag{6.1}$$

Energy transferred in these interactions or the energy lost by the charged particle to the scattering centers of the target medium would be

$$dE(Q) = QN \ d\sigma(E, Q) \ dx \tag{6.2}$$

The energy loss rate ($-dE/dx$) in the target medium after taking into account all possible values of Q, is determined by integrating over values of Q between Q_{min} and Q_{max}.

$$-\frac{dE}{dx} = N \int_{Q_{min}}^{Q_{max}} Qd\sigma(E, Q) \tag{6.3}$$

The exact mathematical nature of $d\sigma(E,Q)$ would depend upon the form of interaction potential between the incident ion and the scattering centers.

6.2.2.2.1 Electronic energy loss

When a swift heavy ion traverses through any absorber material, it is one of the important modes of energy transfer to the absorber material occurs mainly via coulomb interactions with the atomic electrons. Because of the small mass of atomic electrons, the transfer of kinetic energy to these electrons may result in the excitation and ionization of the target atoms.

A simplified classical treatment for the evaluation of electronic energy loss rate $(dE/dx)elec$, based on Bohr's assumptions, is as follows (shown in Figure 6.4b): Consider an incident ion having effective charge, mass M, velocity v, and kinetic energy E, transfers an amount of energy Q to an electron, of the target atom at impact parameter b, which is considered to be free and stationary. It is clear from this figure that the coulomb force F on the electron, which changes its direction continuously with time, can be resolved in two components, F_x and F_y, along and perpendicular, respectively, to the direction of motion of the incident ion. The impulse, $\int F_x dt$ will not contribute to the energy transfer as it is zero by symmetry. This is simply because of the reason that for each position of the incident heavy ion in the negative x direction, there is a corresponding position in the positive x direction, which makes an equal and opposite contribution to the x component of the momentum. Therefore, the impulse will only contribute because throughout the

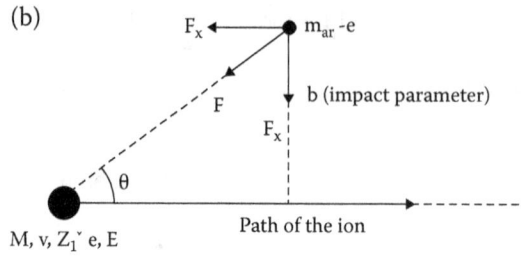

FIGURE 6.4B   Coulomb interaction between heavy ions of effective charge with a target electron.

passage of the incident ion, the direction of F_y, remains same. Simple calculations lead to the fact that the momentum transferred Δp to the electron is

$$\Delta p = \frac{2Z_1^* e^2}{bv} \tag{6.4}$$

Accordingly, the kinetic energy (Q) transferred by the ion to the target electron having rest mass m_0, in non-relativistic velocity region, is given by

$$Q = \frac{\Delta p^2}{2m_0} = \frac{2Z_1^{*2} e^4}{b^2 m_0 v^2} \tag{6.5}$$

The differential cross-section, $d\sigma\,(Q)$, for an energy transfer Q and Q+dQ is given by

$$d\sigma\,(Q) = -2\pi b db \tag{6.6}$$

and the energy loss per unit path length, $-dE/dx$, is

$$\left(-\frac{dE}{dx}\right)_{elec} = NZ_2 \int_{Q_{min}}^{Q_{max}} Q d\sigma(Q) \tag{6.7}$$

where N is the number of target atoms per unit volume and Z_2 is the number of electrons per target atom.

In terms of impact parameter 'b', Equation (6.7) becomes:

$$\left(-\frac{dE}{dx}\right)_{elec} = NZ_2 \int_{b_{min}}^{b_{max}} 2\pi\ b\ Q db \tag{6.8}$$

Substituting the value of Q from Equation (6.5) in Equation (6.8), we get:

$$\left(-\frac{dE}{dx}\right)_{elec} = \frac{4\pi NZ_2 Z_1^{*2} e^4}{m_0 v^2} \ln \frac{b_{max}}{b_{min}} \tag{6.9}$$

The values of b_{min} and b_{max} cannot be taken to be zero and infinite, respectively, because then the predicted energy loss will be infinite in contradiction to the reality. Therefore, some finite values of

b_{max} and b_{min} are to be chosen. The values of b_{max} and b_{min} can be chosen as those values of impact parameter, which correspond to minimum and maximum energy transfer to the target electron, respectively. If the minimum energy transfer is considered the mean ionization potential 'I' of the target atom, then in the light of Equation (6.5):

$$b_{max} = \frac{2Z_1^* e^2}{\sqrt{2m_0 v^2 I}} \tag{6.10}$$

Similarly, in case the maximum energy, which can be transferred to a stationary electron by an ion, is taken to be $2m_0 v^2$, then the corresponding value of impact parameter i.e., b_{min} in the light of the Equation (6.5) can be taken as:

$$b_{min} = \frac{Z_1^* e^2}{m_0 v^2} \tag{6.11}$$

On substituting the values of b_{max} and b_{min} from Equations (6.10) and (6.11), respectively, in Equation 6.9), we obtain:

$$\left(-\frac{dE}{dx}\right)_{elec} = \frac{2\pi\ NZ_2\ Z_1^{*2} e^4}{m_0 v^2} \ln \frac{2m_0 v^2}{I} \tag{6.12}$$

Equation (6.12) is derived after taking into account only the direct collisions of heavy ions with the electrons in the target atoms. In addition to this, there is one more term of comparable magnitude due to distant resonant energy transfer collisions.

Therefore, the total energy loss rate becomes twice that given by Equation (6.12) and finally:

$$\left(-\frac{dE}{dx}\right)_{elec} = \frac{4\pi NZ_2 Z_1^{*2} e^4}{m_0 v^2} \ln \frac{2m_0 v^2}{I} \tag{6.13}$$

The above expression is similar to Bethe's expression for incident ions at energies in the non-relativistic region.

In the SI system of units, Equation (6.13) can be expressed as:

$$\left(-\frac{dE}{dx}\right)_{elec} = \frac{NZ_2 Z_1^{*2} e^4}{4\pi\varepsilon_0^2 m_0 v^2} \ln \frac{2m_0 v^2}{I} \tag{6.14}$$

where ε_0 is the permittivity of the free space.

Incorporating relativistic shell and density effect corrections, Equation (6.13) becomes:

$$\left(-\frac{dE}{dx}\right)_{elec} = \frac{4\pi\ Z_1^{*2} e^4}{m_0 v^2} NZ_2 \left[\ln \frac{2\ m_0 v^2}{I} - \ln(1-\beta^2) - \beta^2 - \delta - U \right] \tag{6.15}$$

where β is the velocity of the incident ion relative to the velocity of the light, δ is the correction factor for the polarization of the medium, generally referred as density correction and U is the shell correction; Equation (6.15) is generally referred as the Bethe–Bloch formula.

6.2.3   DENSITY CORRECTION

The electric field of the incident ion also tends to polarize the target atoms along its path. As a result of this polarization, the outer-lying electrons in the absorber material are shielded from the full electric field intensity. Therefore, these electrons will contribute only partially towards the energy loss and accordingly a correction factor is included in Equation (6.15). This effect depends on the density of the material, since the induced polarization will be greater in condensed materials than in gaseous absorbers.

6.2.3.1   Shell correction

The shell correction accounts for effects, which arise when the velocity of the incident ion is comparable or smaller than the orbital velocity of the bound electrons. At such energies, the assumption that the electron is stationary with respect to the incident ion is not valid and therefore a correction factor U in Equation (6.15) is included.

6.2.3.2   Other corrections

Some more correction factors like Barkas correction, finite nuclear size correction, Bloch and Mott correction, projectile structure correction, etc. are also required to be incorporated but their contribution is quite insignificant.

6.2.3.3   Nuclear energy loss

In this case, the incident ion transfers its energy to the absorber material as a consequence of elastic collisions between the incident heavy ion and the target nuclei, while traversing through it, in contrast to that of electronic energy loss where the interaction takes place between the incident ion and the electrons of the target atoms. The expression for the nuclear energy loss can be derived by adopting the similar procedure as that for the electronic energy loss and is represented as:

$$\left(-\frac{dE}{dx}\right)_{nucl} = \frac{4\pi N Z_2^2 Z_1^{*2} e^4}{M_2 v^2} \ln \frac{b_{max}}{b_{min}} \qquad (6.16)$$

where Z_2 and M_2 are the atomic number and the mass number of the target nucleus, respectively, and N is the number of target atoms per unit volume.

In this case, the maximum energy transfer is given by the expression:

$$Q_{max} = \frac{4M_1 M_2}{(M_1 + M_2)} E \qquad (6.17)$$

and the minimum energy transferred to the target atom can be taken to be equal to the displacement energy (I_{dis}) required to remove an atom. Consequently, the values of b_{max} and b_{min} can be calculated.

6.2.3.3.1   Relative comparison of electronic and nuclear energy loss

The difference between the electronic energy loss rate $(-dE/dx)_{elec}$ and nuclear energy loss rate $(-dE/dx)_{nucl}$ can be evaluated by comparing the expressions given in Equations (6.9) and (6.16) derived for electronic energy loss rate $(-dE/dx)_{elect}$ and nuclear energy loss rate $(-dE/dx)_{nucl}$, respectively. The major difference between the two is that in case of nuclear energy loss, the mass of the target atom M_2 appears instead of the mass of the electron m_0. In addition, NZ_2 (the number of target electrons per unit volume) appeared in the case of electronic energy loss is replaced by the number of target atoms per unit volume N.

For a proton, neglecting the ratio of ln terms, the ratio of nuclear to electronic energy loss rate is:

$$\frac{\left(-\frac{dE}{dx}\right)_{nucl}}{\left(-\frac{dE}{dx}\right)_{elec}} \cong \frac{Z_2}{M_2}m_0 \cong \frac{1}{3600}$$

with $M_2 \cong 2Z_2m_p$ and m_p is the mass of the proton that is equal to $\sim 1800m_0$.

Thus, for MeV heavy ions, the nuclear energy loss is negligible in comparison to electronic energy loss. However, for energies below 1 MeV, when the incident ion behaves as a neutral atom, the energy loss process occurs via hard sphere scattering and now the nuclear energy loss becomes significant.

It is now quite clear that for MeV heavy ions in non-relativistic region, the electronic energy loss is the only dominant mode of energy loss and thus all other modes of energy loss can simply be neglected.

6.3 RADIATION INTERACTION MECHANISM

The properties of material can be modified through ionizing as well as non-ionizing radiation treatments like ultraviolet, IR, gamma, X-rays, ion/electron beam irradiation, etc. As underlying interaction phenomenon is complex, the overall damage cannot be directly probed; instead, individual effects of radiation damage like changes in mechanical, optical, electrical properties, etc. can be studied. The knowledge of interaction mechanism between ionizing radiations and matter is necessary for studying the phenomenon governing the observed changes. The ionizing radiations include all kinds of electromagnetic and corpuscular radiations, e.g., UV rays, X–rays, gamma rays, electron/ion beam, etc. with energies appreciably greater than the dissociation energy of the bonds present in the material.

Any charged particle having a mass with sufficient kinetic energies can easily ionize atoms by fundamental interaction employing coulomb force. However, electromagnetic or neutral particle radiations, while traversing through a material medium, transfer their energy to the medium through following three mechanisms: photoelectric absorption, Compton scattering, and pair production processes [Makuuchi and Cheng, 2012, Attix, 1986, Turner, 2004]. Instead of direct transfer of energy via ionization, the energy is lost by photon absorption via any of the above-mentioned processes, resulting in the creation of fast-moving electrons. These fast-moving electrons further interact with target medium and produce secondary electrons and so on. As energy of these photons gets attenuated in successive interactions rather than showing a continuous loss; therefore, electromagnetic interaction modifies a complete bulk structure. As a result of energy transfer by electromagnetic rays to polymers, molecular-level changes such as chain–scissoring, cross-linking, free radical formation, elimination of volatile species, reordering the chemical bonds, etc. occur in polymers that ultimately results in modification of their properties [Grupen C., 2010].

When an energetic ion is impinged on a material medium, energy lost causes several changes to the medium and also on the ion itself [Bernas H., 2010]. The induced changes in the target medium arise at the cost of energy transferred by the charged particle to the target medium. The energy is lost via one of the following processes:

a. Excitation and ionization

In this mode, the energy from the incident particle is transferred to target electrons through inelastic collisions, leaving target atoms in excited or ionized states. The energy loss through this process is commonly known as electronic energy loss.

b. Nuclear collisions

At ion energies less than 1 MeV, the energy is transferred by a projectile to a target nuclei through elastic collisions and the target atoms get displaced. Energy loss in such collisions is nuclear energy loss.

c. Generation of photons

A charged particle in the vicinity of the target nucleus undergoes acceleration or deceleration and results in the emission of electromagnetic radiations known as Bremsstrahlung process. Also, at velocities greater than phase velocity of light in medium, again there is a loss of energy of the incident particle through the emission of Cherenkov radiations.

The relative contribution of energy loss processes mentioned above depends upon the nature of the incident particle and target material. The electronic energy loss as a result of excitation and ionization dominates for MeV heavy ions while nuclear energy loss becomes important for keV heavy ions. The Bremsstrahlung mode is significant for light charged particles like electrons and emission of Cherenkov radiations occurs at relativistic energies. Therefore, for the case of non-relativistic heavy ions, total energy loss can be considered a mixture of two nearly independent processes: (i) elastic collisions of incident ion with nuclei of target atoms, usually referred to as nuclear energy loss $(-dE/dx)_n$ and (ii) inelastic collisions of incident ion with atomic electrons of target atoms, generally expressed as electronic energy loss $(-dE/dx)_e$.

$$\left(-\frac{dE}{dx}\right)_{total} = \left(-\frac{dE}{dx}\right)_n + \left(-\frac{dE}{dx}\right)_e$$

Further, electron beam interactions, in general, are of two types: (1) elastic scattering of incident electrons by atomic nuclei and (2) inelastic scattering of incident electrons by atomic electrons. The energy transferred by high-energy electrons onto the target medium generally dissipated in the form of heat and in some cases it may leads to rise in temperature of the specimen, resulting in mechanical instability of the specimen. Another consequence of inelastic scattering is the chemical change in the specimen mainly because most polymers are electrically insulators and there is no ready source of free electrons to saturate the radiation-induced free radical. However, soft materials such as polymers with high concentrations of aromatic moieties are more radiation resistant than fully saturated polymers because of the delocalization associated with highly conjugated bonds that provide a source of electron charge and the means to distribute the adsorbed energy over a larger volume of material. The present chapter deals with radiation-induced effects in polymers; therefore, we are discussing briefly polymer types and their application in the next section. Thereafter, we again migrate to the radiation-induced effects in polymers [Durrani and Bull, 1987; Leo, 1994] Figure 6.5.

6.4   OVERVIEW OF POLYMERS

Polymers are very long chain macromolecules in which a large number of fundamental units called 'monomers' are clubbed together through covalent bonds, shown in Figure 6.6. Polymerization is a process through which monomers are connected together and is termed *polymerization* and the degree of polymerization represents the number of repeating units in a polymer chain. The average molecular weight of a polymer is the product of molecular weight of the monomer and the average degree of polymerization. The molecular weight of polymers is normally so high that for all practical purposes they are almost non-volatile [Fink, 2004; Gedde, Mikael, and Hedenqvist, 2019].

Several factors interplay in determining the enormous and intriguing range of properties of polymers. A few of them are chemical compositions of polymers, number of repeating units present, extent of elongation, molecular weight and distribution, spatial organization of repeating

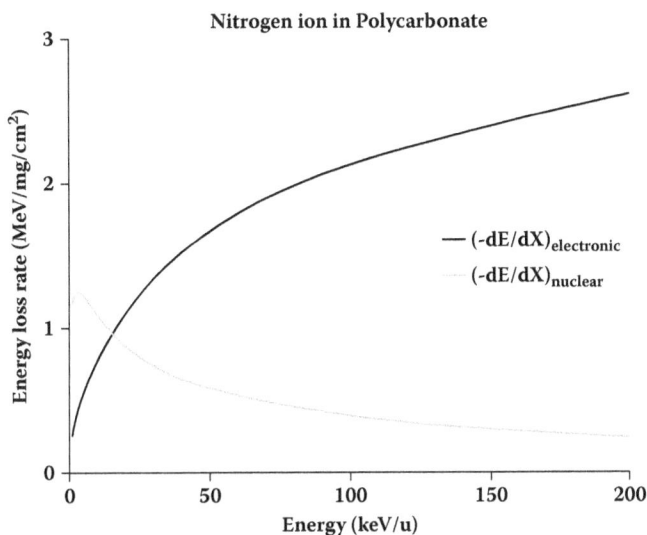

FIGURE 6.5 Comparison of electronic and nuclear energy loss rate as a function of energy for nitrogen ion in polycarbonate absorber.

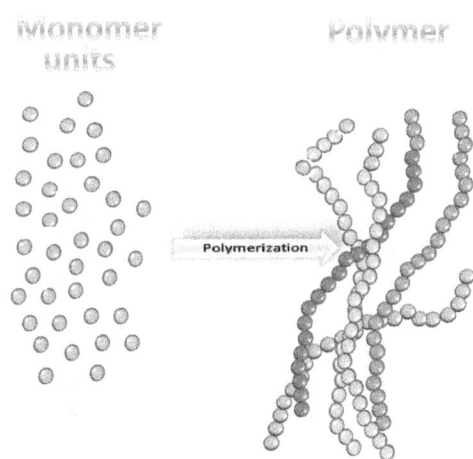

FIGURE 6.6 Schematic of monomer units joined together via polymerization to form a polymer.

units whether zig-zag or helical configuration, parameters for polymers processing, extent of impurities from stabilizers, etc. Further, polymeric chains generally are not linear in the continuum; rather, they are wrapped or folded, leaving large empty spaces in between. Therefore, they may have a straight thread-like structure, branched, or cross-linked morphology, which further may be entangled with each other, giving rise to amorphous or crystalline arrangements therein.

Depending upon the origin of polymers, they can be classified into two groups i.e., natural and synthetic polymers. DNA, RNA, proteins, natural rubber, etc. are a few examples of natural polymers. Synthetic polymers are made in the laboratory from low molecular weight compounds. They can be classified in several categories depending on their skeletal structure, chemistry of synthesis, reaction to temperature, and backbone. Further, the most basic requirement for the polymerization process is the capability of linkage between the molecules of individual monomer units through chemical reaction [Madkou, 1999; Fink, 2004; Gedde, Mikael, and Hedenqvist, 2019]. The monomer units may remain the same or may vary due to the release of small molecules during polymerization and structure is strongly dependent upon the method of polymerization like addition and condensation. Polymers are also categorized as thermosetting and thermoplastic depending upon their tendency to regain their original state after some treatment. Additionally, depending upon the backbone chain nature, they are also categorized as organic and inorganic polymers if they have organic functionality attached with the backbone.

6.5 PROPERTIES OF POLYMERS REQUIRED FOR BIOLOGICAL SYSTEM

Because of extraordinary properties exhibited by polymers like flexibility, mechanical strength, stability, etc. they are being widely considered in various scientific and technological applications. These days, polymers with biocompatibility, biodegradability, and biological activity are of great interest as using these types of polymers can help in mimicking the biological system and their structural parts [Sun et al., 2010, Ratner, 2013, Shen et al., 2015, Joshua et al., 2019, Schottler et al., 2016, Fletcher et al., 2018]. Further, such polymers are easily degradable after use, which automatically solves their decomposition issue and solves the environmental hazard issue. They are environmentally friendly, non toxic, and low cost. The field of polymers as biocompatible materials for tissue engineering, coronary stents, shape memory materials, etc. has shown a drastic development in the last decade [Vrancken, Buma, and van Tienen, 2013, Schottler et al., 2016]. Biodegradable polymers play an important role in orthopedics and can be degraded by hydrolysis and enzymatic activity, thus having a range of mechanical and physical properties that can be engineered to suit a particular application. Their biodegradability depends on several parameters like molecular structure, crystallinity, and copolymer ratio. But for commercialization of these polymers in medical applications, their biocompatibility is still to be checked. Therefore, research emphasis should be paid in modifying the physical and chemical properties of biodegradable polymer so that they will be well compatible with the body [Ratne, 2013, Schottler et al., 2016].

The synthetic material are foreign material for natural biological system and they do not accept the foreign body although generating the self-defense against them; therefore, it must be noted that the synthetic material when used for biological purpose must have compatible biological and bio-adoptive properties. The interdisciplinary research and scientific research have been continuously in progress to produce the synthetic material or modifying the natural material for production of body implants, biomedical devices, sterilization, non-venomous biosensors, failure of material, etc. The biocompatibility and bio-adoptability of the polymer-based implanted biological system must be investigated in-vivo and in-vitro before their use. Also, the desired surface and bulk properties are needed for their optimum performance in a specific biological purpose. Further, different biological applications require different properties. For example, polymers used for contact lenses and artificial bones must have flexibility and very high wettability, while polymers used for artificial teeth require low wettability [Lina, Ali, and Ahmed, 2019, CalóVitaliy and Khutoryanskiy, 2015, Stephen, Musgrave, and Fang, 2019, Xu J. et al., 2018, Musgrav, Nazarov, and Bazin, 2017]. Furthermore, for artificial skin application, the upper side of the skin should have low blood compatibility while the inner side must have high blood compatibility and the overall system must have high porosity. Therefore, use of polymers for the proper biological system is very challenging, and such challenges include the following features:

- Toxicity and eco-friendly nature
- Wettability
- Biocompatibility, biodegradability, and bio-adoptability
- Biological response to foreign material
- Porosity
- Cell/tissue/protein adhesion
- Micro-pattering
- Surface and chemical functionality

- Cross-linked film formation
- Anti-bacterial and anti-microbial properties

As every material does not possess all these properties inherently however, polymers are found to be most suitable. After dictating basic properties and features of few biocompatible polymers we migrate to review changes in properties of PVA through different radiation exposures and biological/biomedical applications.

Poly(methyl methacrylate) (PMMA) is a transparent thermoplastic also known as acrylic, acrylic glass, or Perspex. Its monomer unit and structure is shown in Figure 6.7a. It is lightweight and durable; thus, it is used as a shatter-resistant alternative to glass. It offers very good compatibility with human tissue and was first reported as a synthetic polymer for medical use by a British ophthalmologist, Sir Nicholas Harold, in 1949 for making rigid intraocular lenses in the treatment of cataracts. It is also used in orthopedic surgery, bone cement of PMMA i.e., grout of PMMA powder with liquid methyl methacrylate (MMA). It is used to affix implants and to remodel lost

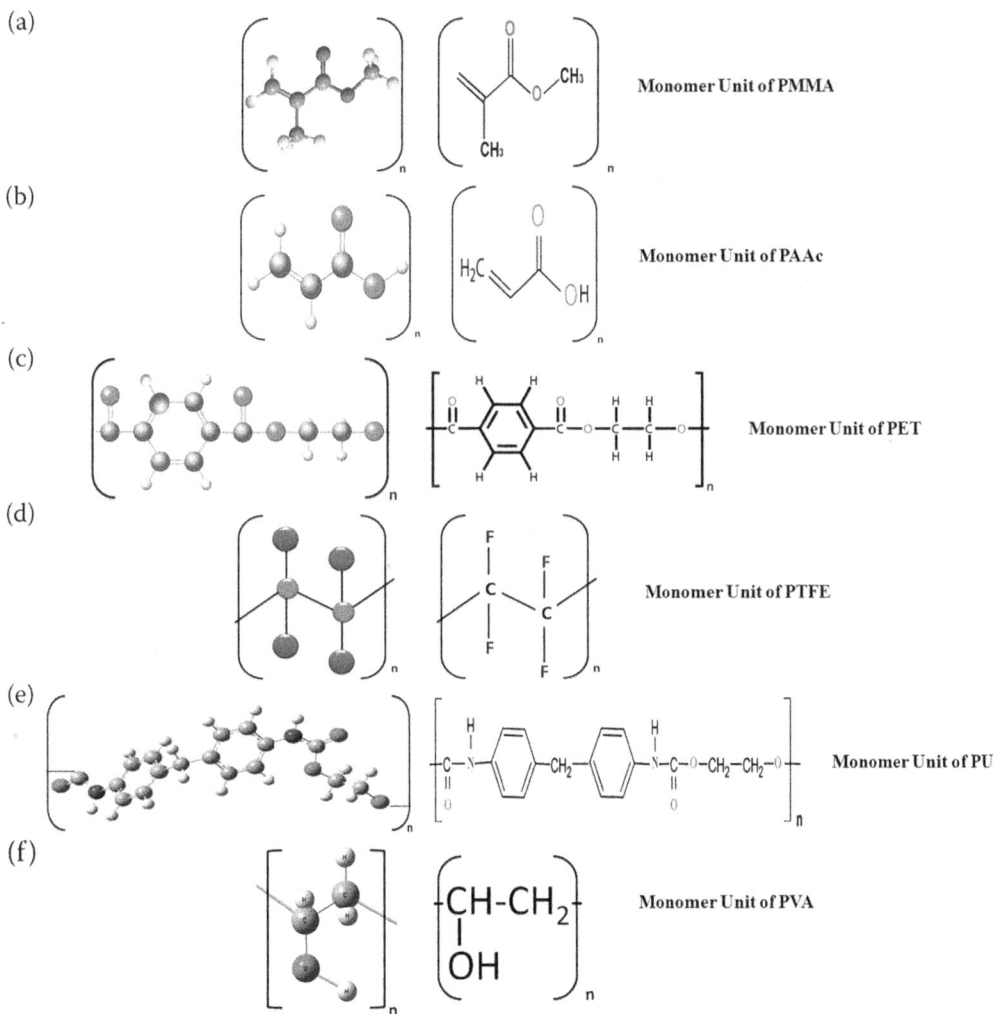

FIGURE 6.7 (a) Monomer structure of PMMA; (b) monomer structure of PAAc; (c) monomer structure of PET; (d) monomer structure of PTFE; (e) monomer structure of PU; (f) monomer structure of PVA.

bone. It is also used in cosmetic surgery to reduce wrinkles or scars permanently. For this purpose, microspheres of PMMA suspended in some biological fluid are injected as soft-tissue filler under the skin. Currently, it is highly used to create micro-fluidic lab-on-a-chip devices.

Polyacrylic acid (PAAc), a derivative of acrylic acid ($CH_2=CHCO_2H$) (shown in Figure 6.7b), it is an acid-base and water-attracting polymer. It has the ability to absorb and retain water and swell. It is one of the mucoadhesive polymers used for bioadhesive drug delivery systems.

Poly (ethylene terephthalate) (PET), is the most common thermoplastic polymer resin of the polyester family. Its monomer structure is shown in Figure 6.7c. It is a colorless, semi-crystalline resin, lightweight in its natural state, and can be rigid or semi-rigid based on processing. Its fibers are used for vascular prostheses or good blood flow.

Polytetrafluoroethylene (PTFE) is another synthetic polymer, and its monomer structure is shown in Figure 6.7d. It is a strong, tough, waxy, nonflammable synthetic resin produced by the polymerization of tetrafluoroethylene. It is well known by its trade name, Teflon. It consists of a chain of carbon atoms with two fluorine atoms bonded to each carbon $\{-CF_2-CF_2-\}$. For its synthesis, tetrafluoroethylene monomers (small, single-unit molecules) are suspended or emulsified in water and then polymerized (linked into giant, multiple-unit molecules) under high pressure in the presence of free-radical initiators. The foam of PTFE is used for vascular prostheses and cardiopulmonary bypass surgery.

Polyurethene (PU) is also used in biomedical applications. Its momomer structure is shown in Figure 6.7e. Its segmented porous foam is used in cardiovascular surgery. Such polymers are used to oxygenate blood, and thus should be made sure that they operate without blood damage.

Polysiloxane is a synthetic polymer made up of silicon, oxygen, carbon, and hydrogen. It is a mixture of semi-inorganic polymeric molecules consisting of an array of polydimethylsiloxane $[(CH_3)_2\text{-}SiO]$ monomers chains of different length. It is commonly called a silicone polymer. Its elastomers are available in a wide range of hardness, have good UV resistance, excellent thermal, and chemical resistance. It allows sterilization using steam, autoclave, or gamma radiation. It is widely used in the biomedical industry. It is prone to microbial infections when exposed and interactions with the host tissue, thus specific antimicrobial treatment is essential before using it for medical inserts. It is also used as speciality contact lenses, drains and shunts, urinary catheters, reconstructive gel fillers, craniofacial prosthesis, nerve conduits, and metatarsophalangeal joint implants.

Semipermeable membrane biopolymers (cellulose), with less antigenicity used in artificial kidney for hemodialysis is futuristic with reduced or no use of immunosuppressive drugs.

Among various polymers, we restrict the discussion to poly (vinyl alcohol) (PVA), as it is one of the most biological acceptable polymers [Jiang, Liu, and Feng, 2011; Qiu and Netravali, 2012; Muppalaneni and Omidian, 2013, Abramowska et al., 2015, Kamoun et al., 2015]. Poly (vinyl alcohol) is considered as one of most capable polymers having great perspectives in development of novel biomaterials. Its water solubility, non-toxicity, and ease in synthesis making it quite popular for making hydrogels in pharmaceutical and biomedical industry for applications like drug delivery, tissue engineering, burn and wound dressing, artificial cornea and contact lenses, etc. [Baker et al., 2012; Binetti, Fussell, and Lowman, 2014; Teodorescu, Bercea, and Morariu, 2018]. Its chemical structure is shown in Figure 6.7f and consists of the main backbone of carbon atoms and pendant hydroxyl groups. PVA is synthesized from poly (vinyl acetate) (PVAc) employing saponification reaction in which ester groups of vinyl acetate are replaced by hydroxyl ones. While the majority of vinylic polymers are obtained by polymerization of vinyl alcohol monomer, which itself is unstable and rearranges itself to acetaldehyde. As saponification reaction itself is not a complete one, therefore the result product is a copolymer of PVA and PVAc. The number of the residual acetate groups determines the degree of hydrolysis and, in addition to polymerization degree, polydispersity index, and chain tacticity, affect the physicochemical properties of the re-sulted polymer (e.g., solubility, crystallinity, mechanical strength, adhesion, diffusivity, etc.). Further properties of PVA-based materials can be easily modulated according to desired applications. The modulation strategies involve either physical methods, such as applying successive freezing-thawing cycles, heat treatment, irradiation or annealing, blending with other polymers, or chemical modifications of OH group and copolymerization. All the above-mentioned things have gained widespread interest in meticulous research in PVA and PVA-based materials for various applications. However, these days, irradiation induced effect in PVA for biomedical application have been explored very much [Muppalaneni and Omidian, 2013, Abramowska et al., 2015, Qiu and Netravali, 2012, Soo-Tueen Bee et al., 2014, Ehab and Shaimaa, 2018].

Further, there are various methods through which the properties of polymer or synthetic material can be tuned to above-mentioned features. However, various radiation-induced modifications in properties of polymers are subjectively chosen widely. In the next section, we briefly understand the irradiation-induced effects in polymers especially in PVA via different radiation exposure.

6.6 RADIATION-INDUCED EFFECTS IN POLYMERS FOR BIOLOGICAL AND BIOMEDICAL APPLICATION

As in this chapter we mainly concentrate on the UV, IR (thermal), gamma, and swift heavy ion-induced changes in polymers. Therefore, we have divided the discussion in two major parts first is related to non-ionizing radiation and second, ionizing radiation induced effects in polymers. Irradiations comprises of various kinds of electromagnetic or corpuscular radiations having energies considerably greater than bond dissociation energy. The energy transferred through radiation–matter interaction and the structure of polymer determines the amount of changes observed in structure of polymers.

Broadly, upon irradiation, various chemical changes take place in the polymeric medium which may be categorized as:

- Cross-linking which may be defined as the formation of different bonds between molecules.
- Chain-scissoring, which is the irretrievable cleavage of bonds resulting in disintegration of molecular chains.
- Free radical formations, which are the unsaturated groups behaving as chemically reactive sites.

These modifications at molecular level alter the structural properties of target material which again become responsible for the alterations in optical, electrical, thermal, mechanical, etc. behavior of the polymeric material.

6.6.1 NON-IONIZING RADIATION-INDUCED EFFECTS

The ultraviolet and IR radiation absorb by naturally and synthetic polymers and undergoes into photochemical changes like photolytic, photo-oxidative, and thermo-oxidative reactions that causes the degradation of the material [Yousif and Haddad, 2013, Cahn, Haasen, and Kramer, 2013, Emad Yousif et al., 2015]. Almost all synthetic as well as natural polymers undergo physical processes, such as polymer re-crystallization, or denaturation of protein structures under UV and IR irradiation. Chemical processes may either increase average molar by cross-linking inside the polymer or decrease it through macromolecular chain bond scission.

6.6.1.1 UV irradiation

UV irradiation may activate the process of photolysis in polymers based on the mobility of radicals present in polymers and their bimolecular recombination. While talking about radicals interactions with polymers, they interact by either combining or removing hydrogen atoms from polymer chains. Among all radicals, hydrogen radicals have large mobility inside polymer matrix while the phenyl ones have limited mobility. When a polymer is irradiated with UV irradiation, then it causes breakage in polymer chains, production of radicals, reduction of molecular weight collectively called photooxidative degradation of polymers. All these processes ultimately deteriorates the properties of polymer and after some fickle amount of time, polymer may get converted to a useless material [Sheela et al., 2014; Yoona et al., 2017; Zidan et al., 2018]. Prior to usage, biomaterials needed to be sterilized. However, many sterilization procedures such as steam autoclave or heat sterilization are known to strongly affect polymer properties. UV irradiation is used as an alternative sterilization method in many laboratories; however, potential alterations of polymer properties have not been extensively considered.

Fischbach et al., (2001) has provided a very good review on the subject. Does UV irradiation affect polymer properties relevant to tissue engineering? They imply the paramount importance of sterilization time control. As extended time of UV sterilization may increase the degradation of the polymer and cell adhesion properties.

Thang et al., (2013) prepared ultraviolet (UV) irradiated series of poly (vinyl alcohol)/ chitosan (PVA/CTS) hydrogel thin films. They studied swelling behaviors, intermolecular chemical bonds, molecular structures, thermal behaviors, degrees of crystallinity, and morphologies of the surfaces and internal structure of the prepared composite films. They confirm the chemical cross-links formed in hydrogel films via free-radical reactions initiated via UV irradiation. They found improved pH stability and controlled the degree of swelling with retaining a high swelling rate and proposed them as promising material for wound dressing and other biomedical applications.

6.6.1.2   Thermal radiation

The principle mode of thermal energy (via IR and microwave radiation) assimilation in polymers is by increase in vibrational energy of its constituent atoms or molecules. The atomic vibrations of adjacent atoms are coupled through atomic bonding. With the increase in the temperature, the relative amplitude of these atomic vibrations increases which results in an enhancement in segmental motion and vibrant bond stretching of the polymeric chains. Thus, these polymer chains rearrange themselves more easily into a more closely packed configuration reducing free volume. This ultimately forms a new polymer with distinguished physical and chemical properties from the original one. The extent of such changes mainly depends on molecular structure of polymer and energy being transferred to medium through heating.

Natalia et al., (2005), study the microwaves irradiation induced effects on structure of PVA. The irradiation causes the stepping up of chemical reactions in the presence of microwave irradiation. In addition to heating effect, the microwave field also produces specific effects on the reactants and resulting in changes of reaction rate constant. These results indicate that microwave irradiation may be a promising tool to alter the structure of polyvinyl alcohol.

Petrova et al., (2005) studied the possibility of cross-linking in thin polyvinyl alcohol films upon microwave irradiation. They presented a comparison of effect of microwave irradiation and heat treatment on cross-linking property of polymers. The cross-linking of polyvinyl alcohol is a result of intermolecular etherification reactions involving hydroxyl groups in both cases, but cross-linking is more in the case of microwave irradiation.

Bernal et al., (2012) studied the effect of microwaves radiation exposure on poly (vinyl alcohol) dissolved ethylene glycol and determine the possible degradation. UV-visible absorption spectroscopy indicates that microwave irradiation results in formation of unsaturated conjugated bond and loss of hydroxyl groups in PVA but this kind of degradation is small. The size exclusion chromatography suggests that microwave radiation exposure do not change significantly the molar mass of PVA and neither chain cleavage nor cross-linking reactions are observed. Thereby, suggesting that MWs hardly degrade polymer solution for prospective applications.

Nguyen et al., (2019) studied the synthesis of silver nanoparticle by microwave heating on in polyvinyl alcohol (PVA)) as a wide-range antibacterial agent. The different concentration of PVA in PVA/Ag$^+$ solution was exposed to microwave radiations for a short period. The structural analysis carried by FTIR reveals that composite samples go through chemical changes depending on the exposure times and concentration of PVA in PVA/Ag^+. It was also confirmed that the Ag^+ ion was reduced within PVA after microwave exposure. Also, a reduction has been observed in the rate of hydrolysis with the addition of Ag^+ in PVA and ether bridges are limited by the spatial structure.

6.6.2 IONIZING IRRADIATION INDUCED EFFECTS

6.6.2.1 Gamma radiations

Gamma rays, while traversing through the medium, transfer their energy through various processes, such as:

- Photoelectric effect
- Compton scattering
- Pair production

All of the above-mentioned processes result in energy transfer to target electrons that further produce secondary electrons as a consequence of the ionization process. Instead of losing energy in a continuum, gamma photons simply attenuate; therefore, gamma irradiation modifies the complete bulk structure of material. Gamma radiation is primarily used in the medical field for sterilization of medical instrument/devices in order to eliminate the problem associated with the pathogen attachment (Kumar et al., 2015].

Cristina et al., (2009) have studied the effect of gamma radiation on various concentrations of porous polymer alloy membranes of poly (hydroxy-urethane) (PHU) and poly (vinyl alcohol) (PVA). They have observed that it is the alloying concentrations of the membranes that affect their hydrophilic character, surface energy, resilience, and initial elastic module based. With increasing doses of gamma irradiation, porosity and hydrophilic properties of membranes also improved.

Stoica-Guzun et al., (2012) explored the use of UV-irradiated poly (vinyl alcohol)-bacterial cellulose (PVA-BC) composites for food packaging applications. They irradiated the prepared composite with 254 nm (using a mercury lamp, Philips TUV-30), for 1 to 10 hrs. They found modification in transparency and swelling characteristics of the composite film.

Yamdej et al., (2017) fabricated cross-linked poly (vinyl alcohol)/sericin hydrogels with different sericin concentrations using gamma irradiation. They studied the physicochemical and biological properties of the poly (vinyl alcohol)/sericin hydrogels and suggested the use of these hydrogels for wound dressings for the treatment of dry and low-exudate wounds.

Ahmed M. Elbarbary and Naeem M. El-Sawy, (2017) prepared polyvinyl alcohol/chitosan/silver (PVA/CS/Ag) nanocomposite membranes and studied the effect of γ-ray irradiation for promising antimicrobial and biomedical applications. They found good antibacterial activity that causes significant reduction in the growth of microbials. These membranes have a non-thrombogenicity effect and slightly hemolytic potential, suggesting their promising use in biomedical applications.

Recently, Swaroop Kumaraswamy et al., (2020) prepared radiolytically synthesized silver/polyvinyl hydrogel nanocomposites for wound dressing applications, Wael H. Eisa et al., (2011) prepared gamma-irradiation-assisted seeded growth of Ag nanoparticles within PVA matrix and Aleksandra Radosavljevic et al., (2020) prepared PVA/Ag nanocomposite hydrogels by gamma irradiation. They found that gamma rays are playing the role of reducing agent for Ag^+ ions and converting them to metallic silver (Ag^0) inside a PVA matrix and also increased the cross-linking of the PVA matrix. The nanocomposites' potential as an antibacterial surface has been observed for both Gram-positive and Gram-negative bacteria while a nil/minimal cytotoxic effect has been observed on human melanoma (SK-MEL-2) and mouse melanoma (B16-F1) cell lines [Ghasemzadeh and Ghanaat, 2014, Helmiyati, Abbas, and Budianto, 2019]. Overall, the conclusion was that gamma-irradiation-induced silver/polyvinyl alcohol hydrogel nanocomposites can be served as marvelous materials for wound dressing applications but with some optimized conditions.

6.6.2.2 Charge particle radiations

The major processes through which an incident-charged particle (electron/heavy ion) loses its energy to the target medium are:

- Elastic energy loss
- Inelastic energy loss

The elastic energy loss occurs as a result of excitation and ionization processes cropping up in the target medium as a consequence of inelastic collisions between the incident-charged particle and the target electrons. This type of interaction dominates for MeV energy charge particles. For the incident-charged particles having energies in keV, the incident ion collides elastically with the target nucleus and displaces it from its mean position. Such energy losses are termed *nuclear energy losses*. In both types of energy loss, the incident ion loses energy continuously along its trajectory. Depending upon incident ion parameters like dose, charge, energy, etc., the near surface layers or deeper layers or the entire bulk may gets modified. These parameters decide to use the particular charged particle irradiation of different biomedical applications.

Wu et al., (2001) studied the effect of irradiation of cross-linked, poly (vinyl alcohol) blended hydrogel for wound dressings. The results of gamma irradiation on blends confirm that irradiation dose and monomer concentration are major factors that govern the gel fraction of the hydrogel films. The gamma-irradiated PVA-blended hydrogel can be used for wound dressing applications without producing any toxic effect. The blends also have an additional advantage of fastening healing of wounds.

Dai et al., (2014) study the effect of electron beam irradiation on the cross-linking of PVA films in the presence of N, N'-methylene bisacrylamide (MBA). They observed that thermal stability of the films was improved after the irradiation and the FTIR study reveals that the cross-linking in PVA chains was increased after the irradiation. They observed that it is the crystallinity, not the crystal form of PVA, that is altered by irradiation of gamma rays. With increasing gamma dose, crystallinity first rises, then shows a drop. The gamma irradiation results in cross-linking among PVA/MBA films, which found advantages in increased thermal stability.

Inamura and Mastro, (2015) prepared PVA-gelatin composites and studied the electron beam irradiation effect. The mechanical properties, color measurements, water absorption, moisture, and film solubility were analyzed after the electron beam irradiation at different doses. After irradiation, the water absorption properties of the PVA and PVA-doped gelatin was found to decrease with an increase in irradiation dose. The decrease in water absorption can be ascribed to radiation-induced cross-linking. The ionizing radiation and the PVA content affect the properties of films must be taken in account whenever gelatin derivative materials were employed in a biomaterial arrangement.

Jahanabadi et al., (2019) studied the annealing effect of electron beam irradiation on different weight ratios of PVA-chitosan blends. They observed that with increasing irradiation dose, the gel contents of the blend films increased and swelling decreased. Finally, the films were prepared through a solution casting method and structural modification after irradiation could be used for biomedical applications.

6.7 SUMMARY

Polymers are one of the largest and most promising classes of materials being extensively used on a large scale in biomedical applications. The type of polymer material produced is mainly dependent upon the kind of radiation-induced product. This chapter discusses the types of radiation and their effect on polymers, being further useful for a variety of applications in biomedical science. The term *biomaterials* corresponds to technology required to boost and implicate recent developments in related strategies like tissue engineering, and the delivery of bioactive agents for treating, repairing, and restoring function of tissues. The major classifications of biomaterials include: metals, ceramics, natural or synthetic polymers, and composites. The radiation processing of polymers found applications in various fields like packaging for food irradiation, sterilization of medical devices, polymer modifications, etc. The radiations discussed are of various types, including the ionizing radiation that comprises ions produced from atomic and nuclear origin while the non-ionizing radiation comprises UV radiations, optical and ionizing radiations, and radio frequency radiations. All of these are essential for the production of biomaterials with various enthralling properties. The mechanism of ionization has been discussed in detail in the chapter along with the types of polymers used in the biomedical field. It has been found that usually hydrogels are the material of choice, which is tailored using the radiation and further used for applications. The effects of radiation upon these materials have also been explained in detail. The effect of different types of irradiation on polymer nanocomposites have been discussed throughout the chapter. These hold a lot of potential in the applications for future purposes as well if their properties are exploited in a productive way.

DEDICATION

Chapter six is dedicated to Late Prof. (Dr.) Shyam Kumar (26/07/1960–20/06/2022). He has made significant contributions to the field of fundamentals of ion beam interaction with matter and material modification through ion beam irradiation for a wide range of applications. He was known as a gentleman who was also a great mentor, scientist, and always passionate about work. He is a beacon in the darkness and a source of hope for the students, giving them the courage and strength to face life's challenges. Sir, you will always be missed.

REFERENCES

Abramowska A., (2015). The influence of ionizing radiation on the properties of starch-PVA films. *Nukleonika* 60(3), 669–677.

Abramowska, A., Cieśla, K. A., Buczkowski, M. J., Nowicki, A., and Głuszewski, W. J. (2015). The influence of ionizing radiation on the properties of starch-PVA films. *Nukleonika* 60 (3), 669–677.

Ahlbom A., and Feychting M. (2003). Electromagnetic radiation. *Br Med Bull* 68, 157.

Ahmed M., Elbarbary, and Naeem M. E. (2017). Radiation synthesis and characterization of polyvinyl alcohol/chitosan/silver nanocomposite membranes: Antimicrobial and blood compatibility studies. *Polymer Bulletin* 74, 195–212.

Attix F. H. (1986). *Introduction to radiological physics and radiation dosimetry*. New York: Wiley & Sons.

Baker, M. I., Walsh, S. P., Schwartz, Z., and Boyan, B. D. A. (2012). Review of polyvinyl alcohol and its uses in cartilage and orthopedic applications. *J. Biomed. Mater. Res. Part B Appl. Biomater.* 100B, 1451–1457.

Banoriya D., Purohit R., and Dwivedi R. K. (2017). Advanced application of polymer-based biomaterials. *Materials Today: Proceedings* 4, 3534–3541.

Bee S. T. et al. (2014). Effects of electron beam irradiation on mechanical properties and nanostructural–morphology of montmorillonite added polyvinyl alcohol composite. *Composites Part B: Engineering* 63, 141–153.

Bernal A., Ivo K., Kasparkova V., and Saha P. (2012). The effect of microwave irradiation on poly (vinyl alcohol) dissolved in ethylene glycol. *J of Applied polymer science* 128, 175–180.

Bernard M. et al. (2018). Biocompatibility of polymer-based biomaterials and medical devices – regulations, in vitro screening and risk-management. *Biomaterial Science* 6(8), 2025–2053.

Bernas H. (2010). *Material Science with ion Beams*, Berlin Heidelberg, Germany: Springer-Verlag.

Binetti, V. R., Fussell, G. W., and Lowman, A. M. (2014). Evaluation of two chemical crosslinking methods of poly(vinyl alcohol) hydrogels for injectable nucleus pulposus replacement. *J. Appl. Polym. Sci.* 131(19), 40843- 1–8.

Bodner G. M., and Rhea T. A. (1978). Natural sources of ionizing radiation, 7.

Brenner D. J., and Hall E. J. (2007). *N Engl J Med* 357, 2277.

Brun C. P., Delobelle M. F., Berger F., Chambaude A., and Jaffiol F., (2001). Mechanical properties determined by nanoindentation tests of polypropylene modified by He+ particle implantation. *Mater. Sci. Eng., A* 315 (1-2), 63–69.

Cahn R. W., Haasen P., Kramer E. J. (2013). *Materials science and technology: Degradation and stabilization of polymers*, Wiley-VCH Verlag GmbH & Co. KGaA.

CalóVitaliy E., and Khutoryanskiy V. (2015). Biomedical applications of hydrogels: A review of patents and commercial products. *European Polymer Journal* 65, 252–267.

Cao D., Yang G., Bourham M., and Moneghan D. (2020). Gamma radiation shielding properties of poly (methyl methacrylate)/Bi2O3 composites. *Nuclear Engineering and Technology* 52, 2613–2619.

Chahal R. P., Mahendia S., Tomar A. K., and Kumar S. (2011). Effect of ultraviolet irradiation on the optical and structural characteristics of in-situ prepared PVP-Ag nanocomposites. *Digest Journal of Nanomaterials and Biostructures* 6, 299–306.

Chahal R. P., Mahendia S., Tomar A. K., and Kumar S. (2012). γ-Irradiated PVA/Ag nanocomposite films: Materials for optical applications. *J. Alloys Comp* 538, 212–219.

Chahal R. P., Mahendia S., Tomar A. K., and Kumar S. (2015). UV irradiated PVA–Ag nanocomposites for optical applications. *Appl. Surf. Sci.* 343, 160–165.

Chahal R. P., Mahendia S., Tomar A. K., and Kumar S. (2016). SHI irradiated PVA/Ag nanocomposites and possibility of UV blocking. *Optical Materials* 52, 237–241.

Cieśla K., Abramowska A., Boguski J., and Drewnik J. (2017). The effect of poly (vinyl alcohol) type and radiation treatment on the properties of starch-poly (vinyl alcohol) films. *Radiation Physics and Chemistry* 141, 142–148.

Constantin C. P., Aflori M., Damian R. F., and Rusu R. D. (2019). Biocompatibility of Polyimides: A Mini-Review. *Materials (Basel)* 12 (19), 3166.

Cristina-Delia N., Dana-Ortansa D., and Constantin C. (2009). The influence of gamma radiations on physicochemical properties of some polymer membranes. *Rom J Phys.* 54, 349–359.

Croisier F., and Jerome C. (2013). Chitosan-based biomaterials for tissue engineering. *Eur Polym J* 49 (4), 780–792.

Dai G., Xiao H., Zhu S., and Shi M. (2014). Effect of electron beam irradiation cross-linking on PVA films. *Advanced Materials Research* 852, 304–308.

Durrani S. A., and Bull R. K. (1987). *Solid State Nuclear Track Detection, Principles, Methods and Applications.* Oxford, England: Pergamon Press.

Ehab E. K., and Shaimaa M. N. (2018). Synthesis and characterization of antimicrobial nanocomposite hydrogel based on wheat flour and poly (vinyl alcohol) using γ-irradiation. *Advances in polymer technology*, 37, 3252–3261.

Eid M., and Hegazy D. (2012). Electron beam synthesis and characterization of poly vinyl alcohol/poly acrylic acid embedded Ni and Ag nanoparticles. *J Inorg Organomet Polym* 22(5), 985–997.

El Salmawi K. M. (2007). Gamma radiation-induced cross-linked pva/chitosan blends for wound dressing. *Journal of Macromolecular Science, Part A* 44, 541–545.

El-Ahdal M. A. (2001). Radiation effect on the molecular structure of dyed poly (vinyl alcohol). *International Journal of Polymeric Materials and Polymeric Biomaterials* 48, 17–28.

Fahmy T. (2007). Investigation of iodine-doped poly (ethyl methacrylate) relaxation by tsdc technique in the vicinity of the glass transition temperature. *Int. J. Polym. Mater.* 56(3), 291–306.

Fink D. (2004). *Fundamentals of Ion-Irradiated Polymers.* Berlin, Heidelberg: Springer-Verlag.

Fischbach C., et al. (2001). *Does UV irradiation affect polymer properties relevant to tissue engineering*, 491, 333–345.

Fletcher N. L., Houston Z. H., Simpson J. D., Veedu R. N., and Thurecht K. J. (2018). Designed multifunctional polymeric nanomedicines: Long-term biodistribution and tumour accumulation of aptamer-targeted nanomaterials. *Chem. Commun.* 54, 11538–11541.

Gajšek P., Ravazzani P., Wiart J., Grellier J., Samaras T., and Thuroczy G. (2013). *J Expo Sci Environ Epidemiol* 1.

Gedde U. W., Mikael S., Hedenqvist (2019). *Fundamental of Polymer science*, (2nd edition) published by Switzerland AG: Springer Nature.

Ghasemzadeh H., and Ghanaat F. (2014). Antimicrobial alginate/PVA silver nanocomposite hydrogel, synthesis and characterization. *J Polym Res* 21, 355–369.

Grandfield K., and Engqvist H. (2012). Focused ion beam in the study of biomaterials and biological matter, advances in materials science and engineering volume, Article ID 841961, 6 pages.

Grupen C. (2010). Interaction of ionizing radiation with matter. In: *Introduction to Radiation Protection. Graduate Texts in Physics.* Berlin, Heidelberg: Springer, Page 31–56.

Gupta R. (2011). Effect of γ-irradiation on thermal stability of CR-39 polymer. *Advances in Applied Science Research* 2, 248–254.

Gupta D. P., Kumar S., Kalsi P. C., Manchanda V. K., and Mittal V. K. (2015). γ-Ray modifications of optical/chemical properties of polycarbonate polymer. *World Journal of Condensed Matter Physics* 05, 129–137.

Hameed A., and Akhtar N. (2019). The skin melanin: An inhibitor of vitamin-D3 biosynthesis: With special emphasis with structure of skin. A mini review. *Dermatol Case Rep* 4, 1.

Helmiyati N. G., Abbas G. H., and Budianto E. (2019) Nanocomposite hydrogel-based biopolymer modified with silver nanoparticles as an antibacterial material for wound treatment. *Journal of Applied Pharmaceutical Science* 9(11), 1–9.

Holzer A. M., and Elmets, C. (2010). The other end of the rainbow: Infrared and skin. *Journal of Investigative Dermatology* 130(6), 1496–1498.

Inamura P. Y., and Mastro N. L. D. (2015). Electron beam irradiation and addition of poly (vinyl alcohol) affect gelatin based-films properties. *International Nuclear Atlantic Conference – INAC.*

International commission on non-ionizing radiation protection (ICNIRP) (2013). ICNIRP guidance on limits of exposure to incoherent visible and infrared radiation. *Health Physics* 105(1), 74–96.

Ionizing Radiation, Part 1: X- and Gamma (γ)-Radiation, and Neutrons, IARC Monographs on the Evaluation of Carcinogenic Risks to Humans Volume 75.

Jahanabadi R., Sheikh N., Mahdavi H., and Bagheri R. (2019). Effect of electron-beam irradiation followed by annealing on the physical properties of poly (vinyl alcohol)–chitosan blend films at different weight ratios. *J. Appl. Polym. Sci.* doi: 10.1002/APP.47820.

Jiang S., Liu S., and Feng W. (2011). PVA hydrogel properties for biomedical application. *Journal of Mechanical Behavior of Biomedical Materials* 4, 1228–1233.

Kaczmarek H., and Podgorski A. (2007). The effect of UV-irradiation on poly (vinyl alcohol) composites with montmorillonite, *Journal of Photochemistry and Photobiology A: Chemistry* 191, 209–215.

Kamoun E. A., Chen X., Eldin M. S., and Kenawy E. R. (2015). Cross-linked poly (vinyl alcohol) hydrogels for wound dressing applications: A review of remarkably blended polymers. *Arab. J. Chem.* 8, 1–14.

Kumar Jaganathan Saravana, Balaji Arunpandian, Vignesh Vellayappan Muthu, Priyadarshni Subramanian Aruna, Priyadarshni Subramanian Aruna, Aruna John Agnes, Kumar Asokan Manjeesh, Supriyanto Eko, Radiation-induced surface modification of polymers for biomaterial application, 2015, 50, *J Mater Sci.*,2007–2018.

Kumar S. et al. (2015). Radiation-induced surface modification of polymers for biomaterial application. *J Mater Sci* 50, 2007–2018.

Kumaraswamy S., Patil S. L., and Mallaiah S. H. (2020). In vitro biocompatibility evaluation of radiolytically synthesized silver/polyvinyl hydrogel nanocomposites for wound dressing applications. *Journal of Bioactive and Compatible Polymers* 35, 435–450.

Laskarakis A., Gravalidis C., and Logothetidis S. (2004). FTIR and Vis-FUV real time spectroscopic ellipsometry studies of polymer surface modifications during ion beam bombardment. *Nucl. Inst. Meth. Phys. Res.* B216, 131–136.

Lee T. -W., Kim Y. M., and Kim Y. C. (2007). Effect of thermal annealing on the charge carrier mobility in a polymer electroluminescent device. *Mol. Cryst. Liq. Cryst.* 462, 241–248.

Leo W. R. (1994). *Techniques for Nuclear and Particle Physics Experiments- A How-to Approach.* Berlin: Springer-Verlag.

Li, D., Liu, T., Yu, X., Wu, D., and Su, Z. (2017). Fabrication of graphene-bio macromolecule hybrid materials for tissue engineering application. *Polym. Chem.*, 8, 4309–4321.

Lina M. S., Ali H Al-H., and Ahmed A Al-A. (2019). Plastic materials for modifying the refractive index of contact lens: Overview. *Res Dev Material Sci.* 11(2). RDMS.000760.2019.

Liu J., Guo T. F., and Yang Y. (2002). Effects of thermal annealing on the performance of polymer light emitting diodes. *J. Appl. Phys.* 91, 1595–1600.

Madkour T. M. (1999). *In Polymer Data Handbook*, edited by Mark J. E., Cincinnati, USA: Oxford University Press, Inc, pp. 363–367.

Makuuchi K., and Cheng S. (2012). *Radiation processing of polymer materials and its industrial applications.* Hoboken, N J: John Wiley & Sons.

Martin J. E. (2013). *Physics for radiation protection*, 3rd edn. USA: Wiley.

Miller A. B. et al. (2019). Risks to health and well-being from radio-frequency radiation emitted by cell phones and other wireless devices. *Front Public Health* 7, 223.

Mishra S., Bajpai R., Katare R., and Bajpai A. K. (2002). Radiation induced crosslinking effect on semi-interpenetrating polymer networks of poly (vinyl alcohol). *eXPRESS Polymer Letters* 1, 407–415.

Mitrousis, N., Fokina, A., and Shoichet, M. S. (2018). Biomaterials for cell transplantation. *Nat. Rev. Mater.*, 3, 441–456.

Muppalaneni S., and Omidian H. (2013). Polyvinyl alcohol in medicine and pharmacy: A perspective. *J Develop Drugs* 2 (3) 1000112.

Musgrave C. S. A., Nazarov W., and Bazin N. (2017). The effect of para-divinyl benzene on styrenic emulsion-templated porous polymers: A chemical Trojan horse. *J. Mater. Sci.* 52, 3179–3187.

Natalia V. et al. (2005). Effect of microwave irradiation on the structuring of polyvinyl alcohol. *Mendeleev Communication* 15, 170–172.

Nations U., and Programme E. (2016). Photochemical photobiology. *Sci.* 15, 14.

Ng K. H. (2003). *Proc Int Conf Non-Ionizing Radiat UNITEN*, 1.

Nguyen et al. (2019). Effect of microwave irradiation on polyvinyl alcohol as a carrier of silver nanoparticles in short exposure time. *International Journal of Polymer Science*, Article ID 3623907, 4 pages.

Nguyen N. T., and Liu J. H. (2013). Fabrication and characterization of poly (vinyl alcohol)/chitosan hydrogel thin films via UV irradiation. *European Polymer Journal*, 49, 4201–4211.

Nouh S. A. (2012). Thermal, structural, and optical properties of c-irradiated poly(vinyl alcohol)/Poly (ethylene glycol) thin film. *Journal of Applied Polymer Science* 124, 654–660.

Petrova V. N. et al. (2005). Effect of microwaves irradiation on the cross-linking of polyvinyl alcohol. *Macromolecular chemistry and Polymeric Materials* 78, 1178–1182.

Podgorsak E. B. (2005). *Radiation oncology physics: A handbook for teachers and students/editor.* Vienna, Austria: International Atomic Energy Agency Publishing.

Qiu K., and Netravali A. N. (2012). Fabrication and characterization of biodegradable composites based on micro fibrillated cellulose and polyvinyl alcohol. *Compos. Sci. Technol.* 72, 1588–1594.

Qiu Y., and Park K. (2012). Environment-sensitive hydrogels for drug delivery. *Adv Drug Deliv Rev* 64, 49–60.

Radiation IARC Monographs on the Evaluation of Carcinogenic Risks to Humans Volume 100D

Radosavljevic A., et al. (2020). *Nanocomposite Hydrogels Obtained by Gamma Irradiation, Cellulose-Based Superabsorbent Hydrogels; Polymers and Polymeric Composites: A Reference Series book series (POPOC)* pp. 601–623: Springer Nature Switzerland AG.

Ratner B. D. (2013). The history of biomaterials, in: B. D. Ratner, A. S. Hoffman, F. J. Schoen, J. E. Lemons (Eds.), *Biomaterials Science: An Introduction to Materials in Medicine*, 3rd ed., Waltham, MA, USA: Academic Press, Elsevier.

Schottler, S., Becker, G., Winzen, S., Steinbach, T., Mohr, K., Landfester, K., Mailander, V., and Wurm, F. R. (2016). Protein adsorption is required for stealth e_ect of poly (ethylene glycol)- and poly (phosphoester)-coated nanocarriers. *Nat. Nanotechnol.* 11, 372–377.

Sharma K., Chahal R. P., Mahendia S., Tomar A. K., and Kumar S. (2013). Optical behaviour of swift heavy ions irradiated poly(vinyl alcohol) films. *Radiation Effects and Defects in Solids* 168, 378–384.

Sheela T. et al. (2014). Effect of UV irradiation on optical, mechanical and microstructural properties of PVA/ NaAlg blends. *Radiation Physics and Chemistry* 103, 45–52.

Shen, W., Luan, J., Cao, L., Sun, J., Yu, L., Ding, J. (2015). Thermogelling polymer–platinum (IV) conjugates for long-term delivery of cisplatin. *Bio macromolecules* 16, 105–115.

Shintani H. (2017). Ethylene oxide gas sterilization of medical devices. *Biocontrol Sci* 2017(22), 1–16.

Simpson J. D., Smith S. A., Thurecht K. J., and Such G. (2019). Engineered polymeric materials for biological applications: Overcoming challenges of the bio-nano interface. *Polymers (Basel)* 11(9), 1441.

Singh R., Singh D., and Singh A. (2016). Radiation sterilization of tissue allografts: A review. *World J Radiol* 8, 355–369.

Singh R., Nath R., Mathur A. K., and Sharma R. S. (2018). Effect of radiofrequency radiation on reproductive health. *Indian J Med Res* 148, S92–S99.

Stephen C., Musgrave A., and Fang F. (2019). Contact lens materials: A materials science perspective. *Materials (Basel)* 12(2), 261.

Stoica-Guzun A., et al. (2012). The effect of UV-irradiation on poly (vinyl alcohol) composites with bacterial cellulose. *Macromol. Symp.* 315, 198–204.

Sun, H., Guo, B., Li, X., Cheng, R., Meng, F., Liu, H., Zhong, Z. (2010). Shell-sheddable micelles based on dextran-SS-poly(ε-caprolactone) diblock copolymer for efficient intracellular release of doxorubicin. *Biomacromolecules* 11, 848–854.

Tamada M. (2018). Radiation processing of polymers and its applications. In: Kudo H. (eds) *Radiation Applications. An Advanced Course in Nuclear Engineering, 07.* Singapore: Springer.

Teodorescu M., Bercea M., and Morariu S. (2018). Biomaterials of poly (vinyl alcohol) and natural polymers. *Polymer Reviews* 58, 247–287.

Thang, N. N., and Liu J. H. (2013). Fabrication and characterization of poly (vinyl alcohol)/chitosan hydrogel thin films via UV irradiation. *European Polymer Journal* 49, 4201–4211.

Thompson J. M. T. (Ed.) (2001). *Visions of the Future: Physics and Electronics.* U. K.: Cambridge University Press.

Turner J. E. (2004). Interaction of radiation with matter. *Health Phys.* 86, 228–252.

UNSCEAR. (2008). *Sources and effects of ionizing radiation.* United Nation.

Vrancken A. C., Buma P., and van Tienen T. G. (2013). Synthetic meniscus replacement: A review. *Int. Orthop.* 37, 291–299.

Wael H. E. et al. (2011). Gamma-irradiation assisted seeded growth of Ag nanoparticles within PVA matrix. *Materials Chemistry and Physics* 128, 109–113.

Wu M., Bao B., Yoshii F., and Makuuchi K. (2001). Irradiation of crosslinked, poly (vinyl alcohol) blended hydrogel for wound dressing. *Journal of Radioanalytical and Nuclear Chemistry* 250(2), 391–395.

Xu J. et al (2018). A comprehensive review on contact lens for ophthalmic drug delivery. *J. Control. Release.* 281, 97–118.

Yamdej R., Pangza K., Srichana T., and Aramwit P. (2017). Superior physicochemical and biological properties of poly (vinyl alcohol)/sericin hydrogels fabricated by a non-toxic gamma-irradiation technique. *Journal of Bioactive and Compatible Polymers* 32, 32–44.

Yoona S. D., Kim Y. M., Kim B., Je J. Y. (2017). Preparation and antibacterial activities of chitosan-gallic acid/polyvinyl alcohol blend film by LED-UV irradiation. *Journal of Photochemistry and Photobiology B: Biology* 176, 145–149.

Yousif E. et al. (2015). Poly (vinyl chloride) derivatives as stabilizers againstphotodegradation. *Journal of Taibah University for Science* 9, 203–212.

Yousif E., and Haddad R. (2013). Photodegradation and photostabilization of polymers, especially polystyrene: review. *Springer plus* 2, 398.

Zidan H. M., El-Ghamaza N. A., Abdelghany M., and Walyc A. L. (2018) Photodegradation of methylene blue with PVA/PVP blend under UV light irradiation. *Spectrochimica Acta Part A: Molecular and Bimolecular Spectroscopy* 199, 220–227.

7 Radiation-induced degradation and grafting of cellulosic substrates

A.K. Rana*

Department of Chemistry, Sri Sai University, Palampur, India

A.S. Singha

Department of Chemistry, National Institute of Technology, Hamirpur, India

*Corresponding author: ranaashvinder@gmail.com; ranaashvinder2020@gmail.com

CONTENTS

7.1   INTRODUCTION

Due to the present environmental issues, one of the most challenging tasks, that researchers are facing, is to provide inexpensive and scalable pathways in the direction of sustainable development. In this context, renewable resources are appealing alternatives to partially/completely substitute synthetic polymers in industrial and domestic applications (Thakur, Rana, and Thakur 2014).

Cellulose is the one of the most abundant renewable polymers obtained from the biosphere (Ashvinder K. Rana, Mishra, et al. 2021). The primary source of cellulose is the existing lignocellulosic materials available in forests, with wood as the chief source. Other cellulose-containing materials are agriculture residues, water plants, grasses, and other plant substances (Singha and Rana 2012g; Ashvinder K. Rana and Thakur 2021). These materials, in addition to cellulose, also contain hemicelluloses, lignin, pectin, and a small amount of extractives. There are numerous methods for isolation of cellulose from agriculture residues that use alkaline treatment (Bhattacharya, Germinario, and Winter 2008), ultrasound treatment (Sun et al. 2004), enzymes technology (Janardhnan and Sain 2006), and dilute acid pre-treatment (Esteghlalian et al. 1997; Ashvinder Kumar Rana, Frollini, and Thakur 2021); and the selection of method depends on the dimensions (long, short and nano form) of the fibers needed. Cellulose, in various dimensions ranging from long to macro or micro to nano fibers/crystal form, can provide attractive structure characteristics, water-absorbing capacity, and channelling of fluids (Ashvinder K. Rana, Gupta, et al. 2021). It is composed of D-anhydrous glucose ($C_6H_{10}O_5$), repeating units joined by ß-1, 4-glycosidic linkages at C_1 and C_4 position (Kraessig 1987; Bledzki, Reihmane, and Gassan 1996). Each D-anhydrous glucose unit contains three hydroxyl groups and can form different types of hydrogen bonds; intramolecular, intermolecular as well as with hydroxyl groups from the air, which plays a key role in governing the physical properties of cellulose. Cellulosic fibers or natural fibers have a lot of attractive properties over the synthetic fibers inclusive of low density, low cost, biodegradable, recyclable, biocompatible, presence of reactive hydroxyl groups, and cap potential to form superastructures. Despite the appealing properties, cellulosic fibers find their utility to a limited volume in industrial applications because of difficulties associated with the surface interactions (Singha and Rana 2012a).

Since cellulose is biocompatible and has high tensile strength, it can be used in various fields such as in nanotechnology, water treatment industries, pharmaceutical industry, food industry, polymer composites industries, textile and paper industry, drug-delivery systems, etc., in virgin form or after surface functionalization of cellulosic materials.

In composite industries, their utility can be enhanced through surface modification by some hydrophobic monomer as inherently polar and hydrophilic cellulosic fibers show poor compatibility with nonpolar thermoplastics. However, their role in water purification industries can be enhanced by surface tailoring through suitable monomers (may be hydrophilic/hydrophobic monomer) or by depositing some nano-metals, depending upon the type of impurities (dye or heavy metals) one may wish to handle. There are numerous methods available in literature for carrying out grafting onto cellulose, such as chemical (Singha and Rana 2012b), enzymatic (Cruz and Fangueiro 2016), and physical methods (Kumar and Anbumalar 2015); however, our focus in the present chapter will only be on physical methods, which include gamma, microwave, electron beam, etc., induced graft copolymerization onto cellulose.

7.2 CELLULOSIC FIBERS AND THEIR CLASSIFICATION

Based upon the type of origin, natural fibers can be classified into the following three categories: i) vegetable fibers: the major constituent of vegetable fibers is cellulose. So we generally called them cellulosic fibers. It consists of cotton, linen, jute, flax, ramie, coir, sisal, and hemp fibers (Cristaldi et al. 2010).

All these fibers, in addition to their use in textiles, also find their utility in the manufacturing of paper, ropes, mats, fabrics, etc. Vegetable cellulosic fibers, depending upon the type of sources, may be further sub-classified into bast, leaf, and seed fibers: ii) animal fibres: They are protein dominant fibres and examples are silk and wool. iii) Third category of natural fibers is mineral fibers and its best example is asbestos, which has been used notably for manufacturing industrial products. However, their utility has been now been phased out due to carcinogenic impacts. The presence of excessive cellulosic contents in vegetable fibers makes it an ideal base material for synthesis of numerous cellulose derivatives such as cellulose ester, cellulose ether, cellulose nitrate, carboxymethyl cellulose, methyl cellulose, hydroxyl propyl cellulose, cellulose xanthate, etc. and can be subsequently used for various industrial applications (Ibarra, Sendón, and de Quirós 2016).

7.3 RADIATION-INDUCED GRAFT COPOLYMERIZATION

Irradiation is one of the simple methods and has been used for a long time for modifying the cellulose or lignocellulosics materials for different purposes, e.g., better compatibility in polymer composites, fire retardancy, abrasion resistance, better diffusion or reactivity in subsequent reactions, ion-exchange materials, sensors, stimuli-responsive materials, etc. (Barsbay and Güven 2019). There are numerous surface modification methods available in literature and among them graft polymerization is exceptionally attractive as it has made a leading influence towards the fabrication of desired cellulose surface (Ashvinder K. Rana, Potluri, and Thakur 2021). Graft copolymerization can be accompanied by various techniques such as by chemical treatment, UV-irradiation, plasma-irradiation, photo-irradiation, high-energy radiation, etc. (Kang, Liu, and Huang 2015). Radiation-induced graft copolymerization is not a new mode for the surface functionalization of polymers, but still it is preferred over conventional chemical approaches because it is budget friendly, a simple and easy-to-control approach, operable at room temperature, and has a tendency to control material composition within the structure (Nasef and Güven 2012). The radiation-mediated grafting reactions onto cellulose are quite rapid and generate long- as well as short-lived free radicals. On paper, more than 20 distinct cellulose-free radicals can be designed and have been distinguished by researchers (Wach, Mitomo, and Yoshii 2004; Wencka et al. 2007). The C_1–H and C_4–H bonds are the most unstable bonds and susceptible to hydrogen abstraction in cellulose molecules during irradiation, and thus lead to the formation of radicals and further graft copolymerization. In addition, Klemm et al. reported that C_2 and C_3 are also presumable to form radicals and graft copolymerization chain (Figure 7.1a) (Ershov 1998; Klemm et al. 1998).

Temperature controls the life span of radicals formed and their substituent reactions (Takács et al. 2007). They are very unstable at room temperature and undergo a thermal mutation; for example, a radical at C_4 forms an allyl radical (Wojnárovits, Földváry, and Takács 2010) (Figure 7.1b).

The breakdown of radicals additionally depends on the supramolecular shape i.e., within the amorphous zones radical decays very fast, whilst in crystalline areas last for a prolonged time (Vismara et al. 2009). It has been also reported in literature that irradiation has no effect on crystalline structure of cellulose and in the ratio of crystalline to amorphous regions up to several hundred kGy irradiation doses (Wojnárovits, Földváry, and Takács 2010).

7.3.1 MECHANISM OF GRAFTING UNDER UV RADIATION

UV radiations are not of sufficiently high energy to rupture C–C or C–H bonds, so generally a photo-sensitizer is added to the system to produce free radicals on the cellulose backbone. The photo initiator in the presence of UV radiation undergoes excitation (having high energy) and transfers its energy to cellulose samples and thus forms free radicals onto the cellulose backbone by hydrogen abstraction and glucosidic bond scission. These cellulose macro radicals then form the graft copolymers as follows (M. A. Khan et al. 2015) (Figure 7.1a).

7.3.2 MECHANISM OF GAMMA RADIATION–INDUCED GRAFTING

There are two methods for grafting onto cellulose in the presence of γ-radiation, namely simultaneous and pre-irradiation grafting. In the former method, cellulosic materials are irradiated in the presence of the vinyl monomer to be grafted. This technique has numerous advantages, such as high reaction rate, low dose of radiation required, easy to carry out, and the presence of monomer protects the cellulose material from degradation. However, certain drawbacks such as excessive homopolymer formation and lack of control on graft copolymers chain length, limit its role in graft copolymerization. In the pre-irradiation grafting method, the cellulose material is pre-irradiated with high-energy radiation followed by the addition of the monomer to be grafted at a predefined temperature in air/or inert atmosphere. The pre-irradiation grafting generally carried out in air or oxygen medium because in this case long-lived peroxy radicals are formed on the cellulose spine, which slowly convert to relatively long-lasting hydroperoxides, Cellulose-OOH- and Cellulose-OO-cellulose-type peroxides, and thus have high efficiency (Figure 7.2).

However, efficiency in the case of pre-irradiation in an inert atmosphere is quite low because in this case radicals formed are of a shorter life span and generally avoided (Roy et al. 2009; Kakati, Bora, and Deka 2015). The peroxy products have an extended lifetime and might be suitably used for the initiation of grafting in an oxygen atmosphere. Mechanism of gamma-irradiated grafting onto cellulosic fibers is very much similar to grafting under UV, except that this technique does not require any photo-initiator (as in case of UV) to induce the grafting as gamma rays are high-energy radiation. The mechanism of cellulosic fiber surface modification by gamma radiation may be summarized as follows, which is based upon the principle of free radical polymerization.

(a)

Chain Initiation

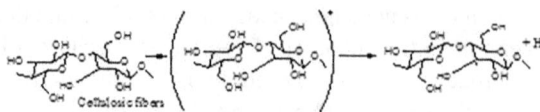

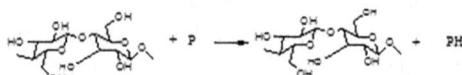

Monomer (M) ——— M* ——→ Ṁ + Ḣ

Photo Initiator ——hv—→ P* ——→ Ṗ + Ḣ

M + P ——→ M + Ṗ

Chain Propagation

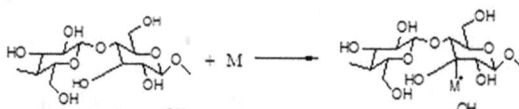

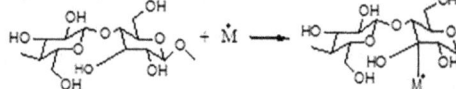

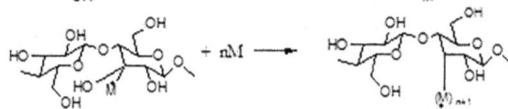

Ṁ + nM ——→ Ṁ_{n+1}

Chain Termination

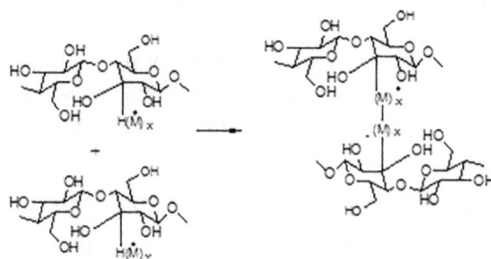

Ṁ_{n+1} + Ḣ ——→ M_{n+1}H

Ṁ + Ṁ ——→ MM

Ṁ + H· ——→ MH

(b)

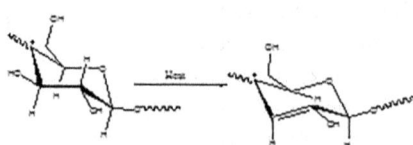

FIGURE 7.1 (a) Possible mechanism for graft copolymerization onto cellulose backbone in the presence of UV radiation (Klemm et al. 1998) and (b) for the formation of allyl radicals during the reaction at room temperature (Wojnárovits, Földváry, and Takács 2010).

Chain Initiation:

Cellulosic fibers

Chain Propagation:

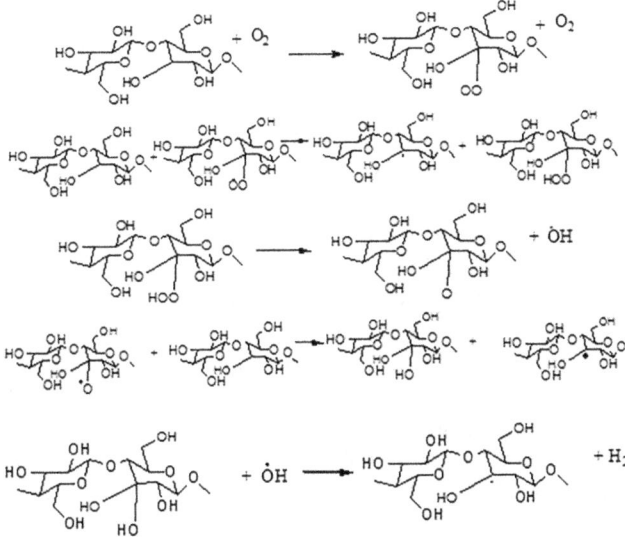

Chain Termination:

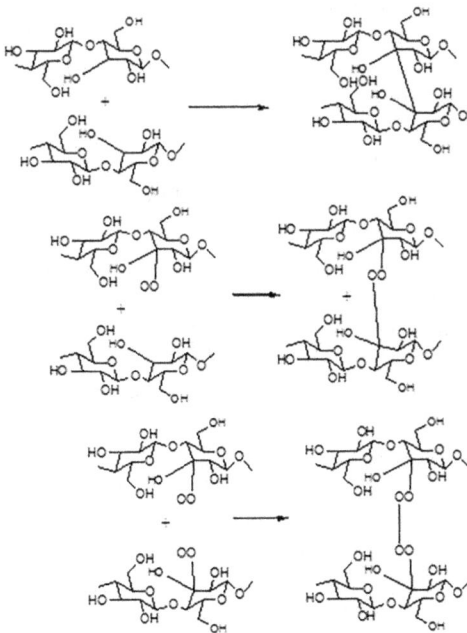

FIGURE 7.2 Possible mechanism for graft copolymerization onto cellulose backbone in the presence of gamma radiations (M. A. Khan et al. 2015).

7.3.3 Gamma radiation–induced graft copolymerization

Through radiation technology, reactions can be carried out at room temperature with high potential for the tailoring the properties of polymeric materials (Omichi et al. 2019; Hosny et al. 2020; I. A. Khan et al. 2020). Lertsarawut et al. carried out radiation-induced graft polymerization of acrylic acid onto carboxymethyl cellulose with an aim to remove dye from wastewater (Lertsarawut et al. 2019). They optimized the gamma radiation dose and reported maximum grafting efficiency at 20 kGy radiation dose. However, grafting efficiency was found to decrease when the radiation dose was increased beyond 30 KgY. Grafting of acrylonitrile (AN) onto cellulosic filter paper beneath γ-irradiation followed by a reaction with hydroxylamine was carried out (Badawy 2017) in order to develop functional cellulosic filter papers containing chelating amidoxime groups. The obtained functional filter papers were characterized by different techniques, such as Fourier transform infrared analysis (FTIR), X-ray diffraction (XRD), scanning electron microscopy (SEM), and thermogravimetric analysis (TGA) and reported nil effects orientation and crystallinity of the cellulosic fibers after graft copolymerization.

Madrid and Abad (Madrid and Abad 2015) have used the simultaneous irradiation method for graft copolymerization of glycidylmethacrylate (GMA) onto microcrystalline cellulose (MCC) in a methanol solvent by using gamma irrradiation under a nitrogen atmosphere. They studied the effect of different factors such as monomer concentration, type of solvent, and absorbed gamma radiation dose for the degree of grafting, Dg. Similar to the study of Badawy, they found a liitle bit of effect on the crystalline region of MCC after graft copolymerization. The thermo gravimetric analysis (TGA) further confirms that grafted MCC is more stable in comparison to the base MCC polymer. Madrid et al. (Madrid, Nuesca, and Abad 2013) also carried out gamma radiation–induced graft copolymerization of GMA onto water hyacinth fibers under a nitrogen atmosphere in various solvents. They optimized the different reactions parameters such as type/solvent ratio, absorbed dose, dose rate, and concentration of monomer. which were found to be 1:3(volume/volume) water-methanol solvent, 10 kGy, 8 kGy/h dose rate, and 5% volume/volume glycidyl methacrylate, respectively. Under optimum conditions, 58% degree of grafting was reported by them. Successful grafting of GMA onto water hyacinth fibers was confirmed by using FTIR-ATR, SEM, and EDX.

Ibrahim et al. (Ibrahim, Mousaa, and Ibrahim 2014) prepared unirradiated and irradiated plasticized carboxymethyl cellulose (CMC)/cum arabic blends by adding glycerol at different concentrations to the blend as a plasticizer. They investigated different properties (gel percent, swelling fraction, etc.) of blend samples before and after exposing different doses (5, 10, and 20 kGy) of gamma irradiation and reported an increase in gel fraction with an increase in irradiation dose, while swelling of blend samples was found to decrease with enhanced irradiation doses. Further, they studied the capacity of different blend samples for removal of dyes.

Ghaffar et al. (Abdel Ghaffar et al. 2016) used a direct radiation grafting technique for the synthesis of carboxymethyl cellulose (CMC) hydrogels, which were synthesized by grafting CMC with different concentrations of acrylamide (AAm) and methacrylic acid (MAAc) individually at a 20 KgY dose. It was reported that for both poly(CMC/AAm) and poly(CMC/MAAc) hydrogels, the grafting yield and grafting ratio increased with the increase in monomer concentration. However, the grafting ratio and grafting yield of the former one was found to be higher than that of the latter one. The effect of monomer concentrations on gel (%) and swelling behavior of cellulose (CMC) hydrogels [poly(CMC/AAm) and poly(CMC/MAAc)] was also evaluated. The gel percent of these hydrogels was found to increase with an increase in concentration of monomers (AAm/MAAc). However, swelling behavior for poly(CMC/AAm) hydrogels decreased with an increase in AAm concentration due to high cross-linking, while for poly(CMC/MAAc) the swelling behavior increased up to 1:25 wt% with an increase in AAc concentration. After the study of swelling kinetics and diffusion mechanism, it was found that water penetration obeys the non-Fickian transport mechanism.

Hemvichian et al. (Hemvichian, Chanthawong, and Suwanmala 2014) have synthesized the superabsorbent polymer (SAP) by gamma radiation-induced graft copolymerization of acrylamide

(AM) onto carboxymethyl cellulose (CMC) in the presence of a cross-linking agent, N,N'-methylenebisacrylamide (MBA), at a radiation dose rate of 0.14 kGy/min ranging from 0.5–10 kGy. The various reaction parameters such as radiation dose, CMC amount, and concentration of AM & MBA were optimized and the effect of ionic strength of the salt solution of Na_3PO_4, Na_2CO_3 and NaCl on the swelling ratio was also investigated. They reported that the optimized radiation dose is 2 kGy and beyond it the swelling ratio was found to decrease. These SAP materials were developed with an aim to study their controlled release behavior of agrochemicals, so SAP was loaded with potassium nitrate (KNO_3) and its potential for controlled release of KNO_3 was evaluated. After the study, it was found that their tendency to release KNO_3 amount increases with an increase in loading percentage of SAP. Pesticide-absorbing strength of cellulose fiber after its grafting by GMA under gamma radiation was found to be improved considerably against phenol and 2,4-dichlorophenoxyacetic acid (Erzsebet Takacs et al. 2012). Ethyl cellulose-g-AN graft copolymers were synthesized by simultaneous irradiation of a mixture of cellulose substrate and AN by using gamma radiation. The nitrile groups were further converted to amidoxime in order to analyze their uranium recovery strength (Başarır and Bayramgil 2012). Rice straw cellulose-g-AAm graft copolymers were synthesized by the simultaneous irradiation method and the grafting efficiency was found higher in the case of bleached cellulose than unbleached one (Swantomo et al. 2013). Various blends of cellulose acetate (CA)/polyethylene oxide (PEO) were developed by using N,N′Methylene bis-acrylamide (MBAAm) as cross-linkers under gamma irradiation (Kamal, Abd-Elrahim, and Lotfy 2014). The water contents of CA cross-linked MBAAm were found to be improved after blending with PEO; however, there was no any effect on chemical stability of CA cross-linked MBAAm after blending with PEO. Thermal stability does not vary so significantly, but melting behavior was found to be improved after blending with POE. Salmawi et al. (Salmawi, El-Naggar, and Ibrahim 2018) prepared the novel superabsorbent polymer by grafting carboxymethyl cellulose with acrylic acid (AAc) under gamma irradiation. They used N,N-Methylene bisacrylamide (MBAM) as a cross-linker and the general mechanism proposed for the synthesis of a superabsorbent polymer (SAP) is given as given in Figure 7.3(a).

CMC-based hydrogels were also prepared by irradiating the mixture of sodium carboxymethyl cellulose (CMC) with two active vinyl monomers {sodium styrene sulfonate (SSS) and Bis[2-(methacryloyloxy)ethyl] phosphate (BMEP)} at a 10–100 kGy gamma radiation dose (T. T. Hong et al. 2019) (Figure 7.3b). The concentration of BMEP was varied form 0.0 to 12.9 m mol/L and its effect on swelling capacities was evaluated. The optimal dose of radiation to get an eco-friendly hydrogel, which has 40% gel fraction and 500 g/g swelling ratio, was 40 kGy. In addition, a hydrogel prepared at optimal coditions showed not only higher compressive modulus 3.4 (kPa) and Poisson's ratio of 0.45, but also conserved better multi-metal adsorption efficiency.

7.3.4 Microwave radiation–induced graft copolymerization

With the increase in interest of clean and green environment chemistry, the researchers prefer the microwave-irradiation method over conventional methods for grafting cellulose substrates. Microwave-irradiated grafting reduces the reaction time, as well as the use of toxic solvents, and thus ensures high percent yields, product selectivity, and product formations (Figure 7.4).

It has been also reported in literature that microwave-synthesized cellulose copolymers exhibit better properties than their conventionally synthesized counterparts for commercial utilization (Singha and Rana 2012e; Singh, Kumar, and Sanghi 2012). Microwaves are radio waves with a frequency in the range of 300 MHz to 300 GHz. On exposing the reaction mixture to microwaves, the polar or charge particles align themselves with electric field components of the microwaves. However, this electric field component of microwave radiation changes its direction at such a fast rate (at the rate of 2.4×10^9/s at 2.45 GHz microwave frequency) that the charged or polar particles in a reaction medium fail to align themselves along the direction of the electric field component and thus create friction and generate a high temperature (Galema 1997).

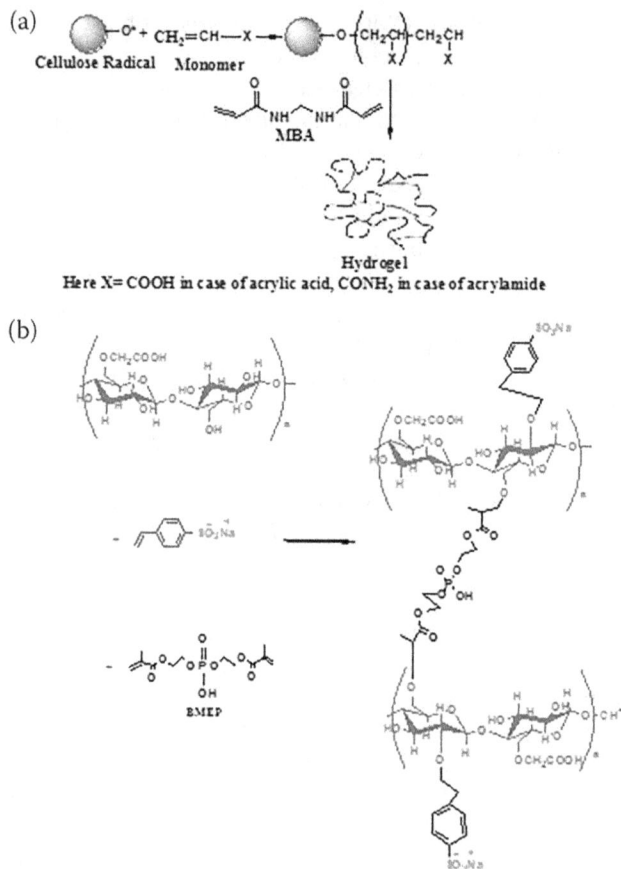

FIGURE 7.3 Proposed mechanism for the synthesis of (a) superabsorbent polymer by grafting carboxymethyl cellulose with acrylic acid under gamma irradiation using N,N-Methylene bisacrylamide (MBAM) as a cross-linker (Salmawi, El-Naggar, and Ibrahim 2018) and (b) hydrogel from CMC using two vinyl monomers (T. T. Hong et al. 2019).

Microwave reactions, just like other irradiation methods, can also be carried out in a solution or in a dry medium. In the dry medium reactions, homogeneous electric field of microwaves is created to produce localized superheating zones, which are called hot spots with dimensions of about 900–1,000 m and temperatures higher (~100–200 K) than the bulk temperature. These hot spots speed up the reactions and make the reactions more fertile than solution phase reactions (Hayward 1999; Kappe 2004). The microwave-irradiated graft copolymerization can be carried out either through microwave-assisted or through microwave-initiated methods. In microwave-assisted grafting, an external redox initiator system is added to the reaction mixture, which produces free radicals under the influence of microwave dielectric heating and thus facilitates the grafting reactions (Neas and Collins 1988; Singh et al. 2007). However, in microwave-initiated grafting reactions, no such redox initiator system is required.

Mishra et al. (Mishra, Rani, and Sen 2012) used microwave radiations to initiate the grafting of polyacrylic acid onto carboxymethyl cellulose (CMC-g-PAA). Reactions parameters such as irradiation time and monomer (AA) concentration were optimized to obtain maximum percent grafting.

Cellulosic canabis indica fibers were grafted with AAc monomer to form AAc-graft-cellulose copolymers (Singha and Rana 2012d). The process was carried out under microwave conditions using a Ce^{3+}-HNO_3 redox initiator system. Various reaction parameters such as time, microwave radiation power, monomer concentration, initiator, and nitric acid concentration were optimized to obtain maximum grafting efficiency (9.86%), which was found at 110 microwave radiation power.

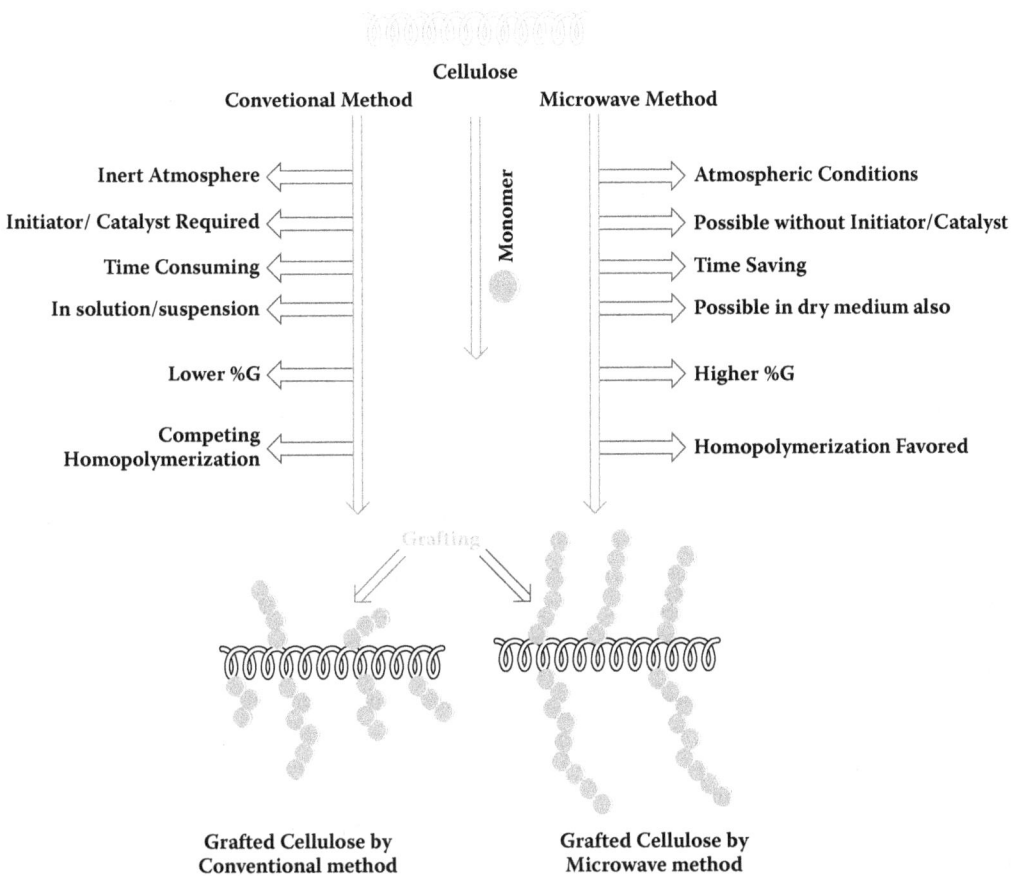

FIGURE 7.4 A comparative view between conventional and microwave insisted graft copolymerization onto cellulose [Figure adapted from Reference (Singh, Kumar, and Sanghi 2012) with permission from Elsevier Copyright (2012)].

The grafted copolymers showed an increase in swelling and moisture absorbance behavior. This provides a proficient method to broaden the applications of biofibers for use as superabsorbents. A similar method was applied for synthesizing poly(AN)-graft-cannabis indica fiber (Singha and Rana 2012c). Contrarory to AAc-graft-cellulose copolymers, the poly(AN)-graft-cannabis indica graft copolymers have been found to be more resistant to swelling and also showed better resistance to chemical and thermal deterioration and thus support their potential application as reinforcement in biocomposites. In addition, graft copolymerization of AN onto Grewia optiva fibers under microwave radiation was also reported by Singha and Rana. They found an increase in thermal stability and water-repelling tendency of resulting graft copolymers (Singha and Rana 2012f). Pathania et al. (Pathania, Sharma, and Sethi 2017) carried out microwave-induced graft copolymerization of binary monomers methacrylate/acrylic acid (MA/AAc) onto luffa cylindrica (LC) fiber for the removal of congo red dye from wastewater. They reported that LC-g-poly(MA/AA) exhibits excellent chemical resistance and dye absorbance behavior over raw fiber.

7.3.5 UV RADIATION–INDUCED GRAFT COPOLYMERIZATION

Among all irradiation methods, UV radiation–induced graft copolymerization is considered the most secure as it has the least tendency to alternate backbone properties because of its low strength.

Hong et al. incorporated benzophenone (Photo-active moiety) onto cotton fabrics by using butyl tetracarboxylic acids (BTCA) as a cross-linker and then grafted copolymerization of acrylamide monomers onto cotton fabrics containing benzophenone chromophoric groups as reactive sites was carried out on long exposure of reaction mixture to UV irradiation (K. H. Hong, Liu, and Sun 2009). Then, grafting of polyacrylamide on the cotton fabrics was further confirmed by SEM, FTIR, XRD, and TGA studies. Mikaela et al. studied the effect of three UV-reactive molecules, namely N-(hydroxymethyl)acrylamide (HMAA), hexanediol diacrylate (HDDA), and trimethylol-propane triacrylate (TMPTA) grafted onto paper sheets with and without acryl ester functionalization on their strength and stiffness (Tsujimoto, Nishiumi, and Kobayashi 2015). 1-hydroxycyclohexyl phenyl ketone was used as photo initiator in concentration of 3–5 wt% of sheets, while HMAA, HDDA, and TMPTA were added to 20 wt%, 15 wt%, and 5–20 wt% of sheets, respectively. For grafting UV reactive molecules onto paper sheets with and without acryl ester functionalization, the reaction mixture was UV irradiated by Black-Ray XX-15M Bench Lump (15W, lambda = 302 nm) at an ambient temperature for 30 min and at a distance of 135 nm. An increase in stiffness after was reported after irradiation.

Bongiovanni et al. (Bongiovanni et al. 2011) carried out UV radiation triggered grafting of a perfluoropolyether urethane methacrylate monomer onto cellulose substrates by using a photo initiator, benzophenone. Various reaction parameters such as the benzophenone concentration, the monomer concentration, and the UV irradiation time were varied during grafting. The grafting was carried out by dipping cellulose substrate into acetone solution containing a known concentration of monomer and initiator followed by irradiation with UV radiation for different intervals of time (60–240 s). The graft copolymers thus synthesized were further characterized by XPS study and during the study they suggested that modification takes place in the region especially on the first layers of the cellulose substrate, which is around 30 A° thickness. Shukla and Athalye (Shukla and Athalye 1992) carried out grafting of 2-hydroxyethyl methacrylate onto cotton cellulose by using benzoin ethyl ether as a photoinitiator in a 10% methanol/acetone solution. Nitrazine yellow (NY) dye was modified by methacrylate (MA) to form a photoreactive monomer (MA-NY), which was subsequently grafted on cotton fabric through UV irradiation in the presence of benzophenone under inert atmosphere (Kianfar et al. 2020). The time of dipping of fabric into a dye solution was optimized by evaluating wet pickup. The fabricated product was further investigated by FT-IR spectroscopy and its halochromic and pH response was studied by CIELAB and UV spectra. Ultraviolet (UV) radiations can also be used for differential dyeing onto cotton fabrics (Migliavacca, Ferrero, and Periolatto 2014). In addition, surface modification of cotton fabrics with the aid of UV curing/grafting with desired chemicals was also carried out in order to develop oil and/or water repellent characteristics (Ferrero et al. 2008). Further, water-resistant antimicrobial activity of cotton fabrics can be enhanced by grafting chitosan through UV irradiation, that is an environmentally friendly method and causes low polymer addition (Ferrero, Periolatto, and Ferrario 2015). Gashti and Almasian (Gashti and Almasian 2013) stabilized multiwall carbon nanotubes (MWCNTs) onto cotton by using a flame retardant cross-linking monomer polyvinylphosphonic acid under UV irradiation. Benzophenone was used as an initiator and the synthesized product was characterized via FTIR, which truly showed cross-linking among OH groups of backbone and monomer to form linkages in the presence of MWCNTs. Khan and Kronfli (F. Khan, Ahmad, and Kronfli 2004) grafted MMA onto jute cellulosic fibers through UV irradiation for a specific time in the air to obtain maximum graft yield. They further reported a maximum graft yield on UV irradiation for 7 min. Physico-mechanical properties of UV cured jute cellulosic fiber had been found to improve through urethane acrylate. The function of plasticizers and water uptake behavior had been also investigated. A decrease in water and moisture uptake was reported after treatment (Hossan Shahid Shohrawardy 2011). The effect of UV curing on physico-mechanical properties of EDGMA-grafted jute fibers were also evaluated by Mollah and co-workers (Mollah et al. n.d.). An increase in mechanical strength has been reported by them.

7.3.6 Electron Beam–Initiated Graft Copolymerization onto Cellulose

Electron beam radiations are of very excessive energy radiations and their rate of imparted energy is also exceptionally higher than that of X-rays, UVs, or γ-rays (Kashiwagi and Hoshi 2012). In addition, these radiations also possess very high processing performance during grafting onto cellulose or on other cellulosic derivatives because electron irradiation directly injects energy in the chemical reaction. The industrial electron accelerators had been categorized into low (80–300 keV), medium (300 keV–5 MeV), and high-energy accelerators (above 5 MeV) (Henniges et al. 2013); and for cellulose functionalization we generally prefer medium-range electron accelerator as higher-energy irradiation may cause cellulose chain scission (Henniges et al. 2012). Grafting can be carried out either in liquid or in vapor phase medium. Grafting in a liquid medium degrades or lowers the mechanical strength of fiber so vapor-phase grafting is generally preferred. It possess some advantages over liquid phase grafting, such as there is no need for a solvent and also grafting efficiency can be increased as there is no loss of monomers through homo-polymerization (Lawal and Wallace 2014; Yagüe et al. 2013).

Moawia et al. (Moawia et al. 2016) used a pre-irradiation technique for graft copolymerization onto raw/bleached cellulosic flax fibres with glycidyl methacrylate (GMA). The reaction was carried out in GMA/water emulsion medium containing polyoxyethylene-sorbitan monolaurate (Tween 20) as surfactant under electron beam irradiation with dose ranging from 0.5 to 50 kGy. Various reaction parameters were optimized and the maximum (148%) degree of grafting (DOG) was found when bleached fiber irradiated with 20 kGy in 5% GMA emulsion medium containing 0.5% polyoxyethylene-sorbitan monolaurate (Tween 20) surfactant at 40°C for 1 h. The synthesis of flax fibers-g-GMA copolymers were further confirmed by Fourier transform infrared spectroscopy (FTIR) and scanning electron microscopy (SEM).

Glycidyl methacrylate grafted bleached kenaf (GMA-g-Kenaf) fibers were also prepared by pre-irradiation grafting technique in GMA/water emulsion system (Sharif et al. 2013). The maximum degree of grafting (>250%) was found at an irradiation dose of 50 kGy, 3h reaction time, 40°C temperature, and 5% of monomer concentration (Figure 7.5a).

Othman et al. (Othman, Selambakkannu, and Abdullah 2020) carried out grafting of GMA onto kenaf fibers in vapor phase using a radiation-induced graft copolymerization technique. Electron beam radiation dose was varied from 10 to 100 kGy; however, maximum grafting was reported at an irradiation dose of 50 kGy. The graft copolymerization was further confirmed by FTIR spectroscopy analysis (Figure 7.5b).

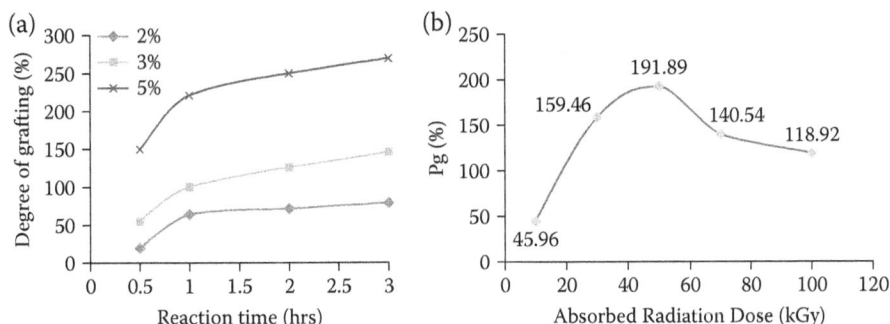

FIGURE 7.5 (a) Variation of degree of grafting with glycidyl methacrylate monomer concentration onto pre-treated (0.7% NaClO$_2$) kenaf fibers, irradiated at 50 kGy (optimized value) at grafting temperature 40°C for 3 h [Figure reprinted from ref (Sharif et al. 2013) with permission from Elsevier Copyright (2013)] and (b) Effect of absorbed dose on percentage of grafting of GMA onto Kenaf fibers (Reaction time: 45 min, temperature: 40°C, Pressure: 0.05 MPa) [Figure reprinted from reference (Othman, Selambakkannu, and Abdullah 2020) with permission from Elsevier Copyright (2020)].

Chen at al. (T. Chen et al. 2019) used two-step methods for synthesis of novel poly (AAc-aniline)-grafted MCC conducting composite. In the first step, AAc was graft copolymerized onto MCC by irradiating with electron beam followed by a second step in which aniline was further grafted onto MCC-AA matrix through oxidative-radical copolymerization technique in ammonium peroxydisulphate (APS) acidic media. Senna et al. (Senna et al. 2016) prepared blends of plasticized starch/cellulose acetate/carboxymethyl cellulose by direct electron beam irradiation of their mixture (different relative ratio) with dose ranging from 0–50 kGy. Physicochemical properties such as gel fraction and swelling characters of irradiated and unirradiated blends were found to be improved after beam irradiation. In addition, study of thermogravimetric analysis (TGA) and differential scanning calorimetry (DSC) curve also showed increase in thermal stability of blends after irradiation and was reported maximum at irradiation dose of 30 kGy (Figure 7.6). Mitra et al. studied the antibacterial activity of silver acrylate–grafted cotton fabric, which were synthesized by graft copolymerization of AAc onto cotton fabrics using electron beam irradiation followed by a reaction with silver nitrate (Mitra et al. 2000).

Sekine et al. (Sekine et al. 2010) carried out electron beam radiation-induced graft polymerization of GMA onto nonwoven cotton fabrics in an emulsion (2% GMA stabilized by Tween 20). These fabrics were subsequently chemically modified by ethylenediamine (EDA) or diethylenetriamine (DETA) to obtain a good Hg ion adsorbent. The optimal irradiation dose for initiation of the grafting of GMA onto fabric, without affecting fiber strength, was found to be 10 KgY. However, fiber strength has been found to reduce to half at an irradiation dose of 50 kGy. The resulting amine-type adsorbents were evaluated for batch and continuous adsorption of Hg, and were found to exhibit extremely high performance for Hg adsorption.

7.3.7 RADIATION-INITIATED RAFT-MEDIATED GRAFT COPOLYMERIZATION

The evolution of controlled free-radical polymerization (CRP) methods has led to the preparation of graft copolymers with well-described traits and properties (Malmström and Carlmark 2012). Among different CRP methods, reversible addition fragmentation chain transfer (RAFT) polymerization is the most valuable and effectively carried out approach to develop well-managed brushes from polymer surfaces to adjust the surface characteristics/properties of base materials in conjunction with radiation-induced grafting approach (Barsbay et al. 2007). γ-radiation triggered RAFT polymerization outcomes in new synthesized materials with well defined or narrow molecular weight distribution, and additionally it supplied a possibility to graft a number of monomers (Barner et al. 2003; Quinn et al. 2007). A RAFT-mediated graft copolymerization bring about excessive grafting frequencies compared to conventional techniques, and it is taken into consideration as a very effective method to tailor cellulosic surfaces with well-defined characteristics. Chenxi et al. (Xu et al. 2020) applied gamma radiation–induced RAFT polymerization to graft poly (glycidymmethacrylate) onto cellulose triacetate surface, using cyanoisopropyl dithiobenzoate (CPDB) as chain transfer agent. Various reaction parameters were optimized to obtain maximum percent grafting and was found maximum (41%) at 10–12 kGy gamma radiation dose with 30% of GMA and molar ratio 1/400 of RAFT agent and GMA. They reported that radiation initiated RAFT polymerization is easy to control in comparison to traditional methods. Barsbay et al. (Barsbay et al. 2009) also used a combination of γ-radiation induced grafting and RAFT polymerization methods to graft copolymerized sodium-4-styrene sulfonate from cellulose. Kodama et al. (Kodama, Barsbay, and Güven 2014) used a similar technique to graft copolymerized HEMA onto cellulose fibers, and maximum grafting 92% (w/w) was reported by them. In another study, Barsbay et al. (Barsbay and Güven 2014) functionalized cellulose with epoxy groups by gamma-initiated RAFT-mediated grafting of GMA. They used Cumyl dithiobenzoate and 2-cyanoprop-2-yl dithiobenzoate RAFT agents. The grafting of cellulose by the radiation-initiated RAFT technique gives a variety of possibilities that still need to be further investigated.

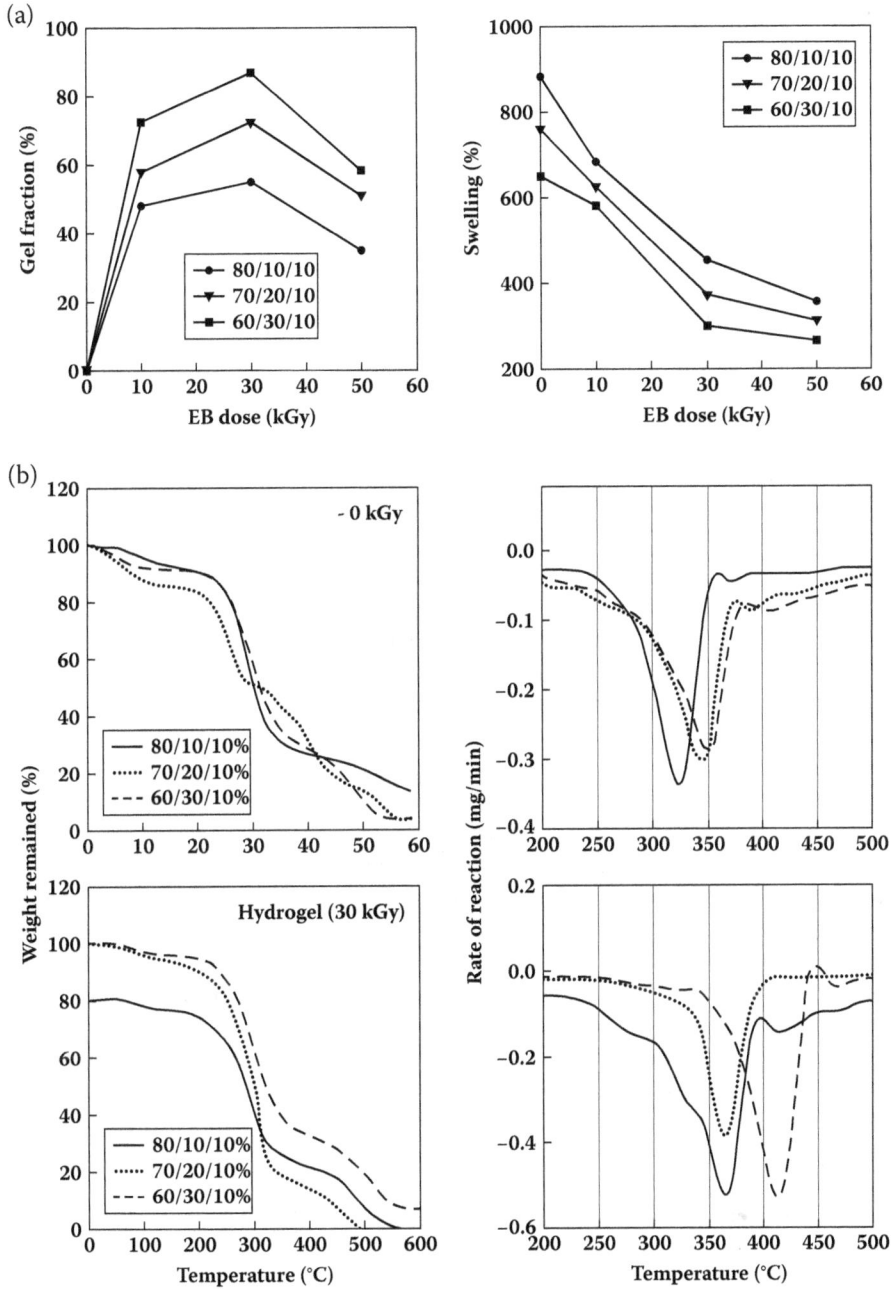

FIGURE 7.6 Effect of EB irradiation dose on (a) gel fraction (%) and swelling (%) in water at pH 7 for plasticized starch/cellulose acetate/carboxymethyl cellulose blend hydrogels irradiated with different EB dosage (b) TGA and the corresponding rates of the thermal decomposition reaction curves of different ratios plasticized starch/cellulose acetate/carboxymethyl cellulose blends and their hydrogels, prepared by EB irradiation at a dose of 30 kGy. [Figure reprinted from reference (Senna et al. 2016) with permission from Elsevier Copyright (2016)].

7.4 DEGRADATION/DISSOLUTION OF CELLULOSE INTO NANO-CELLULOSE THROUGH IRRADIATION

Interactions of cellulose with high-energy electron beam causes depolymerization as a result of chain scission/oxidation (Henniges et al. 2013). The chain scission/oxidation proceeds through the formation of free radicals either by hydrogen abstraction from a cellulose backbone moiety or by the cleavage of glycosidic bonds followed by degradation process after the EBI treatment (Alberti et al. 2005). It has been reported in literature that EB irradiation at low doses (<10 kGy) initiates a cross-linking reaction rather than chain scission/oxidation. However, chain scission/oxidation predominates when EBI doses increased beyond 10 kGy, which may be due to splitting of glycosidic bonds (E. Takacs et al. 2000; Bouchard, Methot, and Jordan 2006). Researchers have found that the conversion yield of cellulose to glucose increases upto 4% after 24 hrs and 6% after 48 hours with 20 kGy of irradiation dose (Duarte et al. 2012; Kristiani et al. 2015; Jeun et al. 2015). Cellulose dissolution has been reported to be increased up to 10% by pre-treatment of the cellulose with ionic liquids [1-ethyl-3-methylimidazolium chloride (EMIM)(Cl mim)] for 1 hour before their exposure to the electron beam (Pezoa et al. 2010; An et al. 2015).

Kim et al. (H. G. Kim et al. 2019) prepared cellulose nanofibers by irradiating Solidago altissima L. pulp for different doses strength 50, 100, 200, and 300 kGy of electron beam at an incremental rate of 6.67 kGy/sec at room temperature in atmospheric air.

The EBI-treated cellulosic fibers were immersed in distilled water and were manually grounded for 1 minute using a pestle mortar for preparation of nanofibers. They labeled the samples as P-0 (raw pulps), P-50, P-100, P-200, and P-300 for different electron beam irradiation strength and reported finer separation of fibers with increasing EBI dose. An increase in electron beam doses reduces the diameter of fiber and was found a minimum (160 nm) in the case of P-300 samples (Figure 7.7). In another research, Kim et al. (D.-Y. Kim et al. 2016) prepared nanocellulose from bleached kenaf core pulp by hydrolyzing EBI-treated pulp in a 63 wt% sulphuric acid concentrated solution.

Lee et al. (Lee et al. 2018) used other techniques such as high-pressure homogenization and alkaline treatment or/and further oxidation/cationization treatment on EBI-treated (~3,000 kGy) cellulose pulp. They have used a two-step process for the preparation of nano-cellulose from cellulose pulp. In the first step, cellulose pulp was irradiated for a short time by electron beam and in the second step it was homogenized by using a high-pressure homogenizer. The degree of polymerization (DP) of cellulose pulp after the EBI irradiation at 100 kGy has been found to reduce from 998 (raw) to 156, whereas the carboxylate contents were found to increase by oxidative activation from 0.04 to 0.84 mmol g^{-1} when the EBI dose was varied from 200–3,000 kGy. The combined EBI and high-pressure homogenization method offer a promising and more efficient approach for the production of cellulose nano-crystals with low environmental impact.

However, Shin et al. (Shin et al. 2012), just like Kim et al. (H. G. Kim et al. 2019) have isolated cellulose fiber using only water, in their process, from EBI-pretreated kenaf bast fibers. Kawaski et al. (Kawasaki et al. 2020) degraded the cellulose by using a greener approach i.e., by irradiating the cellulose by infrared free electron lasers IR-FEL. During irradiation, glucoside bonds were cleaved, and glucose, cellobiose, and low-molecular-weight oligomeric sugars were produced. One key benefit of this method is that the laser system can be operated at room temperature and needs no specific conditions such as acidic or alkaline solution, organic solvents, high pressures, and high temperatures (Bodachivskyi, Unnikrishnan Kuzhiumparambil, and Williams 2019). Chen et al. (Q.-Y. Chen et al. 2017) have used the EBI for dissolving bamboo pulp and reported that when the irradiating dose was increased, a decrease in DP and alpha cellulose had been observed.

Zhang and Liu (Zhang and Liu 2018) prepared nano-cellulose by first irradiating eucalyptus wood at 800 kGy using 60 Co-γ as a radiation source with the intensity of 9.99 × 10^{15} Bq and

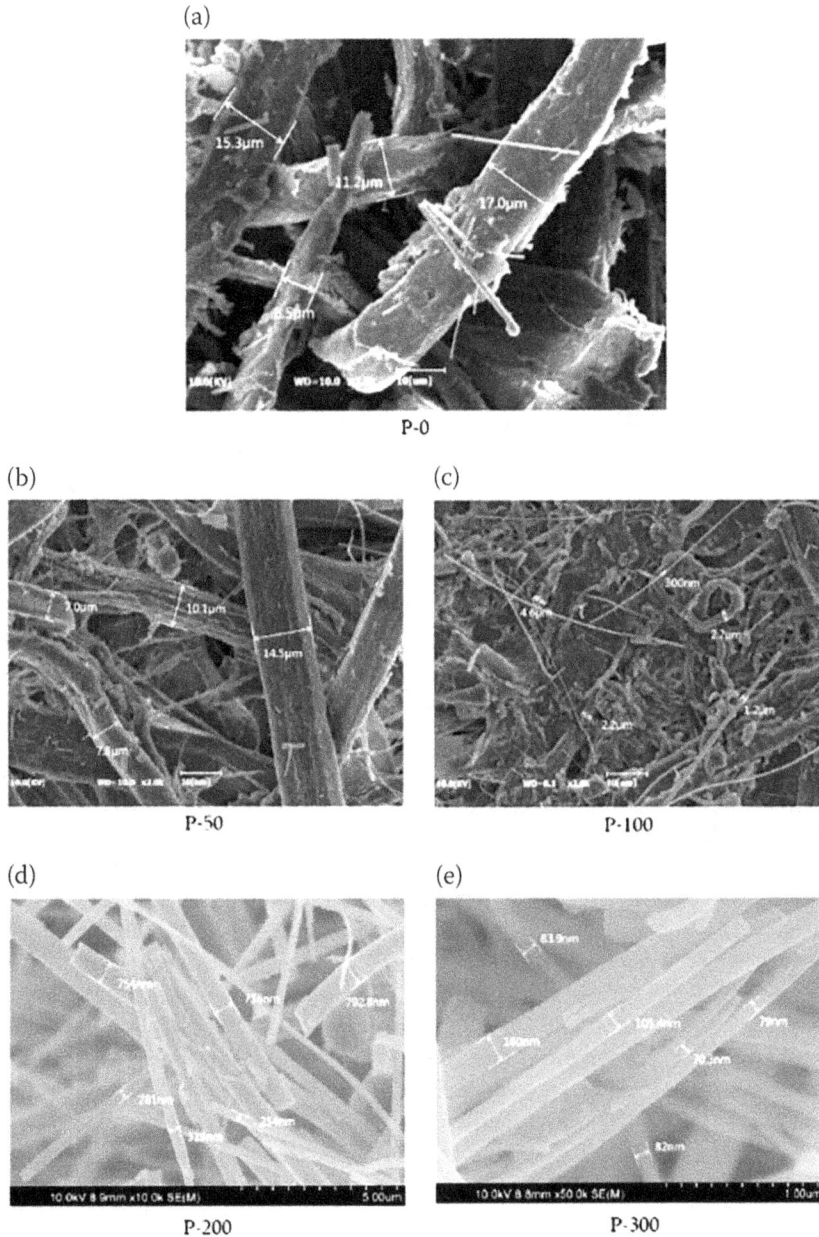

FIGURE 7.7 SEM images showing degree of separation of (a) P-0 (b) P-50 (c) P-100 (d) P-200, and (e) P-300 [Image was copied from research article Published by Kim et al. (H. G. Kim et al. 2019) in Scientific Reports, *Published under a CC BY 4.0 license*].

subsequently subjecting it to a three-step process: cellulose fractionation in organosolv mixture solutions, NaClO bleaching, and ultrasonic disintegration process. In the first step, a residue solid (RS) of cellulose fraction was obtained. In the second step, bleached cellulose (BC) was achieved by bleaching RS with a mixture solution containing NaClO and NaOH. In the third step, BC samples were subjected to ultrasonic disintegration in water to obtain a uniform dispersion, and finally cellulose nano-crystals (CNC) were obtained. They compared two types of eucalyptus, irradiated and untreated feedstocks, during CNC preparation.

7.5 RADIATION-INDUCED GRAFT COPOLYMERIZATION ONTO NANO-CELLULLOSE/NANO-CRYSTALS

The study of functionalization of nano-cellulosic materials (NC) and their properties is an attractive field of research nowadays, as these materials have considerably improved physico-chemical properties that enhance their role in the field of materials science and research. Just like macro cellulosic materials, these nano-cellulosic materials are also hydrophilic by nature due to the presence of OH groups on their surface; therefore, surface chemistry/properties of these NC materials can be tailored as per requirement through chemical, physical (Huang et al. 2019), or biological approaches. Surface modification of NC materials, which can be carried out during or after their production (Wei et al. 2017), led to attaining desirable properties and thus enhanced their potency for a given application (Abushammala and Mao 2019; Liang et al. 2019; Tao et al. 2020). In addition, these cellulose nano-crystals (CNC) have high reactivity in comparison to their macro/micro cellulosic materials counterpart as they possess a high surface area to volume ratio. Lin et al. (Lin, Huang, and Dufresne 2012) noticed that nano-cellulose fibers/crystals reinforced polymer composites with enhanced thermal and mechanical performances can be made through the surface functionalization of cellulosic nano-crystals/fibers.

According to literature, there are numerous physical as well as chemical initiation methods available for surface functionalization of cellulose nano-crystals. However, physical methods for surface functionalization have the edge over chemical methods as they are a simplified process, no residual by-products are formed, and the cost of production is also low. The physical initiators that are mostly employed by researchers are ozone (Partouche, Waysbort, and Margel 2006), γ-rays (Bucio, Skewes, and Burillo 2005), electron beam (Vahdat et al. 2007), plasma (Okubo et al. 2008), corona discharge (Lei, Shi, and Zhang 2000), and ultraviolet irradiation (Stannett 1990). The various methods for surface modification include sulfonation, oxidation, esterification, etherification, silylation, carbamation, etc. (George and Sabapathi 2015; Afrin and Karim 2017; Daud and Lee 2017; Huang et al. 2019). Chen et al. (G. Chen et al. 2009) successfully prepared the first thermoformable bionanocomposite of PCL-grafted CNC under microwave irradiation. The bionanocomposites have been found to have improved hydrophobic character after grafting due to the protective effect of a polycaprolactone (PCL) chain onto the CNC surface, which significantly reduced the number of active -OH groups. Further, injection-molded PCL-grafted CNC bionanocomposite showed better tensile strength and Young's modulus than its compression-molded counterpart. Boujemaoui et al (Boujemaoui et al. 2015) carried out surface modification of cellulose nanopaper (NP) under microwave radiation by placing a known amount of cellulose substrate placed in an E-flask equipped with a magnetic stirrer together with the ε-caprolactone (ε-CL) monomer (20.6 g, 181 mmol) followed by the addition of a catalytic amount of the catalyst Ti(On-Bu)^4 (35.4 mg, 0.10 mmol). The mixture was then degassed for 30 min and conditioned under microwave irradiation (MH) of 180 W using a domestic microwave oven. The samples were continuously withdrawn to determine the extent of monomer conversion by 1H-NMR and were further characterized through FTIR.

Poly(methyl methacrylate) (PMMA) was graft copolymerized onto cellulose nano-fibers without using organic solvents by direct irradiation of nano-fibers with microwave radiation at a constant temperature and stirring speed at a maximum wavelength (λmax) of 365 nm (Yang et al. 2019). The CNF-g-PMMA fibers were characterized through FTIR and SEM techniques. The SEM images of virgin CNF, CNF-g-PMMA0.56, CNF-g-PMMA1.88, and CNF-g-PMMA4.84 samples, where CNF-g-PMMA0.56, CNF-g-PMMA1.88, and CNF-g-PMMA4.84 samples represent the grafted CNFs with low, medium, and high DG have been depicted in Figure 7.8. Jeun et al. (Jeun et al. 2019) hydrophobized cellulose nano-crystals' (CNC) surface by grafting of 1H,1H,2H,2H-perfluorodecyl acrylate (PFDA) using electron beam irradiation. To carry out grafting, CNC suspensions were prepared first and then a PFDA/dichloromethane solution (10 wt%) was added to CNC suspensions in polyethylene bags, which were subsequently purged with nitrogen for 10 min

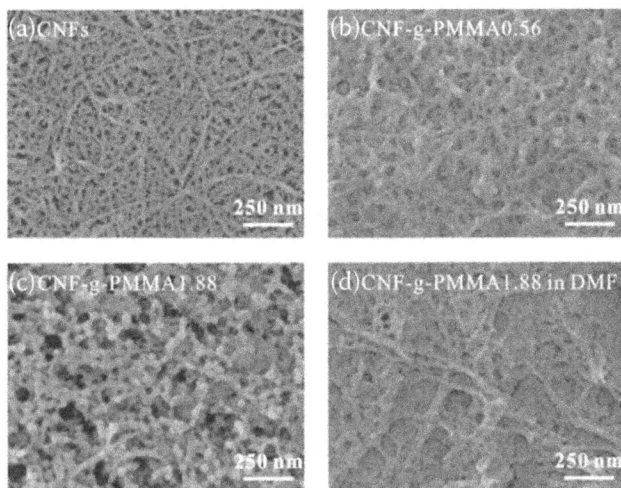

FIGURE 7.8 SEM images of (a) CNFs, (b) CNF-g-PMMA0.56, (c) CNF-g-PMMA1.88, and (d) CNF-g-PMMA1.88 dispersed in DMF [Reproduced from ref. (Yang et al. 2019) with permission from the Royal Society of Chemistry Copyright (2019)].

to remove oxygen. Each sample was irradiated by electron beam radiation at a dose rate of 5 kGy/scan and the irradiation dose of each sample was 10, 20, and 30 kGy. The irradiated samples were washed with THF or chloroform to remove homopolymer and any residual monomer.

Alanis et al. (Alanis et al. 2019) carried out plasma-induced graft copolymerization of three monomers i.e., caprolactone, styrene, and farnesene onto cellulose nanocrystals. The surface functionalized nano-crystals were evaluated by different techniques including XPS, FTIR, and STEM. They reported that plasma-induced graft copolymerization allows a homogenous functionalization, and does not alter the rod-like shape of the nano-crystals, and thus preserves their anisotropic behavior. The modified nano-crystals were used as reinforcement in polymer composites, and an increase in mechanical properties have been observed after grafting the nano-crystals.

7.6 OUTLOOK

This chapter summarizes the various physical methods available in literature for carrying out graft copolymerization onto cellulose. It is difficult to make a comparative study on various physical methods as each one has its own importance and limitations. Also, it is clear from the above study that these methods have an edge over the chemical methods. A lot of work has gone into grafting but physical methods still need to be further explored by researchers.

REFERENCES

Abdel Ghaffar, A. M., M. B. El-Arnaouty, A. A. Abdel Baky, and S. A. Shama. 2016. "Radiation-Induced Grafting of Acrylamide and Methacrylic Acid Individually onto Carboxymethyl Cellulose for Removal of Hazardous Water Pollutants." *Designed Monomers and Polymers* 19 (8): 706–718.

Abushammala, Hatem, and Jia Mao. 2019. "A Review of the Surface Modification of Cellulose and Nanocellulose Using Aliphatic and Aromatic Mono-and Di-Isocyanates." *Molecules* 24 (15): 2782.

Afrin, Sadaf, and Zoheb Karim. 2017. "Isolation and Surface Modification of Nanocellulose: Necessity of Enzymes over Chemicals." *ChemBioEng Reviews* 4 (5): 289–303. 10.1002/cben.201600001

Alanis, Andrés, Josué Hernández Valdés, Neira-Velázquez María Guadalupe, Ricardo Lopez, Ricardo Mendoza, Aji P. Mathew, Ramón Díaz de León, and Luis Valencia. 2019. "Plasma Surface-Modification of Cellulose Nanocrystals: A Green Alternative towards Mechanical Reinforcement of ABS." *RSC Advances* 9 (30): 17417–17424. 10.1039/C9RA02451D

Alberti, A., S. Bertini, G. Gastaldi, N. Iannaccone, D. Macciantelli, G. Torri, and Elena Vismara. 2005. "Electron Beam Irradiated Textile Cellulose Fibres: ESR Studies and Derivatisation with Glycidyl Methacrylate (GMA)." *European Polymer Journal* 41 (8): 1787–1797.

An, Yan-Xia, Min-Hua Zong, Hong Wu, and Ning Li. 2015. "Pretreatment of Lignocellulosic Biomass with Renewable Cholinium Ionic Liquids: Biomass Fractionation, Enzymatic Digestion and Ionic Liquid Reuse." *Bioresource Technology* 192: 165–171.

Badawy, Sayed M. 2017. "Functional Cellulosic Filter Papers Prepared by Radiation-Induced Graft Copolymerization for Chelation of Rare Earth Elements." *Cellulose Chemistry and Technology* 52 (5–6): 551–558.

Barner, Leonie, John F. Quinn, Christopher Barner-Kowollik, Philipp Vana, and Thomas P. Davis. 2003. "Reversible Addition–Fragmentation Chain Transfer Polymerization Initiated with γ-Radiation at Ambient Temperature: An Overview." *European Polymer Journal* 39 (3): 449–459.

Barsbay, Murat, and Olgun Güven. 2014. "Modification of Cellulose by RAFT Mediated Graft Copolymerization." *Hacettepe Journal of Biology and Chemistry* 42 (1): 1–7.

Barsbay, Murat, and Olgun Güven. 2019. "Surface Modification of Cellulose via Conventional and Controlled Radiation-Induced Grafting." *Radiation Physics and Chemistry* 160 (July): 1–8. 10.1016/j.radphyschem. 2019.03.002

Barsbay, Murat, Olgun Güven, Thomas P. Davis, Christopher Barner-Kowollik, and Leonie Barner. 2009. "RAFT-Mediated Polymerization and Grafting of Sodium 4-Styrenesulfonate from Cellulose Initiated via γ-Radiation." *Polymer* 50 (4): 973–982.

Barsbay, Murat, Olgun Güven, Martina H. Stenzel, Thomas P. Davis, Christopher Barner-Kowollik, and Leonie Barner. 2007. "Verification of Controlled Grafting of Styrene from Cellulose via Radiation-Induced RAFT Polymerization." *Macromolecules* 40 (20): 7140–7147.

Başarır, Seyhan Şener, and Nursel Pekel Bayramgil. 2012. "The Uranium Recovery from Aqueous Solutions Using Amidoxime Modified Cellulose Derivatives. I. Preparation, Characterization and Amidoxime Conversion of Radiation Grafted Ethyl Cellulose-Acrylonitrile Copolymers." *Radiochimica Acta* 100 (12): 893–900.

Bhattacharya, Deepanjan, Louis T. Germinario, and William T. Winter. 2008. "Isolation, Preparation and Characterization of Cellulose Microfibers Obtained from Bagasse." *Carbohydrate Polymers* 73 (3): 371–377.

Bledzki, A. K., Sahmir Reihmane, and Julio Gassan. 1996. "Properties and Modification Methods for Vegetable Fibers for Natural Fiber Composites." *Journal of Applied Polymer Science* 59 (8): 1329–1336.

Bodachivskyi, Iurii, Prof Unnikrishnan Kuzhiumparambil, and D. Bradley G. Williams. 2019. "High Yielding Acid-Catalysed Hydrolysis of Cellulosic Polysaccharides and Native Biomass into Low Molecular Weight Sugars in Mixed Ionic Liquid Systems." *ChemistryOpen* 8 (10): 1316.

Bongiovanni, R., E. Zeno, A. Pollicino, P. M. Serafini, and C. Tonelli. 2011. "UV Light-Induced Grafting of Fluorinated Monomer onto Cellulose Sheets." *Cellulose* 18 (1): 117–126.

Bouchard, Jean, Myriam Methot, and Byron Jordan. 2006. "The Effects of Ionizing Radiation on the Cellulose of Woodfree Paper." *Cellulose* 13 (5): 601–610.

Boujemaoui, Assya, Surinthra Mongkhontreerat, Eva Malmström, and Anna Carlmark. 2015. "Preparation and Characterization of Functionalized Cellulose Nanocrystals." *Carbohydrate Polymers* 115: 457–464.

Bucio, Emilio, Phill Skewes, and Guillermina Burillo. 2005. "Synthesis and Characterization of Azo Acrylates Grafted onto Polyethylene Terephthalate by Gamma Irradiation." *Nuclear Instruments and Methods in Physics Research Section B: Beam Interactions with Materials and Atoms* 236 (1–4): 301–306. 10.1016/j.nimb.2005.03.262

Chen, Guangjun, Alain Dufresne, Jin Huang, and Peter R. Chang. 2009. "A Novel Thermoformable Bionanocomposite Based on Cellulose Nanocrystal-Graft-Poly (ε-Caprolactone)." *Macromolecular Materials and Engineering* 294 (1): 59–67.

Chen, Qiu-Yan, Xiao-Juan Ma, Jian-Guo Li, Qing-Xian Miao, and Liu-Lian Huang. 2017. "Effect of the Utilization of Electron Beam Irradiation on the Reactivity of Bamboo Dissolving Pulp." *BioResources* 12 (3): 6251–6261.

Chen, Tao, Yiping Ke, Hongquan Wang, Xinyue Liu, Ziyan Tan, Libing Zhang, and Yuan Zhao. 2019. "Preparation of Cellulose-Polyaniline Composite Microspheres via Electron Beam Irradiation Grafting and It's Properties." In *IOP Conference Series: Materials Science and Engineering*, 493: 012111. IOP Publishing.

Cristaldi, Giuseppe, Alberta Latteri, Giuseppe Recca, and Gianluca Cicala. 2010. "Composites Based on Natural Fibre Fabrics." *Woven Fabric Engineering* 17: 317–342.

Cruz, Juliana, and Raul Fangueiro. 2016. "Surface Modification of Natural Fibers: A Review." *Procedia Engineering* 155: 285–288. 10.1016/j.proeng.2016.08.030

Daud, Jannah B., and Koon-Yang Lee. 2017. "Surface Modification of Nanocellulose." In *Handbook of Nanocellulose and Cellulose Nanocomposites*, edited by Hanieh Kargarzadeh, Ishak Ahmad, Sabu Thomas, and Alain Dufresne, 101–122. Weinheim, Germany: Wiley-VCH Verlag GmbH & Co. KGaA. 10.1002/9783527689972.ch3

Duarte, C. L., M. A. Ribeiro, H. Oikawa, M. N. Mori, C. M. Napolitano, and C. A. Galvao. 2012. "Electron Beam Combined with Hydrothermal Treatment for Enhancing the Enzymatic Convertibility of Sugarcane Bagasse." *Radiation Physics and Chemistry* 81 (8): 1008–1011.

Ershov, Boris G. 1998. "Radiation-Chemical Degradation of Cellulose and Other Polysaccharides." *Russian Chemical Reviews* 67 (4): 315–334.

Esteghlalian, Alireza, Andrew G. Hashimoto, John J. Fenske, and Michael H. Penner. 1997. "Modeling and Optimization of the Dilute-Sulfuric-Acid Pretreatment of Corn Stover, Poplar and Switchgrass." *Bioresource Technology* 59 (2–3): 129–136.

Ferrero, Franco, M. Periolatto, M. Sangermano, and M. Bianchetto Songia. 2008. "Water-Repellent Finishing of Cotton Fabrics by Ultraviolet Curing." *Journal of Applied Polymer Science* 107 (2): 810–818.

Ferrero, Franco, Monica Periolatto, and Stefano Ferrario. 2015. "Sustainable Antimicrobial Finishing of Cotton Fabrics by Chitosan UV-Grafting: From Laboratory Experiments to Semi Industrial Scale-Up." *Journal of Cleaner Production* 96: 244–252.

Galema, Saskia A. 1997. "Microwave Chemistry." *Chemical Society Reviews* 26 (3): 233–238.

Gashti, Mazeyar Parvinzadeh, and Arash Almasian. 2013. "UV Radiation Induced Flame Retardant Cellulose Fiber by Using Polyvinylphosphonic Acid/Carbon Nanotube Composite Coating." *Composites Part B: Engineering* 45 (1): 282–289.

George, Johnsy, and S. N. Sabapathi. 2015. "Cellulose Nanocrystals: Synthesis, Functional Properties, and Applications." *Nanotechnology, Science and Applications* 8: 45.

Hayward, Davidá O. 1999. "Apparent Equilibrium Shifts and Hot-Spot Formation for Catalytic Reactions Induced by Microwave Dielectric Heating." *Chemical Communications* 11: 975–976.

Hemvichian, Kasinee, Auraruk Chanthawong, and Phiriyatorn Suwanmala. 2014. "Synthesis and Characterization of Superabsorbent Polymer Prepared by Radiation-Induced Graft Copolymerization of Acrylamide onto Carboxymethyl Cellulose for Controlled Release of Agrochemicals." *Radiation Physics and Chemistry* 103: 167–171.

Henniges, Ute, Merima Hasani, Antje Potthast, Gunnar Westman, and Thomas Rosenau. 2013. "Electron Beam Irradiation of Cellulosic Materials—Opportunities and Limitations." *Materials* 6 (5): 1584–1598.

Henniges, Ute, Satoko Okubayashi, Thomas Rosenau, and Antje Potthast. 2012. "Irradiation of Cellulosic Pulps: Understanding Its Impact on Cellulose Oxidation." *Biomacromolecules* 13 (12): 4171–4178.

Hong, Kyung Hwa, Ning Liu, and Gang Sun. 2009. "UV-Induced Graft Polymerization of Acrylamide on Cellulose by Using Immobilized Benzophenone as a Photo-Initiator." *European Polymer Journal* 45 (8): 2443–2449.

Hong, Tran Thu, Hirotaka Okabe, Yoshiki Hidaka, Brian A. Omondi, and Kazuhiro Hara. 2019. "Radiation Induced Modified CMC-Based Hydrogel with Enhanced Reusability for Heavy Metal Ions Adsorption." *Polymer* 181: 121772.

Hosny, Alaa El-Dien M.S., Hala A. Farrag, Omneya M. Helmy, Soheir A. A. Hagras, and Amr El-Hag Ali. 2020. "*In-Vitro* Evaluation of Antibacterial and Antibiofilm Efficiency of Radiation-Modified Polyurethane–ZnO Nanocomposite to Be Used as a Self-Disinfecting Catheter." *Journal of Radiation Research and Applied Sciences* 13 (1): 215–225. 10.1080/16878507.2020.1719328

Hossan Shahid Shohrawardy, Mohammed. 2011. "Effect Of•-RAY Irradiation on Mechanical and Thermal Properties of Jute Fabrics Reinforced Polypropylene Composites." M.Phil Thesis submitted in Department of Physicsavailable at Bangladesh University of Engineering and Technology (BUET), Bangladesh. available at http://lib.buet.ac.bd:8080/xmlui/handle/123456789/4106

Huang, Jin, Xiaozhou Ma, Guang Yang, and Dufresne Alain (eds). 2019. Introduction to Nanocellulose. *Nanocellulose: From Fundamentals to Advanced Materials*, 1–20. Wiley-VCH.

Ibarra, V. García, R. Sendón, and A. Rodríguez-Bernaldo de Quirós. 2016. "Antimicrobial Food Packaging Based on Biodegradable Materials." In Barros-Velázquez, V. *Antimicrobial Food Packaging*, 363–384. Academic Press Elsevier.

Ibrahim, Sayeda M., Issa M. Mousaa, and Mervat S. Ibrahim. 2014. "Characterization of Gamma Irradiated Plasticized Carboxymethyl Cellulose (CMC)/Gum Arabic (GA) Polymer Blends as Absorbents for Dyestuffs." *Bulletin of Materials Science* 37 (3): 603–608.

Janardhnan, Sreekumar, and Mohini M. Sain. 2006. "Isolation of Cellulose Microfibrils–an Enzymatic Approach." *Bioresources* 1 (2): 176–188.

Jeun, Joon-Pyo, Byoung-Min Lee, Jin-Young Lee, Phil-Hyun Kang, and Jung-Ki Park. 2015. "An Irradiation-Alkaline Pretreatment of Kenaf Core for Improving the Sugar Yield." *Renewable Energy* 79: 51–55.

Jeun, Joon-Pyo, Yujun Lee, Phil-Hyun Kang, and d Du-Yeon Lee. 2019. "Surface Modification for the Hydrophobization of Cellulose Nanocrystals Using Radiation-Induced Grafting." *Journal of Nanoscience and Nanotechnology* 19 (10): 6303–6308. 10.1166/jnn.2019.17033

Kakati, Dilip Kumar, Montu Moni Bora, and Chayanika Deka. 2015. "Cellulose Graft Copolymerization by Gamma and Electron Beam Irradiation." In Thakur, V. K. *Cellulose-Based Graft Copolymers*, 98–107. CRC Press.

Kamal, H., F. M. Abd-Elrahim, and S. Lotfy. 2014. "Characterization and Some Properties of Cellulose Acetate-Co-Polyethylene Oxide Blends Prepared by the Use of Gamma Irradiation." *Journal of Radiation Research and Applied Sciences* 7 (2): 146–153.

Kang, Hongliang, Ruigang Liu, and Yong Huang. 2015. "Graft Modification of Cellulose: Methods, Properties and Applications." *Polymer* 70: A1–A16.

Kappe, C. Oliver. 2004. "Controlled Microwave Heating in Modern Organic Synthesis." *Angewandte Chemie International Edition* 43 (46): 6250–6284.

Kashiwagi, Masayuki, and Yasuhisa Hoshi. 2012. "Electron Beam Processing System and Its Application." *SEI Tech. Rev* 75: 47–53.

Kawasaki, Takayasu, Takeshi Sakai, Heishun Zen, Yoske Sumitomo, Kyoko Nogami, Ken Hayakawa, Toyonari Yaji, Toshiaki Ohta, Koichi Tsukiyama, and Yasushi Hayakawa. 2020. "Cellulose Degradation by Infrared Free Electron Laser." *Energy & Fuels* 34 (7): 9064–9068.

Khan, Ferdous, S. R. Ahmad, and E. Kronfli. 2004. "UV-Radiation–Induced Preirradiation Grafting of Methyl Methacrylate onto Lignocellulose Fiber in an Aqueous Medium and Characterization." *Journal of Applied Polymer Science* 91 (3): 1667–1675.

Khan, Ijaz A., Hazrat Hussain, Tariq Yasin, and Muhammad Inaam-ul-Hassan. 2020. "Surface Modification of Mesoporous Silica by Radiation Induced Graft Polymerization of Styrene and Subsequent Sulfonation for Ion-Exchange Applications." *Journal of Applied Polymer Science* 137 (26): 48835.

Khan, Mubarak A., Md Saifur Rahaman, A. Al-Jubayer, and J. M. M. Islam. 2015. "Modification of Jute Fibers by Radiation-Induced Graft Copolymerization and Their Applications." Thakur, V. K. (ed). *Cellulose-Based Graft Copolymers: Structure and Chemistry*, 209–235. CRC Press, Taylor and Francis Group, Boca Raton.

Kianfar, Parnian, Molla Tadesse Abate, Valentina Trovato, Giuseppe Rosace, Ada Ferri, Roberta Bongiovanni, and Alessandra Vitale. 2020. "Surface Functionalization of Cotton Fabrics by Photo-Grafting for PH Sensing Applications." *Frontiers in Materials* 7: 39.

Kim, Du-Yeong, Byoung-Min Lee, Dong Hyun Koo, Phil-Hyun Kang, and Joon-Pyo Jeun. 2016. "Preparation of Nanocellulose from a Kenaf Core Using E-Beam Irradiation and Acid Hydrolysis." *Cellulose* 23 (5): 3039–3049.

Kim, Hong Gun, U. Sang Lee, Lee Ku Kwac, Sang Ok Lee, Yong-Sun Kim, and Hye Kyoung Shin. 2019. "Electron Beam Irradiation Isolates Cellulose Nanofiber from Korea 'Tall Goldenrod' Invasive Alien Plant Pulp." *Nanomaterials* 9 (10): 1358.

Klemm, Dieter, B. Philpp, Thomas Heinze, Ute Heinze, and Wolfgang Wagenknecht. 1998. *Comprehensive Cellulose Chemistry. Volume 1: Fundamentals and Analytical Methods.* Wiley-VCH Verlag GmbH.

Kodama, Yasko, Murat Barsbay, and Olgun Güven. 2014. "Radiation-Induced and RAFT-Mediated Grafting of Poly (Hydroxyethyl Methacrylate)(PHEMA) from Cellulose Surfaces." *Radiation Physics and Chemistry* 94: 98–104.

Kraessig, Hans. 1987. "Cellulose Chemistry and Its Applications, T. P. Nevell and S. H. Zeronian, Eds., Halsted Press, John Wiley, New York, 1985, 552 pp." *Journal of Polymer Science Part C: Polymer Letters* 25 (2): 87–88. 10.1002/pol.1987.140250212

Kristiani, Anis, Nurdin Effendi, Yosi Aristiawan, Fauzan Aulia, and Yanni Sudiyani. 2015. "Effect of Combining Chemical and Irradiation Pretreatment Process to Characteristic of Oil Palm's Empty Fruit Bunches as Raw Material for Second Generation Bioethanol." *Energy Procedia* 68: 195–204.

Kumar, S. Sendhil, and V. Anbumalar. 2015. "Selection and Evaluation of Natural Fibers–A Literature Review." *International Journal of Innovative Science, Engineering & Technology* 2: 929–939.

Lawal, Abdulazeez T., and Gordon G. Wallace. 2014. "Vapour Phase Polymerisation of Conducting and Non-Conducting Polymers: A Review." *Talanta* 119: 133–143.

Lee, Minwoo, Min Haeng Heo, Hyunho Lee, Hwi-Hui Lee, Haemin Jeong, Young-Wun Kim, and Jihoon Shin. 2018. "Facile and Eco-Friendly Extraction of Cellulose Nanocrystals via Electron Beam Irradiation Followed by High-Pressure Homogenization." *Green Chemistry* 20 (11): 2596–2610.

Lei, Jingxin, Meiwu Shi, and Jianchun Zhang. 2000. "Surface Graft Copolymerization of Hydrogen Silicone Fluid onto Fabric through Corona Discharge and Water Repellency of Grafted Fabric." *European Polymer Journal* 36 (6): 1277–1281. 10.1016/S0014-3057(99)00169-X

Lertsarawut, P., K. Hemvichian, T. Rattanawongwiboon, and P. Suwanmala. 2019. "Dye Adsorbent Prepared by Radiation-Induced Graft Polymerization of Acrylic Acid onto Carboxymethyl Cellulose." In *Journal of Physics: Conference Series*, 1285: 012023. IOP Publishing.

Liang, Luna, Samarthya Bhagia, Mi Li, Chen Huang, and Arthur J. Ragauskas. 2019. "Cross-Linked Nanocellulosic Materials and Their Applications." *ChemSusChem* 13 (1): 78–87.

Lin, Ning, Jin Huang, and Alain Dufresne. 2012. "Preparation, Properties and Applications of Polysaccharide Nanocrystals in Advanced Functional Nanomaterials: A Review." *Nanoscale* 4 (11): 3274–3294.

Madrid, Jordan F., and Lucille V. Abad. 2015. "Modification of Microcrystalline Cellulose by Gamma Radiation-Induced Grafting." *Radiation Physics and Chemistry* 115: 143–147.

Madrid, Jordan F., Guillermo M. Nuesca, and Lucille V. Abad. 2013. "Gamma Radiation-Induced Grafting of Glycidyl Methacrylate (GMA) onto Water Hyacinth Fibers." *Radiation Physics and Chemistry* 85 (April): 182–188. 10.1016/j.radphyschem.2012.10.006

Malmström, Eva, and Anna Carlmark. 2012. "Controlled Grafting of Cellulose Fibres–an Outlook beyond Paper and Cardboard." *Polymer Chemistry* 3 (7): 1702–1713.

Migliavacca, Gianluca, Franco Ferrero, and Monica Periolatto. 2014. "Differential Dyeing of Wool Fabric with Metal-Complex Dyes after Ultraviolet Irradiation." *Coloration Technology* 130 (5): 327–333.

Mishra, Sumit, G. Usha Rani, and Gautam Sen. 2012. "Microwave Initiated Synthesis and Application of Polyacrylic Acid Grafted Carboxymethyl Cellulose." *Carbohydrate Polymers* 87 (3): 2255–2262.

Mitra, D., K. P. Rawat, S. Sabharwal, and A. B. Majali. 2000. "Antibacterial Activity of Cotton Fabric Grafted with Silver Acrylate by Electron Beam Irradiation." In *Proceedings of Trombay Symposium on Radiation and Photochemistry. Pt. 2: Preprint Volume of Posters and Invited Talks*.

Moawia, Rihab Musaad, Mohamed Mahmoud Nasef, Nor Hasimah Mohamed, and Adnan Ripin. 2016. "Modification of Flax Fibres by Radiation Induced Emulsion Graft Copolymerization of Glycidyl Methacrylate." *Radiation Physics and Chemistry* 122: 35–42.

Mollah, M. Z. I., T. R. Choudhury, M. A. Khan, P. Ali, A. M. S. Chowdhury, and A. I. Mustafa.2015."Effect of Monomer (EGDMA) on the Mechanical Proper-Ties of Jute Fiber by Using Additives: Photocuring with UV Radiation."*ChemXpress* 8(4): 285–295.

Nasef, Mohamed Mahmoud, and Olgun Güven. 2012. "Radiation-Grafted Copolymers for Separation and Purification Purposes: Status, Challenges and Future Directions." *Progress in Polymer Science* 37 (12): 1597–1656. 10.1016/j.progpolymsci.2012.07.004

Neas, E. D., and M. J. Collins. 1988. "Microwave Heating: Theoretical Concepts and Equipment Design." *Introduction to Microwave Sample Preparation, Theory and Practice*. Washington, DC: American Chemical Society.

Okubo, Masaaki, Mitsuru Tahara, Noboru Saeki, and Toshiaki Yamamoto. 2008. "Surface Modification of Fluorocarbon Polymer Films for Improved Adhesion Using Atmospheric-Pressure Nonthermal Plasma Graft-Polymerization." *Thin Solid Films* 516 (19): 6592–6597. 10.1016/j.tsf.2007.11.033

Omichi, Masaaki, Yuji Ueki, Noriaki Seko, and Yasunari Maekawa. 2019. "Development of a Simplified Radiation-Induced Emulsion Graft Polymerization Method and Its Application to the Fabrication of a Heavy Metal Adsorbent." *Polymers* 11 (8): 1373.

Othman, Nor Azillah Fatimah, Sarala Selambakkannu, and Tuan Amran Tuan Abdullah. 2020. "Grafting Yield Determination of Glycidyl Methacrylate Vapor on Radiated Kenaf Fiber via FTIR Spectroscopy." *Materials Today: Proceedings* 29: 207–211.

Partouche, Eran, Daniel Waysbort, and Shlomo Margel. 2006. "Surface Modification of Crosslinked Poly (Styrene-Divinyl Benzene) Micrometer-Sized Particles of Narrow Size Distribution by Ozonolysis." *Journal of Colloid and Interface Science* 294 (1): 69–78. 10.1016/j.jcis.2005.07.007

Pathania, Deepak, Arush Sharma, and Vandana Sethi. 2017. "Microwave Induced Graft Copolymerization of Binary Monomers onto Luffa Cylindrica Fiber: Removal of Congo Red." *Procedia Engineering* 200: 408–415. 10.1016/j.proeng.2017.07.057

Pezoa, R., V. Cortinez, S. Hyvärinen, M. Reunanen, J. Hemming, M. E. Lienqueo, O. Salazar, R. Carmona, A. Garcia, and D. Murzin. 2010. "Use of Ionic Liquids in the Pretreatment of Forest and Agricultural Residues for the Production of Bioethanol."*Cellulose Chemistry & Technology* 44 (4): 165.

Quinn, John F., Thomas P. Davis, Leonie Barner, and Christopher Barner-Kowollik. 2007. "The Application of Ionizing Radiation in Reversible Addition–Fragmentation Chain Transfer (RAFT) Polymerization: Renaissance of a Key Synthetic and Kinetic Tool." *Polymer* 48 (22): 6467–6480.

Rana, Ashvinder Kumar, Elisabete Frollini, and Vijay Kumar Thakur. 2021. "Cellulose Nanocrystals: Pretreatments, Preparation Strategies, and Surface Functionalization." *International Journal of Biological Macromolecules* 182: 1554–1581.

Rana, Ashvinder K., Vijai Kumar Gupta, Adesh K. Saini, Stefan Ioan Voicu, Magda H. Abdellattifaand, and Vijay Kumar Thakur. 2021. "Water Desalination Using Nanocelluloses/Cellulose Derivatives Based Membranes for Sustainable Future." *Desalination* 520: 115359.

Rana, Ashvinder K., Yogendra Kumar Mishra, Vijai Kumar Gupta, and Vijay Kumar Thakur. 2021. "Sustainable Materials in the Removal of Pesticides from Contaminated Water: Perspective on Macro to Nanoscale Cellulose." *Science of The Total Environment* 797 (November): 149129. 10.1016/j.scitotenv.2021.149129

Rana, Ashvinder K., Prasad Potluri, and Vijay Kumar Thakur. 2021. "Cellulosic Grewia Optiva Fibres: Towards Chemistry, Surface Engineering and Sustainable Materials." *Journal of Environmental Chemical Engineering* 9(5):106059.

Rana, Ashvinder K., and Vijay Kumar Thakur. 2021. "The Bright Side of Cellulosic Hibiscus Sabdariffa Fibres: Towards Sustainable Materials from the Macro- to Nano-Scale." *Materials Advances* 2(15): 4945-4965. 10.1039/D1MA00429H

Roy, Debashish, Mona Semsarilar, James T. Guthrie, and Sébastien Perrier. 2009. "Cellulose Modification by Polymer Grafting: A Review." *Chemical Society Reviews* 38 (7): 2046–2064.

Salmawi, Kariman M. El, Amal A. El-Naggar, and Sayeda M. Ibrahim. 2018. "Gamma Irradiation Synthesis of Carboxymethyl Cellulose/Acrylic Acid/Clay Superabsorbent Hydrogel." *Advances in Polymer Technology* 37 (2): 515–521.

Sekine, Ayako, Noriaki Seko, Masao Tamada, and Yoshio Suzuki. 2010. "Biodegradable Metal Adsorbent Synthesized by Graft Polymerization onto Nonwoven Cotton Fabric." *Radiation Physics and Chemistry* 79 (1): 16–21.

Senna, Magdy M., Abo El-Khair B. Mostafa, Sanna R. Mahdy, and Abdel Wahab M. El-Naggar. 2016. "Characterization of Blend Hydrogels Based on Plasticized Starch/Cellulose Acetate/Carboxymethyl Cellulose Synthesized by Electron Beam Irradiation." *Nuclear Instruments and Methods in Physics Research Section B: Beam Interactions with Materials and Atoms* 386: 22–29.

Sharif, Jamaliah, Siti Fatahiyah Mohamad, Nor Azilah Fatimah Othman, Nurul Azra Bakaruddin, Hasnul Nizam Osman, and Olgun Güven. 2013. "Graft Copolymerization of Glycidyl Methacrylate onto Delignified Kenaf Fibers through Pre-Irradiation Technique." *Radiation Physics and Chemistry* 91: 125–131.

Shin, Hye Kyoung, Joon Pyo Jeun, Hyun Bin Kim, and Phil Hyun Kang. 2012. "Isolation of Cellulose Fibers from Kenaf Using Electron Beam." *Radiation Physics and Chemistry* 81 (8): 936–940.

Shukla, S. R., and A. R. Athalye. 1992. "Ultraviolet-Radiation Induced Graft-Copolymerization of Hydroxyethyl Methacrylate onto Cotton Cellulose." *Journal of Applied Polymer Science* 44 (3): 435–442.

Singh, V., P. Kumar, and R. Sanghi. 2012. "Use of Microwave Irradiation in the Grafting Modification of the Polysaccharides–A Review." *Progress in Polymer Science* 37 (2): 340–364.

Singh, V., A. Tiwari, S. Pandey, and S. K. Singh. 2007. "Peroxydisulfate Initiated Synthesis of Potato Starch-Graft-Poly (Acrylonitrile) under Microwave Irradiation." *Express Polymer Letters* 1 (1): 51–58.

Singha, A. S., and Ashvinder K. Rana. 2012a. "Effect of Graft Copolymerization on Mechanical, Thermal, and Chemical Properties of Grewia Optiva/Unsaturated Polyester Biocomposites." *Polymer Composites* 33 (8): 1403–1414.

Singha, A. S., and Ashvinder K. Rana. 2012b. "Effect of Silane Treatment on Physicochemical Properties of Lignocellulosic C. Indica Fiber." *Journal of Applied Polymer Science* 124 (3): 2473–2484.

Singha, A. S., and Ashvinder K. Rana. 2012c. "A Study on Benzoylation and Graft Copolymerization of Lignocellulosic Cannabis Indica Fiber." *Journal of Polymers and the Environment* 20 (2): 361–371. 10.1007/s10924-011-0370-9

Singha, Amar S., and Ashvinder K. Rana. 2012d. "Ce (IV) Ion–Initiated and Microwave Radiation–Induced Graft Copolymerization of Acrylic Acid onto Lignocellulosic Fibers." *International Journal of Polymer Analysis and Characterization* 17 (1): 72–84.

Singha, Amar Singh, and Ashvinder Kumar Rana. 2012e. "A Comparative Study on Functionalization of Cellulosic Biofiber by Graft Copolymerization of Acrylic Acid in Air and under Microwave Radiation." *BioResources* 7 (2): 2019–2037.

Singha, Amar S., and Ashvinder K. Rana. 2012f. "Effect of Surface Modification of Grewia Optiva Fibres on Their Physicochemical and Thermal Properties." *Bulletin of Materials Science* 35 (7): 1099–1110. 10. 1007/s12034-012-0400-9

Singha, A. S., and Ashvinder K. Rana. 2012g. "Preparation and Characterization of Graft Copolymerized Cannabis Indica L. Fiber-Reinforced Unsaturated Polyester Matrix-Based Biocomposites." *Journal of Reinforced Plastics and Composites* 31 (22): 1538–1553.

Stannett, Vivian T. 1990. "Radiation Grafting—State-of-the-Art." *International Journal of Radiation Applications and Instrumentation. Part C. Radiation Physics and Chemistry* 35 (1–3): 82–87. 10.1016/ 1359-0197(90)90062-M

Sun, J. X., X. F. Sun, H. Zhao, and R. C. Sun. 2004. "Isolation and Characterization of Cellulose from Sugarcane Bagasse." *Polymer Degradation and Stability* 84 (2): 331–339.

Swantomo, D., Rochmadi Rochmadi, K.T. Basuki, and R. Sudiyo. 2013. "Synthesis and Characterization of Graft Copolymer Rice Straw Cellulose-Acrylamide Hydrogels Using Gamma Irradiation." *Atom Indonesia* 39 (2): 57. 10.17146/aij.2013.232

Takács, E., H. Mirzadeh, L. Wojnárovits, J. Borsa, M. Mirzataheri, and N. Benke. 2007. "Comparison of Simultaneous and Pre-Irradiation Grafting of N-Vinylpyrrolidone to Cotton-Cellulose." *Nuclear Instruments and Methods in Physics Research Section B: Beam Interactions with Materials and Atoms* 265 (1): 217–220. 10.1016/j.nimb.2007.08.098

Takacs, E., L. Wojnarovits, CS Földváry, P. Hargittai, J. Borsa, and I. Sajo. 2000. "Effect of Combined Gamma-Irradiation and Alkali Treatment on Cotton–Cellulose." *Radiation Physics and Chemistry* 57 (3–6): 399–403.

Takacs, Erzsebet, Laszlo Wojnarovits, Eva Koczog Horvath, Tamas Fekete, and Judit Borsa. 2012. "Improvement of Pesticide Adsorption Capacity of Cellulose Fibre by High-Energy Irradiation-Initiated Grafting of Glycidyl Methacrylate." *Radiation Physics and Chemistry* 81 (9): 1389–1392.

Tao, Han, Nathalie Lavoine, Feng Jiang, Juntao Tang, and Ning Lin. 2020. "Reducing End Modification on Cellulose Nanocrystals: Strategy, Characterization, Applications and Challenges." *Nanoscale Horizons* 5 (4): 607–627.

Thakur, Manju Kumari, Aswinder Kumar Rana, and Vijay Kumar Thakur. 2014. Lignocellulosic Polymer Composites: A Brief Overview. *Lignocellulosic Polymer Composites: Processing, Characterization, and Properties*, 1–15. Wiley.

Tsujimoto, Takashi, Yohei Nishiumi, and Shiro Kobayashi. 2015. "Synthesis and Curing Behaviors of Reactive Cellulose Esters from Renewable Resources." *Synthesis* 4 (6): 3909–3916.

Vahdat, A., Hajir Bahrami, N. Ansari, and F. Ziaie. 2007. "Radiation Grafting of Styrene onto Polypropylene Fibres by a 10MeV Electron Beam." *Radiation Physics and Chemistry* 76 (5): 787–793. 10.1016/ j.radphyschem.2006.05.009

Vismara, Elena, Lucio Melone, Giuseppe Gastaldi, Cesare Cosentino, and Giangiacomo Torri. 2009. "Surface Functionalization of Cotton Cellulose with Glycidyl Methacrylate and Its Application for the Adsorption of Aromatic Pollutants from Wastewaters." *Journal of Hazardous Materials* 170 (2–3): 798–808.

Wach, R., H. Mitomo, and F. Yoshii. 2004. "ESR Investigation on Gamma-Irradiated Methylcellulose and Hydroxyethylcellulose in Dry State and in Aqueous Solution." *Journal of Radioanalytical and Nuclear Chemistry* 261 (1): 113–118.

Wei, Liqing, Umesh P. Agarwal, Kolby C. Hirth, Laurent M. Matuana, Ronald C. Sabo, and Nicole M. Stark. 2017. "Chemical Modification of Nanocellulose with Canola Oil Fatty Acid Methyl Ester." *Carbohydrate Polymers* 169: 108–116.

Wencka, Magdalena, Kinga Wichlacz, Henryk Kasprzyk, Stefan Lijewski, and Stanislaw K. Hoffmann. 2007. "Free Radicals and Their Electron Spin Relaxation in Cellobiose. X-Band and W-Band ESR and Electron Spin Echo Studies." *Cellulose* 14 (3): 183–194.

Wojnárovits, L., Cs M. Földváry, and E. Takács. 2010. "Radiation-Induced Grafting of Cellulose for Adsorption of Hazardous Water Pollutants: A Review." *Radiation Physics and Chemistry* 79 (8): 848–862.

Xu, Chenxi, Minghan Cao, Jing Peng, Jiuqiang Li, and Maolin Zhai. 2020. "Modification of Cellulose Triacetate Membranes with Glycidyl Methacrylate via γ-Ray Initiated Controlled Grafting." *Chinese Journal of Applied Chemistry* 37 (3): 293.

Yagüe, Jose Luis, Anna Maria Coclite, Christy Petruczok, and Karen K. Gleason. 2013. "Chemical Vapor Deposition for Solvent-Free Polymerization at Surfaces." *Macromolecular Chemistry and Physics* 214 (3): 302–312.

Yang, Xianpeng, Ting-Hsuan Ku, Subir K. Biswas, Hiroyuki Yano, and Kentaro Abe. 2019. "UV Grafting: Surface Modification of Cellulose Nanofibers without the Use of Organic Solvents." *Green Chemistry* 21 (17): 4619–4624. 10.1039/C9GC02035G

Zhang, Renli, and Yun Liu. 2018. "High Energy Oxidation and Organosolv Solubilization for High Yield Isolation of Cellulose Nanocrystals (CNC) from Eucalyptus Hardwood." *Scientific Reports* 8 (1): 1–11.

8 Radiation-initiated tailored membranes for ready fit

Mayank Saxena

Membrane Science and Separation Technology Division, Council of Scientific and Industrial Research-Central Salt and Marine Chemicals Research Institute (CSIR-CSMCRI), Bhavnagar Gujarat, India

A. Bhattacharya*

Membrane Science and Separation Technology Division, Academy of Scientific and Innovative Research, Council of Scientific and Industrial Research-Human Resource Development Centre Campus, Ghaziabad, Uttar Pradesh, India

*Corresponding author: amit@csmcri.res.in

CONTENTS

8.1 INTRODUCTION

The hurdles to invent new materials have driven polymer scientists to explore the different possibilities of modifying the existing polymeric materials. The modification through graft polymerization is quite attractive for introducing desired properties into the polymer matrices.

The graft copolymer is a branched one composed of a leading chain of a polymer backbone onto which side chains are covalently attached. The grafted polymer's chemical structure and composition differ from the polymer backbone (Herman, 2007). The grafting process is a unique way to combine the properties of two incompatible monomers/polymers. Graft polymerization is a procedure where an active chemical site in the pre-existing polymer is created first. This site could be a free radical or a functional group, which may initiate the polymerization of a monomer, leading to a copolymer structure. Problems related to chemical grafting are the control of the extent of grafting, forming homopolymer, and tackling toxic initiators.

DOI: 10.1201/9781003321910-8

Radiation-induced grafting is a useful method in the preparation of functionalized membranes. The process is versatile as numerous polymer substrates and monomers are available. Moreover, the grafting parameters could be capable of tailoring and tuning the properties as required. The radiation grafting technique is preferred, as the polymers desired properties, retaining most of the inherent properties, could be achieved. The radiation grafting does not need additives or catalysts to initiate it. It does not leave any residue like chemical grafting. Thus, it is a simple and clean one. The graft distribution is achieved throughout the modified polymeric matrix because of the uniform radiation activation of the film. The radiation technique avoids the local heating effect on the system (Bhattacharya, 2000, Bhattacharya et al., 2008). Moreover, one can tune the degree of grafting by merely adjusting the radiation dose and dose rate. A suitable attenuator can easily administer the radiation dose rate (Chapiro, 1962). The adaptability of this method in this membrane arena is because it can make a couple of hydrophobic backbones with hydrophilic side chains and vice versa. It can also utilize commercial membrane/films, thereby avoiding casting or other difficulties associated with it (Jetsrisuparb et al., 2014, Dargaville et al., 2003, Nasef and Hegazy, 2004, Gubler and Scherer, 2009). Thus, a wide choice of monomers and base membranes are available to develop specially tailored membranes.

Atomic nuclei have energy levels similar to discrete energy levels in which electrons rotate. The configurations of the protons and neutrons determine the energy-level structures. In the nuclear transition from a high to lower energy level, the excess energy emits in photons. The nuclear energy-level differences correspond to photon wavelengths in the gamma-ray region. The fraction of photons that do not interact with a finite thickness of the membrane material are transmitted with their original energies and directions. Thus, the radiation dose rate may easily be controlled using a suitable attenuator without influencing the photon energy, an essential aspect of radiation-initiated grafting (Gursel et al., 2008).

The energy bombardment breaks the polymer substrate's chemical bond and forms fragments with an unpaired electron, i.e., free radical. The free radical initiates other monomers as a neat liquid, vapor, or solution. In the radiation technique, energy absorption initiates free radical formation in the polymer system and does not need temperature. Thus, it is a zero activation energy process. The high penetration ability of radiation makes it favourable for solid substrate grafting. It is a heterogeneous system and the grafting technique evolved. The characteristic of radiation grafting is that the dimensions of the grafted product can be contained easily by proper choice of the irradiation dose and intensity (Chapiro, 1962). Thus, it is feasible to produce different membranes with a wide range of properties.

The difference between UV-photo and γ-radiation is there. The low-energy photon for the photochemical reaction can interact with a single molecule 'photo-initiator' that initiates grafting. The γ-radiation is of high-energy; it does not require any initiators but can produce free radicals to initiate a grafting system. This chapter introduces grafting as a powerful technique for modifying polymeric membranes. The possibilities of using grafted membranes in different applications are presented.

8.2 DIFFERENT TECHNIQUES

The grafting reactions can be performed by different pathways viz. chemical, photo-irradiation, and enzymatic. The steps follow their mechanistic pathways according to their features. In the radiation technique, generally, three types of mechanistic pathways occur viz. pre-irradiation, peroxidation, and simultaneous (Bhattacharya and Mishra, 2004).

a. Pre-irradiation technique: The polymer substrate is first irradiated, usually in vacuum or in inert gas, to produce somewhat stable free radicals by splitting of hydrogen atoms, which are then reacted with the monomer in neat liquid vapor, or solution, usually at elevated temperature. An essential advantage of this technique is that less homopolymer is formed.

The simple mechanistic pathways from initiation to termination is as follows:

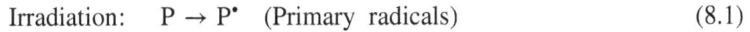

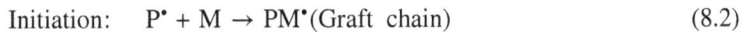

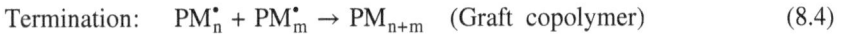

$$\text{—H} \xrightarrow{\quad\sim\sim\sim\quad} \text{—•} + \text{H}^{\bullet}$$

$$\text{—•} + \text{M} \longrightarrow \text{—M}^{\bullet}$$

Irradiation: $P \rightarrow P^{\bullet}$ (Primary radicals) (8.1)

Initiation: $P^{\bullet} + M \rightarrow PM^{\bullet}$ (Graft chain) (8.2)

Propagation: $PM^{\bullet} + nM \rightarrow PM_{n+1}^{\bullet}$ (Graft growing chains) (8.3)

Termination: $PM_{n}^{\bullet} + PM_{m}^{\bullet} \rightarrow PM_{n+m}$ (Graft copolymer) (8.4)

b. Peroxidation technique: In this technique, the method is similar to the previous one, except it irradiates in the presence of air or oxygen. It forms peroxides and hydroperoxides. It then reacted with a monomer at an elevated temperature to form the grafted polymer substrate. In this technique, high irradiation doses are required to form a sufficient number of hydroperoxides to accomplish reasonable graft levels. The simple reaction mechanism is as follows:

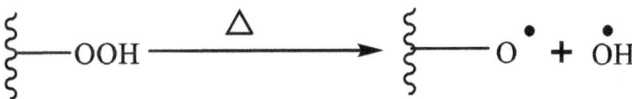

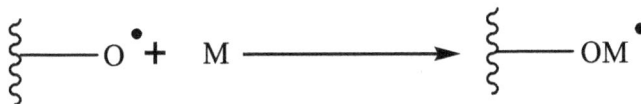

$$\text{—H} + O_2 \xrightarrow{\quad\sim\sim\sim\quad} \text{—OOH}$$

$$\text{—OOH} \xrightarrow{\quad\Delta\quad} \text{—O}^{\bullet} + {}^{\bullet}\text{OH}$$

$$\text{—O}^{\bullet} + M \longrightarrow \text{—OM}^{\bullet}$$

Propagation: $POM^{\bullet} + nM \rightarrow POM_{n+1}^{\bullet}$ (8.5)

c. Simultaneous: The polymer substrate and monomer is irradiated in the same system. The monomer may be present as a vapor, liquid, or solution state. The grafting follows a similar pre-irradiation mechanistic pathway, but homopolymer formation is possibly parallel to the grafting reaction. It leads to monomer wastage; thus, a low level of grafting efficiency in the system results.

The simple mechanism of homopolymer formation is as follows:

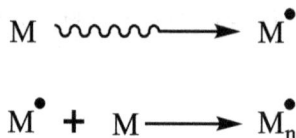

$$M \rightsquigarrow M^{\bullet}$$

$$M^{\bullet} + M \longrightarrow M_n^{\bullet}$$

There is also a high possibility for the mutual recombination of the polymer backbone. It is represented as follows:

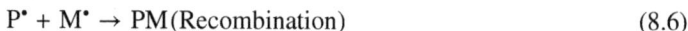

$$P^{\bullet} + M^{\bullet} \rightarrow PM \, (\text{Recombination}) \tag{8.6}$$

The typical cationic/anionic grafting mechanistic pathways are also probable in the simultaneous technique's monomer system. The cationic mechanistic path is in the following:

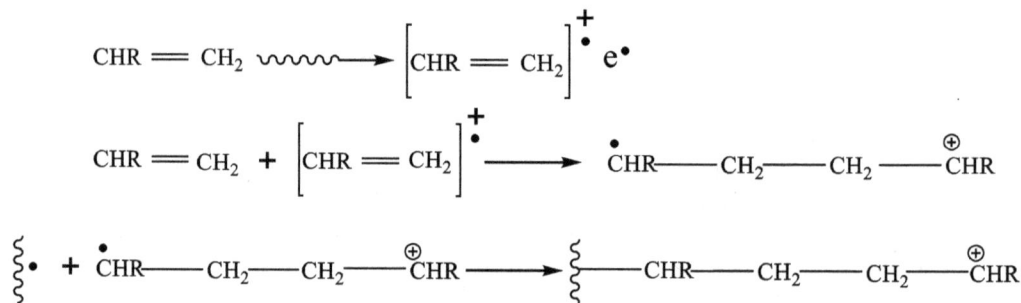

$$CHR = CH_2 \rightsquigarrow \left[CHR = CH_2 \right]^{\overset{+}{\bullet}} e^{\bullet}$$

$$CHR = CH_2 + \left[CHR = CH_2 \right]^{\overset{+}{\bullet}} \longrightarrow \overset{\bullet}{C}HR - CH_2 - CH_2 - \overset{\oplus}{C}HR$$

$$\xi \bullet + \overset{\bullet}{C}HR - CH_2 - CH_2 - \overset{\oplus}{C}HR \longrightarrow \xi - CHR - CH_2 - CH_2 - \overset{\oplus}{C}HR$$

8.3   GRAFTING PARAMETERS

The extent of grafting is expressed by the degree of grafting or grafting yield. Mathematically, it is described as the percentage mass of the grafted moiety within the copolymer matrix. The grafting efficiency defines the percentage conversion of the monomer into the grafted moiety concerning the total monomer conversion.

The degree of grafting (DG) and grafting efficiency (GE) is mathematically represented by:

$$\% \; grafting = \frac{Wg - Wo}{Wo} \times 100$$

$$\% \; GE = \frac{W_g}{W_m - W_r} \times 100$$

Here, W_g, the grafted polymer's weight, W_0; the weight of the ungrafted polymer, W_m; weight of blank monomer, W_r, weight of residual monomer.

8.4   GRAFTING APPROACHES

There are two approaches to grafting, i.e., 'grafting from' and 'grafting to'. The first one utilizes active species on the polymeric surface. It can initiate the polymerization of monomers from the surface. In another method, i.e., 'grafting to' preformed polymer chains carrying the reactive functionalities at the end or the side chains are covalently coupled to the polymer surface (Zhao and Brittain, 2000).

The simplified graft polymeric structure is composed of polymeric chain substrate and a side-grafted chain. It can be done for a wide range of polymer substrates (viz. polysulfone, poly-ethylene, poly (vinyl alcohol)), and monomers (viz. acrylic acid, acrylamide, styrene, methyl acrylates).

The DG, GE, and grafting distribution control the properties of the grafted membranes. The influencing factors are (Bhattacharya, 2000, Gupta et al., 2004):

i. Irradiation parameters (radiation source, dose, dose rate)
ii. Grafting components (polymers and monomers)
iii. Type of monomers and concentration
iv. Nature of polymer matrix (chemical as well as physical) thickness, the life of free radical in terms of pre-irradiation
v. The radiation medium also influences the mechanistic pathways as well as the product nature
vi. Presence of solvent, acid, and cross-linkers
vii. Grafting medium temperature and atmosphere

8.5 MEMBRANE AND GRAFTING

The membrane has become the necessary one to society over the years. In the architect's words, the membrane material is 'unbelievably network and texture and phenomenally varied'. They are materials made of by joining smaller units (i.e., monomers) together in long chains. They have unique properties, depending on the type of molecules associated with them. Staudinger, the father of modern polymer development, coined two terms (viz. polymerization and macromolecules) associated to polymers. Polymeric materials play a significant role in membrane formation. The outstanding processibility, abundance, and low cost make them so attractive. The properties of polymers (viz. chains rigidity, chain interactions, stereo regularity, and polarity of their functional groups) control the network and three-dimensional (3D) membrane morphology. Many polymers (synthetic and natural) are available worldwide. But, the choice of the polymers is not a trivial task. The selection of polymers into membranes is based on the compatibility with membrane fabrication and their use in targeted applications. Developing a polymeric membrane for a clearly defined application is not easy, as it needs complete control over the physical/chemical properties to attain the best possible performance. The grafting on the polymeric membrane influences the properties of membranes. Some of the polymers and monomers with their structures are depicted in Tables 8.1 and 8.2.

The graft copolymers display a distinctive concept of merging desirable properties of two or more moieties. As mentioned earlier, the traditional methods of grafting (viz. chemical, enzymatic, photo-irradiation, plasma treatment etc.) have their advantages and limitations (Bhattacharya et al., 2008). The grafting has opened exciting prospects for membrane development, as it can deliver new membranes without finding new polymers.

Radiation grafting has many advantages, including spatial and temporal control of graft polymerization, control over reaction kinetics, and flexible tuning of polymerization conditions. The idea of polymer membrane development by the radiation grafting technique is promising. Radiation-induced graft copolymerization was initiated in the 1950s. The grafted polymer retains most of its inherent characteristics and acquires additional properties needed for the specific applications. The irradiation can activate the whole polymer matrix. Therefore, it offers a unique way to combine the properties of two highly incompatible polymers.

TABLE 8.1

List of Common Monomers Used for Radiation Grafting on Various Membranes and Their Molecular Structures

Monomer	Formula
Acrylic monomers	
Acrylic acid	$CH_2=CH\text{-}COOH$
Methyl acrylate	$CH_2=CH\text{-}COOCH_3$
Acrylamide	$CH_2=CH\text{-}CONH_2$
Acrylonitrile	$CH_2=CH\text{-}CN$
Methacrylic acid	$CH_2=C(CH_3)\text{-}COOH$
Methylmethacrylic acid	$CH_2=C(CH_3)\text{-}COOCH_3$
Vinyl acetate	$CH_2=CH\text{-}COOCH_3$
Glycidyl methacrylate	$CH2=C(CH3)\text{-}COOCH_2CH\text{---}CH_2$

Vinyl monomers

Styrene

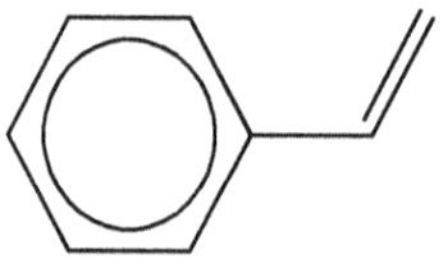

α-Methyl-styrene

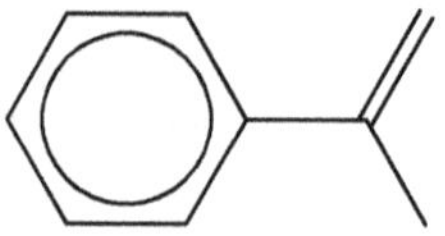

α-β,β-Trifluorostyrene

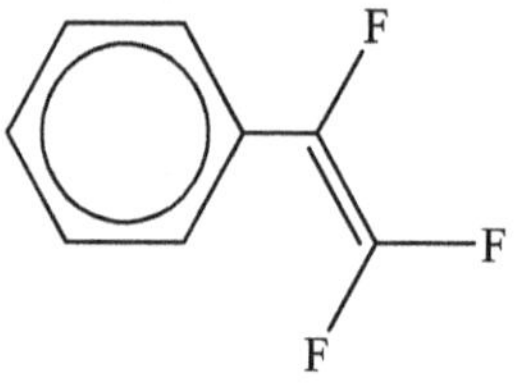

Substituted trifluorostyrene
(R= SO$_2$F, Me, MeO, PhO.)

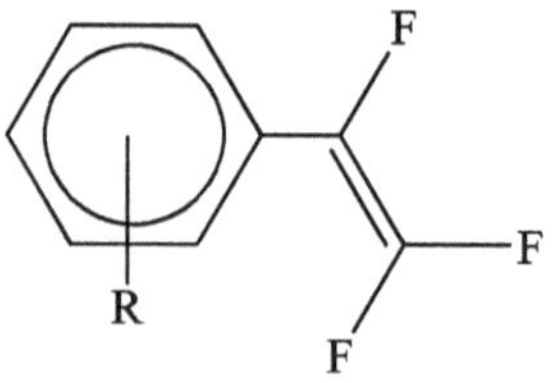

2-Vinyl pyridine

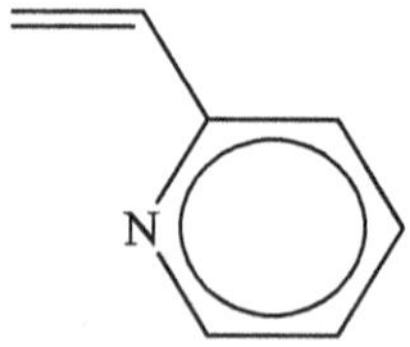

TABLE 8.1 (Continued)
List of Common Monomers Used for Radiation Grafting on Various Membranes and Their Molecular Structures

Monomer	Formula
4-Vinyl pyridine	
N-Vinylpyrrolidone	
N-Vinylimidazole	

8.6 APPLICATIONS

Radiation-induced graft copolymerization has been used to prepare various types of membranes. The remarkable feature is that the functional monomers could be straightforwardly grafted onto prebuilt films; a process is tough to perform using typical grafting. The grafted membranes have shown encouraging futures in many applications viz. metal ion absorption/adsorption, biological orientation, electrochemical, and membrane separation. The radiation-grafted membranes replace conventional polymer membranes. The results of research activities in these areas are in the different ensembles in the following.

8.6.1 Metal ion absorption/adsorption

The traditional metal ion solvent extraction is a high capacity, selective, and fast method. Still, it is not suitable for dilute metal ion solution, as more extractant is needed. The grafted polymer in that aspect is the viable direction. The metal ion adsorption/absorption depends on the physically attractive force or complexation/ chelation. The polymeric membrane matrices should have those functionalities that hold the metal ions. The direction is to design polymer adsorbents capable of capturing metal ion adsorbents. Grafting techniques can incorporate functionalities onto the polymeric membrane. In this regard, various examples are there where radiation was employed.

Low-density polyethylene (LDPE)-g-(styrene/methacrylic acid) films were sulfonated first. Then, alkaline treatment was done to their respective grafted monomers to impart reactive cationic/anionic character. The matrices developed the potentiality to recover various metal ions viz. Ni (II), Co(II), Cu(II), Cd(II), Mg (II), Zn(II), Mn(II), and Cr(III) from their aqueous solutions (Hegazy et al., 2012).

Cationic ((LDPE)-g-poly(acrylic acid) (PAAc)) and cationic/anionic (LDPE-g-P(AAc/4-vinyl pyridine (4VP)) membranes were experimented for the metal uptake (Hegazy et al., 1997). As the membranes were pH responsive, the metal uptake increased significantly as pH (especially ≤ 5.3). The desorption could be done, i.e., by treating the adsorbed membrane with 0.1M HCl for 2 hours at room temperature. The cationic/anionic membranes showed maximum metal uptake efficiency compared to the cationic ones.

The exchange properties and/or complexation significantly improved by introducing anionic and cationic behaviour in the grafted polymeric membrane. The stability constant and bond

TABLE 8.2

List of Common Polymers Used for Radiation Grafting on Various Membranes and Their Molecular Structures

Polymer	Abbreviation	Repeating Unit
Perfluorinated polymers		
Polytetrafluoroethylene	PTFE	$\left[\!\!\left[CF_2\!-\!CF_2 \right]\!\!\right]_n$
Poly(tetrafluoroethylene-co-hexafluoropropylene)	FEP	$\left[\!\!\left[CF_2\!-\!CF_2 \right]\!\!\right]_n \left[\!\!\left[CF_2\!-\!\underset{\underset{CF_3}{\mid}}{CF} \right]\!\!\right]_m$
Poly(tetrafluoroethylene-co-perfluoropropyl vinyl ether)	PFA	$\left[\!\!\left[CF_2\!-\!CF_2 \right]\!\!\right]_n \left[\!\!\left[CF_2\!-\!\underset{\underset{OC_3F_7}{\mid}}{CF} \right]\!\!\right]_m$
Partially fluorinated polymers		
Polyvinylidene fluoride	PVDF	$\left[\!\!\left[CF_2\!-\!CH_2 \right]\!\!\right]_n$
Poly(vinylidene fluoride-co-hexafluoropropylene)	PVDF-co-HFP	$\left[\!\!\left[CF_2\!-\!CH_2 \right]\!\!\right]_n \left[\!\!\left[CF_2\!-\!\underset{\underset{CF_3}{\mid}}{CF} \right]\!\!\right]_m$
Poly(ethylene-alt-tetrafluoroethylene)	ETFE	$\left[\!\!\left(CH_2\!-\!CH_2 \right)\!\!\left(CF_2\!-\!CF_2 \right)\!\!\right]_n$

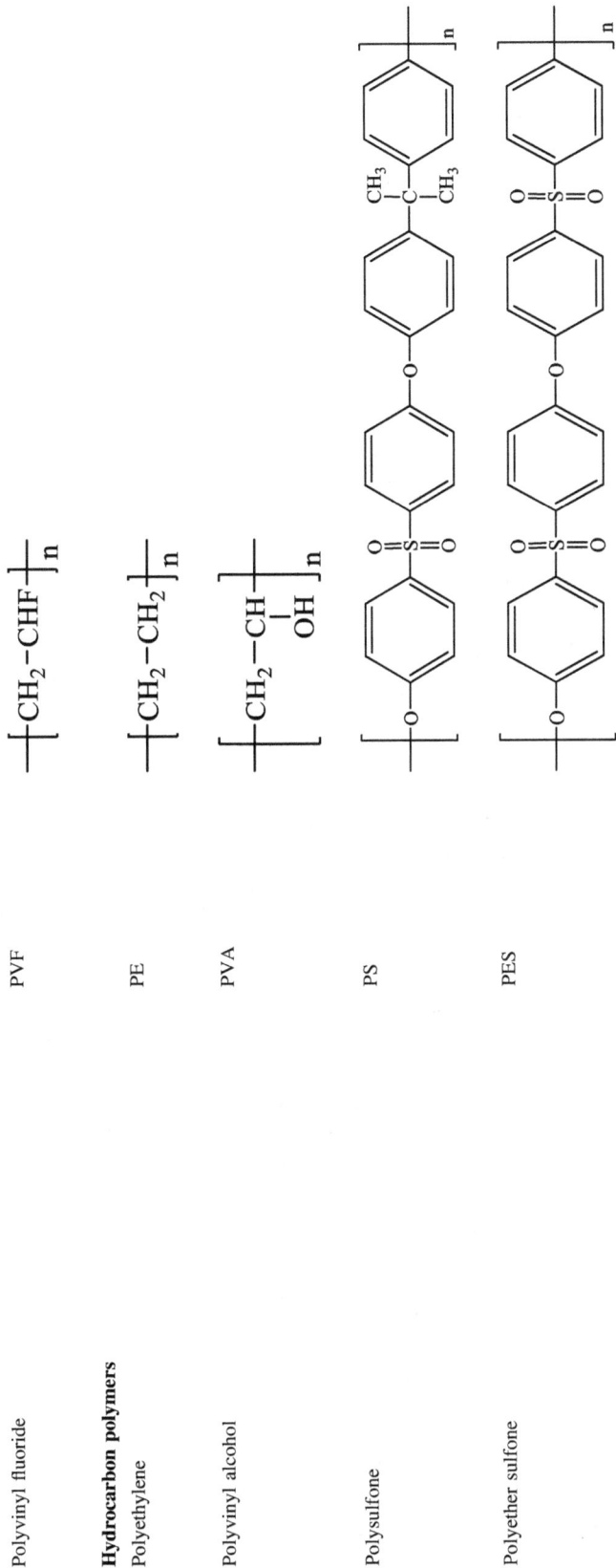

Polyvinyl fluoride

PVF

$$\left[\begin{array}{c} CH_2-CHF \end{array}\right]_n$$

Hydrocarbon polymers

Polyethylene

PE

$$\left[\begin{array}{c} CH_2-CH_2 \end{array}\right]_n$$

Polyvinyl alcohol

PVA

$$\left[\begin{array}{c} CH_2-CH \\ | \\ OH \end{array}\right]_n$$

Polysulfone

PS

Polyether sulfone

PES

strength of complexation between the metals and functionalities on the membranes increased. It was due to incorporating pyridine moieties, as pyridine has a lone pair of electrons on the nitrogen atoms, which could easily form quaternary pyridinium metal salts or chelate. The membranes showed high uptake efficiency towards Fe(III) ions.

Polyethylene (PE)-g-(GMA/AN) (glycidyl methyl acrylate/acrylonitrile) hollow fibers with post-treatment attached chelating imino diacetate (IDA) and amidoxime (AO). The IDA-containing grafted membranes showed higher uptake compared to AO. The saturation level was 0.13 to 1.3 mol Co(II)/kg of the membrane depending on the grafting yield of poly GMA 20% to 260% (Tsuneda et al., 1991).

PE-g-GMA flat sheet membrane functionalized with IDA showed the potential to remove Cu(II) from $CuCl_2$ solution. The epoxy ring into IDA groups removed Cu(II) from its solutions (Yamashiro et al., 2007). PE-g-(GMA) membrane treated with different amines showed different efficiencies of Pd(II). The metal adsorption efficiency followed diethyl triamine > hexamethylene diamine > ethylene diamine > trimethylamine (Choi and Nho, 1999a). Amine-treated PE-g-GMA hollow fiber membranes are suitable adsorbents for removing Pb(II) and Pd(II) ions (Zhi-Li, 1987, Choi and Nho, 1999a). The chelating PE hollow fiber membrane modified with imino diacetic acid (IDA) adsorbed much more Pb(II) ions compared to Pd (II) (Choi and Nho, 1999a). The first step is radiation grafting, whereas the second step is post-treatment.

PE-g-GMA hollow fiber membrane post-treated with IDA showed the selectivity of Cu(II) from Cu(II)/Co(II) reaction mixture (Konishi et al., 1996). PE-g-GMA hollow fiber membranes activated with IDA and succinic acid (SA) groups could adsorb Co(II) and Cs(I) from solutions (Choi and Nho, 1999b, Choi and Nho, 1999c). Imino-diacetic acid (IDA) groups have very high selectivity for cobalt from a mixture of metal ions (Choi and Nho, 1999c). PE-g-GMA hollow fiber membranes functionalized with phenylalanine showed their potential in separating Cu(II) ions from the solution (Kiyohara et al., 1996). PE-g-GMA hollow fiber membranes functionalized with sodium bisulfite showed their capabilities in capturing Pb(II) ions from the solution (Kim and Saito, 2000).

LDPE-g-styrene/maleic anhydride films gained their functionality by alkaline hydrolysis and the treatment with different alkylamines (El-Rehim et al., 2000). The grafted films treated with thio semicarbazide preferentially adsorbed Cu(II). The hydroxylamine hydrochloride treatment preferred Cr(III), sodium hydroxide treatment picked Fe(III) in a mixture of the three Cu(II), Cr(III), and Fe(III) ions at 70°C. The adsorption depended strongly on the pH of the medium (El-Rehim et al., 2000).

A post-treated PE-g-(AN/GMA) hollow fiber membrane showed their uranyl ions sorption depended on post-treated chemicals. The sorption order followed with the treated chemicals amidoxime > IDA > amine. PE-g-(AA/VP) showed their metal ions (viz. Co, Ni, Zn, and Cd) adsorption efficiency satisfactorily. The metal uptake followed the reverse order of the atomic radii of the metal ion. The 4-vinylpyridine (4VP) is a more effective chelating agent than the 2-vinylpyridine (VP) in the grafted membrane (Hegazy et al., 2000). PP/PE-g-vinyl ether/monoethanolamine membranes showed their affinity towards Pb(II), Cd(II), Zn(II), Fe(II), Cu(II), Ni(II), Co(II), and Ag(I). The maximum adsorption of Cu(II), Ni(II), and Co(II) occurred at pH 5 (Mun et al., 2005). LDPE-g-(AAc/styrene) cation exchange membranes had the potential to adsorb Fe(III), Cd(II) and Pb(II) ions from the waste solution (Hegazy et al., 1999). PE-g-(AAc) showed Hg(II) removal efficiency of 99% for 200 mg/L of Hg(II) (Gupta and Anjum, 2002). PE-g-(AAc/4VP) showed separation efficiency Pb(II),Cu(II), Cd(II), Fe(II), Ni(II), Co(II), Zn(II), Mn(II) ions. The Cu(II) uptake depended on the degree of grafting. The binary grafting of two monomers showed better Cu(II) uptake than individual AAc grafting (Choi and Nho, 2000a, Sugiyama et al., 1993). The cationic membrane in its R-CONH-OH form showed high selectivity towards zirconium from a zirconium-uranium mixture (Hegazy et al., 2001). PE-g (AAm) film develops high Hg(II) sorption abilities (Gupta and Anjum, 2003). The film showed its reusability nature. PP-g-(acrylic acid/vinyl pyrrolidone) developed the selectivity of Pb(II) from Hg(II) or Cd(II) ions mixture. The uptake of metal ions depended on the grafting degree. PE-g-(acrylic acid/N-vinyl imidazole) membrane also showed their potential to uptake metal ions (viz. Pb(II), Cd(II), Ni (II), Co(II)) (Ajji, 2012).

The strong acid cation-exchange membrane with sulphonic groups was prepared using p-vinyl benzene sulphonic acid sodium salt hydrate (SSS) grafting on the PE-g-AAc grafted membrane (Reddy et al., 2005). The ion-exchange capacity of the SSS grafted membrane was 5.5–5.8 milli equivalent per gram (meq/g), which showed very promising and better than most commercially available membranes in this orientation. It established that strong acid groups could be incorporated into polyethylene (PE) by grafting hydrophilic acrylic acid(AAc) and then by the strong acid group.

Polysulfone (PSf)-g-2-acrylamido-2-methyl propane sulfonic acid (AMPS) nanofiltration (NF) hollow fiber membranes showed their abilities towards Mithe removal of Cr(VI). The presence of AMPS developed the negative charge density on the membrane and had the separation ability of Cr (VI) ions with relatively large pore size. The study showed that the separation performance of Cr(VI) ions did not vary above pH 8 by sodium chloride and sodium sulphate (Xu et al., 2014).

Poly(vinyl alcohol)(PVA)-g-(sty/AAc) showed high permeation selectivity for Ni(II) from Ni(II)/ Co (II) mixture solutions. Ni(II) ions were transported through the membranes, whereas Co(II) did not transport as they chelate with the membrane functional groups. Co(II) prefers Ni(II) to chelate with the grafted membrane's –COOH and -OH functional groups. The rate of transport of Ni(II) ions increases with the degree of grafting (El-Rehim et al., 1999). PE-g-(sty/maleic anhydride) developed their potential by activating carboxylic acid and amino groups by alkaline hydrolysis and different reactions with alkylamines (El-Rehim et al., 2000). The membrane treated with thiosemicarbazide showed a high affinity towards Cu(II). On the other hand, hydroxylamine hydrochloride treatment showed high Cr(III) selectivity. The metal ion absorption depended on the pH of the medium. PVA-g-(AAc/VIm) (acrylic acid/N-vinyl imidazole) membranes used for the adsorption of Cu(II) ions from contaminated water (Ajji and Ali, 2005). PVA-g-(AAc/VIm) retained Fe(III) from Cu(II)/ Fe(III) mixture. Cu(II) diffusion depended on grafting yield and pH in the feed solution (Ajji and Ali, 2010). The Cu(II) ions concentration increased with time for all the grafted membranes. The rate and concentration of transported Cu(II) ions increased inversely with the degree of grafting. The higher cross-linked network structure of the membrane, which restricted the diffusion of ions through it. Thus, the rate of the Cu(II) ions decreased, and much duration was needed for its transportation at higher grafting yield. On the other hand, Fe(III) ions were not transported; they might be adsorbed on the membrane. LDPE-g-acrylonitrile/vinyl pyrrolidone (AN/N-VP) membrane turned into amidoxime and treated with hydroxylamine hydrochloride. The modified membranes showed their potential for adsorption of Cr(III), Ni(II), Cd(II, Ag(I), and Pb(II) metallic ions. The adsorption capacity depended on the grafting yield and pH of the medium (El-Sawy, 2000).

PE-g(AM)(acrylamide) membranes showed their chelating ability of Hg(II) ions from the aqueous solution (Gupta et al., 2002). The Hg(II) binding capacity of the membranes strongly depended on the extent of grafting. The membrane with 590% degree of grafting uptake 6.2 mmol/g mercury. The membranes showed their potential in separating Hg(II) from a solution of as low a concentration as 200 mg/L.

The PE-g-AN/AAc (acrylonitrile/acrylic acid) film made into amidoxime end showed its potential in uranyl ion (UO_2^{2+}) separation (Choi and Nho, 2000a). The maximum adsorption of UO_2^{2+} ions was observed in the AN/AA (50/50, mol%) composition. In another study, PE-g- AN/ AAc, PE-g- AN/MAAc (acrylonitrile/methyl acrylic acid) membranes were post modified into amidoxime groups have the potential to separate UO_2^{2+} (Choi and Nho, 2000b). The carboxyl, amidoxime, and carboxyl /amidoxime groups onto the support membrane have the chelating site of metal ions. Sekiguchi et al. reported convincing results of amidoxime grafted hollow fiber membranes for separation of UO_2^{2+} (Sekiguchi et al., 1994).

8.6.2 IN THE BIOLOGICAL FIELDS

There is considerable progress regarding the grafted membranes in the orientation of biology. Grafted membranes showed their potential in terms of an artificial kidney. PE-g-polyvinyl alcohol/N-vinyl pyrrolidone (PVA/NVP) grafted membranes could remove toxic low molecular weight organics viz.

urea, uric acid, creatinine, and phosphate (Nasef and Hegazy, 2004). PVA-g-(AAc/MAc) membranes were permeable to the particularly mentioned solutes (Shantora and Huang, 1981). The permeation process was temperature-dependent. It could be expressed as $\log(permeation\ rate) = 1/T$. The permeability of the grafted membranes increased as the degree of grafting, though no appreciable selectivity developed. The PVA-g-(AAc) membranes developed better improvement in permeability than the PVA-g-(MAc) membranes. The PP-g-(HEMA, hydroxyl ethyl methacrylate) membrane developed the permeability of urea and uric acid (Yuee and Tianyi, 1988). PE-g-Sty/MAn, (styrene/maleic anhydride) films showed their potential to permeate urea, uric acid, creatinine, glucose, and phosphate salts. It was also seen that the permeability increases with the degree of grafting (El-Rehim et al., 1999).

PE-g-(VAc/MAn) (vinyl acetate/maleic anhydride) membranes post-treated with different chemicals (viz., sodium hydroxide, hydrochloric acid, ammonium hydroxide, sulfamic acid, or sulfamic acid, aminopyridine) showed their abilities to permeate toxic organics like urea, creatinine, and uric acid. The chemicals used in post-treatment developed the characteristic transport property and low protein adsorption (El-Rehim, 2005). PE-g-AAc membranes transformed into corresponding acrylates with different metal salts showed permeability for the solutes (glucose, urea) (El-Awady, 2004).

Another important direction is the immobilization of urease enzymes on grafted membranes. Polymeric membranes are functionalized through surface grafting. It provided substrates suitable for immobilization. The grafted surface with biologically active biomolecules can undergo specific interactions that mimic the physiological interactions. The immobilized enzymes hydrolyzed urea during the permeation. Kobayashi et al. approach this technique for post functionalized amines for hollow fibre membranes for PE-g-GMA (glycidyl methacrylate) (Kobayashi et al., 2003). In membrane chromatography, the binding of proteins was the primary mechanistic pathway. The separation mode forced, and convection through the pores enhanced the mass transfer of proteins to affinity ligands and ion exchange groups.

The grafting of vinyl monomers reduced the steric hindrance associated with large molecules' binding on the rigid membrane surfaces (Koguma et al., 2000, Dessouki et al., 1998, Dessouki et al., 1990, Hegazy et al., 1984, Hegazy et al., 1990). PE-g-GMA amine-functionalized hollow fiber membranes showed strong potential for protein adsorption (Godjevargova and Dimov, 1995). PE-g-GMA hollow fiber membranes coupled with tryptophan, or L-phe ligand showed Bovine- γ-globulin adsorption (Kim et al., 1991a, 1991b). PE-g-GMA hollow fiber membranes treated with phenol, n-butyl alcohol, aniline, or n-butyl amine showed their ability in the binding rate of BSA because of the negligible diffusional mass transfer resistance of the protein (Kubota et al., 1997). Cellulose hydrate grafted with selective ligands shows chromatographic purification of human serum albumin from desalted human plasma (Gebauer et al., 1997). PE-g-GMA hollow fiber membranes incorporated diethyl amino groups were used to adsorption gelsolin using $CaCl_2$ solution (Hagiwara et al., 2005). Hollow fiber membranes grafted with GMA followed by hydrolysis of the grafted epoxide group into diol converted with sulfopropyl (SP) functionality. It showed negligible diffusional resistance of lysozyme, the compact protein of 129 amino acids (Shinano et al., 1993). The binding capacity of lysozyme to the fibers is independent of the SP group density (Tsuneda et al., 1994).

PE-g-GMA hollow fiber membranes with Ag^+ immobilized had the potential to purify docosahexaenoic acid ethyl ester (DHA-Et). The adsorption was caused by the selective interaction between Ag^+ ion and C=C of DHA-Et (Ozawa et al., 2000). The matrix developed the matrix in PE-g-GMA hollow fiber membranes with subsequent amination with diethyl amino functionality immobilized with amino acylase. It showed the permeation of acetyl-DL-methionine, an endogenous metabolite (Kawai et al., 2001).

Graft polymer membranes have received attention regarding the selective affinity of ligands in terms of antigens and antibodies. The affinity membranes were used in biomolecular separation and purification (Charcosset, 1998). The graft membrane-based matrices overcome the limitations of the bead and micro/ultrafiltration membranes based on sieving. Moreover, the affinity membranes could run the process at low pressure and large volume compared to beads (Elias, 2000, Nasef and Guven, 2012).

The most used polymer substrates in radiation grafting PE, PVDF, and PS were used in this application, as they are radiation-resistant. The use of γ -irradiation is quite significant compared to the electron beam, as it showed better penetration to modify the interlayer layers (Nasef and Hegazy, 2004). PE-g-GMA post-treated with different chemicals turn into ligand active membranes with imino di acetate, amidoxime, L-phenyl alanine, chlorovinyl pyridine, and acrylic acid were reported. Apart from these, the functionality with amino acid, immobilized metal as pseudo-biospecific affinity ligand, sulfonic acid, diethylamino, succinimide, and amino sulfonic acid group was incorporated on the radiation-induced grafted substrate (Kim et al., 1991a, Tsuneda et al., 1994, Iwata et al., 1991, Tsuneda et al., 1995a, Kiyohara et al., 1997, Miyoshi et al., 2005).

PE-g-GMA hollow fiber membranes post-treated with ammonia, followed by immobilization of red dye (HE-3B) through spacers formed by amino groups. These membranes showed an affinity towards targeted ions or molecules (Wolman et al., 2007 , Asai, 2005). The radiation-grafted polymeric hollow fiber membranes, depending on affinity, metal chelating, ion exchange, and hydrophobic groups, could have high protein binding ability (Koguma et al., 2000, Kubota et al., 1997, Ito et al., 2001, Sasagawa et al., 1999). The protein adsorption depends on various interactions between protein molecules and membranes. It includes Van der Waals, electrostatic, hydrophobic, dipole-dipole, and H-bonding (Hamza et al., 1997).

Hydroxy ethyl methacrylate (HEMA), vinyl acetate/acrylamide, and GMA-grafted polyethylene microporous membranes were used for enzyme and protein immobilization (Gupta and Anjum, 2003). The grafted membranes were transformed into hydroxyl derivatives; thus, it was fit for bovine-γ- globulin immobilization (Kim et al., 1994, Tsuneda et al., 1995b, Kim et al., 1991c). The amount of immobilization was independent of chosen grafted vinyl monomers. PE-g-sodium p-styrene sulfonate/HEMA membranes post-treated with sodium sulfite could adsorb lysozyme. The adsorption depended on the sulfonic acid group density in the treated membrane (Shinde and Salovey, 1985). PE-g-GMA hollow fiber modified with sulfuric acid, post-treated with tannin, to prepare tannin functionality with the epoxide group. Bovine serum albumin (BSA) immobilization was better concerning a purely hydrolyzed one on this tannin formation surface (Kim et al., 1993).

The anti-thrombosis dialytic membrane based on PE-g-HEMA/Sty was reported by Wu et al. (Wu et al., 2000). The adsorption of plasma proteins and subsequent adherence of platelets did not adhere, and the thrombus formation was inhibited. The grafted membrane showed an excellent permeability coefficient. Moreover, the permeability of the grafted membrane was higher as the HEMA content. In another study, Neffe et al. showed that protein adsorption and thrombocyte adhesion depends on the chain length and end groups of linear PEG grafted on poly(ether imide) (PEI) membranes. The shorter OEG (oligo ethylene glycol) resulted in higher grafting densities than the longer PEG (polyethylene glycol). It showed a better surface shielding effect (Neffe et al., 2013).

The introduction of the microbial cell onto grafted surfaces is quite beneficial. PE-g-GMA membranes post-treated with DEA (diethyl amine) and EA (ethyl amine) exhibited *S. aureus* cell-capturing ability (Lee et al., 1996, Lee et al., 1997). The bacterial cell retain their shape on the grafted surface. The brush-type grafted surface containing graft chains adsorbs the bacterial cell 1-000-fold faster than the cross-linked surface. The grafted membrane material has the potential in bacterial recovery for water purification.In another study, the epoxy ring of GMA of the PE-g-GMA grafted matrix opened by introducing diethylamine (DEA) or sodium sulfite (SS) (Terada et al., 2005). The bacterial adhesion onto the grafted membrane was due to electrostatic interaction and increased surface area. The initial bacterial adhesion led the possibility to influence subsequent biofilm formation. The bacterial adhesion is beneficial as it could stimulate efficient degradation of specific substances in a bioreactor.

For haemodialysis application LDPE-g(Sty/ MAn) membrane on post-treatment with acid/base treatment incorporated thiosemicarazide, hydroxamic acid, 2-amino pyridine, and aspartic acid. The membranes grafted with different functionalities show their separation potential from other basic metabolites (glucose, creatinine, urea, vitamin B_{12}, bovine albumin). The alkali-treated membranes show higher permeation than membranes containing the carboxylic acid group (Lee et al., 1996, Terada et al., 2005).

The surface modification of expanded poly (tetrafluro ethylene) (ePTFE) membranes with the single as well as mixture of glass ionomers (viz. AA, AA, and itaconic acid (IA), AA, and hydroxyl ethyl methacrylate, HEMA) was feasible in the radiation grafting technique (Hidzir et al., 2012, Hidzir et al., 2015, Radzali et al., 2020). Thus, it was made to swell in water. The water uptake was significant and modified the highly hydrophobic nature of the ePTFE to a lesser one. The modified one had the potential to enhance its bone-integration properties.

8.6.3   IN ELECTROCHEMICAL APPLICATIONS

Electrochemical energy systems are of two types: energy conversion and energy storage devices (Nasef et al., 2016). Polymer electrolyte membranes have not only been used as electrode separators; they mediate the electrochemical reaction (Kreuer and Portale, 2013). They have generated considerable interest as solid polymer electrolytes in fuel cells, power sources. The invention of fuel cells was occurred in the middle of the 19th century. The application began in the 1960s during space exploration. The development of PEMFC was done during the launching of the Gemini space vehicle of the National Aeronautics and Space Administration (NASA) (Nasef, 2014).

The fuel cell is an eco-friendly replacement for energy production. It is in the lead of the attempt towards green and sustainable energy production. A fuel cell is similar to conventional power plants, where the chemically stored energy of a fuel is continuously transformed into electrical energy. The fuel cell features the advantage of the same transformation in one step. In a fuel cell, the permanent consumption of a supplied fuel is the primary difference from secondary batteries, where recharging via applying an external voltage is required (Hoogers, 2003, Hayre et al., 2006).

The stupendous research effort directed to developing membranes by introducing ionic functionalities into a broad range of polymers. In the fuel cell, there is the direction to find low-cost raw materials for the base polymer and graft monomer. The radiation grafting technique opens up the possibilities to find tailor-made and cost-effective membranes in the polymer electrolyte fuel cell (PEFC). The grafted polymer membranes can display a distinctive ionic nature, water absorption, and high conductivity. The central part of the PEFC consists of a membrane-electrode assembly (MEA), compacted in sandwich configuration between two conducting plates. They are of three types viz. proton exchange fuel cells (low and high temperature), direct methanol fuel cell (DMFC), and anion exchange membrane fuel cell (AEMFC) (Nasef et al., 2016).

In the PEFC, the electrode reaction is as follows:

Anode reaction (hydrogen oxidation)

$$H_2 \rightleftharpoons 2H^+ + 2e$$

Cathode reaction (oxygen reduction)

$$2H^+ + 2e + \tfrac{1}{2}O_2 \rightleftharpoons H_2O$$

The overall chemical reaction

$$H_2 + \tfrac{1}{2}O_2 \rightleftharpoons H_2O$$

The proton exchange membranes are used in the PEMFCs and DMFCs. Proton exchange membrane fuel cells were invented in the 1960s by General Electric (GE). PEMFC and AEMFC schematic diagrams are depicted in Figures 8.1 and 8.2. In contrast, the anion exchange membrane is an in-built

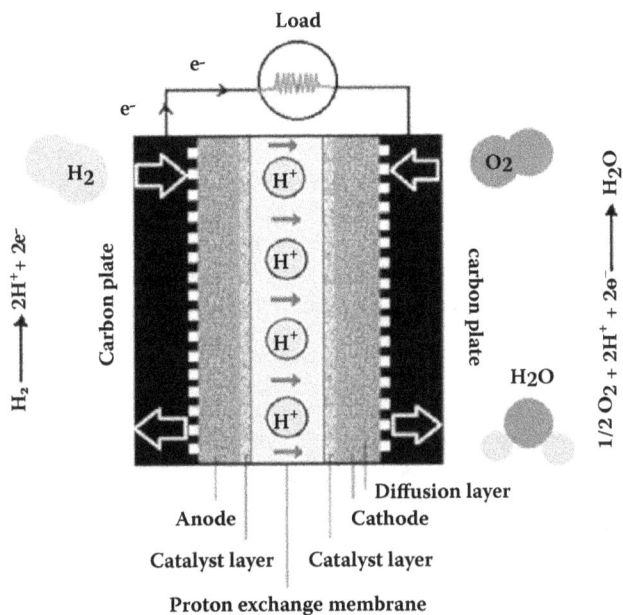

FIGURE 8.1 Schematic diagram of basic unit of proton exchange membrane fuel cell (PEMFC) [Reproduced with permission from *Progress in Polymer Science,* 63, 1–41, (2016)].

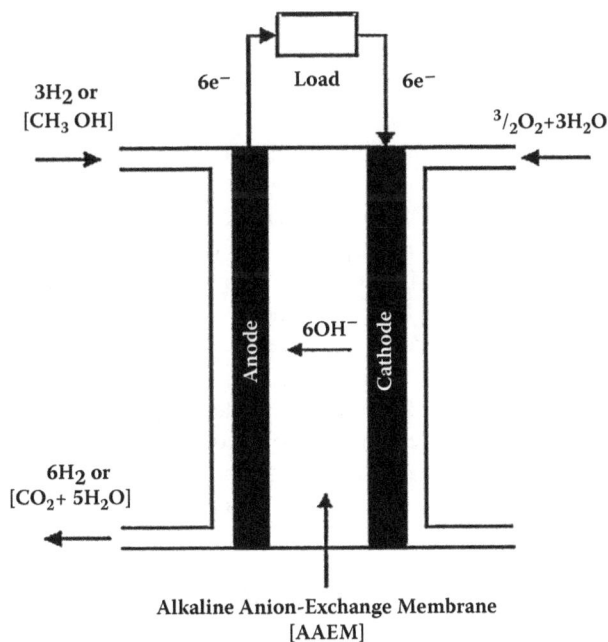

FIGURE 8.2 Schematic diagram of anion exchange membrane fuel cell (AEMFC) [Reproduced with permission from *Progress in Polymer Science,* 63, 1–41, (2016)].

component in anion exchange membrane fuel cell (AEMFCs). Apart from low-cost criteria, the proton exchange membrane has high proton conductivity, low permeability to fuel, thermal and chemical stability, and good mechanical strength (Gubler and Scherer, 2010, Peighambardoust et al., 2010, Gubler et al., 2004). Moreover, radiation can effectively control the reaction parameters to attain desirable properties. The polymers viz. perfluorinated, partially fluorinated, and purely hydrocarbon are excellent performers in this application (Wokaun et al., 2004). The base polymer poly (ethylene-alt-tetrafluoroethylene) (ETFE) showed promising results, but poly(vinylidene fluoride) (PVDF), as well as poly (tetra fluoroethylene-co-hexafluoropropylene)(TEFLON-FEP), are not far behind. Cross-linkers are essential as they provide stability. Moreover, low grafting is preferred as it allows for better chemical and mechanical strength (Gubler et al., 2005a).

The most commonly available monomer is styrene, as it possesses an aromatic ring in the nucleus. The sulfonation to the aromatic ring is available is a feasible approach to impart proton conductivity in the membranes. The different cross-linkers (viz. divinylbenzene, bis (vinyl phenyl ethane) are also used. The substituted styrene with a protected α position shows better alternatives with higher chemical stability (Nasef, 2004, Gupta et al., 2004).

The proton conductivity incorporation is made by sulfonating the grafted styrene monomer as it exhibits $pk_a < -1$, the sulfonic acid dissociates in an aqueous medium, and mobile protons result in conductivity (Gubler et al., 2005a). Typical sulfonating agents are sulfonic acid, oleum, sulfur trioxide, chlorosulfonic acid, and acetyl sulfate.

The base polymer provides stability in the radiation-grafted proton exchange membrane, whereas the sulfonated part delivers the water proton conductivity. The hydration of the membrane develops conductivity. Hydration is associated with sulfonic acid. The number of water molecules per sulfonic acid quantifies the hydration. The grafted membranes reflect hydrophobic and aqueous domains compared to the ungrafted membrane (Kallio et al., 2002).

PVDF-g-Sty, PVDF-co-HFP-g-Sty (Kallio et al., 2002, 2003), LDPE-g-Sty (Horsfall and Lovell, 2001, 2002). FEP-g-sty (Brack et al., 2000), ETFE-g-Sty (Gubler et al., 2005b), PTFE-g-meSty (Assink et al., 1991), PFA-g- Sty (Nasef and Saidi, 2002), and PES-g-sty (Filho and Gomes, 2006) are a few examples of radiation-grafted polymeric membranes in this arena. PVDF-g-PSSA membranes are prepared by sulfonation of the PVDF-g-sty matrices. They showed helium and hydrogen gas permeabilities. The permeability increased with the grafting content. Because of the incompatibility of the PVDF and PS, the PS graft content is phase-separated from the base polymer matrix thus, the permeabilities of PS are higher than of PVDF. The gas permeabilities are quite different for dried and wet grafted membranes. The gas permeation in the wet grafted membrane follows diffusion pathways of the gas in water. The permeabilities for grafted membranes reached the same order as commercial Nafion® 117 membrane. PVDF-g- PSSA membranes are an excellent candidate for use in electrochemical cells (Hietala et al., 1999).

The co-monomer grafting is also trending to get better membranes viz. ethylene tetrafluroethylene (ETFE)-g- meSty/tBuSty (Chen et al., 2006) and fluorinated ethylene propylene (FEP)-g-meSty/AN (Gubler et al., 2006). Novel composite membranes based on ETFE-g(styrene) followed by sulfonation and subsequent polypyrrole, finally reported platinum deposition (Dogan and Gursel, 2011). PE-g- sty and ETFE-g-Sty membranes are also used for direct methanol fuel cells (DMFC) as they exhibit a low methanol diffusion coefficient. It is potentially useful in reducing methanol crossover from anode to cathode (Scott et al., 2000).

Recently, pore-filled membranes have also made attention in direct methanol fuel cells because of certain characteristics like controlled swelling, low methanol crossover, high proton conductivity, and low cost. PVDF-g-styrene, followed by sulfonation with a chlorosulfonic acid/dichloromethane mixture, is used in this application (Nasef et al., 2006a, 2006b).

Alkaline anion exchange membranes (AEMs) are used as electrolytes in AEMFCs. The reduced methanol permeability (in the context of methanol permeability), effective water management (water generated at the anode and used at cathode), and improved electrokinetics made the feature of the alkaline membrane (Varcoe et al., 2007). (PVDF, ETFE, FEP)-g-vinyl benzyl chloride and subsequent amination or quaternization with 1-methylimidazole and alkalization made it fit for this type of fuel cell (Varcoe and Slade, 2006, Yoshimura et al., 2014). ETFE-g(VBC) membrane followed by quaternization and cross-linking with 1,4-diazabicyclo [2,2,2] octane, alkylation with p-xylylenedichloride, and quaternization again with trimethylamine (Varcoe et al., 2007, Fang et al., 2012, Biancolli et al., 2021) PE-g(VBC) and quaternization again with trimethylamine are fit for this application (Biancolli et al., 2021).

In battery applications, membranes are used as separators between the two electrodes. Apart from its chemical and electrochemical stabilities, membranes should have the property to conduct the ions but electronic insulators. Radiation-induced grafting is one of the most suitable techniques to prepare the battery separator membranes (Nasef and Hegazy, 2004). PE-g-(AAc), PE-g (MeAAc), PP-g-(AAc) show their potential in this application. With the increase in grafting, conductivity increases with requisite properties (viz. water uptake, ion exchange) (Choi et al., 2000, Goel et al., 2009).

PE-g-GMA, PVDF-g-styrene (Ko et al., 2004, Nasef et al., 2004, 2007) and PE-g-MMA (Gwon et al., 2009) are also used as battery separators for lithium batteries with better properties (viz. recyclability, ionic conductivity, stability). PVDF-g-sty/DMAEMA (dimethyl amino ethyl methacrylate) (Qiu et al., 2009), PVDF-g-AMS/DMAEMA, PVDF-g-PMAODEMAC-co-PAMSSA, (Hu et al., 2012), and Nafion-g-DMAEMA (Ma et al., 2013) are a few examples in terms of vanadium redox battery separators.

In supercapacitors, the high-energy storage battery radiation grafted membranes are also useful. There are a few examples in this regard, FEP-g- Sty/AAc, FEP-g- AAc-SO$_3$H (Sivaraman et al., 2006), and PVDF-g-2, 3, 4, 5, 6 penta fluorostyrene (Dumas et al., 2014).

8.6.4 IN SEPARATION ARENAS

Membrane science and technology have progressed a lot during the last few decades. The processes have replaced many conventional separation processes. The physicochemical nature of membranes drives them into the proper separation application arena. The membrane's bulk properties (permeability and selectivity) are essential critical factors for this application. The grafted membranes alter the bulk properties. The phenomenon 'fouling' hinders the spontaneous functioning process of membranes. Fouling membranes result from adsorption and deposition of foulants (organic or/and inorganic compounds) on the surface and within pores. The physicochemical interactions between membranes and foulants are the reasons behind it. It disturbs the performance, increasing operating costs and the life cycle of membranes (Dang et al., 2010). The modification is one of the approaches to counter fouling. The hydrophilic surface modification on membranes has low fouling tendency than the hydrophobic surface. There are two types of modification approaches (viz. grafting on membranes and grafting on materials).

Polysulfone (Psf)-g-2-acrylamido-2-methylpropanesulfonic acid (AMPS) nanofiltration hollow fiber membrane possessed a high negative charge density and relatively high permeability for separating Cr(VI) (Xu et al., 2014). The performance increased with the grafting. The negative charge density increased with the degree of grafting (DG). Thus, the electrostatic interactions between different ions increased. The electrostatic effect operated for the separation of Cr(VI).

Polyethersulfone-g-polyethylene glycol (PES-g-PEG) hollow ultrafiltration membrane showed ~67% higher flux than unmodified PES membrane during the filtration of porcine albumin solution (Mok et al., 1994). PVDF-g-HEA (hydroxy ethyl acrylate) membrane offered an antifouling ability during bovine serum albumin (BSA) filtration (Shen et al., 2017). The grafted membrane showed that the flux of the bovine serum albumin (BSA) solution was more than that of pure water in three cycles of filtration. The grafted membrane's flux rate decrease was lower than the unmodified membrane.

PDMAA (NN dimethyl acrylamide) graft chains in the PVDF-g-PDMAA membrane improved hydrophilicity and reduced protein adsorption. The water recovery percentage of the grafted (DG: 17.1%) membranes was 2.6 times higher than the unmodified after surface fouling

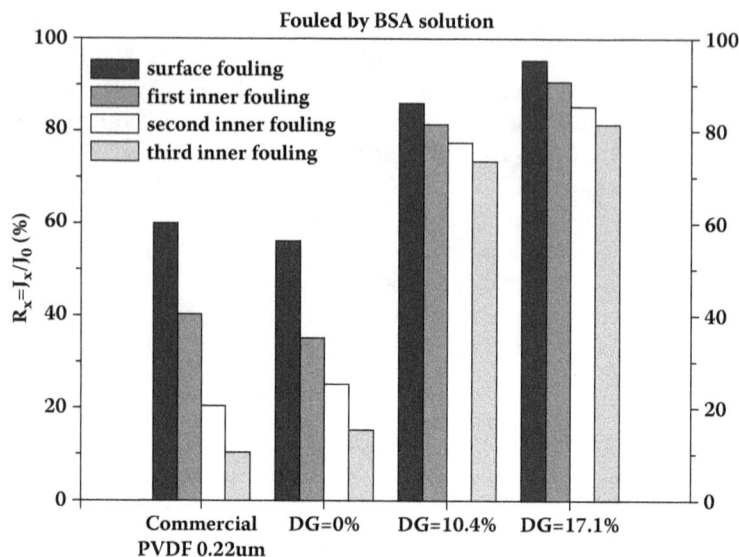

FIGURE 8.3 Variation of water flux recovery percentage of PVDF-g-PDMAA membranes at different degree of grafting and commercial membrane after the surface and inner fouling with BSA solution. [Reproduced with permission from *Journal of Materials Chemistry*, 21, 11908–11915, (2011)].

by lysozyme solution. The BSA solution experimented with the low fouling behavior of the PVDF-g-PDMAA grafted membrane. The water flux recovery of membranes after the surface and internal fouling for different grafted membranes are depicted in Figure 8.3. The flux recovery increased with the grafting percentage (0% to 17.1%), and it is better than the commercial PVDF membrane (0.22 μm thickness) (Yang et al., 2011). PVDF-g-AAc/MeAAc membrane showed similar antifouling properties. Moreover, the grafted membranes exhibited a 20% better flux than ungrafted ones (Deng et al., 2010). PA-g-VIm (vinyl imidazole) membranes showed 55% better water flux than the unmodified membrane (Reis et al., 2017). The performance of polyamide composite membranes and surface modification by hydrophilic acrylic acid was one of the approaches to counter fouling (Freger et al., 2002).

PVDF-g-PVP was prepared by grafting PVP (poly vinyl pyrrolidone) on PVDF materials and casting it into membrane form. PVP is an uncharged hydrophilic the water molecules immobilize in the vicinity of the amide group. The grafted membrane was hydrophilic (Qin et al., 2013, Chen et al., 2013). The hydrophilic layer can immobilize water molecules on the grafted membrane surface. The hydrophilicity of those grafted membranes increased with the DG. Thus, protein (viz. BSA) adsorption did not quickly occur and showed antifouling behavior (Hautojarvi et al., 1996). The hydrophilic grafted surface immobilizes water on it and thus repels proteins. Evidence is also there regarding binary grafting of N, N, dimethyl acrylamide (DMAA), and N-vinyl pyrrolidone (NVP) on PVDF powder. The phase separation technique was employed to prepare the membrane from the grafted material (Yang et al., 2011, Chen et al., 2013).

Grafting can be one of the approaches to preparing a stimuli-responsive membrane. The membranes can change their properties with the changes in the conditions (viz. pH, temperature). There are reports regarding PVDF-g-HEA (hydroxyl ethyl acrylate), PVDF-g-AA, PVDF-g-PAM, IPP-g-HEMA (hydroxyl ethyl methacrylate) membranes (Shen et al., 2017, Hautojarvi et al., 1996, Yang et al., 2010, Rattan and Sehgal, 2012). PVDF-g-HEA membrane exhibited improved dependence water flux in low pH. With the increase in pH, chemical potential changes and swelling of the grafted matrix occurred; thus, water flux decreased (Shen et al., 2017). PVDF-g(AA) and PES-g-(AA) membranes showed pH-dependent permeability.

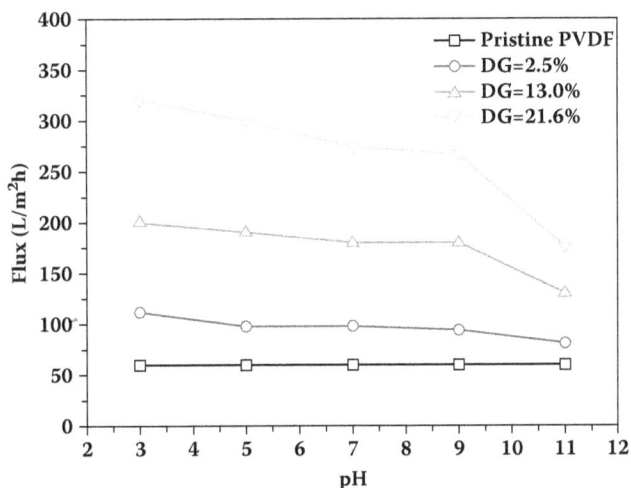

FIGURE 8.4 Variation of water flux with pH for pristine and PVDF-g-PAM membranes [Reproduced with permission from *Journal of Membrane Science*, 362(1–2), 298–305, (2010)].

The more grafted membranes showed a more significant change in permeability (Hautojarvi et al., 1996, Deng et al., 2008). The open-coil structure of the grafted moiety in high and low pH. The pore opening followed as the conformation of the grafted chains. The grafted moiety's open structure closed the membrane's pores, thus having low permeability. On the other hand, the coiled structure opened the pores.

PVDF-g-PAM (poly acrylamide) membranes showed the water flux behavior of on the variation of pH (Figure 8.4). The pristine PVDF and low grafted (2.5%) membrane showed no flux variation with pH. The higher grafting percentage (13% and 21.6%) led to the pH flux variation (Yang et al., 2010). Another interesting observation was that BSA flux was higher for grafted membranes than water flux. It was due to electric double layer compression on the polymeric swelling (Shen et al., 2017). The microfiltration membrane from PVDF-g-PAM powder showed better water flux performance than PVDF membrane (Rattan and Sehgal, 2012). The pH dependence permeability of the modified membrane was reproducible during the 12-time cyclic test. Through radiation-induced graft polymerization, Liu et al. successfully prepared temperature-sensitive PVDF-g-NIPAAm (N-isopropyl acrylamide) membrane (Liu et al., 2007). The NIPAAm is one of the intelligent polymers that changes their conformation with temperature.

The study exhibited the temperature-dependent permeation of water and the electric current through the membrane. The behaviour was explained by conformation variation of the NIPAAm chains when temperature varied through lower critical solution temperature (LCST) (Curti et al., 2005). The NIPAAm polymer side chains are became hydrophilic below the LCST, and thus polymer swelled by absorbing water. The swollen grafted poly-NIPAAm in the pores decreased the pore diameter, and thus the permeation rate and electric current through the modified membrane were hampered. On the other hand, above the LCST, as the grafted NIPAAm polymer chains became hydrophobic, it shrank, and thus a pore opening occurred. A similar observation was also seen in PET-g-NIPAAm membranes (Shtanko et al., 2000).

LDPE-g-DMAEMA/ PEGMEMA) membrane showed the stimuli-responsive behavior. It exhibited a low critical solution temperature of 40°C and critical pH ~ 7.5. PDMAEMA and PEGMEMA both polymers are pH-sensitive and thermo-sensitive (Martínez and Bucio, 2009) Gel membranes of polyacryloyl-L-proline methyl ester (A-ProOMe) grafted with 15 mol% poly(acrylic acid) showed the thermos and pH-responsive behavior (Hasegawa et al., 2005).

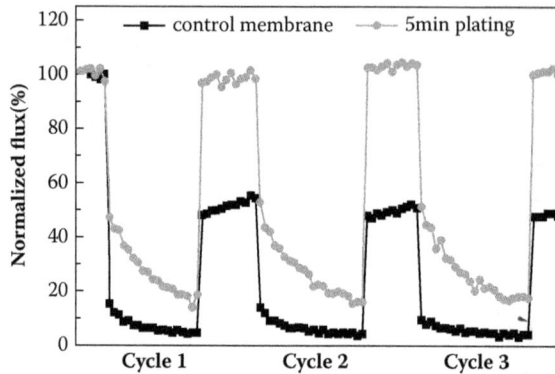

FIGURE 8.5 Normalized flux profiles of the control membrane and the PVDF-g-PAA-Ag@Ni membrane at electroless plating time of 5 min when filtrating SA solution [Reproduced with permission from *Journal of Colloid Interface Science*, 543, 64–75, (2019)].

The selective permeation of Li-ion compared to Co and Ni ions results at pH 6.0 and 30°C. In this condition, the thermo-sensitive unit shrinks, whereas pH-sensitive unit swells. A novel strategic approach to modify hydrophobic PVDF ultrafiltration membrane was adopted by Shen et al. (Shen et al., 2019). It combines radiation grafting of poly(acrylic acid)(PAA) and electroless nickel plating on PVDF ultrafiltration membrane. At first, the PVDF membrane was radiation grafted with acrylic acid and then immersed into AgNO$_3$ solution to absorb Ag metal ions. The resulting membranes were placed in an electroless plating bath where Ag$^+$ could be reduced to Ag micro cores, promoting Ni deposition. The modified PVDF-g-PAA-Ag@Ni membranes developed significant improvements in surface wettability and hydrophilicity. The modified membranes showed 100% flux recovery and reduced fouling propensity after a three-cycle filtrated sodium alginate (SA) solution. Figure 8.5 depicts the observation.

Grafting of hydroxyethyl acrylate (HEA) on the PVDF membrane showed the improved antifouling ability for sodium alginate (SA) and bovine serum albumin (BSA) solution (Shen et al., 2018).

The interesting feature of the grafted membrane had a lower flux decline rate than the unmodified one and higher BSA solution flux than pure water flux. The preparation of cationic and anionic membranes by grafting vinyl and acrylic monomers showed promise in reverse osmosis applications of desalinating water (Hegazy and Dessouki, 1986) prepared anionic membranes by radiation grafting of 4-vinylpyridine (4VP) onto polyethylene (PE) films followed by quaternization of pyridine rings with suitable quaternizing agents (viz. methyl iodide, allyl bromide, and HCl).

The preparation of cationic and anionic membranes by grafting vinyl and acrylic monomers showed promise in reverse osmosis applications of desalinating water (Hegazy and Dessouki, 1986) prepared anionic membranes by radiation grafting of 4-vinylpyridine (4VP) onto polyethylene (PE) films followed by quaternization of pyridine rings with suitable quaternizing agents (viz. methyl iodide, allyl bromide, and HCl). The membrane showed application in desalination. The cationic membranes PE-g-AAc (polyacrylic acid) by grafting acrylic acid onto low-density polyethylene films needed post-treatment with alkali showed good desalination application (Hegazy et al., 1989).

Thin-film polyamide composite (TFC) membrane is a milestone in developing membrane science. It is of three-layer structure. The polyamide coated on asymmetric polysulfone, which was supported by non-woven polyester. The polyamide was made up of phenylenediamine and 1, 3, 5 trimesoyl chloride through interfacial polymerization, mainly water and hexane. Figure 8.6 depicts the thin film composite membrane (TFC). The membrane suffered some limitations viz. chlorine susceptibility, prone to biofouling. The chlorine exposure partially destructs the rigid polyamide

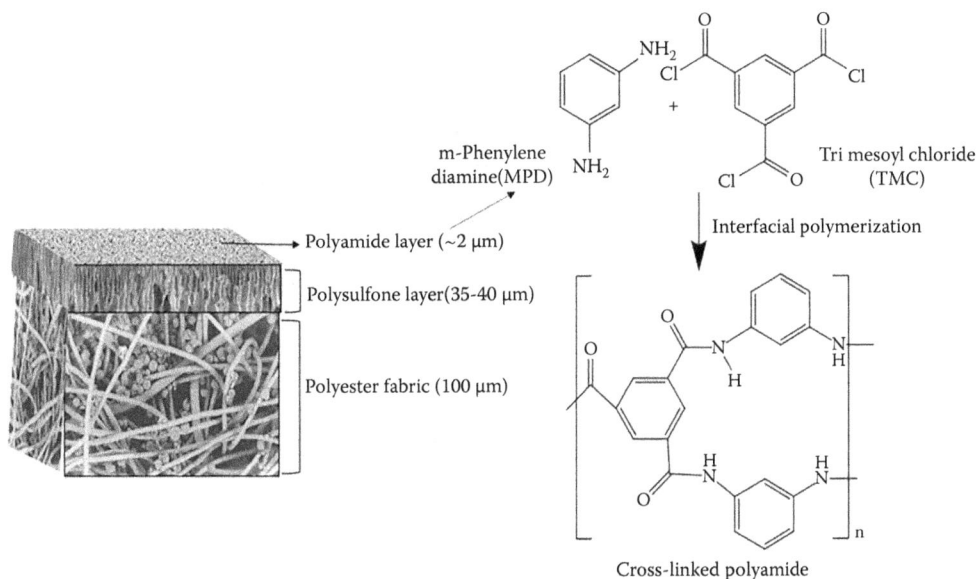

FIGURE 8.6 Schematic presentation of thin film composite membrane.

structure, and thus the collapse of the polymer chain (Kwon and Leckie, 2006). Modification of TFC membrane by grafting N-Isopropyl acrylamide (NIPAAM) and ZnO nanoparticles developed better chlorine and biofouling resistance properties (El-Arnaouty et al., 2018). Figures 8.7a and 8.7b depict the improvement of chlorine resistance and biofouling. The salt rejection profile of the modified membrane and water flux was steady after 4,000 ppm chlorine exposure. Figure 8.8 shows the comparison of water flux in pre- and post-fouling conditions. It shows that the unmodified membranes sacrificed water flux, whereas modified membranes developed antifouling properties. The enhanced chlorine resistance of the grafted modified membrane might be the presence of NIPAAM, which hindered the replacement of hydrogen by chlorine on the amide functionality (Wu et al., 2010). The grafted layer also acts as the protective and sacrificial layer.

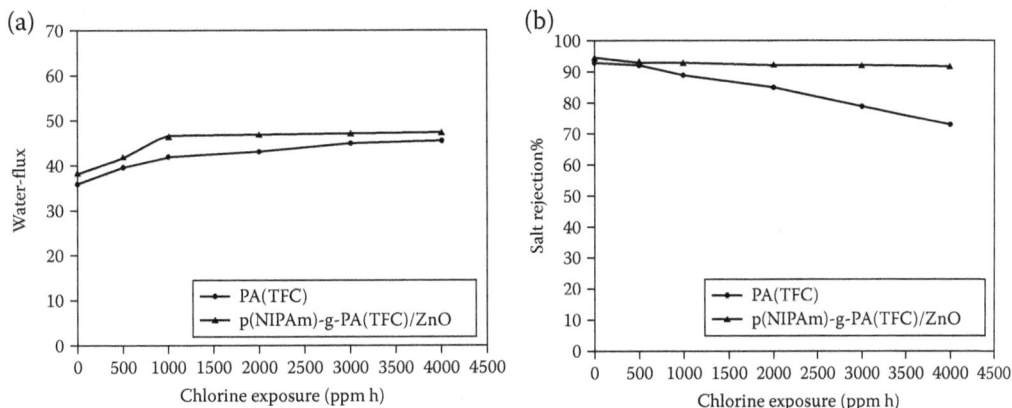

FIGURE 8.7 (a) Water flux and (b) salt rejection (%) on the chlorine exposure concentration for virgin PA(TFC) and ZnO NPs (1.25) grafted TFC(PA)-g- poly(N-isopropyl acrylamide) [Reproduced with permission from *Journal of Radiation Research and Applied Sciences*, 11(3), 204–216, (2018)].

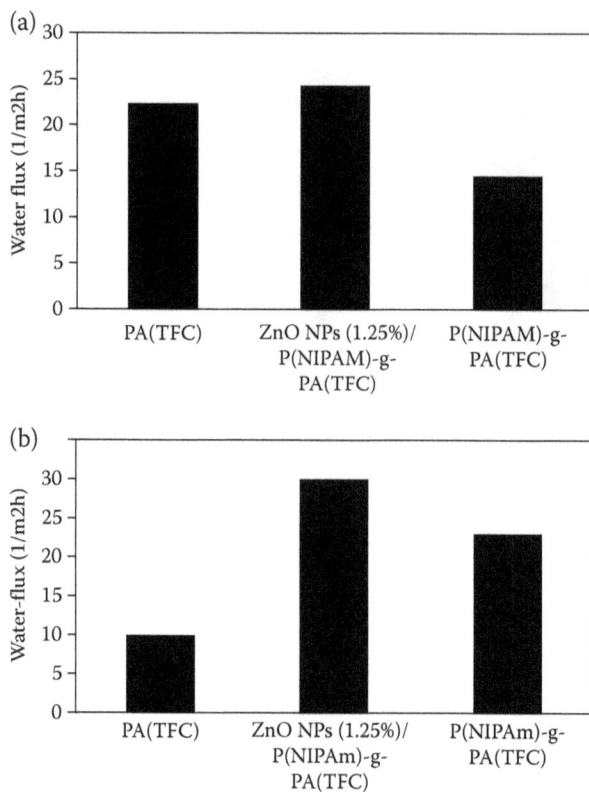

FIGURE 8.8 Water fluxes for the virgin PA(TFC) and the grafted membranes before and after fouling. [Reproduced with permission from *Journal of Radiation Research and Applied Sciences*, 11(3), 204–216, (2018)].

Pervaporation is a separation process in which azeotropic liquids are separated from their mixtures. It is operated by partial vaporization through a porous/nonporous membrane. Radiation grafted pervaporation membranes are well reported in the literature. Poly tetrafluoroethylene (PTFE)–g–P4VP and PTFE-g-PVP membranes successfully separate various liquid mixtures viz., chloroform-hydrocarbons, ethyl ether, and water-alcohols (Patel et al., 1972, 1974, Tealdo et al., 1981, Niemoller et al., 1988, Semenova et al., 1997). Poly(ethylene terephthalate)-g-polystyrene (PET-g-PSt) membranes were useful for the pervaporation of toluene/methanol mixtures (Kheyet et al., 2005). The grafted membrane was toluene selective and performed at degree of grafting (35%) was better than the ungrafted base PET membrane, though permeation flux decreased. The permeation flux might be the larger thickness of the grafted layer. Polyimide-g-4VP–based membrane established its selectivity of benzene in benzene/cyclohexane mixture (Yanagishita et al., 2004). PE hollow fibers grafted with GMA, followed by treatment with DEA showed selectivity in permeation of acetone from acetone/water mixture (Tsuneda et al., 1992). Apart from pervaporation, grafted membranes could dehydrate from the organic/water mixture. PVA-g-AAc/MAc membranes were promising in dehydrating methanol/water mixtures (Shantora and Huang, 1981).

The chemical process of separating THF from its water mixture is not friendly. THF reacts readily with oxygen in the presence of air forming an unstable hydroperoxide. Distillation of the combination results in an explosion. THF also forms an azeotrope with water and the mixture of THF–water. Polyvinyl chloride (PVC)-g- 4VP and PVC-g-4VP and vinyl acetate (VA) membranes were used for sorption and pervaporation of 0.5–20 wt% of tetrahydrofuran (THF) from water (Choudhury and Ray, 2021).

Hydrophilic membranes modified to hydrophobic surface membranes have the potentiality in membrane distillation application. Membrane distillation (MD) is a thermally driven non-isothermal process in which vapor molecules transported through porous hydrophobic polymer membranes. The hydrophobicity and porosities of the polymer membranes are the important regulators in this application of membrane distillation. In this direction, Wu et al. (Wu et al., 1992) prepared cellulose acetate-g-(sty) membrane that showed the maximum 99.1% rejection for 0.3M NaCl solution.

Apart from the separation activities, grafted membranes are used as catalyzing the reaction. Poly(tetrafluoroethylene-co-perflurovinyl ether)-g-PSSA (PFA-g-PSSA) acted as catalytic membrane for the following hydrolysis reaction (Nasef et al., 2005):

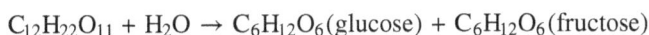

$$C_{12}H_{22}O_{11} + H_2O \rightarrow C_6H_{12}O_6(\text{glucose}) + C_6H_{12}O_6(\text{fructose})$$

The hydrolysis depended on the concentration of sulfonic acid on the membranes and the degree of grafting.

8.7 FUTURE TRENDS

The beauty of the radiation grafting technique is that it is dynamic and moves forward slowly, and recommendations are made based on the best methodology available at that particular time. The adaptability of research process is like putting stones of an old-fashioned balance scale; when enough weight gathers on one side, the scale tips, favoring a particular recommendation.

The impressively remarkable extant studies related to varieties of radiation-grafted polymeric membranes with different functionalities have been documented in this research orientation; yet, the arena's overall potential remains unrealized, and expectations are unmet. It may be of a different mindset to use radiation techniques. Thus, the challenge is humongous. The advantages include the ease of preparation, the superb controllability over the composition, and properties of the graft modified membranes through optimization of the grafting conditions. The obstacles (viz. long irradiation time, the purification step from homopolymer) are also pointed out. Regarding the time obstacles, it may be the direction of electron beam irradiation.

The potential end-use viability scales the success of radiation grafting. Moreover, membranes are always ready to spread their wings in a different arena. In this regard, bio-applications is growing tremendously, as researchers have covered few viz. affinities to biomolecules, immobilization of enzymes, and antifouling property. The development of grafted membrane biomaterials in the drug-eluting medical devices, antifouling biomedical devices, and tissue engineering is very promising. The fuel cell application is another direction in which the grafting of suitable functionalities can serve the property. A few companies (GE, USA Pall-RAI, USA, Ashai Chlorine Engineers, Japan and De Morgan, France, DuPont, USA, Saint-Gobain, France) have begun preparing radiation grafted ion exchange membranes for various applications (Nasef, 2004).

Such advantages include the ease of preparation and controllability over the graft copolymers' composition and properties through optimization of the grafting conditions. This confers the membrane materials' separation capabilities through various mechanisms, including ion exchange, chelation, affinity, and physical entrapment, diffusion, and convective diffusion.

8.8 CONCLUSIONS

Every sector has to look for new and unique ways to keep their orientation more functional in today's unprecedented scenario. The direction needs to be taken on multiple fronts to craft a cohesive strategy. Radiation grafting is a fantastic approach to modifying polymer membranes' properties for achieving specifically desired properties. Radiation processing is recognized as an energy-efficient technology with promising perspectives in different applications. The present chapter describes the radiation grafted membranes application in other arenas viz. electrochemical, metal ion absorption/adsorption, biological, and separation science. The membranes fabricated in this technique offer a cost-competitive option since inexpensive commercial materials are used. The stupendous progress of membrane orientation shows the enormous researchers' confidence in the future relationship between radiation grafting and membrane. Radiation-induced grafting is a potential method for introducing desirable properties into a polymer membrane due to its simplicity in handling and control. This is the approach to take membranes to human well-being. The method opens up the possibilities to use a wide range of polymer–monomer combinations. Radiation processing can be controlled and successfully used to develop novel materials with tailored properties for specific applications.

Overall, radiation grafting-based methods have the versatility to make the spectrum of membranes broader. Continuous efforts are needed to extend radiation-grafted functional copolymers' practical applications by bringing many potential products.

ACKNOWLEDGMENT

The manuscript is having the CSIR-CSMCRI communication number CSIR-CSMCRI–173/2020.

REFERENCES

Ajji, Z. (2012). *IAEA Radiation Tech Report (No. 3) International Atom Energy Agency*, Vienna, 199.

Ajji, Z., and Ali, A. M. (2005). Preparation of poly(vinyl alcohol) membranes grafted with N vinyl imidazole/acrylic acid binary monomers. *Nuclear Instrument and Methods in Physics Research B: Beam Interactions with Materials and Atoms*, 236 (1), 580–586.

Ajji, Z., and Ali, A. M. (2010). Separation of copper ions from iron ions using PVA-g-(acrylic acid/N-vinyl imidazole) membranes prepared by radiation-induced grafting. *Journal of Hazardous Materials*, 173(1–3), 71–74.

Asai, S., Watanabe, K., Sugo, T., and Saito, K. (2005). Preparation of an extractant-impregnated porous membrane for the high-speed separation of a metal ion. *Journal of Chromatography A*, 1094(1–2), 158–164.

Assink, R. A., Arnold, C., and Hollandsworth, R. P. (1991). Preparation of oxidatively stable cation-exchange membranes by the elimination of tertiary hydrogens. *Journal of Membrane Science*, 56(2), 143–151.

Bhattacharya, A. (2000). Radiation and industrial polymers. *Progress in Polymer Science*, 25(3), 371–401.

Bhattacharya, A., and Mishra, B. N. (2004). Grafting: A versatile means to modify polymers: Techniques, factors and applications. *Progress in Polymer Science*, 29(8), 767–814.

Bhattacharya, A., Rawlins, J. W., and Ray, P., Eds. (2008). *Polymer Grafting and crosslinking*, published by A John Wiley and Sons, New Jersey, pp. 1–341.

Biancolli, A. L. G., Barbosa, A. S., Kodama, Y., DeSousa Jr., R. R., Lanfredi, A. J. C., Fonseca, F. C., Rey, J. F. Q., and Santiago, E. I. (2021). Unveiling the influence of radiation-induced grafting methods on the properties of polyethylene-based anion-exchange membranes for alkaline fuel cells. *Journal of Power Sources*, 512, 230484.

Brack, H. P., Buchi, F. N., Huslage, J., Rota, M., Scherer, G. G., Pinnau, I., and Freeman, B. D. (2000). Development of radiation-grafted membranes for fuel cell applications based on poly(ethylene-alt-tetrafluoroethylene), ACS symposium series. *Journal of the American Chemical Society*, 744, 174.

Chapiro, A. (1962). *Radiation chemistry of polymeric systems, High polymers*, John Wiley and Sons, Interscience Publishers, New York, pp. 712.

Charcosset, C. (1998). Review: Purification of proteins by membrane chromatography. *Journal of Chemical Technology and Biotechnology*, 71(2), 95–110.

Chen, J., Asano, M., Yamaki, T., and Yoshida, M. (2006). Preparation and characterization of chemically stable polymer electrolyte membranes by radiation-induced graft copolymerization of four monomers into ETFE films. *Journal of Membrane Science*, 269(1-2), 194–204.

Chen, L., Hou, Z., Lu, X., Chen, P., Liu, Z., Shen, L., Bian, X., and Qin, Q. (2013). Antifouling microfiltration membranes prepared from poly(vinylidene fluoride)-*graft*-Poly(*N*- vinyl pyrrolidone) powders synthesized via pre-irradiation induced graft polymerization. *Journal of Applied Polymer Science*, 128(6), 3949–3956.

Choi, S. H., and Nho, Y. C. (1999a). Adsorption of CO^{2+} by stylene-g-polyethylene membrane bearing sulfonic acid groups modified by radiation-induced graft copolymerization. *Journal of Applied Polymer Science*, 71(13), 2227–2235.

Choi, S. H., and Nho, Y. C. (1999b). Modification of hollow fiber membrane with amidoxime, iminodiacetic acid and diethylene triamine by radiation-induced graft polymerization. *Korea Polymer Journal*, 7(1), 38–45.

Choi, S. H., and Nho, Y. C. (1999c). Adsorption of Co^{2+} and Cs^{1+} by polyethylene membrane with iminodiacetic acid and sulfonic acid modified by radiation-induced graft copolymerization. *Journal of Applied Polymer Science*, 71, 999–1006.

Choi, S. H., and Nho, Y. C. (2000a). Adsorption of UO^{2+} by polyethylene adsorbents with amidoxime, carboxyl, and amidoxime/carboxyl group. *Radiation Physics and Chemistry*, 57(2), 187–193.

Choi, S. H., and Nho, Y. C. (2000b). Radiation-induced graft copolymerization of binary monomer mixture containing acrylonitrile onto polyethylene films. *Radiation Physics and Chemistry*, 58(2), 157–168.

Choi, S. H., Park, S. Y., and Nho, Y. C. (2000). Electrochemical properties of polyethylene membrane modified with carboxylic acid group. *Radiation Physics and Chemistry*, 57(2), 179–186.

Curti, P. S., Moura, M. R. D., Veiga, W., Radovanovic, E., Rubira, A. F., and Muniz, E. C. (2005). Characterization of PNIPAAm photografted on PET and PS surfaces. *Applied Surface Science*, 245(1–4), 223–233.

Choudhury, S., and Ray, S. K. (2021). Poly(4-vinylpyridine) and poly(vinyl acetate –co-4-vinylpyridine) grafted polyvinyl chloride membranes for removal of tetrahydrofuran from water by pervaporation. *Separation and Purification Technology*, 254(1), 117618.

Dang, H. T., Rana, D., Narbaitz, R. M., and Matsuura, T. (2010). Key factors affecting the manufacture of hydrophobic ultrafiltration membranes for surface water treatment. *Journal of Applied Polymer Science*, 116(5), 2626–2637.

Dargaville, T. R., George, G. A., Hill, D. J. T., and Whittaker, A. K. (2003). High energy radiation grafting of fluoropolymers. *Progress in Polymer Science*, 28, 1355–1376.

Deng, B., Li, J., Hou, Z., Yao, S., Shi, L., Liang, G., and Sheng, K. (2008). Microfiltration membranes prepared from polyethersulfone powder grafted with acrylic acid by simultaneous irradiation and their pH dependence. *Radiation Physics Chemistry*, 77(7), 898–906.

Deng, B., Yu, M., Yang, X., Zhang, B., Li, L., Xie, L., Li, J., and Lu, X. (2010). Antifouling microfiltration membranes prepared from acrylic acid or methacrylic acid grafted poly(vinylidene fluoride) powder synthesized via pre-irradiation induced graft polymerization. *Journal of Membrane Science*, 350(1-2), 252–258.

Dessouki, A. M., Taher, N. H., and Boohy, H. A. E. (1990). Radiation-induced graft polymermization of acrylamide onto poly(tetrafluoroethylene/Hexaflouropropylene/vinylidene fluoride) (TFB) films. *International Journal of Radiation Application and Instrumentation Part C: Radiation Physics and Chemistry*, 36(3), 371–375.

Dessouki, A. M., Taher, N. H., and Arnaouty, M. B. (1998). Gamma ray induced graft copolymerization of N-vinylpyrrolidone, acrylamide and their mixtures onto polypropylene films. *Polymer International*, 45(1), 67–76.

Dogan, H. D. C., and Gursel, S. A. (2011). Preparation and characterisation of novel composites based on a radiation grafted membrane for fuel cells. *Fuel Cells*, 11, 361–371.

Dumas, I., Fleury, E., and Portinha, D. (2014). Wettability adjustment of PVDF surfaces by combining radiation-induced grafting of (2, 3, 4, 5, 6)-pentafluorostyrene and subsequent chemoselective "click-type" reaction. *Polymer*, 55(11), 2628–2634.

El-Arnaouty, M. B., Ghaffar, A. M. A., Eid, M., Aboulfotouh, M. E., Taher, N. H., and Solimen, E. S. (2018). Nano-modification of polyamide thin film composite reverse osmosis membranes by radiation grafting. *Journal of Radiation Research and Applied Sciences*, 11(3), 204–216.

El-Awady, N. I. (2004). Dialysis membranes from polyethylene films grafted with acrylic acid. *Journal of Applied Polymer Science*, 91(1), 10–14.

Elias, K. (2000). Affinity membranes: A 10-year review. *Journal of Membrane Science*, 179(1-2), 1–27.

El-Rehim, H. A. A. (2005). Hemodialysis membranes based on functionalized high-density polyethylene. *Journal of Bioactive and Compatible Polymer*, 20(1), 51–75.

El-Rehim, H. A. A., Hegazy, E. A., and Ali, A. M. (1999). Preparation of poly(vinyl alcohol) grafted with acrylic acid/styrene binary monomers for selective permeation of heavy metals. *Journal of Applied Polymer Science*, 74(4), 806–815.

El-Rehim, H. A. A., Hegazy, E. A., and Ali, A. E. H. (2000). Selective removal of some heavy metal ions from aqueous solution using treated polyethylene-g-styrene/maleic anhydride membranes. *Reactive and Functional Polymers*, 43(1-2), 105–116.

El-Sawy, N. M. (2000). Amidoximation of acrylonitrile radiation-grafted onto low-density polyethylene and its physicochemical characterization. *Polymer International*, 49(6), 533–538.

Fang, J., Yang, Y. X., Lu, X. H., Ye, M. L., Li, W., and Zhang, Y. M. (2012). Cross-linked, ETFE-derived and radiation grafted membranes for anion exchange membrane fuel cell applications. *International Journal of Hydrogen Energy*, 37(1), 594–602.

Filho, A. A. M. F., and Gomes, A. S. (2006). Copolymerization of styrene onto polyethersulfone films induced by gamma ray irradiation. *Polymer Bulletin*, 57(4), 415–421.

Freger, V., Gilron, J., and Belfer, S. (2002). TFC polyamide membranes modified by grafting of hydrophilic polymers: an FT-IR/AFM/TEM study. *Journal of Membrane Science*, 209(1), 283–292.

Gebauer, K. H., Thommes, J., and Kula, M. R. (1997). Plasma protein fractionation with advanced membrane adsorbents. *Biotechnology and Bioengineering*, 54(2), 181–189.

Godjevargova, T., and Dimov, A. (1995). Grafting of acrylonitrile copolymer membranes with hydrophilic monomers for immobilization of glucose oxidase. *Journal of Applied Polymer Science*, 57(4), 487–491.

Goel, N. K., Bharwaj, Y. K., Manoharan, R., Kumar, V., Dubey, K. A., Chaudhari, C. V., and Sabharwal, S. (2009). Physiochemical and electrochemical characterization of battery seperator prepared by radiation induced grafting of acrylic acid onto microporous polypropylene membranes. *Express Polymer Letters*, 3(5), 268–278.

Gubler, L., Beck, N., Gursel, S. A., Hajboulouri, F., Kramer, D., Reimer, A., Steiger, B., and Scherer, G. G. (2004). Materials for polymer electrolyte fuel cells. *CHIMIA International Journal for Chemistry*, 58(12), 826–836.

Gubler, L., Gursel, S. A., and Scherer, G. G. (2005a). Radiation grafted membranes for polymer electrolyte fuel cells. *Fuel Cells*, 5(3), 317–335.

Gubler, L., Prost, N., Gursel, S. A., and Scherer, G. G. (2005b). Proton exchange membranes prepared by radiation grafting of styrene/divinylbenzene onto poly(ethylene-alt-tetrafluoroethylene) for low temperature fuel cells. *Solid State Ionics*, 176(39–40), 2849–2860.

Gubler, L., and Scherer, G. G. (2009). Radiation-grafted proton conducting membranes. In Vielstich, W., Lamn, A. , and Gasteiger, H. A. *Handbook of Fuel Cells: Fundamental, Technology and Applications.* John Wiley and Sons, Chichester, pp. 313–321.

Gubler, I., and Scherer, G. G. (2010). Trends for fuel cell membrane development. *Desalination*, 250(3), 1034–1037.

Gubler, L., Slaski, M., Wokaun, A., and Scherer, G. G. (2006). Advanced monomer combinations for radiation grafted fuel cell membranes. *Electrochemistry Communications*, 8(8), 1215–1219.

Gupta, B., and Anjum, N. (2002). Radiation grafted membranes: Innovative materials for the separation of toxic metal ions from industrial effluent. *Indian Journal of Environmental Health*, 44(2), 154–163.

Gupta, B., and Anjum, N. J. (2003). Preparation of ion-exchange membranes by the hydrolysis of radiation-grafted polyethylene-*g*-polyacrylamide films: Properties and metal-ion separation. *Journal of Applied Polymer Science*, 90(14), 3747–3752.

Gupta, B., Anjum, N., and Sen, K. (2002). Development of membranes by radiation grafting of acrylamide into polyethylene films: Properties and metal ion separation. *Journal of Applied Polymer Science*, 85(2), 282–291.

Gupta, B., Anjum, N., Jain, R., Revagade, N., and Singh, H. (2004). Development of membranes by radiation-induced graft polymerization of monomers onto polyethylene films. *Journal of Macromolecular Science, Part C: Polymer Reviews*, C44(3), 275–309.

Gursel, S. A., Gubler, L., Gupta, B., and Scherer, G. G. A. (2008). Radiation Grafted Membranes, In book: Fuel Cells I, Springer-Verlag Berlin Heidelberg. *Chapter in Advances Polymer Science*, 215, 157–217.

Gwon, S. J., Choi, J. H., Sohn, J. Y., Ihm, Y. E., and Nho, Y. C. (2009). Preparation of a new micro-porous poly(methyl methacrylate)-grafted polyethylene separator for high performance Li secondary battery. *Nuclear Instruments and Methods in Physics Research, Section B: Beam Interactions with Materials and Atoms*, 267, 3309–3313.

Hagiwara, K., Yonedu, S., Saito, K., Shiraish, T., Sugo, T., Tojyo, T., and Katayama, E. (2005). High-performance purification of gelsolin from plasma using anion-exchange porous hollow-fiber membrane. *Journal of Chromatography B*, 821(2), 153–158.

Hasegawa, S., Ohashi, H., Maekawa, Y., Katakai, R., and Yoshida, M. (2005). Thermo-and pH-sensitive gel membranes based on poly-(acryloyl-L-proline methyl ester)-graft-poly (acrylic acid) for selective permeation of metal ions. *Radiation Physics and Chemistry*, 72(5), 595–600.

Hautojarvi, J., Kontturi, K., Nasman, J. H., Svarfar, B. L., Viinikka, P., and Vouristo, M. (1996). Characterization of graft-modified porous polymer membranes. *Industrial and Engineering Chemistry Research*, 35(2), 450–457.

Hamza, A., Pham, V. A., Matsuura, T., and Santerre, J. P. (1997). Development of membranes with low surface energy to reduce the fouling in ultrafiltration applications. *Journal of Membrane Science*, 131(1-2), 217–227.

Hegazy, E. A., Abdel-Ehim, H., Hegazy, D., Ali, A. A., Kamal, H., and Sayed, A. (2012). *IAEA Rad. Tech Report (No. 3) International Atomic Energy Agency*, Vienna 49.

Hegazy, E. S. A., Assy, N. B. E., Rabie, A. G. M., Ishigaki, I., and Okamoto, J. (1984). Kinetic study of preirradiation grafting of acrylic acid onto poly(tetrafluoroethylene–perfluorovinyl ether) copolymer. *Journal of Polymer Science: Polymer Chemistry*, 22(3), 597–604.

Hegazy, E. A., and Dessouki, A. M. (1986). Ion-containing reverse osmosis membranes obtained by radiation grafting method. *International Journal of Radiation Applications and Instrumentation. Part C. Radiation Physics and Chemistry*, 28(3), 273–279.

Hegazy, E. A., EL-Assy, N. B., Dessouki A. M., and Shaker (1989). Ion-containing reverse osmosis membranes obtained by radiation grafting method. *Radiation Physics and Chemistry*, 33(1), 13–18.

Hegazy, El-Sayed A., El- Rehim, H. A. A., and Shawky, H. A. (2000). Investigations and characterization of radiation grafted copolymers for possible practical use in waste water treatment. *Radiation Physics and Chemistry*, 57(1), 85–95.

Hegazy, E. A., El-Rehima, H. A. A., Kamal, H., and Kandeel, K. A. (2001). Advances in radiation grafting. *Nuclear Instruments and Methods in Physics Research Section B: Beam Interactions with Materials and Atoms*, 185(1), 235–240.

Hegazy, E. A., El-Rehima, H. A. A., Khalifa, N. A., Atwa, S. M. and Shawky, H. A. (1997). Anionic/cationic membranes obtained by a radiation grafting method for use in waste water treatment. *Polymer International*, 43(4), 321–332.

Hegazy, E. A., Kamal, H., Maziad, N., and Dessouki, A. M. (1999). Membranes prepared by radiation grafting of binary monomers for adsorption of heavy metals from industrial wastes. *Nuclear Instruments and Methods in Physics Research, Section B: Beam Interactions with Materials and Atoms*, 151, 386–392.

Hegazy, E. S. A., Taher, N. H., and Ebaid, A. R. (1990). Preparation and some properties of hydrophilic membranes obtained by radiation grafting of methacrylic acid onto fluorinated polymers. *Journal of Applied Polymer Science*, 41(11-12), 2637–2647.

Herman, F. M. (2007). Graft copolymers. In *Encyclopedia of Polymer Science and Technology. Concise* (Mark Herman F 3rd ed., pp. 526–549). Wiley, New Jersey.

Hidzir, N. M., Hill, D. J. T., Martin, D., and Grøndahl, L. (2012). Radiation-induced grafting of acrylic acid onto expanded poly(tetrafluoroethylene) membranes. *Polymer*, 53(26), 6063–6071.

Hidzir, N. M., Lee, Q., Hill, D. J. T., Rasoul, F., and Grøndahl, L. (2015). Grafting of acrylic acid-*co*-itaconic acid onto ePTFE and characterization of water uptake by the graft copolymers. *Journal of Applied Polymer Science*, 132(7), 41482–41494.

Hietala, S., Skou, E., and Sundholm, F. (1999). Gas permeation properties of radiation grafted and sulfonated poly-(vinylidene fluoride) membranes. *Polymer*, 40, 5567–5573.

Hoogers, G. (2003). *Fuel Cell Technology Handbook*. CRC Press, Chicago, pp. 1–376.

Horsfall, J. A., and Lovell, K. V. (2001). Fuel cell performance of radiation grafted sulphonic acid membranes. *Fuel Cells*, 1(3-4), 186–191.

Horsfall, J. A., and Lovell, K. V. (2002). Comparison of fuel cell performance of selected fluoropolymer and hydrocarbon based grafted copolymers incorporating acrylic acid and styrene sulfonic acid. *Polymers for Advanced Technologies*, 13(5), 381–390.

Hu, G., Wang, Y., Ma, J., Qui, J., Peng, J., Li, J., and Zhai, M. A. (2012). A novel amphoteric ion exchange membrane synthesized by radiation-induced grafting α-methylstyrene and N,N-dimethylaminoethyl methacrylate for vanadium redox flow battery application. *Journal of Membrane Science*, 407–408, 184–192.

Hayre, R. O., Cha, S. W., Colella, W., and Prinz, F. B. (2006). *Fuel Cell Fundamentals*, (3rd Edition). Jhon Wiley and Sons Ltd., New York, pp. 1–608.

Ito, H., Nakamura, M., Saito, K., Sugita, K., and Sugo, T. (2001). Comparison of L-tryptophan binding capacity of BSA captured by a polymer brush with that of BSA adsorbed onto a gel network. *Journal of Chromatography A*, 925(1-2), 41–47.

Iwata, H., Saito, K., Furusaki, S., Sugo, T., and Okamoto, J. (1991). Adsorption characteristics of an immobilized metal affinity membrane. *Biotechnology Progress*, 7(5), 412–418.

Jetsrisuparb, K., Youcef, B. H., Wokaun, A. and Gubler, L. (2014). Radiation grafted membranes for fuel cells containing styrene sulfonic acid and nitrile comonomers. *Journal of Membrane Science*, 450, 28–37.

Kallio T., Jokela, K., Ericson, H., Serimaa, R., Sundholm, G., Jacosson, P., and Sundholm, F. J. (2003). Effects of a fuel cell test on the structure of irradiation grafted ion exchange membranes based on different fluoropolymers. *Journal of Applied Electrochemistry*, 33, 505–514.

Kallio, T., Lundstrom, M., Sundholm, G., Walsby, N., and Sundholm, F. (2002). Electrochemical characterization of radiation-grafted ion-exchange membranes based on different matrix polymers. *Journal of Applied Electrochemistry*, 32(1), 11–18.

Kawai, T., Nakamura, M., Sugita, K., Saito, K., and Sugo, T. (2001). High conversion in asymmetric hydrolysis during permeation through enzyme-multilayered porous hollow-fiber membranes. *Biotechnology Progress*, 17(5), 872–875.

Kheyet, M., Nasef, M. M., and Mengual, J. I. (2005). Radiation grafted Poly(ethylene terephthalate)-graft-polystyrene pervaporation membranes for organic/organic separation. *Journal of Membrane Science*, 263(1-2), 77–95.

Kim, M., Kojima, J., Saito, K., Furusaki, S., and Sugo, T. (1994). Reduction of nonselective adsorption of proteins by hydrophilization of microfiltration membranes by radiation-induced grafting. *Biotechnology Progress*, 10, 114–120.

Kim, M., and Saito, K. (2000). Radiation-induced graft polymerization and sulfonation of glycidyl methacrylate on to porous hollow fiber membranes with different pore sizes. *Radiation Physics and Chemistry*, 57(2), 167–172.

Kim, M., Saito, K., Furusaki, S., Sato, T., Sugo, T., and Ishigaki, I. (1991a). Adsorption and elution of bovine gamma-globulin using an affinity membrane containing hydrophobic amino acids as ligands. *Journal of Chromatography A*, 585(1), 45–51.

Kim, M., Saito, K., Furusaki, S., Sato, T., Sugo, T., and Ishigaki, I. (1991b). Protein adsorption capacity of a porous phenylalanine-containing membrane based on a polyethylene matrix. *Journal of Chromatography A*, 586(1), 27–33.

Kim, M., Saito, K., Furusaki, S., Sugo, T., and Okamoto, J. (1991c). Water flux and protein adsorption of a hollow fiber modified with hydroxyl groups. *Journal of Membrane Science*, 56(3), 289–302.

Kim, M., Saito, K., Furusaki, S., and Sugo, T. (1993). Comparison of BSA adsorption and Fe sorption to the diol group and tannin immobilized onto a microfiltration membrane. *Journal of Membrane Science*, 85(1), 21–28.

Kiyohara, S., Sasaki, M., Saito, K., Sugita, K., and Sugo, T. (1996). Radiation-induced grafting of phenylalanine-containing monomer onto a porous membrane. *Reactive and Functional Polymer*, 31(2), 103–110.

Kiyohara, S., Kim, M., Toida, Y., Saito, K., Sugita, K., and Sugo, T. (1997). Selection of a precursor monomer for the introduction of affinity ligands onto a porous membrane by radiation-induced graft polymerization. *Journal of Chromatography A*, 758(2), 209–215.

Ko, J. M., Min, B. G., Kim, D. W., Ryu, K. S., Kim, K. M., Lee, Y. G., and Chang, S. H. (2004). Thin-film type Li-ion battery, using a polyethylene separator grafted with glycidyl methacrylate. *Electrochimica Acta*, 50(2–3), 367–370.

Kobayashi, S., Yonezu, S., Kawakita, H., Saito, K., Sugito, K., Tamada, M., Sugo, T., and Lee, W. (2003). Highly multilayered urease decomposes highly concentrated urea. *Biotechnology Progress*, 19(2), 396–399.

Koguma, I., Sugita, K., Saito, K., and Sugo, T. (2000). Multilayer binding of proteins to polymer chains grafted onto porous hollow-fiber membranes containing different anion-exchange groups. *Biotechnology Progress*, 16(3), 456–461.

Konishi, S., Saito, K., Furusaki, S., and Sugo, I. (1996). Binary metal-ion sorption during permeation through chelating porous membranes. *Journal of Membrane Science*, 111(1), 1–6.

Kreuer, K. D., and Portale, G. A. (2013). A critical revision of the nano-morphology of proton conducting ionomers and polyelectrolytes for fuel cell applications. *Advanced Functional Materials*, 23(43), 5390–5397.

Kubota, N., Kounosu, M., Saito, K., Sugita, K., Watanabe, K., and Sugo, T. (1997). Protein adsorption and elution performances of porous hollow-fiber membranes containing various hydrophobic ligands. *Biotechnology Progress*, 13(1), 89–95.

Kwon, Y. N., and Leckie, J. O. (2006). Hypochlorite degradation of crosslinked polyamide membranes. II. Changes in hydrogen bonding behavior and performance. *Journal of Membrane Science*, 282(1–2), 456–464.

Lee, W., Furusaki, S., Saito, K., Sugo, T., and Makuuchi, K. (1996). Adsorption kinetics of microbial cells onto a novel brush-type polymeric material prepared by radiation-induced graft polymerization. *Biotechnology Progress*, 12(2), 178–183.

Lee, W., Saito, K., Furusaki, S., and Sugo, T. (1997). Capture of microbial cells on brush-type polymeric materials bearing different functional groups. *Biotechnology and Bioengineering*, 53(5), 523–528.

Liu, Q., Zhu, Z., Yang, X., Chen, X., and Song, Y. (2007). Temperature-sensitive porous membrane production through radiation co-grafting of NIPAAm on/in PVDF porous membrane. *Radiation Physics and Chemistry*, 76(4), 707–713.

Ma, J., Wang, S., Peng, J., Yuan, J., Yu, C., Li, J., Ju, X., and Zhai, M. (2013). Covalently incorporating a cationic charged layer onto Nafion membrane by radiation-induced graft copolymerization to reduce vanadium ion crossover. *European Polymer Journal*, 49(7), 1832–1840.

Martínez, A. R. H., and Bucio, E. (2009). Novel pH- and temperature-sensitive behavior of binary graft DMAEMA/PEGMEMA onto LDPE membranes. *Designed Monomers and Polymers*, 12(6), 543–552.

Miyoshi, K., Saito, K., Shiraaishi, T., and Sugo, T. (2005). Introduction of taurine into polymer brush grafted onto porous hollow-fiber membrane. *Journal of Membrane Science*, 264(1-2), 97–103.

Mok, S., Worsfold, D. J., Fouda, A., and Matsuura, T. (1994). Surface modification of polyethersulfone hollow-fiber membranes by γ-ray irradiation. *Journal of Applied Polymer Science*, 51(1), 193–199.

Mun, G., Nurkeeva, Z., and Sergaziev, A. (2005). IAEA-TecDoc-1465, International Atomic Energy Agency, Vienna International Atomic Energy Agency, Vienna 91.

Nasef, M. M. (2014). Radiation-grafted membranes for polymer electrolyte fuel cells: Current trends and future directions. *Chemical Reviews*, 114(24), 12278–12329.

Nasef, M. M., Gursel, S. A., Karabelli, D., and Guven, O. (2016). Radiation-grafted materials for energy conversion and energy storage applications. *Progress in Polymer Science*, 63, 1–41.

Nasef, M. M., and Guven, O. (2012). Radiation-grafted copolymers for separation and purification purposes: Status, challenges and future directions. *Progress in Polymer Science*, 37(12), 1597–1656.

Nasef, M. M., and Hegazy, E. S. A. (2004). Preparation and applications of ion exchange membranes by radiation-induced graft copolymerization of polar monomers onto non-polar films. *Progress in Polymer Science*, 29(6), 499–561.

Nasef, M. M., and Saidi, H. (2002). Post-mortem analysis of radiation grafted fuel cell membrane using X-ray photoelecton spectroscopy. *Journal of New Materials for Electrochemical Systems*, 5(3), 183–190.

Nasef, M. M., Saidi, H., Dahlan, K. Z. M. (2007). Preparation of composite polymer electrolytes by electronbeam-induced grafting: Proton- and lithium ion-conducting membranes. *Nuclear Instruments and Methods in Physics Research, Section B: Beam Interactions with Materials and Atoms*, 265, 168–172.

Nasef, M. M., Saidi, H., Senna, M. M. (2005). Hydrolysis of sucrose by radiation grafted sulfonic acid membranes. *Chemical Engineering Journal*, 108(1), 13–17.

Nasef, M. M., Suppiah, R. R., and Dahlan, K. Z. M. (2004). Preparation of polymer electrolyte membranes for lithium batteries by radiation-induced graft copolymerization. *Solid State Ionics*, 171(3–4), 243–249.

Nasef, M. M., Zubir, N. A., Ismail, A. F., Dahlan, K. Z. M., Saidi, H., Khayet, M. (2006b). Preparation of radiochemically pore-filled polymer electrolyte membranes for direct methanol fuel cells. *Journal of Power Sources*, 156(2), 200–210.

Nasef, M. M., Zubir, N. A., Ismail, A. F., Khayet, M., Saidi, H., Rohani, R., Ngah, T. I. S., and Sulaiman, N. A. (2006a). PSSA pore-filled PVDF membranes by simultaneous electron beam irradiation: Preparation and transport characteristics of protons and methanol. *Journal of Membrane Science*, 268(1), 96–108.

Neffe, A. T., Lange, M. V. R., Braune, S., Luetzow, K., Roch, T., Richau, K., Jung, F., and Lendlein, A. (2013). Poly(ethylene glycol) grafting to poly(ether imide) membranes: Influence on protein adsorption and thrombocyte adhesion. *Macromolecular Bioscience*, 13(12), 1720–1729.

Niemoller, A., Scholz, H., Gotz, B., and Ellinghorst, G. (1988). Radiation-grafted membranes for pervaporation of ethanol/water mixtures. *Journal of Membrane Science*, 36, 385–404.

Ozawa, I., Saito, K., Sugita, K., Sato, K., Akiba, M., and Sugo, T. (2000). High-speed recovery of germanium in a convection-aided mode using functional porous hollow-fiber membranes. *Journal of Chromatography A*, 888(1-2), 43–49.

Patel, P., Cuny, J., Jozefowicz, J., Morel, G., and Neel, J. (1972). Liquid transport through membranes prepared by grafting of polar monomers onto poly(tetrafluoroethylene) films. I. Some fractionations of liquid mixtures by pervaporation. *Journal of Applied Polymer Science*, 16(5), 1061–1076.

Patel, P., Cuny, J., Jozefowicz, J., Morel, G., and Neel, J. (1974). Liquid transport through membranes prepared by grafting of polar monomers onto poly(tetrafluoroethylene) films. II. Some factors determining pervaporation rate and selectivity. *Journal of Applied Polymer Science*, 18(2), 351–364.

Peighambardoust, S. J., Rowshanzamir, S., and Amjadi, M. (2010). Review of the proton exchange membranes for fuel cell applications. *International Journal of Hydrogen Energy*, 35(17), 9349–9384.

Qin, Q., Hou, Z., Lu, X., Bian, X., Chen, L., Shen, L., and Wang, S. (2013). Microfiltration membranes prepared from poly(N-vinyl-2-pyrrolidone) grafted poly(vinylidene fluoride) synthesized by simultaneous irradiation. *Journal of Membrane Science*, 427, 303–310.

Qiu, J. Y., Zhang, J., Chen, J., Peng, J., Xu, I., Zhai, M., Li, J., and Wei, G. S. (2009). Amphoteric ion exchange membrane synthesized by radiation-induced graft copolymerization of styrene and dimethylaminoethyl methacrylate into PVDF film for vanadium redox flow battery applications. *Journal of Membrane Science*, 334(1-2), 9–15.

Radzali, N. A. M., Hidzir, N. M., Mokhtar, A. K., and Rahman, I. A. (2020). ^{60}Co-induced grafting of dual polymer (acrylic acid-co-HEMA) onto expanded poly(tetrafluoroethylene) membranes. *AIP Conference Proceedings*, 2295, 020009.

Rattan, S., and Sehgal, T. (2012). Stimuli-responsive membranes through peroxidation radiation-induced grafting of 2-hydroxyethyl methacrylate (2-HEMA) onto isotactic polypropylene film (IPP). *Journal of Radioanalytical and Nuclear Chemistry*, 293(1), 107–118.

Reddy, P. R. S., Agathian, G., and Kumar, A. (2005). Preparation of strong acid cation-exchange membrane using radiation-induced graft polymerization. *Radiation Physics and Chemistry*, 73(3), 169–174.

Reis, R., Duke, M. C., Tardy, B. L., Oldfield, D., Dagastine, R. R., Orbell, J. D., and Dumee, L. F. (2017). Charge tunable thin-film composite membranes by gamma-ray triggered surface polymerization. *Scientific Reports*, 7, 4426.

Sasagawa, N., Saito, K., Sugita, K., Kumori, S. I., and Sugo, T. (1999). Ionic crosslinking of SO$_3$H-group-containing graft chains helps to capture lysozyme in a permeation mode. *Journal of Chromatography A*, 848(1–2), 161–168.

Scott, K., Taama, W. M., and Argyropoulos, P. (2000). Performance of the direct methanol fuel cell with radiation-grafted polymer membranes. *Journal of Membrane Science*, 171(1), 119–130.

Sekiguchi, K., Serizawa, K., Konishi, S., Saito, K., Furusaki, S., and Sugo, T. (1994). Uranium uptake during permeation of seawater through amidoxime-group-immobilized micropores. *Reactive Polymers*, 23(2–3), 141–145.

Semenova, S. I., Ohya, H., and Soontarapa, K. (1997). Hydrophilic membranes for pervaporation: An analytical review. *Desalination*, 110(3), 251–286.

Shantora, V., and Huang, R. Y. M. (1981). Separation of liquid mixtures by using polymer membranes. III. Grafted poly(vinyl alcohol) membranes in vacuum permeation and dialysis. *Journal of Applied Polymer Science*, 26(10), 3223–3243.

Shen, L., Feng, S., Li, J., Chen, J., Li, F., Lin, H., and Yu, G. (2017). Surface modification of polyvinylidene fluoride (PVDF) membrane via radiation grafting: Novel mechanisms underlying the interesting enhanced membrane performance. *Scientific Reports*, 7, 2721.

Shen, L., Wang, H., Zhang, Y., Li, R., Fabien, B., Yu, G., Lin, H., and Liao, B. Q. (2018). New strategy of grafting hydroxyethyl acrylate (HEA) via γ ray radiation to modify polyvinylidene fluoride (PVDF) membrane: Thermodynamic mechanisms of the improved antifouling performance. *Separation and Purification Technology*, 207, 83–91.

Shen, L., Zhang, Y., Yu, W., Li, R., Wang, M., Gao, Q., Li, J., and Lin, H. (2019). Fabrication of hydrophilic and antibacterial poly(vinylidine fluride) based separation membranes by a novel strategy combining radiation grafting of poly(acrylic acid)(PAA) and electroless nickel plating. *Journal of Colloid Interface Science*, 543, 64–75.

Shinde, A., and Salovey, R. (1985). Irradiation of ultrahigh-molecular-weight polyethylene. *Journal of Polymer Science*, 23(8), 1681–1689.

Shinano, H., Tsuneda, S., Saito, K., Furusaki, S., and Sugo, T. (1993). Ion exchange of lysozyme during permeation across a microporous sulfopropyl-group-containing hollow fiber. *Biotechnology Progress*, 9(2), 193–198.

Shtanko, N. I., Kabanov, V. Y., Apel, P. Y. Yoshida, M., and Vilenskii, A. I. (2000). Preparation of permeability controlled track membranes on the basis of smart polymers. *Journal of Membrane Science*, 179(1-2), 155–161.

Sivaraman, P., Rath, S. K., Hande, V. R., Thakur, A. P., Patri, M., and Samui, A. B. (2006). All-solid-supercapacitor based on polyaniline and sulfonated polymers. *Synthetic Metals*, 156(16-17), 1057–1064.

Sugiyama, S., Tsuneda, S., Saito, K., Furusaki, S., Sugo, T., and Makuuchi, K. (1993). Attachment of sulfonic acid groups to various shapes of polyethylene, polypropylene and polytetrafluoroethylene by radiation-induced graft polymerization. *Reactive Polymers*, 21(3), 187–191.

Tealdo, G. C., Canepa, P., and Munari, S. (1981). Water-ethanol permeation through grafted PTFE membranes. *Journal of Membrane Science*, 9(1–2), 191–196.

Terada, A., Yuasa, A., Tsuneda, S., Hirata, A., Katakai, A., and Tamada, M. (2005). Elucidation of dominant effect on initial bacterial adhesion onto polymer surfaces prepared by radiation-induced graft polymerization. *Colloids and Surfaces B: Biointerfaces*, 43(2), 99–107.

Tsuneda, S., Kagawa, H., Saito, K., and Sugo, T. (1995a). Hydrodynamic evaluation of three-dimensional adsorption of protein to a polymer chain grafted onto a porous substrate. *Journal of Colloid and Interface Science*, 176(1), 95–100.

Tsuneda, S., Saito, K., Furusaki, S., Sugo, T., Ishigaki, I. (1992). Water/acetone permeability of porous hollow-fiber membrane containing diethylamino groups on the grafted polymer branches. *Journal of Membrane Science*, 71(1-2), 1–12.

Tsuneda, S., Saito, K., Furusaki, S., Sugo, T., and Okamoto, J. (1991). Metal collection using chelating hollow fiber membrane. *Journal of Membrane Science*, 58(2), 221–234.

Tsuneda, S., Saito, K., Sugo, T., and Makuuchi, K. (1995b). Protein adsorption characteristics of porous and tentacle anion-exchange membrane prepared by radiation-induced graft polymerization. *Radiation Physics and Chemistry*, 46, 239–245.

Tsuneda, S., Shinano, H., Saito, K., Furusaki, S., and Sugo, T. (1994). Binding of lysozyme onto a cation-exchange microporous membrane containing tentacle-type grafted polymer branches. *Biotechnology Progress*, 10(1), 76–81.

Varcoe, J. R., and Slade, R. C. T. (2006). An electron-beam-grafted ETFE alkaline anion-exchange membrane in metal-cation-free solid-state alkaline fuel cells. *Electrochemistry Communications*, 8(5), 839–843.

Varcoe, J. R., Slade, R. C. T., Yee, E. L. H., Poynton, S. D., Driscoll, D. J. and Apperley, D. C. (2007). Poly (ethylene-*co*-tetrafluoroethylene)-derived radiation-grafted anion-exchange membrane with properties specifically tailored for application in metal-cation-free alkaline polymer electrolyte fuel cells. *Chemistry of Materials*, 19(10), 2686–2693.

Wokaun, A., Rajesh, B., and Thampi, K. R. (2004). Materials for polymer electrolyte fuel cells. *Chimia*, 58, 826.

Wolman, F. J., Maglio, D. G., Grasselli, M., and Cascone, O. (2007). One-step lactoferrin purification from bovine whey and colostrum by affinity membrane chromatography. *Journal of Membrane Science*, 288(1–2), 132–138.

Wu, Y., Kong, Y., Lin, X., Liu, W., and Xu, J. (1992). Surface-modified hydrophilic membranes in membrane distillation. *Journal of Membrane Science*, 72(2), 189–196.

Wu, M., Bao, B., and Chen, J. (2000). Reduction of the thrombogenicity of polyethylene membranes by radiation grafting. *Journal of Radioanalytical and Nuclear Chemistry*, 246(2), 457–461.

Wu, D., Liu, X., Yu, S., Liu, M., and Gao, C. (2010). Modification of aromatic polyamide thin-film composite reverse osmosis membranes by surface coating of thermo responsive copolymers P(NIPAM-co-Am). I. Preparation and characterization. *Journal of Membrane Science*, 352(1), 76–85.

Xu, H. M., Wei, J. F., and Wang, X. L. (2014). Nanofiltration hollow fiber membranes with high charge density prepared by simultaneous electron beam radiation-induced graft polymerization for removal of Cr(VI). *Desalination*, 346(1), 122–130.

Yamashiro, K., Miyoshi, K., Ishihara, R., Umeno, D., Saito, K., Sugo, T., Yamada, S., Fukunaga, H., and Nagai, M. (2007). High-throughput solid-phase extraction of metal ions using an iminodiacetate chelating porous disk prepared by graft polymerization. *Journal of Chromatography A*, 1176, 37–42.

Yang, X., Deng, B., Liu, Z., Shi, L., Bian, X., Yu, M., Li, L., Li, J., and Lu, X. (2010). Microfiltration membranes prepared from acryl amide grafted poly(vinylidene fluoride) powder and their pH sensitive behaviour. *Journal of Membrane Science*, 362(1-2), 298–305.

Yang, X., Zhang, B., Liu, Z., Deng, B., Yu, M., Li, L., Jiang, H., and Li, J. (2011). Preparation of the antifouling microfiltration membranes from poly (N, N-dimethylacrylamide) grafted poly (vinylidene fluoride) (PVDF) powder. *Journal of Materials Chemistry*, 21, 11908–11915.

Yanagishita, H., Arai, J., Sandoh, T., Negishi, H., Kitamoto, D., Ikegami, T., Haraya, K., Idemoto, Y., Koura, N. (2004). Preparation of polyimide composite membranes grafted by electron beam irradiation. *Journal of Membrane Science*, 232(1–2), 93–98.

Yoshimura, K., Koshikawa, H., Yamaki, T., Shisitani, H., Yamamoto, K., Yamaguchi, S., Tanaka, H., and Maekawa, Y. (2014). Imidazolium cation based anion-conducting electrolyte membranes prepared by radiation induced grafting for direct hydrazine hydrate fuel cells. *Journal of Electrochemical Society*, 161(9), F889–F893.

Yuee, F., and Tianyi, S. (1988). Polypropylene dialysis membrane prepared by cobalt-60 gamma-radiation-induced graft copolymerization. *Journal of Membrane Science*, 39(1), 1–9.

Zhao, B., and Brittain, W. J. (2000). Polymer brushes: Surface-immobilized macromolecules. *Progress in Polymer Science*, 25, 677–710.

Zhi-Li, X. (1987). Preparation of NF ion-exchange membranes via radiation-induced grafting and their application. *Desalination*, 62, 259–264.

9 Radiation-grafted ion exchange membranes (RGIEMs) for fuel cell applications

Prachi Singhal, Safiya Nisar, Bharath Govind, and Sunita Rattan*
Amity Institute of Applied Sciences, Amity University, Uttar Pradesh, Noida, India

*Corresponding author: srattan@amity.edu

CONTENTS

9.1   INTRODUCTION

Fuel cells are clean, efficient electrochemical power generating devices, capable of converting chemically bonded energy of fuel (as hydrogen, methanol, etc.) directly into electrical energy that can create a radical shift in the field of electricity generation. Fuel cells can be grouped as solid oxide fuel cells, phosphoric acid fuel cells, molten carbonate fuel cells, alkaline fuel cells, polymeric ion exchange fuel cells based on components used in the cell such as the type of electrolyte, operating temperature, type of fuel, etc. Among them, the development of polymeric ion exchange membrane-based fuel cells (PIEMFCs) has accomplished global interest to come up with a potential source of power for stationary and portable applications. In the last few decades, significant attempts are being made to commercialize PIEMFCs that can achieve DOE's (Department of Energy, US) durability (life span greater than 5000 h) and cost ($45/kW) targets by 2025. The PIEMs are being used as an electrolyte in PIEMFCs, as its key component. Performance of such PIEMFCs depends largely on the physiochemical nature of the developed membranes and their stability in the hostile environment (Yang et al., 2020).

DOI: 10.1201/9781003321910-9

PIEMs are semi-permeable polymeric membranes, delivering two major functions, first acting as a medium to conduct ions through ion conducting channels and second, acting as a barrier for the anodic and cathodic compartment of the fuel cell preventing the connection between the oxidant and the fuel. They are composed of ionic pendant groups attached to polymeric chains and can be categorized into anion exchange membranes (AEMs) and cation/proton-exchange membranes (PEMs). PEMs contain negatively charged groups, such as $-SO_3^-$, $-COO^-$, $-PO_3^{2-}$, $-PO_3H^-$, $-C_6H_4O^-$, etc., fixed onto the membrane backbone and allows the passage of cations only, while AEMs contains positively charged groups, such as $-NH_3^+$, $-NRH_2^+$, $-NR_2H^+$, $-NR_3^+$, $-PR_3^+$, $-SR_2^+$, etc., fixed to the membrane backbone and allows the passage of anions only (Hagesteijn et al., 2018; Strathmann et al., 1995; Ortega et al., 2005). Fuel cells with proton conducting and hydroxide conducting membranes are schematically represented in Figure 9.1.

Most of the reported literature is focused on PEMFCs. The fundamental property of PEMs is the proton conduction, thus, consisting of multiple proton-conducting functionalities, allowing required proton movement from one group to another. Fuel cells based on PEMs capable of making use of hydrogen, methanol, or formic acid by harnessing their chemical energy and converting it to electrical energy (efficiency ~60%), thus addressing the need of the hour for the future of the hydrogen economy. They have a revolutionary potential in powering the automotive, transport industry, portable power batteries/supplies, and stationary power plants for residential homes and buildings, advancement in power generation, and providing transferrable power for our growing personal electronic devices.

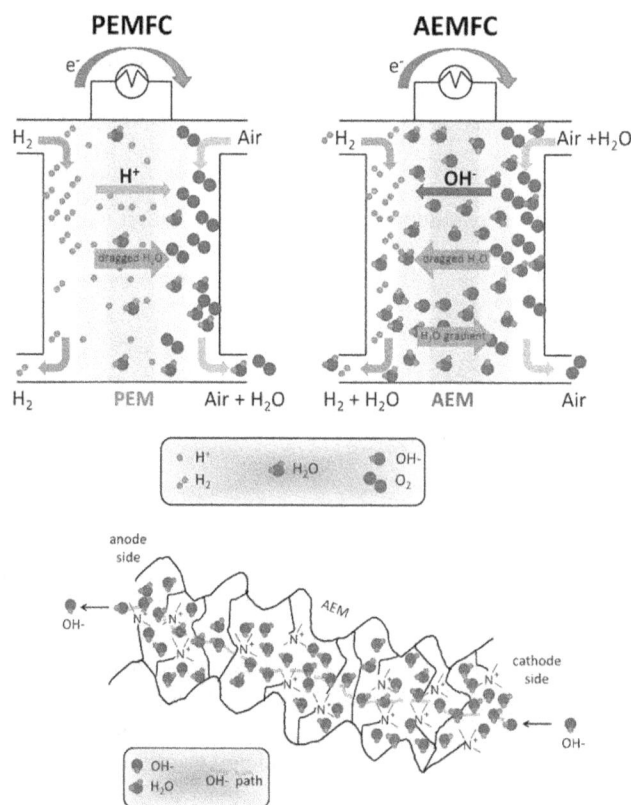

FIGURE 9.1 Schematic of an AEMFC and PEMFC (top scheme), and of an AEM based on a quaternary ammonium pendant functional group (bottom scheme). Reproduced from (Dekel et al., 2018) with permission. copyright @ 2018 Elsevier.

A breakthrough happened in the 1970s in the development of a chemically stable ion-exchange membrane formed, using sulfonated polytetrafluoro-ethylene, commercially known as Nafion, a cation-exchange membrane developed by DuPont. The presence of sulfonic acid groups onto the perfluorinated polymeric backbone makes the membrane proton conductive. Typically, the Nafion membranes have micro/nanophase structures consisting of hydrophobic matrices along with the consistent inter-connected ionic channels that are hydrophilic in nature. The Nafion membranes have many looked-for attributes like high proton conductivity under normal conditions, excellent mechanical stability, and satisfactory chemical stability, thus, making it an excellent candidate and a standard as an ion exchange membrane for IEPMFCs. It offers high proton conductivity of 0.13 S/cm (at 75°C and 100%RH) and durability > 60,000 hrs and good chemical stability (Xu et al., 2008; Alberti et al., 2001). Other top-notch producers have also aimed to develop Nafion-like ionomeric membranes containing perfluorosulphonic acid groups, such as Aciplex® (Asahi Chemical Company) or Flemion® (Asahi Glass), Dow® (Dow Chemical Company) or 3 M® (3 M Company) (Motupally et al., 2000; Peighambardoust et al., 2010).

However, Nafion being the most extensively used proton conducting membrane in PEMFCs, has a lot of disadvantages that require further modification of the membrane or development of some novel products. For ex: it is very costly, ($800–$1,000)/m^2, due to the use of expensive platinum group catalysts and complicated synthetic procedure, the upper temperature limit for the working of this membrane is 80–100°C, exceeding this limit proton conductivity is compromised. The performance of the membrane was further compromised due to relatively low humidity and higher methanol permeability (Casciola et al., 2006; Rikukawa & Sanui et al., 2000). Much research work has been conducted on the possibility of modifying the attributes of Nafion membrane to make its performance more suitable for proton conductivity under elevated temperature and relatively lower humidity conditions with improved chemical and mechanical stability and better water management plans, etc.

During the past few years, many advances have been made in the development of PEMs to overcome the current issues associated with development of PEMs, but there are still techno-economic obstacles in the developed technology related to water management, use of high-cost noble metal catalyst, cooling issues, etc. Much of the research development of PEMs, while promising, may be too far away from commercialization to meet this time frame, and could be the next-generation technology.

In the last few years, the alkaline anion-exchange membranes (AEMs) have drawn increased attention to resolve the challenges and limitations of the conventional proton exchange membranes (PEMs) (Xu et al., 2020). The anion exchange membranes (AEMs) are the crucial components of anion exchange membrane fuel cell (AEMFCs) that offer faster oxygen reduction reaction kinetics under alkaline conditions (instead of acidic conditions in PEMFCs) lowering the activation losses, prospect of acting in an alkaline environment avoids the use of noble platinum group catalysts in fuel cells, permitting the use of non-noble metals as electrocatalysts such as Ni, Co, Fe, etc., lower membrane cost, and cheaper cell components due to less corrosive alkaline environment. Thus, AEMFCs have been considered as the most promising contender to PEMFCs, addressing the cost issues associated with PEMFCs (Wang et al., 2018; Gottesfeld et al., 2018).

Considerable work has been reported on AEMs fabricated using a range of polymers including polystyrene, fluoropolymers, partially fluorinated polymers, polyaromatics, polyimides, poly-ethylene, polypropylene, etc. through halo alkylation of the polymers followed by quaternization with cation pendant groups such as trimethylamine, dimethyl amine, imidazole, guanidine, phosphine, etc. The majority of the work has been conducted on quaternary ammonium poly-electrolyte membranes (Pan et al., 2020. Pasquini and his co-workers (Pasquini et al., 2016) synthesized AEM based on polysulfone modified using trimethylamine as quaternary ammonium groups or 1,4-diazabicyclo [2.2.2] octane (DABCO) as tertiary amines with conductivity obtained in the range of 8-12 mS cm^{-1}. Some other examples of fabrication of such membranes includes the quaternary amination of polysulfones (Yan and Hickner, 2010, Zhou et al., 2009), polyimides, poly-phenylenes (Wang et al., 2009a), polyetherketone, polytetrafluoroethylene (PTFE) and other flouro- and partially fluorinated polymers (Zhang et al., 2010; Xu et al., 2010). Efforts have also been made to use other cationic groups such as imidazolium (Guo et al., 2020), phosphonium (Gu et al., 2009), guanidium (Wang et al., 2010) etc. For example, polysulfone-based AEMs were investigated by Zhang et al. by treating chloromethylatedpolysulfone with imidazole followed by alkalization (Zhang et al., 2011). Some of the developed AEM designs approach the PEMFCs' performance with low loading of precious metal catalysts, but scientific knowledge is still inferior related to AEM durability and short lifetime in comparison to PEMFCs and thus are not suitable to be used at the commercial level. Consequently, the search for new AEM materials with low cost and the required electrochemical characteristics, along with performances matching those of Nafion, is in continuation and has become the most focused research area in the designing of polymer electrolyte fuel cells.

Thus, significant and sincere research attempts are needed for the development of ion exchange membranes (both PEMs and AEMs) with improved material design and more understanding of structural-property relationship. The designed ion exchange membranes (both PEMs and AEMs) should exhibits the following attributes to be eligible for its efficient working in a fuel cell; (1) high conductivity for ions to support and enable high currents with decreased loses due to resistance and negligible electronic conductivity; (2) sufficient stability and high mechanical strength; (3) adequately decreased fuel/oxidant by-pass; (4) stability of the fuel cell under working conditions thermally and chemically; (5) lesser transport of water molecules through diffusion; (6) ability to form membrane electrode assemblies and electro-osmosis (Phadnis et al., 2003). Critical challenges to the efficient design approach of PIEMs are the selection of base materials, optimization of composition between the base polymer and the functionalization, and reaction conditions. For example, it was reported that the number and yield of cationic groups bonded onto the substrate polymer affect the overall property of IEMs (Wang et al., 2009b). Wang et al. studied the effect of temperature and time on the addition of chloromethyl group to the polysulfone as substrate polymer, different quaternization approaches on ionic conductivity of the developed IEM (Wang et al., 2010).

Various synthetic strategies have been reported for the development of IEMs highlighting the relationship between the structure and the resulting properties of the membrane such as functionalization of existing commercial polymers, synthesizing new block polymeric structures through direct polymerization of monomers, heterogeneous polymeric compositions based on blending, reinforcement, or pore filling, using grafting technologies as chemical grafting, plasma grafting, ATRP, and radiation grafting, etc. (Pan et al., 2021; Fan, 2019; Sgreccia et al., 2021; & Suresh et al., 2021). Polymer systems can be directly developed by polymerization of possible monomers followed by the introduction of charge pendant groups, the method provides capability of controlling numbers of pendant groups with controlled crosslinking. Cross-linking methods are used to improve the properties of the developed polymeric system. The sol-gel method is used, which aids in the introduction of inorganic phase into the polymer matrix.

A favorable strategy to design IEMs with required attributes is to fabricate hydrophilic/hydrophobic microphase separation structure in which a range of tailored copolymers have been tethered with cationic or anionic species. The main hydrophobic polymer backbone provides the mechanical stability, gas permeability, and water-repellent character while the hydrophilic part provides ionic mobility, as depicted in Figure 9.2(a). Such heterogeneous polymeric networks can be developed by grafting required functionalization on the base polymeric chain through chemical or irradiation methods (Han et al., 2019; Zhu et al., 2018).

One critical challenge to design such a system is the balance of composition between the base polymer and the grafted part for the optimization of the membrane properties (Wang et al., 2020). As a solution to the designing of ion exchange membranes, radiation grafting is a versatile technique to introduce the desired properties into the material, as such reactions can easily be regulated by adjusting the experimental parameters. This method involves a uniform radical

(a)

Hydrophobic rigid chain

Hydrophilic flexible chain

(b)

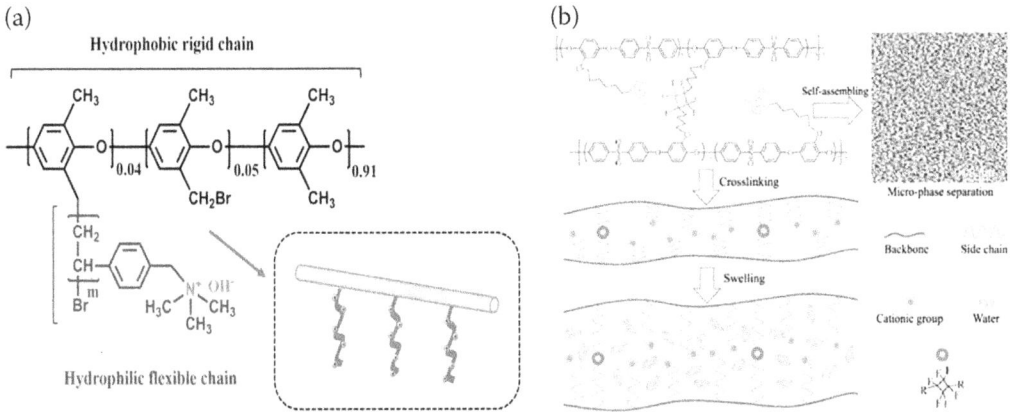

FIGURE 9.2 (a) PPO-*g*-QVBC with hydrophobic base PPO and aminated grafted chain of VBC. Reproduced from (Zhu et al., 2018) with permission. Copyright @2018 Elsevier. (b) Cross-linked anion exchange membrane based on aryl-ether free polymer backbones. Reproduced from (Lin et al., 2017) with Permission.

formation throughout the membrane, without the use of any catalyst or initiator (Nisar et al, 2021; Rattan & Seghal, 2012). Although considerable attempts have been made to develop radiation grafted ion exchange membranes (RGIEMs), but still significant progress is needed in this field related to (a) physiochemical interaction between the base polymer and the monomeric entities, (b) mechanical and chemical stability, (c) membrane cost, (d) optimization with respect to structure and morphological characteristics to get desirable architecture, and (e) conductive mechanism of ion exchange membranes.

9.2   RADIATION-GRAFTED ION EXCHANGE MEMBRANES (RGIEMS)

Radiation-induced grafting is an attractive method to introduce desirable properties into a polymer owing to its simplicity in handling and control over the grafting process. This method offers the promise of polymerization of monomers that are difficult to polymerize by conventional methods without residues of initiators and catalysts (Nisar et al., 2021). Moreover, polymerization can be carried out even at low temperatures, unlike the polymerization with catalysts and initiators. Radiation grafting involves the polymerization of a monomer onto the preformed polymer film, leading to the formation of a branched structure, in which macromolecular chains of the base polymer film act as the backbone and the branches are side chains represented by the grafted moiety.

Some standard methods of radiation grafting are: i) *simultaneous method*, in which the base polymer is irradiated in the presence of the monomer, resulting in the formation of active free radicals in both monomer and base polymer (Zhou et al., 2015); ii) *pre-irradiation method (trapped radicals)*, in which the base polymer is irradiated in vacuum or under an inert atmosphere to generate radicals prior to exposure to the monomer (Gwon et al., 2009); iii) *peroxidation method*, in which the polymer is irradiated in the presence of air or oxygen to form mainly hydroperoxides that are decomposed at elevated temperature to initiate the grafting reaction (Willson et al., 2019); and iv) *ATRP (atom transfer radical polymerization)* is a recent method of polymer modification that helps in controlling the radical polymerization. Radiation-induced ATRP is one of the prominent techniques that involve a fast dynamic equilibrium between active radical species with other dormant species that in turn provide control over the polymerization using radiation like plasma (Zhao et al., 2019; Feng et al., 2016). Different methods of radiation grafting are schematically represented in Figure 9.3(a).

One of the important properties of RGIEMs includes degree of grafting (DG), defined as the percentage of successfully grafted functional groups onto the polymeric base chain that influences

(a)

(b)

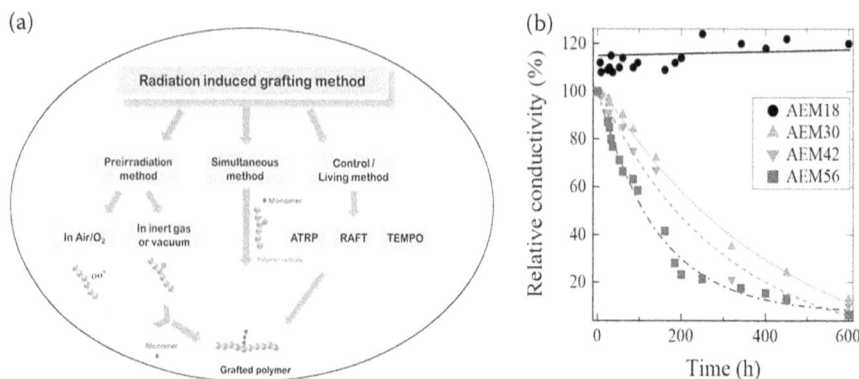

FIGURE 9.3 (a) Different methods of radiation-induced frafting onto polymers to fabricate ion exchange membranes; (b) time-dependence of relative OH conductivity (σ_t/σ_0) of AEMs with different IECs after immersion in a 1 M KOH solution at 80°C. Reproduced from (Zhao et al., 2020) with permission.

the final characteristics of IEMs for fuel cell application. DG can be estimated from the difference in the mass of grafted and ungrafted membranes (Espiritu et al., 2016; Sehgal & Rattan, 2010).

Relative ion exchange capacity, ionic conductivity, and water uptake properties of the developed IEMs depend significantly on DGs, which in turn depends on the number of factors discussed below including radiation source, radiation dose, dose rate, type of solvent, choice of base polymers, monomers or ionic pendant groups, etc. The process of radiation grafting offers huge possibilities for the design of the final IEMs architecture by careful variation of all these factors. For example, in a very recent study, a series of p-(2-imidazoliumyl) styrene-g-ETFE AEMs are prepared and named as AEM 18, AEM30, AEM42, and AEM56 based on a different DG that shows different relative conductivities, as shown in Figure 9.3(b) (Zhao et al., 2020).

a. *Radiation source* – The polymerization reaction is initiated by radicals generated by irradiation of the base polymer. Different electromagnetic radiation sources used for inducing radiation grafting includes gamma rays, X-rays, ultraviolet rays, charged particles as electrons, etc. Energetic electron beams or γ-rays are the main ionizing radiations used for grafting purposes, with average linear energy transfer (LET) of 1 MeV, and [60]Co γ-ray around 0.3 keVμm[−1]. This amount of LET results in homogeneous deposition of energy in the irradiated polymer. Still, the choice of radiation source depends on the target properties and nature of the material used (Golubenko et al., 2020).

b. *Type of solvents* – The graft copolymerization reaction is carried out by bringing the activated base polymer film in contact with the monomer in liquid or vapor form. According to the *grafting front mechanism*, grafting occurs initially at the surface of the film, then proceeds gradually inwards as the grafting zone is swollen by the monomer (Nasef, 2001). The type of solvent and the composition of the monomer/solvent mixture may influence the grafting kinetics, the length of grafted chains, and polymer microstructure. Benzene, toluene, and dichloromethane, and alcohols, in particular methanol, ethanol, or propanol, have been used as solvents for radiation grafting. It was found that polar solvents are generally favorable solvents for pre-irradiation grafting.

c. *Radiation dose, dose rate* – Radiation dose and dose rate influence grafting process significantly with increased density of generated radical sites with increased radiation dose. However, high irradiation doses may harm the polymeric main chain structure leading to loss in mechanical properties of the membranes. Thus, dose rate has to be optimized, varying different parameters of the reaction so as to obtain the maximum conductivity, ion exchange capacity, and mechanical properties of the membranes. Zhao and his co-workers (Zhao et al.,

2020) fabricated styrene and imidazolium grafted ETFE membrane with 50 kGy as total γ-ray dose at dose rate of 10 KGy h^{-1} with optimized grafting percentage of 18% at 72 min reaction time, exhibiting IEC as 0.54 mmolg^{-1} and ionic conductivity > 50 mS cm^{-1}. Wang et al. optimized the irradiation dose to 30 kGy by moderating the VBC concentration and changing the solvent to attain a high IEC of 2.01 mmol g^{-1} along with balanced overall mechanical properties including Young's modulus, elongation at break, etc.

  d. *Temperature* – Grafting temperature is significant as an increased mobility of polymeric chains above glass transition temperature, results in recombination of chains and decreased formation of radical sites, with negative effect on grafting.

  e. *Concentration of monomers, solvent amount, time of grafting reaction* – The reaction parameters such as time of reaction, solvent amount, monomer concentration are optimized by carrying out the grafting at varying concentrations until it approaches a maximum percentage. For example, while optimizing for monomer concentration, grafting is carried out at varied monomer concentrations. Initially, enough free radical sites are available on protein backbone to completely occupy the increasing amount of monomers added. But, as the concentration of monomer increases, homo-polymerization increases due to increased competition for free radical sites between monomer moieties, resulting in decreased grafting percentage. Thus, with a limited number of free radical sites available on the protein backbone, grafting, once it attains a maximum value, decreases thereafter with preferential homopolymer formation at monomer concentration beyond an optimum value.

  f. *Cross-linkers* – Cross-linkers are used in conjunction with the monomers to achieve certain desirable properties in a grafted membrane. Cross-linking agents such as divinylbenzene (DVB) may be introduced into the grafting solution. These are monomers with two or more vinyl groups, such that the grafted polymer chains are bridged and linked together, because of which the tensile modulus and strength of the material increases (Gao et al., 2020). For example, higher 'structural density' of cross-linked membranes is reported with significantly improved chemical stability of FEP- and ETFE-based membranes. However, grafting yield is affected by concentration of cross-linkers, low concentration of the cross-linker increases the degree of grafting due to enhanced number of branching reactions, while high concentration of the cross-linker causes a suppression in the swelling of the graft, an increase in viscosity of the grafting solution and decrease in the diffusion and availability of the monomer due to the formation of network structure. Thus, a balance cross-linking is required in functional IEMs (Lin et al., 2017). A cross-linked anion exchange membrane based on aryl-ether free polymer backbones are represented in Figure 9.2(b).

  g. *Choice of base polymer and monomers and ionic pendant groups:* Different polymers can be used as base material for ion exchange membranes that are then functionalized through radiation grafting using suitable monomers with cationic groups such as ammonium, phosphonium, imidazolium, guanidine, etc. or anionic groups such as sulphonic acid group, carboxylic group, phosphate group, etc. Using radiation grafting techniques, these groups can be uniformly distributed throughout the polymer matrix to get ion exchange membranes with the desired architecture.

9.3 SELECTION OF BASE MATERIALS (POLYMERIC BACKBONE, GRAFTED MONOMERIC MOIETY WITH CATIONIC PENDANT GROUP)

The selection of base materials for the fabrication IEMs for their application in polymeric fuel cells depends on the following basic requirement:

  a. Uniform distribution of hydroxide ion (H^{+}/OH^{-}) conducting groups on the entire polymeric membrane.

b. Fuel crossover across the membrane should be minimum.

c. Mechanical, thermal, and chemical stability of IEMs at required hydration and temperature.

9.3.1 TYPES OF POLYMERIC BACKBONES USED FOR FABRICATION OF IEMs

The choice of polymeric membrane base material plays a very important role in the fabrication of IEMs with required attributes. Required characteristics for the selection of base polymer includes (a) conducting polymeric backbone along with mechanical strength for which choices are restricted, (b) covalent attachment of monomeric units on the polymeric backbone depends on the number of active sites on the polymeric backbone which provides the final ionic conductivity to the fabricated IEMs, (c) chemical nature of the polymeric backbone and its compatibility with the attached monomer and ionic pendant groups, which will be responsible for the chemical and thermal stability of the fabricated IEMs. The degradation of IEMs is the prominent factor that must be considered during their fabrication as this decreases their lifetime and is one of the limiting factors in the development of commercial RGIEM fuel cells.

Different polymeric backbones that have been explored for the synthesis of RGPEMs and RGAEMs that can be broadly classified into fluorinated polymers, partially fluorinated polymers, and non-fluorinated polymeric backbones including both aliphatic and aromatic structures such as polyphenylene oxide (PPO), polystyrene (PS), polyether sulfones (PES), polyimides (PI), polyether ketone (PEK), polyvinyl alcohol (PVA), etc.

9.3.1.1 Fully fluorinated polymer-based RGPEMs and RGAEMs

Fully fluorinated polymeric-based backbones such as poly tetrafluoro ethylene (PTFE), poly tetrafluoro ethylene-co-hexafluoropropylene (FEP), etc. are one of the important classes of polymers that have been used for the fabrication of RGAEMs due to their base-resistant nature and thermal stability. Liu et al. (Liu et al., 2011) carried out radiation-induced graft copolymerization of vinyl benzyl chloride (VBC) onto polytetrafluoroethylene-co-perfluoropropylvinyl ether (PFA) membranes that have been quaternized and alkalinized, resulting into AEM with maximum ionic conductivity and power density as 0.05 Scm^{-1} and 16 $mWcm^{-2}$, respectively, at 60°C, with degree of grafting (DG) of 42.3% using PtRu/C and MnO_2 as electrodes and methanol as direct fuel, as shown in Figure 9.4(a).

VBC grafted fully fluorinated FEP poly (tetrafluoroethylene-co-hexafluoropropylene) AEM membranes are also reported (Varcoe et al., 2007). The membranes are comparatively stable and show conductivity values up to 23 mS cm^{-1} at 50°C. It was indicated that such types of AEMs are suitable for fuel cell application below 60°C (thermal stability limitation temp) (Figure 9.4(b)).

Fluorinated polymer-based AEMs pose some synthesis related disadvantages such as irradiation of fluorinated polymers suffers from main chain scission due to stronger C-F bond in comparison to C-C bond; in addition to that, irradiation-generated radical sites are not stable and decay with time that can affect the grafting degree.

9.3.1.2 Partially fluorinated polymer-based RGAEMs and RGPEMs

To overcome the above-mentioned shortcomings associated with fluorinated polymer-based membranes, partially fluorinated polymers have also been employed for the development of RGPEMs and RGAEMs. Partially fluorinated polymers such as polyethylene tetrafluoroethylene (ETFE), poly (vinylidene fluoride-co-hexapropylene) (PVDF-co-HFP), poly (vinyl fluoride), etc. can be used for the fabrication of AEMs.

Felix N. Bijch et al. prepared membranes by pre-irradiation grafting of styrene/cross-linker mixtures onto 10*10 cm^{-2} FEP-films (Du Pont) and their subsequent sulfonation yielding grafted copolymer membranes consisting of poly(tetrafluoroethylene-co-hexafluoropropylene) and sulfonated polystyrene (Bai et al., 2019). Linfan Li et al. designed a novel approach, involving the preparation of poly(styrene-sulfonic acid) grafted poly(vinylidene fluoride) (PVDF-g-PSSA) membrane

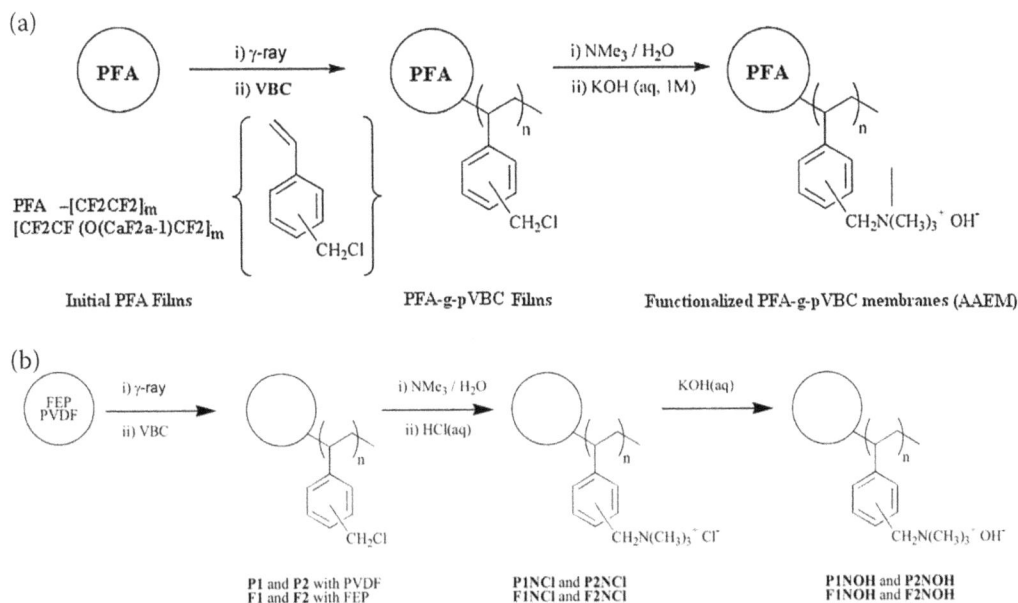

FIGURE 9.4 (a) Radiation-induced graft copolymerization of vinyl benzyl chloride (VBC) on polytetrafluoroethylene-co-perfluoropropylvinyl ether (PFA) membranes. Reproduced from (Liu et al., 2011) with permission. (b) Radiation grafting of VBC onto PVDF and FEP, followed by amination. Reproduced from (**Danks et al., 2002**) with permission.

using successive graft copolymerization of styrene onto PVDF. The IEC (ion exchange capacity) value and proton conductivity of the PEM obtained with a DG of 21.8% is as high as 1.41 meq/g and 10^{-3} S/cm, respectively, whereas the IEC value of Nafion® NRE-212 is 0.91 meq/g. The water uptake of the membrane was too small (Li et al., 2010). ETFE-based AEMs were prepared by Varcoe and his co-workers (Varcoe et al., 2007) through irradiation using a ^{60}Co-radiation source and is functionalized using chloromethyl styrene (CBS) followed by quaternization, resulting into AEMs with IEC reported as 0.92 meqg^{-1} and power density as 110 mWcm^{-2} in H_2/O_2 alkaline fuel cell.

Danks et al. (Danks et al., 2003) reported PVDF radiation-grafted AEM with tri methyl amine (NMe$_3$) as quaternary exchange sites that were found to be brittle in nature due to dehydrofluorination of the backbone polymer. The maximum conductivity of 20–30 mS cm^{-1} was observed at a temperature of 50°C and at 25% DG. PVDF-g-PVBC and FEP-g-PVBC were also synthesized by Danks et al. (Danks et al., 2002) with subsequent amination with NMe$_3$ and ion exchange using KOH, which showed an IEC of 0.7 meq g^{-1} and 1.0 meq g^{-1}, respectively. PVDF-based AEM showed structural instability on treatment with alkali while FEP-based AEMs were analyzed to be thermally stable and resulted in conductivity of 0.01–0.02 S m^{-1} at ambient temperatures. Grafting of chloromethyl styrene (CMS) onto FEP and ETFE also showed promising results (Büchi et al., 1995); schematically, it is represented in Figure 9.5(a).

Partially fluorinated polymers also degraded easily under basic conditions by elimination of HF and formation of C=C bond, leading to main chain scission.

9.3.1.3 Non-fluorinated polymer-based RGAEMs and RGPEMs

Due to disadvantages exhibited by fluorinated and partially fluorinated polymer-based RG IEMs, some non-fluorinated polymeric networks have also been explored for fuel cell applications. Some of the examples of non-fluorinated polymers used for the fabrication of RGPEMs and RGAEMs are given below.

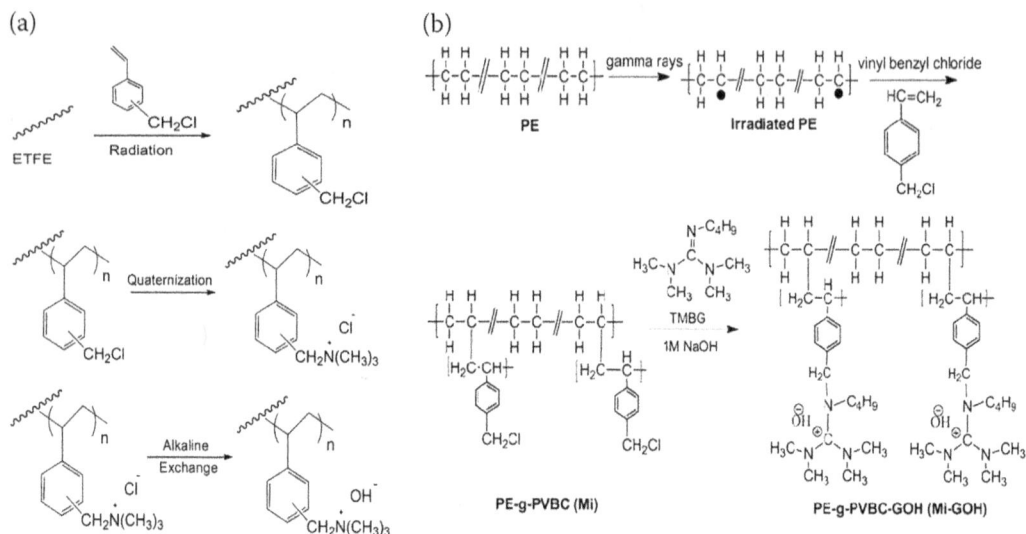

FIGURE 9.5 (a) Preparation of ETFE-g-CMS radiation-grafted membranes. Subsequent amination and ion exchange yields alkaline anion-exchange membranes. Reproduced from (Zhou et al., 2015) with permission. (b) Irradiation grafting of VBC onto non-flourinated polyethylene backbone for fabrication of AEMs. Reproduced from (Sherazi et al., 2015) with permission.

RG PE-g-PSSA (polyethylene-g-polystyrene sulphonic acid) PEM developed with 18% DG and 0.9 mmol/g and 75 mS/cm as ion exchange capacity and conductivity, respectively, better than Nafion but its chemical stability is compromised [56]. Polystyrene has been shown to be stable under high pH conditions. Good polymer backbone alkaline stability is also observed in non-fluorinated polyolefins and polyaromatic membranes. Polyphenylene backbone has attracted much attention due to elimination of weak aryl ether bonds.

Gublerand his co-workers (Gubler et al., 2018) tried polyarylene block copolymer membranes as an option for fluorine-free membranes. In this, polystyrene-grafted membranes cross-linked with DVB were found to have a considerably lower surface energy compared to perfluorinated membranes of the Nafion® type, at all water content values investigated. The acid functionalized polymers exhibit acceptable thermal stability, i.e., a 5% weight loss occurring between 310°C and 324°C in TGA under synthetic air. The mechanical properties of these polymers indicate ductile mechanical behavior with an elongation at a break up to 230%. The water uptakes of films cast from the polymers are between 39 and 49 wt%, which corresponds to 8.8 to 10.0 molecules of H_2O per sulfonic acid group. In practice, however, these films are difficult to handle as solid electrolytes for fuel cell applications (Gubler et al., 2006).

Huijuan Bai et al. (Bai et al., 2019) attempted the grafting of poly(1-vinylimidazole) on polysulfone ((P-g-V-x) for fuel cell applications-ATRP polymerization. P-g-V-3.82/PA shows the highest proton conductivity (127 mS cm^{-1} at 160°C); thus, a single H_2-O_2 HT-PEMFC-V-3.82/PA shows the highest proton conductivity (127 mS cm^{-1} at 160°C). Significantly, the tensile strength of P-g-V-3.82/PA remains at 7.94 MPa and it showed almost no noticeable degradation during a test period of 200 h (150°C). The tensile strengths (mechanical properties) of the P-g-V-x/PA membranes are greater than 7.9 MPa, and this meets the requirement for practical applications (Miyata et al., 2005).

Filho and Gomes (Filho, & Gomes, 2012) reported a composite sulfonated bisphenol-A-polysulfone (SPSF)–based PEMs, containing tungstophosphoric acid (TPA) and modified using electron beam (EB) irradiation and thermal treatment showed similarities with Nafion 117 commercial PEM. Especially the 58.6% sulfonated membrane they made at 100 kGy showed an IEC of 1.199 meqg^{-1}, proton conductivity of 41.4 mS cm^{-1} at 30°C, and 59.7 mS cm^{-1} at 80°C and it had good stability.

The main advantages of using PE as a polymeric backbone include low cost, excellent physio-chemical properties, stability under basic conditions, etc. (Sherazi et al., 2013). Low-density and high-density polyethylene films are being used for the fabrication of AEMs. For example, irradiated polyethylene was used to fabricate AEM by grafting glycidyl methacrylate (GMA) and DVB. Radiation-grafted LDPE-based AEMs provide high ion conductivities and exhibit fast water transport but with limiting mechanical properties. To overcome the limiting aspect of mechanical strength of LDPE-based AEMs, use of HDPE-based radiation-grafted AEM is proposed. Sproll et al. (Sproll et al., 2016) reported radiation-grafted HDPE-based AEMs using low dose rates but that resulted in poor performance due to lower intensity of grafting. A comparison study between LDPE-based and HDPE-based RG AEM membrane was also carried out by Wang et al. Both represent comparable properties w.r.t. ion exchange capacity, ionic conductivity, etc. but HDPE-based membranes showed better mechanical properties, stability, and fuel cell performance (Wang et al. 2019).

Sherazi et al. (Sherazi et al., 2015) focused on radiation-grafted ultra-high molecular weight polyethylene-based AEMs fabrication followed by quaternization and alkalization, represented in Figure 9.5(b). The fabricated membrane exhibited maximum ionic conductivity as 47.5 mS cm^{-1} at 90°C using methanol as a fuel in an alkaline fuel cell. However, the ionic conductivities obtained are still considerably lower than that obtained with commercial Nafion-based PEMs.

Several aliphatic and aromatic polymer-based PEMs and AEMs were also explored such as poly (ether-imide) (PEI), polysulfone (PS), poly (arylene ether sulfone) (PESF), etc. Stoica et al. fabricated poly (epichlorohydrin-allylglycidyl ether)–based AEM (Stoica et al., 2007), poly vinyl alcohol (PVA)–based AEM explored by Nikolic et al. (Nikolic et al., 2011) cross-linked using gamma radiations, etc.

9.3.2 MODIFICATION OF POLYMERIC BACKBONE TO RGPEMs AND RGAEMs (INTRODUCING IONIC GROUPS)

Modification can be done using different ways: grafting of the main polymer base using a potential monomer is the primary functionalization and then to make it specific towards RGAEMs or RGPEMs, which depends on the secondary functionalization using ionic moieties. Introduction of ionic groups has also led to high proton conductivity in RGPEMs and high anion conductivity in RGAEMs, and stability in the aggressive hydrolytic environments of a working fuel cell. Performances of RGAEMS and RGAEMs for fuel cell applications have been summarized in Table 9.1 and Table 9.2.

RGPEMs are usually developed with charged groups attached to the polymer backbone, including the sulfonic group (-SO$_3$H), carboxyl group (-COOH), phosphate group (-PO$_3$H$_2$), and orthophosphorous acid group (-PO$_2$H$_2$), etc. Mostly, membranes with sulfonic groups are studied widely among all kinds of functional groups for their strongest acidic proton exchange ability (Esmaeili et al., 2019; Nasef et al., 2009). Whereas, AEMs polymer backbones stand with nitrogen-containing groups such as quaternary ammonium/tertiary diamines, imidazolium, guanidinium, pyridinium, and nitrogen groups like phosphonium and sulphonium (Sana et al., 2021).

Immobilization of various anionic and cationic head groups as side chains or directly in the polymeric backbones, creates conduction channels to transport cations or anions through the PEMs of AEMs.

9.3.2.1 Introducing anionic pendant groups as cation exchange sites on the PEMs

In PEMs, the most commonly used cation exchange pendant groups include carboxylic acid groups (COOH), sulfonic acid (-SO$_3$H), and phosphoric acid (H$_3$PO$_4$). These acidic groups dissociate in the presence of water, providing mobile protons, responsible for the ionic conductivity of the membrane.

TABLE 9.1

Properties of RGAEMs for Fuel Cell Applications

AEM Membrane	Type of Fuel Cell	Radiation Source and Degree of Grafting	Performance (IEC), Power Density, Ionic Conductivity
3,4-VBC-g-ETFE	H_2/O_2 AEMFC	Electron Beam (20-100 kGy)	Peak performance of 1,350 mW cm^{-2} at 70°C. (Biancolli et al., 2022)
VBC-g-ETFE (aminated with aq. Trimethylamine)	H_2/O_2 AEMFC	Electron Beam 30 kGy (4.5 MeV) DG = 70%	IEC = 2.11 mmol g^{-1} at 70°C; Peak power density of 1.16 W cm^{-2} at 60°C. (Wang et al., 2017a)
ETFE-g-poly(vinylbenzyl-N-methylpyrrolidiunium)	H_2/O_2 AEMFC At 60°C	Electron irradiated (70 kGy)	IEC = 1.66 mmol g^{-1}; Peak power density = 800 mW cm^{-2} (Ponce-González et al., 2016)
ETFE-g-1-(-4-chlorobutyl)-4-vinylbenzene (with butyl spacer)	H_2/O_2 AEMFC at 60°C	Radiation dose 40 kGy DG = 66%	Peak power increased to 1.12 W cm^{-2} at 70°C. (Ponce-González et al., 2018)
LDPE-g-VBC (aminated with benzyl trimethyl ammonium)	–	Gamma ray with radiation dose = 20 kGy DG = 65.6%	IEC = 2.3 mmol g^{-1} (Espiritu et al., 2018)
ETFE-g-VBC RG AEMs	H_2/O_2 Cell	DG = 23.6%	Maximum Power density= 94 mWcm^{-2}; Ion exchange capacity (IEC) = 1.03 e^{-3} molg^{-1} (Varcoe et al., 2007)
PVDF-VBC (aminated with TMA)	Methanol fuel cell	Gamma rays Total dose = 6.3 Mrad DG = 54%	Low IEC. (Danks et al., 2003)
LDPE-RGAEMs	H_2/O_2 AEMFC	–	IEC = 2.54 +_021 mmol g^{-1}; OH- ion conductivity at 80°C (RH = 100%) = 208 ± 6 mS cm^{-1} (Wang et al., 2017b)
HDPE – RG AEMs	H_2/O_2 AEMFC	–	IEC = 2.44 ± 0.04 mmol g^{-1}; OH- ion conductivity at 80°C (RH = 100%) = 214 ± 2 mS cm^{-1} (Wang et al., 2019)
Stlm-RG-ETFE AEMs	Hydrazine-Hydrate fuel	–	IEC = 0.54 mmol g^{-1}; Ion Conductivity => 50 mS cm^{-1}; Low water uptake < 10%; Alkaline stability over 600 h in 1M KOH solution at 80°C. (Zhao et al., 2020)

ETFE-g-VBC (crosslinked with DABCO and amination with triethylamine)	H_2/O_2 AEMFC At 40°C	60Co gamma ray source (90 kGy at dose rate of 4 kGy h^{-1}) DG = 41.6%	48 mW cm^{-2} at current density of 69 mA cm^{-2} (Fang et al., 2012)
VBC-g-Poly (ethylene-co-tetrafluoroethylene) functionalized with DABCO, DCX, TMA	–	γ-ray Irradiation Dose = 90 kGy	Ionic conductivity = 74 mS cm^{-1} at 80°C. (Wang et al., 2017a)
LDPE-g-VBC (aminated with TMA, MPY and MPIP)	–	Electron beam (100 kGy)	IEC = 2.68 mmol g^{-1} (TMA) IEC = 2.40 mmol g^{-1} (MPY) IEC = 2.35 mmol g^{-1} (MPIP) (Meek et al., 2020)
Styrene and 2-methyl-N-vinylimidazole grafted onto ETFE	–	Gamma (^{60}Co γ-ray Source with 80 kGy irradiation dose)	Ionic conductivity = 251 mS cm^{-1}. (Dang and Jannasch, 2015)
Hexafluoropropylene grafted with 4-VP	Methanol fuel cell	Plasma treatment	Maximum Power density = 44 mW/cm^2 (Sudoh et al., 2011)
ETFE-g-VBC (aminated with TMA)	–	Electron beam (Total dose = 7 MRad)	IEC = 20-40 mS cm^{-1} (Kizewski et al., 2013)
LDPE-co-VBC (triethylamineaminated)	–	Gamma irradiated DG = 26%	DG = 26% Peak power Density = 478 mW cm^{-2} (Mamlouk and Scott, 2012)

TABLE 9.2
Properties of RGPEMs for Fuel Cell Applications

PEMs	Type of Fuel Cell	Radiation Source and Degree of Grafting	Performance (IEC), Power Density, Ionic Conductivity
ETFE-g-P1-Vlm/4VP	H$_2$/O$_2$ Cell	Gamma radiation	Power density of 237 mW cm^{-2}. (Rajabalizadeh Mojarradet al., 2019)
Styrene-g-AN/Styrene-g-MAN	H$_2$/O$_2$ Cell	Electron beam	(Jetsrisuparb et al., 2014)
Polystyrene-g-polymethylpentene	–	UV	IEC=GCM-1 (2.9 ± 0.1 meq/g) GCM-2 (1.91 ± 0.06 meq/g) (Golubenko et al., 2021)
Styrene grafted PVA/SiO2 (PVA/SiO2-g-PSS)	–	Gamma; Degree of grafting =40%	IEC = 1.3 (meq/g), proton conductivity = 70 mS/cm (Awad et al., 2021)
FEP-g-PSSA (sulphonated with chlorosulphonic acid)	Methanol fuel cell	–	DG = 27.48%; Peak Power Density = 0.896 W cm^{-2}; Gamma rays (Dose 10–60 kGy min^{-1}) (Li et al., 2022)
PBI doped with phosphoric acid	H$_2$/O$_2$ Cell Pt/C electrodes	Electron Beam	Peak power density of 934 mW cm^{-2}. (Lai et al., 2016)
Styrene -co-acrylic acid (sulphonated)	DMFC	Gamma Radiation	Gamma radiation (10–100 kGy) (Benavides et al., 2019)
Nafion membranes irradiated (PFA-g-PSSA) membranes	DMFC Methanol fuel cell	UV radiation (Dose = 198 mJ cm^{-2}); Gamma radiation; DG = 79.3%	Peak power density = 27.39 mW cm^{-2} (Rao et al., 2019); IEC = 2.78 meq/g; Peak power density – 123 mW cm^{-2} (Kang et al., 2013)
PVDF-g-PSSA	H$_2$/O$_2$ fuel cell (60°C)	Gamma radiation; Radiation dose = 50 kGy; DG = 35%	Peak power density = 250 mW/cm^2 (Sadeghi et al., 2018)

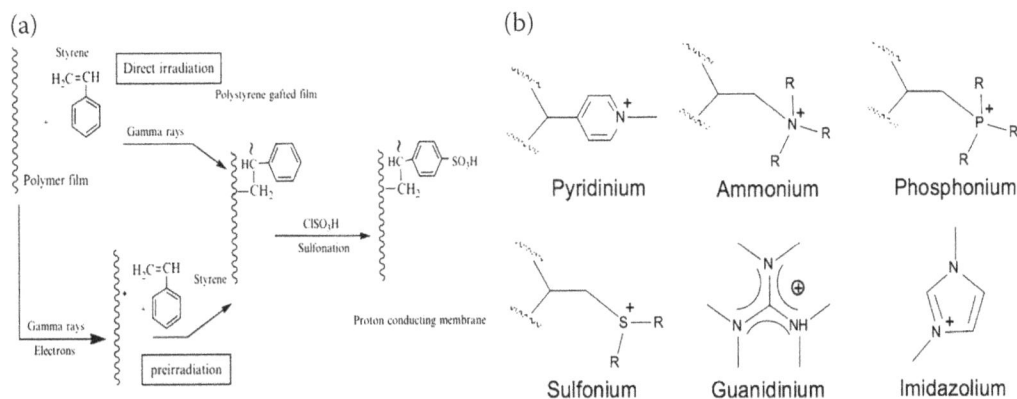

FIGURE 9.6 (a) PEM preparation by direct or pre-irradiation grafting of styrene onto base polymeric chain followed by sulfonation. Reproduced from (Nasef et al., 2009) with permission. (b) Different cationic pendant groups used for the Fabrication of AEMs. Reproduced from (Zhou et al., 2015) with permission.

9.3.2.1.1 Sulfonation

Sulfonation is generally carried out using chlorosulfonic acid, sulfonyl chloride, or concentrated sulfuric acid diluted in inert solvents such as dichloromethane, tetrachloroethane, or carbon tetrachloride, followed by hydrolysis and re-acidification to obtain the $-SO_3H$ functional group. PEMs are commonly prepared by grafting of styrene or its substituents onto polymer films followed by sulfonation reaction as shown in Figure 9.6(a). Sulfonation can be performed on the non-fluorinated polymeric chains such as polystyrene [PS], polyphenylene oxide [PPO], polysulfone [PS], polyether ether ketone [PEEK], polybenzimidazole [PBI], etc.

Sulfonates are the most used proton-conducting groups as its dissociation constant is nearly one, more than the acidic character of other groups as carboxylic group. Li et al. fabricated a FEP-g-PSSA membrane and sulfonated using chlorosulphonic acid exhibited 27.48% degree of grafting, 130.1 mS cm^{-1} proton conductivity, water uptake of 36.1±1.8%, IEC of 1.08±0.02 mmol·g^{-1}, and especially the mass transfer polarization of FEP-g-PSSA reaches up to 0.204 V, far higher than Nafion® 211 (0.084 V) (Li et al., 2022).

Awad et al. (Awad et al., 2021) made PVA/SiO2-g-PSS using gamma irradiation and sulfonated via immersing the grafted film in chlorosulphonic acid. The highest value of grafting obtained was 40% in 3 h of irradiation time. The maximum value of IEC was 1.3 (meq/g), which shows that PVA/SiO$_2$-g-PSS has high protonic form than Nafion and proton conductivity was 70 mS/cm, which is greater than Nafion. Hao et al. explored grafting of styrene onto ETFE and then followed that by sulfonation with chlorosulfonic acid. They found that the membrane surfaces had less accumulation of PSSA grafts. Moreover, among all possible sulfonation products, the sulfur species originated from sulfonic acid groups' predominance.

9.3.2.1.2 Carboxylation

Carboxylic acid group can be added to the base polymeric film through direct grafting of allyl monomers such as AA or MAA, or grafting epoxy acrylate monomers such as glycidyl acrylate and then converting the epoxy groups into carboxylic acid groups. Literature reported the preparation of fluoropolymer film-based PEMs, developed through co-grafting of styrene and maleic anhydride [MA] with two carboxylic groups obtained after hydrolysis, resulting in the high ion-exchange capacity of the grafted film. Phadnis et al. reported radiation grafting of styrene and acrylic acid onto pre-irradiated FEP films followed by sulfonation simultaneously to both carboxylic and sulfonated moieties (Wang et al, 2009b).

9.3.2.1.3   Phosphorylation

The sulfonic acid functionality has been dominating the ion-exchange component of virtually all polymeric materials that have been developed and commercialized for application as membranes in proton exchange membrane fuel cells (PEMFC). Nowadays, phosphoric acid functionalized polymers have been preferred as potential candidates to augment the arsenal of available materials for use in PEMFC.

Mojarradand his co-workers (Mojarred et al., 2020) made HT-PEM by doing metal-salt-enhanced grafting of vinylpyridine and vinylimidazole combinations onto the ETFE film followed by protonation with phosphoric acid. The membranes showed good proton conductivity. One of the prepared membranes showed 80 mS/cm conductivity at 110°C in 60% relative humidity and excellent thermal stability at 300°C.

9.3.2.2   Introducing cation pendant groups as anion exchange sites on the AEMs

Conventionally, the polymeric backbone used for fabrication of AEMs was modified following a sequential process including polymerization, chloromethylation and amination. The process involves the use of chloromethyl ether as the chloromethylation agent, which proves to be hazardous. Thereof, as an alternative, vinyl benzene chloride (VBC) or vinyl pyridine (VP) were copolymerized onto a base polymer as the halo substituted monomers to which quaternary ammonium exchange sites are introduced through amination using trimethylamine (NMe3), 1,4-diazobicyclo [2,2,2] octane (DABCO), etc. Radiation grafting of VBC onto polymeric chains is one of the frequently used grafting monomers that is an effective way to avoid the use of chloromethyl ether.

Different ionic and neutral monomers have been reported in the literature that can be radiation grafted onto polymeric membranes to fabricate AEMs (Gopi et al., 2014; Varcoe, 2007; Herman et al., 2003). Further, a range of cation head groups has been studied that work as anion exchange sites on AEMs such as quaternary ammonium, imidazolium, guanidinium, pyridinium, tertiary sulfonium, pyrrolidinium, imidazolium and phosphonium or metal ions, etc. (shown in Figure 9.6(b)) with alkyl ammonium ions being the most commonly studied anion exchange sites. The chemical structure of the cationic *head groups* attached to AEM polymeric structure greatly influences the various physico-chemical properties of the AEMs such as conductivity, stability, mechanical strength, water uptake, etc.

9.3.2.2.1   Quaternary ammonium group

Quaternary ammonium groups have been the most extensively reported cationic exchange head groups attached to the polymeric chains for the fabrication of AEMs as they can be easily functionalized and show high alkaline stability and hydroxide conductivity.

For example, AEMs based on poly (ETFE-g-VBC) were reported, fabricated by radiation grafting of VBC onto ETFE in the presence of chloroform as solvent using gamma irradiation followed by quaternization using DABCO (Ko et al., 2012). Samaniego and co-workers (Samaniego et al., 2021) designed novel gamma radiation–grafted cellulose acetate-based anion exchange membranes (CA-AEM) functionalized through TMA, as represented in Figure 9.7.

Poynton et al. (Poynton et al., 2014) developed PVDF, ETFE-based AEMs using VBC directly as a grafting monomer, which then undergoes nucleophilic substitution to introduce hydroxide exchange sites on the membrane in the form of cationic functional groups such as quaternary ammonium groups. A stable (polyethylene-co-tetrafluoroethylene)-based anion exchange membrane with through plane conductivities of 0.034±0.004 Scm−1 at 50°C was obtained. However, one of the disadvantages associated with VBC is that it is harmful and expensive so it can be used in limited amounts.

A wide range of tertiary amines are also available which allows the selection of amines to serve the dual action of quaternization as well as of cross-linking. Some of them include DABCO, TMHDA (N, N, N, N-tetramethylhexanediammonium), etc. The major drawback of using tertiary amines for quaternization is their susceptibility to hydroxide ions.

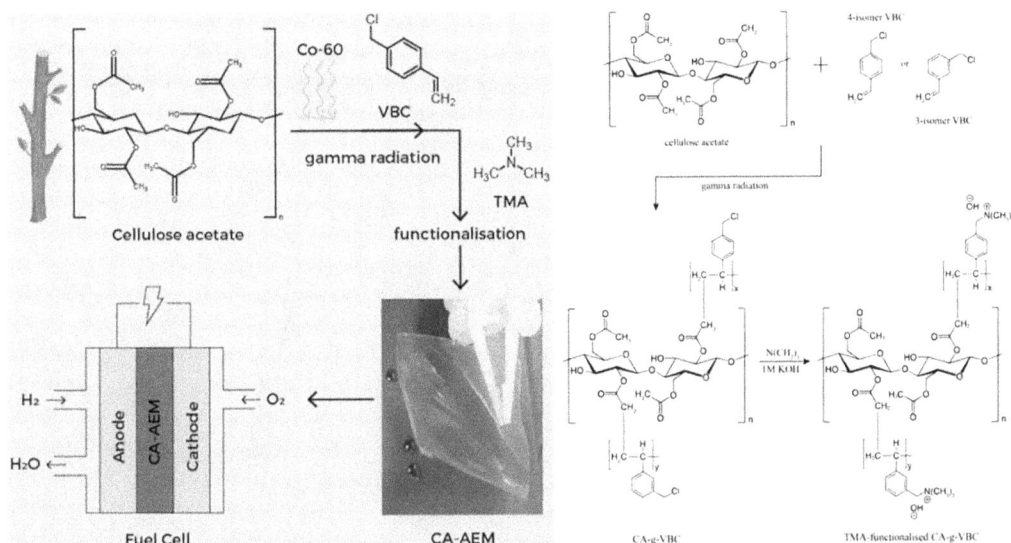

FIGURE 9.7 Schematic and mechanism of aminated CA-g-VBC (cellulose acetate- grafted-vinyl benzyl chloride) synthesis showing grafting and functionalization steps. Reproduced from (Samaniego et al., 2021) with permission.

One of the methods to introduce benzyl trimethyl ammonium groups in the membrane includes radiation-induced grafting of monomers such as styrene onto the base polymer followed by its halo alkylation, amination, and finally alkalization. Chakravorty et al. (Chakravorty et al., 1989) developed radiation-grafted polyethylene-based anion exchange films by grafting styrene and divinylbenzene onto polyethylene films, followed by chloromethylation and amination. Halo alkylation reactions are slow and involve toxic reagents. Another method includes the direct radiation grafting of vinyl benzyl trimethyl ammonium chloride groups onto the base polymer, followed by amination and alkalization. The cationic head groups directly attached to benzylic positions are reported to be susceptible to attack by hydroxide ion resulting into the degradation of benzyl trimethyl ammonium ions to benzyl alcohol and benzyl trimethylamine mediated by nucleophilic substitution reaction (Wang et al, 2009b).

Some other anion exchange groups are also used such as 1,4-diazobicyclo [2,2,2] octane (DABCO) by Ko et al. (Ko et al., 2012). The alkaline stability of the DABCO functionalized membrane is found to be more than quaternary ammonium group functionalized AEMs.

The covalently attached cationic head groups attached to the benzylic positions of the aromatic components of the base polymer proves to be a major drawback due to the sensitivity for substitution by the hydroxide group. It is observed that instability in AEMs arises due to covalent bonding of cationic head -groups to benzylic positions of grafted chains. Such AEMs degrade on treatment with aq. alkali solution at temperature > 70°C as benzyl ammonium groups are susceptible to attack by hydroxide ions. The membranes degrade through ylide formation and Hoffmann elimination. A strategic solution reported to this problem is use of flexible spacers between the benzylic group of the AEM and cationic head group, as shown in Figure 9.8. Some reports are there in which an aliphatic chain as a spacer is used between the benzylic and cationic head group that reduces the degradation of polymeric chains (Yang et al., 2015).

González and his fellow-workers (Ponce-González et al., 2018) used a butyl-spacer between the benzylic group and cationic head group (methyl pyrrolidinium group). VBC along with butyl spacer was grafted onto the ETFE by radiating through an electron beam followed by final amination done with MPY results in enhanced chemical stability of the AEM.

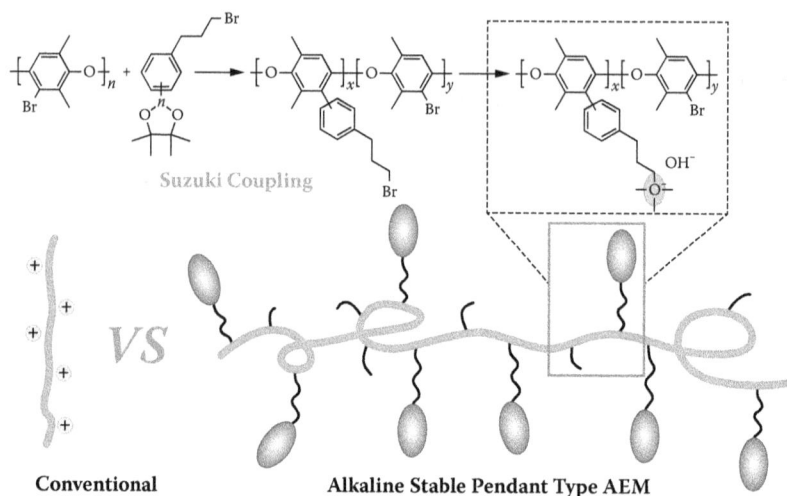

FIGURE 9.8 Synthetic strategy to construct AEMs with flexible spacers between the benzylic group of the AEM and cationic head group. Reproduced from (Yang et al., 2015) with permission.

Dang et al. (Dang and Jannasch, 2015) explored a series of different length cationic alkyl side chains grafted to poly(phenylene oxide)-based AEMs which results in their alkaline stability in 1M NaOH at 80°C. The AEMs developed reach the ion exchange capacity ~1.4 meqg^{-1} and OH$^-$ conductivities of 0.1S cm^{-1} at 80°C.

9.3.2.2.2 Imidazolium moieties

Heterocyclic imidazolium rings with multiple N-atoms were introduced as the cationic pendant groups to fabricate AEMs. Imidazolium anion exchange groups can be introduced using 1-methylimidazole, 1,2-dimethylimidazole, 1-vinyl imidazole, 4-vinyl imidazole, etc. Zhao et al. (Zhao et al., 2020) developed first time imidazolium based AEMs by grafting 2-methyl-1-vinylimidazole/styrene on ETFE films. Yoshimura et al. [86] functionalized VBC-grafted ETFE membrane with imidazolium anion exchange group using 1-methylimidazole. Mojarrad et al. (Mojarred et al., 2020) used 1-vinylimidazole (1-VIm) and 4-vinylpyridine (4VP) monomers to get simultaneously-grafting onto pre-irradiated ETFE (ethylene-*co*-tetrafluoroethylene) films (metal salt as inhibitor). They achieved maximum power density of 237 mW·cm^{-2}.

According to some reports, such imidazolium-based AEMs degrade quickly in an alkaline solution, where degradation is due to β-elimination of acidic proton at β-carbon atom from two adjacent imidazolium rings, which eventually reduces the anionic conductivity of AEMs. For example, AEMs radiation grafted with 1-benzyl-2,3-dimethyl imidazolium as anion exchange groups were found to be less alkali stable than radiation-grafted AEMs with benzyl trimethyl ammonium anion exchange groups. Yan et al. (Yan and Hickner, 2010) showed that AEM degradation is slowed down on using N3 or C2-substituted imidazolium cationic head groups.

Recently, Zhao et al. (Zhao et al., 2020) grafted p-(2-imidazoliumyl) styrene onto the ETFE followed by N-alkylation. Thus, according to this study, a favorable strategy to improve the alkaline stability of AEMs could be increasing the hydrophobicity of the polymer backbone and chemical hindrance around the ionic groups.

Some other cationic head groups had also been explored, such as benzimidazolium, triazoles, guanidinium, triarylsulfonium, pyrrolidinium, pyridinium, morpholinium, etc. (Zhang et al., 2015). There is still room for the improvement of radiation-grafted membranes and there are huge research efforts to develop high-performance polymer membranes by radiation grafting.

9.4 CONCLUSION

Radiation-induced graft copolymerization has proven to be an attractive way for the preparation of both PEMs and AEMs for fuel cell applications. The physicochemical characteristics of the membrane in the IEMs greatly affect the overall performance and efficiency of the fuel cell. Moreover, radiation-induced grafting used for the development of such polymeric fuel cell membranes has become a promising and reliable method with the perks of being cost-effective when it comes to production, considerable control over the degree of grafting and the parameters applied for it, relatively simple and eco-friendly synthesis procedures, devoid of catalytic contamination, devoid of toxic initiators and organic solvents, lower reaction temperatures, uniformity in the generation of active sites for grafting, and the pertinence to any polymer used for the fuel cell application. In addition to that, this technique offers the desired tunability for the use of a variety of polymeric membranes with a variety of monomers. Moreover, radiation-induced grafting enables the control of hydrophilic and hydrophobic domains to ensure good micro/nanophase separation in the AEM to facilitate high ionic conduction. Radiation grafting spacers onto base polymers also improve the chemical stability properties of AEMs. Thus, radiation-induced grafting of ion exchange membranes has become one of the noteworthy alternatives for fuel cells with a great scope for the future of more efficient materials for such applications.

ACKNOWLEDGMENTS

The authors are grateful to Amity University Uttar Pradesh, Noida,Uttar Pradesh for providing support and infrastructure. The authors would also like to thank IAEA, Vienna, Austria for their constant technical support.

REFERENCES

Alberti, G. & Casciola, M. (2001). Solid state protonic conductors, present main applications and future prospects. *Solid State Ionics*, 145, 3–16. 10.1016/S0167-2738(01)00911-0

Awad, S., Alomari, A. H., Abdel-Hady, E. E., & Hamam, M. F. (2021). Characterization, nanostructure, and transport properties of styrene grafted PVA/SiO2 hybrid nanocomposite membranes: Positron lifetime study. *Polymers for Advanced Technologies*, 32(4), 1742–1751. 10.1002/pat.5210

Bai, H., Wang, H., Zhang, J., Zhang, J., Lu, S., & Xiang, Y. (2019). High temperature polymer electrolyte membrane achieved by grafting poly(1-vinylimidazole) on polysulfone for fuels application. *Journal of Membrane Science*, 592, 117395. 10.1016/j.memsci.2019.117395

Biancolli, A. L. G., Bsoul-Haj, S., Douglin, J. C., Barbosa, A. S., de Sousa Jr, R. R., Rodrigues Jr, O., ... & Santiago, E. I. (2022). High-performance radiation grafted anion-exchange membranes for fuel cell applications: Effects of irradiation conditions on ETFE-based membranes properties. *Journal of Membrane Science*, 641, 119879. 10.1016/j.memsci.2021.119879

Benavides, R., Urbano, R., Morales-Acosta, D., Martínez-Pardo, M.E., Carrasco, H., Paula, M.M.S., & da Silva, L. (2019). Effect of gamma radiation on crosslinking, water uptake and ion exchange on polystyrene-co-acrylic acid copolymers useful for fuel cells. International Journal of Hydrogen Energy, 44, 12525–12528. 10.1016/j.ijhydene.2018.08.131

Büchi, F. N., Gupta, B., Haas, O., & Scherer, G. G. (1995). Performance of Differently Cross-Linked, Partially Fluorinated Proton Exchange Membranes in Polymer Electrolyte Fuel Cells. *Journal of Electrochemical Society*, 142, 3044–3048. 10.1149/1.2048683

Casciola, M., Alberti, G., Sganappa, M., & Narducci, R. (2006). On the decay of Nafion proton conductivity at high temperature and relative humidity. *Journal of Power Sources*, 162, 141–145. 10.1016/J.JPOWSOUR.2006.06.023

Chakravorty, B., Mukherjee, R. N., & Basu, S. (1989). Synthesis of ion-exchange membranes by radiation grafting. *Journal of Membrane Science*, 41, 155–161. 10.1016/S0376-7388(00)82398-4

Dang, H. S., & Jannasch, P. (2015). Exploring different cationic alkyl side chain designs for enhanced alkaline stability and hydroxide ion conductivity of anion-exchange membranes. *Macromolecules*, 48(16), 5742–5751. 10.1021/acs.macromol.5b01302

Danks, T. N., Slade, R. C. T., & Varcoe, J. R. (2002). Comparison of PVDF- and FEP-based radiation-grafted alkaline anion-exchange membranes for use in low temperature portable DMFCs. *Journal of Material Chemistry*, 12, 3371–3373. 10.1039/b208627a

Danks, T. N., Slade, R. C. T., & Varcoe, J. R. (2003). Alkaline anion-exchange radiation-grafted membranes for possible electrochemical application in fuel cells. *Journal of Material Chemistry*, 13, 712–721. 10.1039/b212164f

Dekel, D. R. (2018). Review of cell performance in anion exchange membrane fuel cells. *Journal of Power Sources*, 375, 158–169. 10.1016/j.jpowsour.2017.07.117

Esmaeili, N., Gray, E. M. A., & Webb, C. J. (2019). Non-Fluorinated Polymer Composite Proton Exchange Membranes for Fuel Cell Applications – A Review. *ChemPhysChem*, 20, 2016–2053. 10.1002/CPHC.201900191

Espiritu, R., Golding, B. T., Scott, K., & Mamlouk, M. (2018). Degradation of radiation grafted anion exchange membranes tethered with different amine functional groups via removal of vinylbenzyltrimethylammonium hydroxide. *Journal of Power Sources*, 375, 373–386. 10.1016/j.jpowsour.2017.07.074

Espiritu, R., Mamlouk, M., & Scott, K. (2016). Study on the effect of the degree of grafting on the performance of polyethylene-based anion exchange membrane for fuel cell application. *International Journal of Hydrogen Energy*, 41, 1120–1133. 10.1016/j.ijhydene.2015.10.108

Fan, J., Willdorf-Cohen S., Schibli, E. M., Paula, Z., Li, W., Skalski, T. J. G., Sergeenko, A. T., Hohenadel, A., Frisken, B. J., Magliocca, E., Mustain, W. E., Diesendruck, C. E., Dekel, D. R., & Holdcroft, S. (2019). Poly(bis-arylimidazoliums) possessing high hydroxide ion exchange capacity and high alkaline stability. *Nature Communication*, 10, 1–10. 10.1038/s41467-019-10292-z

Fang, J., Yang, Y., Lu, X., Ye, M., Li, W., & Zhang, Y. (2012). Cross-linked, ETFE-derived and radiation grafted membranes for anion exchange membrane fuel cell applications. *International Journal of Hydrogen Energy*, 37(1), 594–602. 10.1016/j.ijhydene.2011.09.112

Feng, K., Liu, L., Tang, B., Li, N., & Wu, P. (2016). Nafion-Initiated ATRP of 1-Vinylimidazole for Preparation of Proton Exchange Membranes. *ACS Applied Material Interfaces*, 8(18), 11516–11525. 10.1021/acsami.6b02248

Furtado Filho, A. A. M. & Gomes, A. S. (2012). Sulfonatedbisphenol-A-polysulfone based composite PEMs containing tungstophosphoric acid and modified by electron beam irradiation. *International Journal of Hydrogen Energy*, 37(7), 6228–6235. 10.1016/j.ijhydene.2011.10.056

Gao, X., Yu, H., Xie, F., Hao, J., & Shao, Z. (2020). High performance cross-linked anion exchange membrane based on aryl-ether free polymer backbones for anion exchange membrane fuel cell application. *Sustainable Energy Fuels*, 4, 4057–4066. 10.1039/d0se00502a

Golubenko, D. V., Gerasimova, E. V., & Yaroslavtsev, A. B. (2021). Proton conductivity and performance in fuel cells of grafted membranes based on polymethylpentene with radiation-grafted crosslinkedsulfonated polystyrene. *International Journal of Hydrogen Energy*, 46(32), 16999–17006. 10.1016/j.ijhydene.2021.01.102

Golubenko, D. V., Van der Bruggen, B., & Yaroslavtsev, A. B. (2020). Novel anion exchange membrane with low ionic resistance based on chloromethylated/quaternized-grafted polystyrene for energy efficient electromembrane processes. *Journal of Applied Polymer Sciences*, 137, 48656. 10.1002/app.48656

Gopi, K. H., Peera, S. G., Bhat, S. D., Sridhar, P., & Pitchumani, S. (2014). 3-Methyltrimethylammonium poly (2, 6-dimethyl-1, 4-phenylene oxide) based anion exchange membrane for alkaline polymer electrolyte fuel cells. *Bulletin of Materials Science*, 37(4), 877–881. 10.1017/s12034-014-0020-7

Gottesfeld, Dekel, D. R., Page, M., Bae, C., Yan, Y., Zelenay, P., & Kim, Y. S. (2018). Anion exchange membrane fuel cells: Current status and remaining challenges. *Journal of Power Sources*, 375, 170–184. 10.1016/j.jpowsour.2017.08.010

Gu, S., Cai, R., Luo, T., Chen, Z., Sun, M., Liu, Y., He, G., & Yan, Y. (2009). A soluble and highly conductive ionomer for high-performance hydroxide exchange membrane fuel cells. *AngewandteChemie International Edition*, 48, 6499–6502. 10.1002/anie.200806299

Gubler, L., Beck, N., Gürsel, S. A., Hajbolouri, F., Kramer, D., Reiner, A., Steiger, BScherer, G. G., Wokaun, A., Rajesh, B., & Thampi, K. R. (2006). Materials for polymer electrolyte fuel cells. *Chimia International Journal of Chemistry*, 58(12), 826–836. 10.2533/000942904777677128

Gubler, L., Nauser, T., Coms, F. D., Lai, Y. H., & Gittleman, C. S. (2018). PerspectiveProspects for Durable Hydrocarbon-Based Fuel Cell Membranes. *Journal of Electrochemical Society*, 165, F3100–F3103. 10.1149/2.0131806jes

Guo, M., Fang, J., Xu, H., Li, W., Lu, X., Lan, C., & Li, C. (2020). Synthesis and characterization of novel anion exchange membranes based on imidazolium-type ionic liquid for alkaline fuel cells. *Journal of Membrane Science*, 362, 97–104. 10.1016/j.memsci.2010.06.026

Gwon, S. J., Choi, J. H., Sohn, J. Y., Lim, Y. M., Nho, Y. C., & Ihm, Y. E. (2009). Battery performance of PMMA-grafted PE separators prepared by pre-irradiation grafting technique. *Journal of Industrial Engineering Chemistry*, 15, 748–751. 10.1016/j.jiec.2009.09.057

Hagesteijn, K. F. L., Jiang, S., & Ladewig, B. P. (2018). A review of the synthesis and characterization of anion exchange membranes. *Journal of Material Science*, 5316(53), 11131–11150. 10.1007/S10853-018-2409-Y

Han, J., Pan, J., Chen, C., Wei, L., Wang, Y., Pan, Q., Zhao, N., Xie, B., Xiao, L., Lu, J., & Zhuang, L. (2019). Effect of Micromorphology on Alkaline Polymer Electrolyte Stability. *ACSApplied Material Interfaces*, 11, 469–477. 10.1021/acsami.8b09481

Hao, L. H., Hieu, D. T. T., Danh, T. T., Long, T. H., Phuong, H. T., Van Man, T., … & Tap, T. D. (2021). Surface features of polymer electrolyte membranes for fuel cell applications: An approach using S2p XPS analysis. *Science and Technology Development Journal*, 24(3), 2100–2109. 10.32508/stdj.v24i3.2556

Herman, H., Slade, R. C., & Varcoe, J. R. (2003). The radiation-grafting of vinylbenzylchlorideonto poly (hexafluoropropylene-co-tetrafluoroethylene) films with subsequent conversion to alkaline anion-exchange membranes: optimisation of the experimental conditions and characterisation. *Journal of Membrane Science*, 218(1–2), 147–163. 10.1016/S0376-7388(03)00167-4

Jetsrisuparb, K., Wokaun, A., & Gubler, L. (2014). Radiation grafted membranes for fuel cells containing styrene sulfonic acid and nitrile comonomers. *Journal of membrane science*, 450, 28–37. 10.1016/j.memsci.2013.08.037

Kang, S., Jung, D., Shin, J., Lim, S., Kim, S. K., Shul, Y., & Peck, D. H. (2013). Long-term durability of radiation-grafted PFA-g-PSSA membranes for direct methanol fuel cells. *Journal of membrane science*, 447, 36–42. 10.1016/j.memsci.2013.07.005

Kizewski, J. P., Mudri, N. H., & Varcoe, J. R. (2013). An empirical study into the effect of long term storage (− 36 ±2 C) of electron-beamed ETFE on the properties of radiation-grafted alkaline anion-exchange membranes. *Radiation Physics and Chemistry*, 89, 64–69. 10.1016/j.radphyschem.2013.04.005

Ko, B. S., Sohn, J. Y., & Shin, J. (2012). Radiation-induced synthesis of solid alkaline exchange membranes with quaternized 1, 4-diazabicyclo [2, 2, 2] octane pendant groups for fuel cell application. *Polymer*, 53(21), 4652–4661. 10.1016/j.polymer.2012.08.002

Lai, S., Park, J., Cho, S., Tsai, M., Lim, H., & Chen, K. (2016). Mechanical property enhancement of ultrathin PBI membrane by electron beam irradiation for PEM fuel cell. *International Journal of Hydrogen Energy*, 41(22), 9556–9562. 10.1016/j.ijhydene.2016.04.111

Li, L., Deng, B., Ji, Y., Yu, Y., Xie, L., Li, J., & Lu, X. (2010). A novel approach to prepare proton exchange membranes from fluoropolymer powder by pre-irradiation induced graft polymerization. *Journal of Membrane Science*, 346, 113–120. 10.1016/j.memsci.2009.09.027

Li, X., Zhang, H., Lin, C., He, Z., & Ramani, V. (2022). Quantitative analysis of proton exchange membrane prepared by radiation-induced grafting on ultra-thin FEP film. *International Journal of Hydrogen Energy*, 47(3), 1874–1887. 10.1016/j.ijhydene.2021.10.234

Lin, C. X., Zhuo, Y. Z., Hu, E. N., Zhang, Q. G., Zhu, A. M., & Liu, Q. L. (2017). Crosslinked side-chain-type anion exchange membranes with enhanced conductivity and dimensional stability. *Journal of Membrane Science*, 539, 24–33. 10.1016/j.memsci.2017.05.063

Liu, H., Yang, S., Wang, S., Fang, J., Jiang, L., & Sun, G. (2011). Preparation and characterization of radiation-grafted poly (tetrafluoroethylene-co-perfluoropropyl vinyl ether) membranes for alkaline anion-exchange membrane fuel cells. *Journal of Membrane Science*, 369, 277–283. 10.1016/j.memsci.2010.12.002

Mamlouk, M., & Scott, K. (2012). Effect of anion functional groups on the conductivity and performance of anion exchange polymer membrane fuel cells. *Journal of Power Sources*, 211, 140–146. 10.1016/j.jpowsour.2012.03.100

Meek, K. M., Reed, C. M., Pivovar, B., Kreuer, K. D., Varcoe, J. R., & Bance-Soualhi, R. (2020). The alkali degradation of LDPE-based radiation-grafted anion-exchange membranes studied using different ex situ methods. *RSC Advances*, 10(60), 36467–36477. 10.1039/d0ra06484j

Miyata, H., Kawashima, Y., Itoh, M., & Watanabe, M. (2005). Preparation of a mesoporous silica film with a strictly aligned porous structure through a sol-gel process. *Chemistry of Materials*, 17, 5323–5327. 10.1021/cm0504992

Motupally, S., Becker, A. J., & Weidner, J. W. (2000). Diffusion of Water in Nafion 115 Membranes. *Journal of Electrochemistry Society*, 147, 3171. 10.1149/1.1393879/XML

Rajabalizadeh Mojarrad, N., Sadeghi, S., Yarar Kaplan, B., Güler, E., & Alkan Gürsel, S. (2019). Metal-Salt Enhanced Grafting of Vinylpyridine and Vinylimidazole Monomer Combinations in Radiation Grafted

Membranes for High-Temperature PEM Fuel Cells. ACS Applied Energy Materials, 3, 532–54010.1021/acsaem.9b01777

Nasef, M. M. (2001). Effect of solvents on radiation-induced grafting of styrene onto fluorinated polymer films. Polymer International, 50, 338–346. 10.1002/pi.634

Nasef, M. M. (2009). Fuel cell membranes by radiation-induced graft copolymerization: current status, challenges, and future directions. In Polymer Membranes for Fuel Cells (pp. 87–114). Springer, Boston, MA. 10.1007/978-0-387-73532-0_5

Nikolić, V. M., Žugić, D. L., Maksić, A. D., Šaponjić, D. P., & MarčetaKaninski, M. P. (2011). Performance comparison of modified poly(vinyl alcohol) based membranes in alkaline fuel cells. International Journal of Hydrogen Energy, 36, 11004–11010. 10.1016/j.ijhydene.2011.05.164

Nisar, S., Pandit, A. H. A. H., Nadeem, M., Rizvi, M. M. A., & Rattan, S. (2021). γ-Radiation induced L-glutamic acid grafted highly porous, pH-responsive chitosan hydrogel beads: A smart and biocompatible vehicle for controlled anti-cancer drug delivery. International Journal of Biological Macromolecules, 182, 37–50. 10.1016/j.ijbiomac.2021.03.134

Ortega, E. A., Singh, A., & Xu, T. (2005). Ion exchange membranes: State of their development and perspective Related papers Removal of Heavy Metal from Drinking Water Supplies through the Ion Exchange Membrane-A … IOSR Journals Ion Exchange Membranes SyarifHidayat Elect ro-Membrane Processes Ion exchange membranes: State of their development and perspective. Journal of Membrane Science, 263, 1–29. 10.1016/j.memsci.2005.05.002

Page, O. M., Poynton, S. D., Murphy, S., Ong, A. L., Hillman, D. M., Hancock, C. A., … & Varcoe, J. R. (2013). The alkali stability of radiation-grafted anion-exchange membranes containing pendent 1-benzyl-2, 3-dimethylimidazolium head-groups. Rsc Advances, 3(2), 579–587. 10.1039/C2RA22331G

Pan, D., Pham, T. H., & Jannasch, P. (2021). Poly(arylenepiperidine) Anion Exchange Membranes with Tunable N-Alicyclic Quaternary Ammonium Side Chains. ACS Applied Energy Materials, 4, 11652–11665. 10.1021/acsaem.1c02389

Pan, J., Lu, S., Li, Y., Huang, A., Zhuang, L., & Lu, J. (2020). High-Performance alkaline polymer electrolyte for fuel cell applications. Advanced Functional Materials, 20, 312–319. 10.1002/adfm.200901314

Pasquini, L., Di Vona, L. M. L., & Knauth, P. (2016). Effects of anion substitution on hydration, ionic conductivity and mechanical properties of anion-exchange membranes. New Journal of Chemistry, 40, 3671–3676. 10.1039/c5nj03212a

Peighambardoust, S. J., Rowshanzamir, S., & Amjadi, M. (2010). Review of the proton exchange membranes for fuel cell applications. International Journal of Hydrogen Energy, 35, 9349–9384. 10.1016/J.IJHYDENE.2010.05.017

Phadnis, S., Patri, M., Hande, V. R., Deb, P. C. (2003). Proton Exchange membranes by grafting of styrene-acrylic acid onto FEP by pre-irradiation technique. I. Effect of synthesis conditions. Journal of Applied Polymer Science, 90(9), 2572–2577. 10.1002/app.12727

Ponce-González, J., Ouachan, I., Varcoe, J. R., & Whelligan, D. K. (2018). Radiation-induced grafting of a butyl-spacer styrenic monomer onto ETFE: the synthesis of the most alkali stable radiation-grafted anion-exchange membrane to date. Journal of Materials Chemistry A, 6(3), 823–827. 10.1039/c7ta10222d

Ponce-González, J., Whelligan, D. K., Wang, L., Bance-Soualhi, R., Wang, Y., Peng, Y., … & Varcoe, J. R. (2016). High performance aliphatic-heterocyclic benzyl-quaternary ammonium radiation-grafted anion-exchange membranes. Energy & Environmental Science, 9(12), 3724–3735. 10.1039/c6ee01958g

Poynton, S. D., Slade, R. C., Omasta, T. J., Mustain, W. E., Escudero-Cid, R., Ocón, P., & Varcoe, J. R. (2014). Preparation of radiation-grafted powders for use as anion exchange ionomers in alkaline polymer electrolyte fuel cells. Journal of Materials Chemistry A, 2(14), 5124–5130. 10.1039/c4ta00558a

Rajabalizadeh Mojarrad, N., Sadeghi, S., Yarar Kaplan, B., Güler, E., & AlkanGürsel, S. (2019). Metal-salt enhanced grafting of vinylpyridine and vinylimidazole monomer combinations in radiation grafted membranes for high-temperature PEM fuel cells. ACS Applied Energy Materials, 3(1), 532–540. 10.1021/acsaem.9b01777

Rao, A. S., Rashmi, K. R., Manjunatha, D. V., Jayarama, A., Prabhu, S., & Pinto, R. (2019). Pore size tuning of Nafion membranes by UV irradiation for enhanced proton conductivity for fuel cell applications. International Journal of Hydrogen Energy, 44(42), 23762–23774. 10.1016/j.ijhydene.2019.07.084

Rattan, S., & Sehgal, T. (2012). Stimuli-responsive membranes through peroxidation radiation-induced grafting of 2-hydroxyethyl methacrylate (2-HEMA) onto isotactic polypropylene film (IPP). Journal of Radioanalytical Nuclear Chemistry, 293, 107–118. 10.1007/s10967-012-1728-8

Rikukawa, M. & Sanui, K. (2000). Proton-conducting polymer electrolyte membranes based on hydrocarbon polymers. Progress in Polymer Science, 25, 1463–1502. 10.1016/S0079-6700(00)00032-0

Rohde, L. E., Clausell, N., Ribeiro, J. P., Goldraich, L., Netto, R., Dec, G. W., ... & Polanczyk, C. A. (2005). Health outcomes in decompensated congestive heart failure: a comparison of tertiary hospitals in Brazil and the United States. *International journal of cardiology*, 102(1), 71–77. 10.1016/j.ijcard.2004.04.006

Sadeghi, S., Şanlı, L. I., Güler, E., & Gürsel, S. A. (2018). Enhancing proton conductivity via sub-micron structures in proton conducting membranes originating from sulfonated PVDF powder by radiation-induced grafting. *Solid State Ionics*, 314, 66–73. 10.1016/j.ssi.2017.11.017

Samaniego, A. J., Arabelo, A. K., Sarker, M., Mojica, F., Madrid, J., Chuang, P. Y. A., ... & Espiritu, R. (2021). Fabrication of cellulose acetate-based radiation grafted anion exchange membranes for fuel cell application. *Journal of Applied Polymer Science*, 138(10), 49947. 10.1002/app.49947

Sana, B., Das, A., Sharma, M., & Jana, T. (2021). Alkaline Anion Exchange Membrane from Alkylated Polybenzimidazole. *ACS Applied Energy Materials*, 4(9), 9792–9805. 10.1021/acsaem.1c01862

Sehgal, T. & Rattan, S. (2010). Synthesis, characterization and swelling characteristics of graft copolymerized isotactic polypropylene film. *International Journal of Polymer Science*, 2010(14758), 1–9. 10.1155/2010/147581

Sgreccia E., Narducci, R., Knauth, P., & Di Vona, M. L. (2021). Silica containing composite anion exchange membranes by sol–gel synthesis: A short review. *Polymers (Basel)*, 13, 1874. 10.3390/polym13111874

Sherazi, T. A., Yong Sohn, J., Moo Lee, Y., & Guiver, M. D. (2013). Polyethylene-based radiation grafted anion-exchange membranes for alkaline fuel cells. *Journal of Membrane Science*, 441, 148–157. 10.1016/j.memsci.2013.03.053

Sherazi, T. A., Zahoor, S., Raza, S., Shaikh, A. J., Naqvi, S. A. R., Abbas, G., Khan, Y., & Li, S. (2015). Guanidine functionalized radiation induced grafted anion-exchange membranes for solid alkaline fuel cells. *International Journal of Hydrogen Energy*, 40, 786–796. 10.1016/j.ijhydene.2014.08.086

Sproll, V., Nagy, G., Gasser, U., Embs, J. P., Obiols-Rabasa, M., Schmidt, T. J., Gubler, L., & Balog, S. (2016). Radiation Grafted Ion-Conducting Membranes: The Influence of Variations in Base Film Nanostructure. *Macromolecules*, 49, 4253–4264. 10.1021/acs.macromol.6b00180

Stoica, D., Ogier, L., Akrour, L., Alloin, F., & Fauvarque, J. F. (2007). Anionic membrane based on poly-epichlorhydrin matrix for alkaline fuel cell: Synthesis, physical and electrochemical properties. *ElectrochimicaActa*, 53, 1596–1603. 10.1016/j.electacta.2007.03.034

Strathmann, H. (1995). Electrodialysis and related processes. *Membrane Science Technology*, 2, 213–281. 10.1016/S0927-5193(06)80008-2

Sudoh, M., Niimi, S., Kurozumi, T., & Okajima, Y. (2011). Anion conductive membrane prepared by plasma polymerization for direct methanol alkaline fuel cell. *ECS Transactions*, 41(1), 1775. 10.1149/ma2011-02/16/867

Suresh, D., Goh, P. S., Ismail, A. F., & Hilal, N. (2021). Surface design of liquid separation membrane through graft polymerization: A state of the art review. *Membranes (Basel)*, 11, 1–40. 10.3390/membranes11110832

Varcoe, J. R. (2007). Investigations of the ex-situ ionic conductivities at 30 C of metal-cation-free quaternary ammonium alkaline anion-exchange membranes in static atmospheres of different relative humidities. *Physical Chemistry Chemical Physics*, 9(12), 1479–1486. 10.1039/B615478F

Varcoe, J. R., Slade, R. C. T., Lam How Yee, E., Poynton, S. D., Driscoll, D. J., & Apperley, D. C. (2007). Poly(ethylene-co-tetrafluoroethylene)-derived radiation-grafted anion-exchange membrane with properties specifically tailored for application in metal-cation-free alkaline polymer electrolyte fuel cells. *Chemistry of Materials*, 19, 2686–2693. 10.1021/cm062407u

Wang, G., Pan, J., Jiang, S. P., & Yang, H. (2018). Gas phase electrochemical conversion of humidified CO2 to CO and H$_2$ on proton-exchange and alkaline anion-exchange membrane fuel cell reactors. *Journal of CO$_2$ Utilization*, 23, 152–158. 10.1016/j.jcou.2017.11.010

Wang, G., Weng, Y., Chu, D., Chen, R., & Xie, D. (2009b). Developing a polysulfone-based alkaline anion exchange membrane for improved ionic conductivity. *Journal of Membrane Science*, 332, 63–68. 10.1016/j.memsci.2009.01.038

Wang, G., Weng, Y., Chu, D., Xie, D., & Chen, G. (2009a). Preparation of alkaline anion exchange membranes based on functional poly(ether-imide) polymers for potential fuel cell applications. *Journal of Membrane Science*, 326, 4–8. 10.1016/j.memsci.2008.09.037

Wang, J., Li, S., & Zhang, S. (2010). Novel hydroxide-conducting polyelectrolyte composed of an poly (arylene ether sulfone) containing pendant quaternary guanidinium groups for alkaline fuel cell applications. *Macromolecules*, 43, 3890–3896. 10.1021/ma100260a

Wang, L., Brink, J. J., Liu, Y., Herring, A. M., Ponce-González, J., Whelligan, D. K., & Varcoe, J. R. (2017b). Non-fluorinated pre-irradiation-grafted (peroxidated) LDPE-based anion-exchange membranes with high performance and stability. *Energy & Environmental Science*, 10(10), 2154–2167. 10.1039/c7ee02053h

Wang, L., Brink, J. J., & Varcoe, J. R. (2017a). The first anion-exchange membrane fuel cell to exceed 1 W cm− 2 at 70 C with a non-Pt-group (O 2) cathode. *Chemical Communications*, 53(86), 11771–11773. 10.1039/c7cc06392j

Wang, L., Peng, X., Mustain, W. E., & Varcoe, J. R. (2019). Radiation-grafted anion-exchange membranes: the switch from low-to high-density polyethylene leads to remarkably enhanced fuel cell performance. *Energy & Environmental Science*, 12(5), 1575–1579. 10.1039/c9ee00331b

Wang, Y., Ruiz Diaz, D. F., Chen, K. S., Wang, Z., & Adroher, X. C. (2020). Materials, technological status, and fundamentals of PEM fuel cells – A review. *Materials Today*, 32, 178–203. 10.1016/j.mattod.2019.06.005

Willson, T. R., Hamerton, I., Varcoe, J. R., & Bance-Soualhi, R. (2019). Radiation-grafted cation-exchange membranes: An initial: Ex situ feasibility study into their potential use in reverse electrodialysis. *Sustainable Energy Fuels*, 3, 1682–1692. 10.1039/c8se00579f

Xu, F., Su, Y., & Lin, B. (2020). Progress of Alkaline Anion Exchange Membranes for Fuel Cells: The Effects of Micro-Phase Separation. *Frontiers in Materials*, 7, 4. 10.3389/fmats.2020.00004

Xu, H., Fang, J., Guo, M., Lu, X., Wei, X., & Tu, S. (2010). Novel anion exchange membrane based on copolymer of methyl methacrylate, vinylbenzyl chloride and ethyl acrylate for alkaline fuel cells. *Journal of Membrane Science*, 354, 206–211. 10.1016/J.MEMSCI.2010.02.028

Xu, T., Woo, J. J., Seo, S. J., & Moon, S. H. (2008). In situ polymerization: A novel route for thermally stable proton-conductive membranes. *Journal of Membrane Science*, 325, 209–216. 10.1016/J.MEMSCI.2008.07.036

Yan, J. & Hickner, M. A. (2010). Anion exchange membranes by bromination of benzylmethyl-containing poly(sulfone)s. *Macromolecules*, 43, 2349–2356. 10.1021/MA902430Y/SUPPL_FILE/MA902430Y_SI_001.PDF

Yang, Z., Peng, H., Wang, W., & Liu, T. (2020). Crystallization behavior of poly(ε-caprolactone)/layered double hydroxide nanocomposites. *Journal of Applied Polymer Science*, 116, 2658–2667. 10.1002/app

Yang, Z., Zhou, J., Wang, S., Hou, J., Wu, L., & Xu, T. (2015). A strategy to construct alkali-stable anion exchange membranes bearing ammonium groups via flexible spacers. *Journal of Materials Chemistry A*, 3(29), 15015–15019. 10.1039/C5TA02941D

Ye, Y. S., Rick, J., & Hwang, B. J. (2012). Water soluble polymers as proton exchange membranes for fuel cells. *Polymers (Basel)*, 4, 913–963. 10.3390/polym4020913

Yoshimura, K., Koshikawa, H., Yamaki, T., Maekawa, Y., Yamamoto, K., Shishitani, H., … & Tanaka, H. (2013). Alkaline durable anion exchange membranes based on graft-type fluoropolymer films for hydrazine hydrate fuel cell. *ECS Transactions*, 50(2), 2075. 10.1149/ma2012-02/13/1611

Zhang, F., Zhang, H., & Qu, C. (2011). Imidazolium functionalized polysulfone anion exchange membrane for fuel cell application. *Journal of Material Chemistry*, 21, 12744–12752. 10.1039/c1jm10656b

Zhang, F., Zhang, H., Ren, J., & Qu, C. (2010). PTFE based composite anion exchange membranes: thermally induced in situ polymerization and direct hydrazine hydrate fuel cell application. *Journal of Material Chemistry*, 20, 8139–8146. 10.1039/C0JM01311K

Zhang, H. W., Chen, D. Z., Xianze, Y., & Yin, S. B. (2015). Anion-Exchange Membranes for Fuel Cells: Synthesis Strategies, Properties and Perspectives. *Fuel Cells*, 15(6), 761–780. 10.1002/fuce.201500039

Zhao, Y., Li, X., Wang, Z., Xie, X., & Qian, W. (2019). Preparation of graft poly(arylene ether sulfone)s-based copolymer with enhanced phase-separated morphology as proton exchange membranes via atom transfer radical polymerization. *Polymers (Basel)*, 11, 1297. 10.3390/polym11081297

Zhao, Y., Yoshimura, K., Mahmoud, A. M. A., Yu, H. C., Okushima, S., Hiroki, A., … & Maekawa, Y. (2020). A long side chain imidazolium-based graft-type anion-exchange membrane: Novel electrolyte and alkaline-durable properties and structural elucidation using SANS contrast variation. *Soft matter*, 16(35), 8128–8143. 10.1039/D0SM00947D

Zhou, J., Unlu, M., Vega, J. A., & Kohl, P. A. (2009). Anionic polysulfoneionomers and membranes containing fluorenyl groups for anionic fuel cells. *Journal of Power Sources*, 190(2009), 285–292. 10.1016/J.JPOWSOUR.2008.12.127

Zhou, T., Shao, R., Chen, S., He, X., Qiao, J., & Zhang, J. (2015). A review of radiation-grafted polymer electrolyte membranes for alkaline polymer electrolyte membrane fuel cells. *Journal of Power Sources*, 293, 946–975. 10.1016/j.jpowsour.2015.06.026

Zhu, M., Zhang, X., Wang, Y., Wu, Y., Wang, H., Zhang, W.,Chen, Q., Shen, Z., & Li, N. (2018). Novel anion exchange membranes based on quaternizeddiblockcopolystyrene containing a fluorinated hydrophobic block. *Journal of Membrane Science*, 554, 264–273. 10.1016/J.MEMSCI.2018.01.055

10 High-temperature thermoplastic elastomeric materials by electron beam treatment – Challenges and opportunities

Aiswarya S. and Pratiksha Awasthi
Department of Materials Science and Engineering, Indian Institute of Technology Delhi, New Delhi, India

Nischay Kodihalli Shivaprakash and A. Wayne Cooke
Mitsubishi Chemical America Performance Polymers Division, Greer-South Carolina, USA

Subhan Salaeh
Department of Rubber Technology and Polymer Science, Faculty of Science and Technology, Prince of Songkla University, Pattani Campus, Pattani, Thailand

Shib Shankar Banerjee*
Department of Materials Science and Engineering, Indian Institute of Technology Delhi, New Delhi, India

*Corresponding author: ssbanerjee@mse.iitd.ac.in

CONTENTS

DOI: 10.1201/9781003321910-10

10.1   INTRODUCTION

In recent decades, electron beam treatment (EBT) has opened up new avenues for material science as well as polymer science research and technology. This technology is an extremely adaptable and versatile tool that is effectively employed in academic and industrial research projects to enable modeling, elemental and structural characterization, additive manufacturing, and effective material modification (Frank and Hirano 1993, Datta, Chaki et al. 1998). All of this is possible at sizes ranging from hundreds of micrometers to nanometers. Because of their particular versatility, electron beams may be employed for fast prototyping as well as nanostructure modification and analysis. In the case of polymers, EBT has various advantagesand is frequently used to modify the physicochemical characteristics to meet the requirements of numerous applications (Kashiwagi 2004, Bhowmick 2008, Banerjee, Janke et al. 2018). The kinetic energy of primary electrons is much higher than the binding and excitation energies of electrons in the atoms (Chaudhary, Singh et al. 2015). Consequently, many excitations and/or ionizations are generated by an electron during its interaction with the polymer material to be modified. The chemical process is referred to as the "electron beam treatment" and accelerated electrons belong to direct "ionizing radiation" (Frank and Hirano 1993, Kashiwagi 2004). Based on the characteristic structure of polymeric material along with the chemical environment under EBT, different reactions such as cross-linking, curing, and imbedding as well as degradation and functionalization are induced (Spenadel and Chemistry 1979, Chaudhary, Singh et al. 2020, Manaila, Craciun et al. 2021). Polymer adaptation by high electron energy is commonly used in the production of novel materials and products with higher affinity and performances.

Thermoplastic elastomers (TPEs) are essential category of polymers used in numerous applications and industries. TPEs endow the functional characteristics of cross-linked rubber with the exceptional thermoplastic polymer processing behavior.Whether based on block copolymers orrubber-plastic blends, the overwhelming common TPEs have a biphasic nature, with the rubber phase distributed throughout the persistent thermoplastic matrices (Drobny 2014, Datta Sarma, Padmanathan et al. 2017). Amid the several kinds of TPEs, those prepared by melting semicrystalline thermoplastics and then mixing with rubber through high shearing and elongational flow have piqued the interest of researchers due to their ease of preparation and ability to achieve a wide range of chemical and physical characteristics. In the case of thermoplastic vulcanizates (TPVs), an elastomeric phase of a blend is fully cross-linked over the melt mixing process and becomes a dispersed phase in a continuous thermoplasticmediumto provide elasticity. This is, in general, referred to as dynamic vulcanization, and it includes vulcanizing the elastomer phase in the vicinity of a small amount of activated phenol-formaldehyde resin or peroxide (so-called reaction initiators) under the dynamic conditions of melt mixing (Ghosh, Chattopadhyay et al. 1994, Banerjee and Bhowmick 2015, Liao, Brosse et al. 2019, Ghumman, Nasef et al. 2021). Dynamic vulcanization relies on temperature-sensitive initiators. Consequently, the utilization of this type of reaction initiator causes a coupling of all processes involved in melt mixing, including flow and blending, heat transmission, as well as the chemical reactions. As a result, it is challenging to sustain and maintain the preferred chemical reactions as well as to regulate the altered viscosity caused by these chemical reactions. Furthermore, shear stresses and rates-of-deformation significantly affect the chemical reaction depending on-the degree of-filling. As a result, extensive knowledge of chemistry, flow models, mixing, process control, and rheology is needed. Nevertheless, conventional reactive processing is difficult to analyze and control. Also, due to these drawbacks, high-temperature TPEs can hardly be prepared by traditional dynamic vulcanization. To overcome this issue, electron -induced reactive processing was developed and tested as an alternative for preparing TPEs and high-temperature TPEs (Mandal, Chakraborty et al. 2014, Banerjee, Janke et al. 2018, Mandal, Chakraborty et al. 2021). In this process, the chemical reactions are initiated by high-energy electrons in a spatially and temporally precise manner at temperatures above the melting temperature of thermoplastic as well as at low shear stresses and rates of deformation. Thus, the accurate process puts in prospect enhanced mechanical properties by reducing undesired side reactions induced by polymer radicals. Furthermore, the electron beam treatment of TPE can be used to enhance a lot of its properties while maintaining its thermoplastic processability. This chapter provides an overview on electron-induced reactive processing to be used to prepare high temperature TPEs, including the EB process framework, process factors, and application technology.

10.2 BASIC PRINCIPLES OF ELECTRON BEAM TREATMENT

Electron beam treatment involves accelerated electrons with an enhanced kinetic energy that is-converted into excitation/ionization in the physical phase of interaction with polymer materials. The electron energy controls the treatment depth in polymers to be modified (Monem, Ali et al. 2005, Kongmon, Kangrang et al. 2016). If the electron energy is less than ~ 0.08 MeV, they cannot penetrate the electron beam exit window to modify polymer materials under atmospheric conditions. Depending on the thickness of the beam exit window, their energy level typically ranges between 0.1 and 10 MeV (Chaudhary, Singh et al. 2020). These electron beams are generated by electron accelerators, mainly consisting of an electron gun, acceleration tube, and scanning device, as presented in Figure 10.1a (Datta, Chaki et al. 1998). A key component of an accelerator is an electron-injector system, or electron gun, which is situated at the top of a cavity. The electron-injector generates a-triode, which includes the cathode, anode, and a control grid. For heating, a heater constructed of tungsten wire is coupled to the cathode. At high voltage, electrons generated in the cathode are propelled through the accelerating gap. Then these electrons are emitted by the electron gun through thermionic emission, then moved via the potential of around 100 KW in the electron beam treatment, and are therefore sent into the target via the window (Datta, Chaki et al. 1998). Some critical parameters associated with EBT are discussed below:

- The energy of electrons: Electronvolts (eV) is a unit of measurement for the energy of an electron. One electron receives one eV of energy when propelled via one volt.
- Power: Power is referred as the amount of electron energy released per second. Thepower of industrial electron accelerators amounts up to several 100 KW (Makhlis 1972).
- Dose: The absorbed energy is defined as a dose where the unit of dose is in gray (Gy). (Bhowmick 2008).
- Dose rate: It is the ionizing energy absorbed within a unit of time (Cleland 2006, Leo, Chulan et al. 2016). This is demonstrated in grays per-unit-time (kGy/s). Using mathematical notation, this is:

$$D_R = K * I/A$$

  where D_R is the rate of dose absorption, I is electron beam current in ampere, A is the area of the radiation field in mm^2, and K is the electrons' stopping power in watt that is decided by the electron energy and density of the material to be treated.
- Electron beam penetration depth: Secondary electrons created by ionizing reactions from an intruding beam of high-energy electrons are directed in space at random. After a minimal distance from the surface of a material being treated in contact with air, spatial "equilibrium" is reached. As a result, the absorbed energy dosage reaches a maximum at a distance of around 2 mm from the surface for 1 M eV electron (this varies with electron

(a)

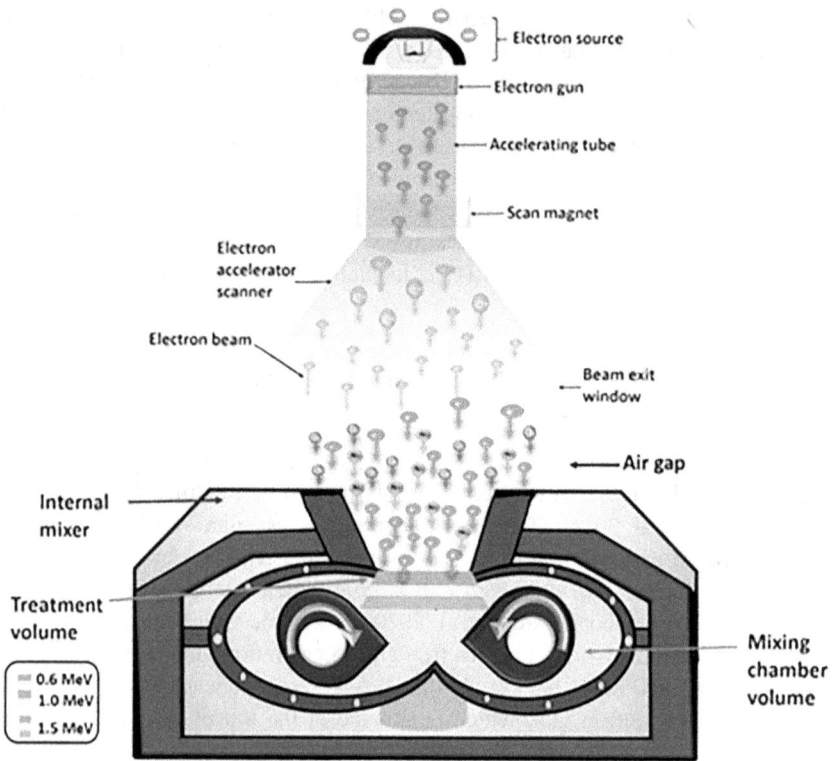

(b)

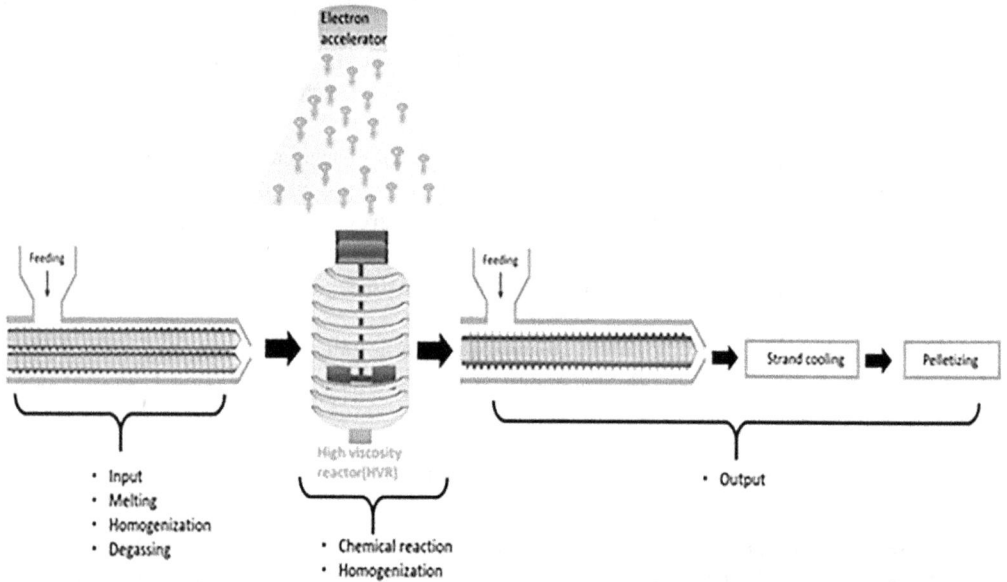

FIGURE 10.1   (a) Setup for reactive electron-induced reactive processing: the combination of a mid-energy electron accelerator with an internal mixer; (b) continuous low-energy electron-induced reactive processing.

energy for particular materials) (Becker, Bly et al. 1979, Cleland 2006). At a finite penetration depth, the energy deposition then drops to zero. This is calculated by:

$$E = (T * d)/3$$

where E represents the energy of electron beam in MeV, T represents the thickness of the layer in mm, and d represents the radiated material's density in/cm^3.

- Estimated yield: The quantity of a certain reaction (cross-linking, degradation, grafting, etc.) generated per 100 eV of absorbed energy is the yield of an EBT (Cleland, Parks et al. 2003). This can be expressed as;

$$\Phi = \frac{No: of\ molecules\ involved\ in\ a\ specific\ process}{The\ total\ amount\ of\ energy\ absorbed\ by\ the\ system}.$$

- The efficiency of electron beam treatment: In order to ensure maximum efficiency, all of the EB power must be captivated by the product so that the desired dose distribution is achieved (Berejka, Cleland et al. 2014). For materials with optimal thickness and treated from one side, the efficiency is reduced to around 50%. However, the efficiency is higher i.e., around 70%–80%, if the material of equal thickness is treated from both sides.

In the case of reactive electron treatment (Figure 10.1a), a typical electron accelerator is directly linked to mixing equipment. The accelerated electrons from the electron accelerator provide a spatial and temporal accurate energy input for the generation of polymer radicals. In contrast to conventional curatives for curing treatments, the radical generation is temperature independent and does not require the use of any additional additive. The penetration depth of electrons in this novel EBT is restricted to a portion of the mixing volume. The overall mixing volume is altered as a result of the modification in polymer content contained by the electron penetration depth throughout the mixing procedure. Consequently, the dose rate and penetration depth are determined by the electron current and electron energy, respectively. Further, the electron energy regulates the ratio of modification volume to the overall volume of the mixing chamber. Finally, the ratio of dose to mixing rate (dose per revolution) is controlled by electron current and revolution used for the melt mixing process.

Naskar et al. used this treatment to inspect the performance characteristics of electron-induced cross-linked polypropylene (PP) ethylene propylene diene monomer rubber (EPDM) blends with specific emphasis on absorbed dosage, electron-treatment period, and electron energy of the electron-induced reactive processing (Naskar, Gohs et al. 2009). These parameters were mainly selected based on the experimental data for peroxide-induced vulcanization of PP/EPDM TPV. Based on experiences in EBT of PP, the total number of radicals for electron-induced reactive processing was reduced to 10% (~100 kGy) compared to the peroxide-induced vulcanization of PP/EPDM TPV. Compared with peroxide cured PP-EPDM TPV, the electron-induced reactive treatment with 1.5 MeV electrons for 15 seconds at a dosage of 100 kGy yielded higher mechanical properties. Additionally, their experimental findings point out that two processes are taking place concurrently, both of which are critical to the progress of mechanical characteristics. One is the insitu compatibilization of PP and EPDM, and the other is the cross-linking in the EPDM phase. Similarly, Banerjee et al. used the same method to prepare high-performance polyamide 6/fluoroelastomer blends (Becker, Bly et al. 1979, Cleland, Parks et al. 2003, Banerjee, Gohs et al. 2016). They prepared the blends with three different absorbed doses (12.5, 17.5, and 25 kGy) while maintaining 1.5 MeV values of electron energy, dose/revolution (17.2 kGy/rpm), at temperature 230°C, and rotor speed of 60 rpm. The range of dose was precisely selected to avoid any cross-linking of the fluoroelastomer. As a result, the modified blends showed outstanding mechanical performance, excellent oil swelling resistance, and high thermal stability.

The coupling of an electron accelerator with an internal mixer is called discontinuous reactive electron processing. The Leibniz Institute of Polymer Research Dresden recently developed a continuous low-energy electron (~0.3 MeV) induced reactive processing (CEIReP) by coupling the reactive extrusion process with a low-energy electron emitter. This process consists of three sections, as displayed in Figure 10.1b (Zschech, Pech et al. 2020). The first section consists of a non-reactive compounding on a typical twin-screw extruder, which helps to melt and physically combine the compound ingredients. Along with this, the second section is intended for the low-energy electron-induced reactive treatmentthat occurs for the duration of melt-mixing in-the presence oflow-energy electrons to precisely initiate the chemical reaction in both times and an edge layer of molten polymer. As a result, a specific high viscosity reactor (HVR) was developed. Based on the operating principle, geometry, and process parameters of the internal mixer employed for the-discontinuous mid-energy electron-inducedprocessing, the overall design of the HVR was designed. The HVR's primary components are tangentially inward counter-rotating-rotors with a constant friction ratio. They were finally followed by the electron-induced reactive process. The third section consists of downstream processing with a second screw extruder, a water bath, and a granulator. The second screw extruder is needed for homogenization and mixing of extra fillers or stabilizers that would interfere with the low-energy electron-induced reaction. This experimental equipment has a maximum outputof 20 kg/h. Due to the use of allow energy electron emitter, the costs could be reduced, but the continuous low-energy electron-induced reactive processing (CEIReP) cannot be used to produce TPVs and TPEs as was done by Naskar et al. 2009 and Banerjee et al. 2018.

Using the continuous mid-energy electron (1.5 MeV) induced reactive processing (CEIReP) based on a 1.5 MeV electron accelerator, Aghjeh et al. investigated the rheological, morphological, and mechanical properties of ethylene octene copolymer (EOC) toughened polypropylene (PP) (Aghjeh, Khonakdar et al. 2016). According to their findings, continuous mid-energy electron (1.5 MeV) induced reactive processing enhances the tensile and impact performance of the EOC-PP blend. Furthermore, the following advantages are associated with the continuous mid-energy electron-induced reactive treatment (Zschech, Pech et al. 2020):

  i. In-line processing is possible
  ii. Continuous process
  iii. No need for reaction promoters
  iv. Combining multiple processes into one

10.2.1 CHARACTERISTIC FEATURES ASSOCIATED WITH REACTIVE ELECTRON PROCESSING

The characteristics mentioned above are associated with electron-induced reactive processing for modifying typical polymeric material (Mondal, Gohs et al. 2013, Banerjee, Janke et al. 2018, Zschech, Pech et al. 2020).

- Energy efficiency is high because the energy required for the chemical reaction is directly injected by accelerated electrons. These accelerated electrons create ionization/excitations, which get transformed into polymer radicals in the physical-chemical phase. The average yield amounts to about three radicals per 100 eV absorbed energy. Because the energy used in the radiation processing ranges from several 10 kGy to 200 kGy, it may also be used for materials with lower heat resistance.
- Third components, such as catalysts or precursors, are not necessary. Chemical reactions are triggered by electron-induced polymer radicals and the precise selection of a chemical environment.
- It's simple to control the reaction by electron current and duration of electron treatment. Because the electron beam is fundamentally a part of an electric system, hence controlling

the chemical process is directly interlinked with the control system. Therefore, initiation/termination is simply a matter of flipping a switch. In thermally induced systems, this level of control is absent.

- The system, particularly the discontinuous one is simple to use and keep up with. These are activated by simply moving the switch and configuring a few settings, and maintenance work are accessible after turning the switch off. This is not possible in continuous processes because of flow and mixing processes. One of the drawbacks of continuous electron-induced treatment is this. In addition, each polymer system to be modified needs its screw design.

10.2.2 ESSENTIAL STAGES OF INTERACTION OF POLYMER THROUGH EBT

The interaction of electron beam with polymeric materials results in energy dissipation into the substance, which affects the development of a variety of techniques, as depicted in Figure 10.2a. Depending on the time domain, these functions could be separated into three stages, as discussed below (Chattopadhyay 2001, Gohs, Dorschner et al. 2006):

10.2.2.1 Ionization and excitation stage

This stage, also known as the physical stage, is the initial step, and it occurs in less than 10^{-15} seconds after electron beam contact with a polymeric material, as shown in Figure 10.2b. Also, a substance treated with an electron beam can show the appearance of an electrical charge. This is the outcome of electrons being extracted from the treated polymer molecules by transmitting a significant amount of energy, resulting in the formation of ions. The binding-energies of the most labile electrons, known as ionization potentials, for polymeric compounds are in the range of 10–15 eV, and as a result, it is the ionisation threshold energy. Because electron beams created by accelerators carry far additional energy, ionization is very probabilistic. As a result, merely a portion of the deposited energy is utilized in ionization, with the balance causing electron activation.

Excitation: excited molecules

$$XY + \Delta E \rightarrow XY^*$$

$$XY^+ + e^- \rightarrow XY^*$$

Ionization: ions or free electrons

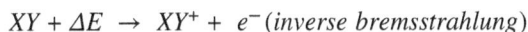

$$XY + \Delta E \rightarrow XY^+ + e^- \, (inverse \; bremsstrahlung)$$

$$XY \rightarrow X^+ + Y + e^- \, (Ionization \; via \; dissociation)$$

$$XY + e^- \rightarrow XY^- \, (Capture \; of \; electrons)$$

$$XY + e^- \rightarrow X^- + Y \, (Electron \; capture \; by \; dissociation)$$

where XY is a typical polymeric molecule, XY^+ is an ion, e^- is an electron, and XY^* is an exciting atom.

10.2.2.2 The free radical formation stage

Free radicals are generated next after the activation of an electron beam in the physical-chemical phase. Free radical generation typically takes 10^{-12} to 10^{-6} s to complete (Figure 10.2b). The electrons extracted from the treated molecular particles, are exposed to the high electric field

(a)

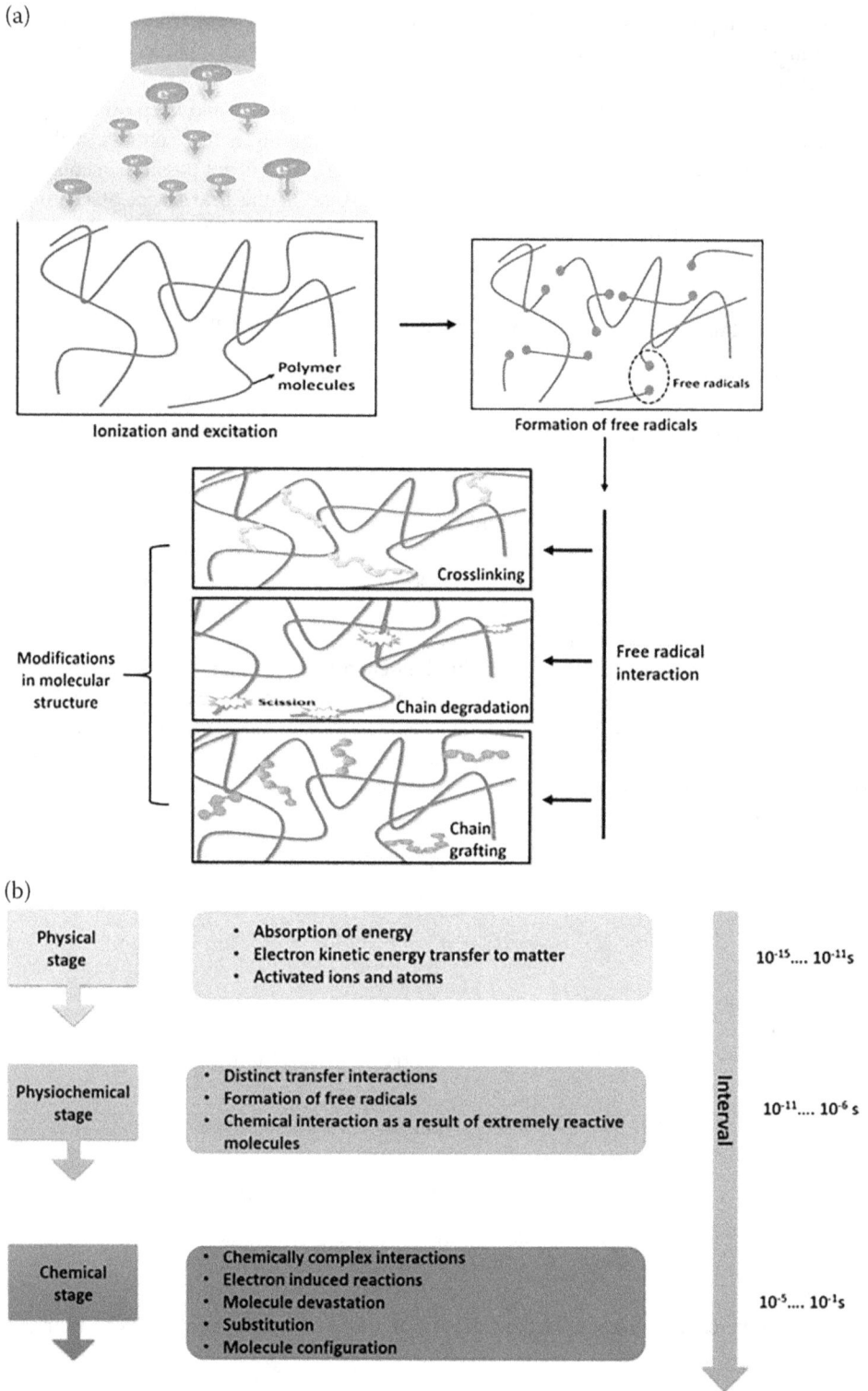

Polymer molecules

Ionization and excitation

Free radicals

Formation of free radicals

Crosslinking

Modifications in molecular structure

Free radical interaction

Scission   Chain degradation

Chain grafting

(b)

| Physical stage | • Absorption of energy
• Electron kinetic energy transfer to matter
• Activated ions and atoms | 10^{-15}.... 10^{-11}s |

| Physiochemical stage | • Distinct transfer interactions
• Formation of free radicals
• Chemical interaction as a result of extremely reactive molecules | 10^{-11}.... 10^{-6} s |

| Chemical stage | • Chemically complex interactions
• Electron induced reactions
• Molecule devastation
• Substitution
• Molecule configuration | 10^{-5}.... 10^{-1}s |

Interval

FIGURE 10.2   (a) The interaction of polymers and electron beams is depicted in this diagram. (b) A time-course of polymeric molecules in contact with electron beams.

created by the ions. As a result, charge recombination is a common phenomenon, either throughout the treatment or as an after-effect of treatment. The ionizing potential of around 10–15 eV is recovered, resulting in extremely excited molecules with an extent of energy far greater than any bond strength. Typical molecular bond strength in polymeric compounds is in the 3–8 eV range. As a result, such extremely agitated polymer molecules will very certainly degrade into free radicals. As ions are broken down, free radicals are also produced. Ions and free radicals can become trapped in polymeric materials for extended periods of time.

$$XY^* \rightarrow X^\blacksquare + Y^\blacksquare$$

$X^\blacksquare$ and $Y^\blacksquare$ are uncharged pieces of XY^* excited molecules or free radicals. It is these active species, i.e., ions and free radicals that contributes to the observed alterations for a typical polymeric molecule when exposed to electron beam treatment.

10.2.2.3   Chemical stage

The last stage of contact is referred to as the chemical stage because chemical reactions occur, owing to the active species formed in the aforementioned stages. After treatment, the chemical stage comprises 10^{-6} to 10^{-1} seconds to complete (Figure 10.2b). In addition, this stage results in molecular structural alteration and the development of polymeric material with improved or deteriorated properties. Such chemical stage is very significant for modifying molecular network, especially at elevated temperature and presence of free radical generated by EBT. This results in cross-linking, grafting, and chain scission.

10.3   THERMOPLASTIC ELASTOMERIC MATERIALS

Thermoplastic elastomers are the class of fundamentally engrossing polymeric materials. TPEs possess biphasic morphology having combined properties of semi-crystalline thermoplastics accompanied with elastic nature of elastomers. Basically, thermoplastic elastomers require a combination of elastomeric features with the processing capabilities of thermoplastic polymers. They provide a two-phased system composed of soft and hard segments. The hard phase is typically thermoplastic, enabling efficient processability and recyclability. On the contrary, the soft phase consists of a flexible elastomeric segment that provides elastic characteristics to TPEs. In addition, this elastomeric phase is dispersed in the continuous thermoplastic matrix, which is typically connected with a rubber ligament of 4–12 nm, as illustrated in Figure 10.3a. (Banerjee, Bhowmick et al. 2017, Watts, Kurokawa et al. 2017, Hochleitner, Fürsattel et al. 2018, Yuan, Fan et al. 2018, Fazli and Rodrigue 2020). During stretching, TPEs are known to deform plastically via the thinnest region of the thermoplastic matrix, whereas the thicker region withholds deformation. When the external force is released, the connected rubber ligaments pull back the plastically deformed thermoplastic matrix, thereby providing rubber-like elasticity in TPEs (Banerjee, Kumar et al. 2015). The typical stress-strain plot of TPE is marked by the yield point absence before failure of the material, as shown in Figure 10.3c. Generally, thermoplastic elastomers are preferred over conventional elastomers because they offer better processing and durability than those of conventional elastomers. In addition, they reduce waste generation because of their recyclability. Similarly, since TPEs processing characteristics are akin to thermoplastic processing, it is easier, faster, and requires fewer steps, allowing them to achieve almost no compounding, low power consumption, and consistent product quality. Likewise, TPE material is less expensive per unit volume since it has a lower density than conventional rubbers (Seymour and Kauffman 1992, Abdou-Sabet, Puydak et al. 1996, Kear 2003). TPEs are mainly classified under six categories like thermoplastic polyolefins (TPO), thermoplastic copolyester (TPC), thermoplastic vulcanizate (TPV), thermoplastic polyamide (TPA), styrene-based block copolymers (TPS), and thermoplastic polyurethane (TPU). Blend of polypropylene (PP) or polyethylene (PE) with ethylene-propylene-diene rubber (EPDM) or acrylonitrile butadiene rubber

(a)

(b)

(c)

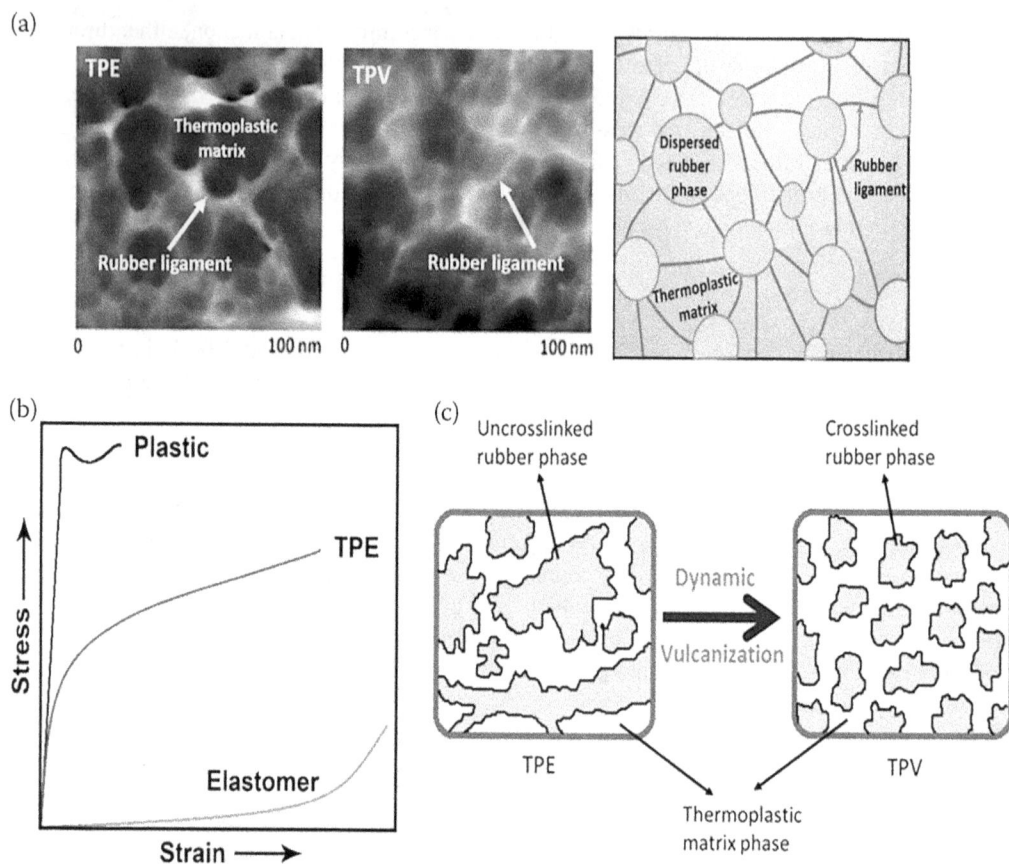

FIGURE 10.3 (a) Dispersed elastomeric phase in a continuous thermoplastic matrix interconnected by nanometric rubber ligaments (~4–12 nm). (b) Stress-strain curve of a typical TPE (c) An illustration of the formation of TPV by dynamic vulcanization.

(NBR) is termed a TPO. TPC is a blend of a hard segment consisting of polybutylene terephthalate and a soft segment of polytetramethylene oxide glycol. TPAs are made up from rigid polyamide blocks and polyether soft blocks.

Thermoplastic vulcanizates (TPVs) are a sub-class of TPEs that are characterized by dynamic vulcanization of rubber phases during melt mixing of semi-crystalline thermoplastic matrix phases at high temperatures. Therefore, the process of dynamic vulcanization is used to produce TPVs (Figure 10.3b). Dynamic vulcanization (DV) is a dynamic cross-linking method that uses shear and elongational forces during the vulcanization of the soft rubber phase in the presence of molten thermoplastic. The melt-mixing of all blend components and cross-linking of the rubber phase occur concurrently in this process. Dynamically vulcanized TPVs contain at least three components; among these are elastomer, thermoplastic, and an appropriate thermal initiator. Moreover, dynamic vulcanization-derived TPVs main advantages are increased tensile strength, high elongation at break, low permanent set, high crack resistance, fluid resistance, and consistent thermoplastics productivity (Banerjee, Kumar et al. 2015, Yuan, Fan et al. 2018, Awasthi and Banerjee 2021).

Both TPEs and TPVs are characterized by large values for recoverable strain (>100%) and lower tension set (<50%) similar to elastomers. Young's modulus, a ratio of stress to strain, is lower than 10^6–10^9 Pa for TPE. This value is in the range of modulus for soft materials. These properties together make the TPE a better choice for various applications and slowly overcome the market of thermoplastics or elastomers (Sbrescia, Ju et al. 2021, Zanchin and Leone 2021). Due to ease of processability and excellent cost performance, TPEs are proving as a boon for industries.

10.3.1 High-temperature Thermoplastic Elastomeric Material

Currently, industrial requirements have undergone drastic changes in a variety of applications, mainly for the automotive field. It requires high-temperature resistance and high durability, which requires an evolution of a superior performance polymeric material at elevatedtemperatures. Therefore, the high-temperature thermoplastic elastomeric material is developed by a melt-mixing technique of high-temperature performing thermoplastics or elastomers. High-temperature TPEs can be classified in accordance with the selection of a suitable type of high heat resistance thermoplastics and elastomers such as (Banerjee, Kumar et al. 2015, Wang, Schlegel et al. 2016, Banerjee, Bhowmick et al. 2017):

i. High-temperature TPEs are constructed from high heat resistance thermoplastics like polyamide (PA), copolyesters, polyethylene terephthalate, and alternate high-performance engineering plastics.
ii. High-temperature TPEs are built from silicone rubber (Q), acrylate rubber (ACM or AEM), fluorinated rubber (FKM), and hydrogenated nitrile rubber (HNBR).
iii. Super TPVs made from blending of both high-temperature resistance thermoplastic and high-temperature resistance elastomer.

The higher the heat resistance level of the selected material, the more will be the high-temperature functional properties of developed TPEs. Conventional TPEs experiences escalating change in their mechanical, chemical, and physical properties on exposure to heat, whereas high-temperature TPEs retain their properties even at high temperature (Banerjee, Bhowmick et al. 2017). Furthermore, it was observed that weathering resistance, and aging properties of TPEs could be enhanced by narrowing the morphology of developed TPE from microstructure to nanostructure and by incorporating nanofillers (Fang, Zhang et al. 2015, Hidayah, Mariatti et al. 2015, Ristić, Krakovsky et al. 2018, Bhattacharya, Chatterjee et al. 2020). The morphology transformation can be done by adopting high shear mixing method for mixing the desired thermoplastic and elastomeric material.

The addition of nanofillers in the matrix of TPE is a highly effective way to enhance the functional and high-temperature properties of thermoplastic elastomeric materials. Properties of the TPE depend on the quality of dispersion and the level of interaction of dispersed nanofiller with surrounding phases.

10.4 DEVELOPMENT OF HIGH-TEMPERATURE TPEs USING ELECTRON BEAM TREATMENT

A growing number of requirements on polymeric materials and sustainable processing, numerous physical methods are used for the physicochemical modification of polymers in order to apt the requirements of various applications and sustainability. In this sense, electron beam processing is a prevailing tool to modify and develop new polymeric materials with superior properties in comparison to the conventional processing method. Thermoplastic elastomer (TPEs) is an applied engineering material that has unique rubber-like elasticity and thermoplastic-like processability. However, conventional TPEs generally have poor high-temperature features, such as low heat resistance, swelling resistance, and poor compression set. There are several challenges to develop high-temperature TPEs by conventional dynamic vulcanization method. In dynamic vulcanization, the chemical reaction is induced by temperature-sensitive initiators or curatives, which leads to a coupling of all processes involved, such as the flow and mixing process, heat transport, and the

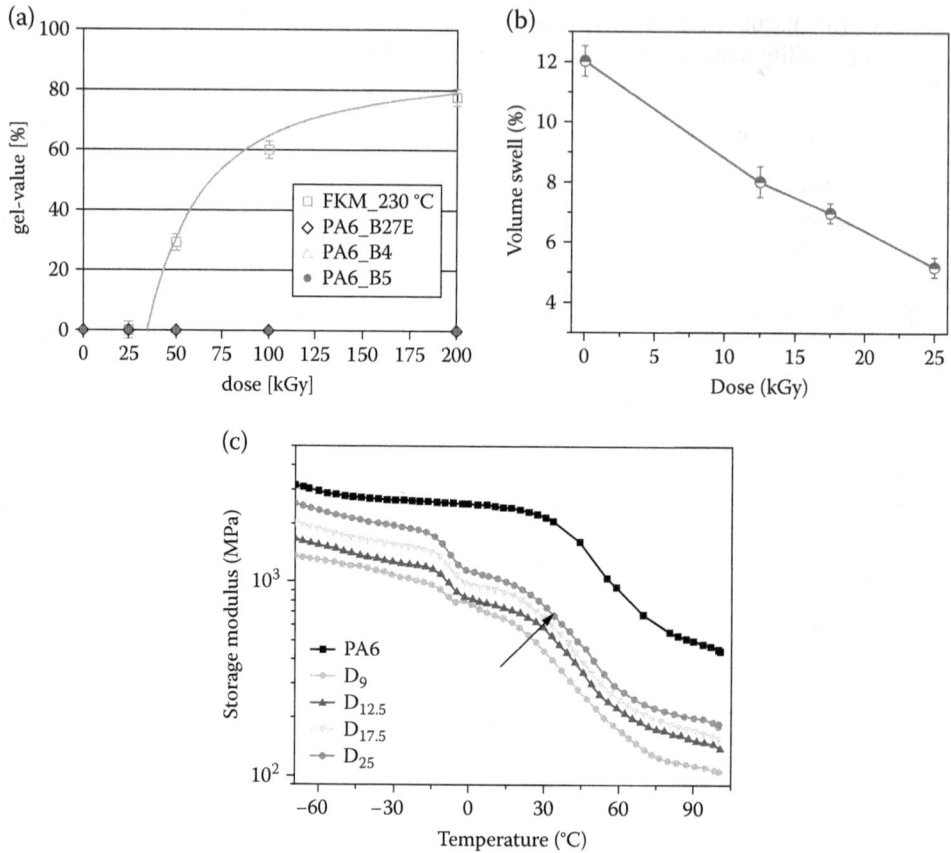

FIGURE 10.4 (a) Gel content of FKM-PA blend with respect to dose (kGy), (b) a volume swelling study of PA/FKM blends in IRM oil at 150°C versus different dose, (c) the storage modulus of EBT modified and unmodified PA/FKM blends with respect to temperature.

chemical reaction. As an outcome, effective controlling the intended chemical reaction and managing the variable viscosity as a result of the chemical reaction is challenging. Furthermore, shear forces and deformation rates impact the chemical process. Consequently, comprehensive experiences in chemistry, flow modeling, as well as mixing, process control, and rheology are required. Nevertheless, it is difficult to analyze and control conventional reactive processing.

Electron beam treatment (EBT) is used for cross-linking, curing, degrading, grafting, and polymerization of polymers (Figure 10.4a). In contrast to conventional reactive processing, EBT modifies polymers and their compounds by using the controlled spatial and temporal supply of electron energy. The main processing parameters of EBT are acceleration voltage, beam current, dose, and dose rate. In electron beam treatment, the influence of the properties of polymers to be modified (constitution and configuration of the polymers, cross-linking additives, and antioxidants) and the chemical environment during EBT (gas atmosphere, water vapor, pH-value, and temperature) is taken into account to enable a tailored chemical modification of polymers (Gohs, Leuteritz et al. 2010). In this section, the recent development of high-temperature TPEs by EBT, their structure, morphology, and properties are discussed.

Based on the basic principles of EBT, Banerjee et al. created a high-performance polyamide 6(PA)/fluoroelastomer (FKM) blends without the addition of any additives. It was found that EBT process parameters such as non-reactive blending time, dose/revolution, and electron-energy have a significant effect on blend morphologyand characteristics. In addition,based on gel-content measurements for FKM-and PA, the dosage parameters to be used were derived, as shown in Figure 10.4a. The blends were created by adjusting the absorbed dosage to 12.5, 17.5, and 25 kGy, while maintaining the electron energy constant at 1.5 MeV, the dose per revolution constant at 17.2 kGy/rpm, the temperature constant at 230°C, and the rotor speed constant at 60 rpm. Also, the gel content for FKM and PA was determined after 24 hours of extraction in boiling methyl-ethyl-ketone and formic-acid, -respectively. As illustrated in Figure 10.4a, from 0 to 25kGy, there is no cross-linking, so three dosage ranges of 12.5, 17.5, and 25 kGy were used. These dosage levels relate to electron-exposure-times of 9, 12, and 18 seconds. According to their findings, the EBT adjusted blends demonstrated excellent mechanical performance, high thermal stability, and high oil swelling resistance. For instance, the high oil resistance of the PA/FKM blends was determined by submersing the blended samples in IRM-903 oil at 150°C for 72 hours. As shown in Figure 10.4b, the volume swell was significantly reduced in electronbeam treated blends. Moreover, during electron treatment, the volume swell of the blends was also reduced when the dosage was increased. As shown in Figure 10.4b, the EBT with 12.5 kGy led to a volume swell of about 8%. After EBT with 25 kGy, the volume swells further decreased to 5%. Therefore, it is understood that the inclusion of highly oil-resistant FKM domains in the least swelling PA medium primarily regulates the swelling resistance of these blends. Furthermore, the decreasing volume swell with increasing dose might be caused by the enhanced electron induced branching as well as decreased particle size and increased volume of reduced mobility region. Due to the constant volume fraction of FKM in the PA medium, the relative volume of reduced mobility enlarged along with increasing dosage. Due to this, the equilibrium volume does not swell as much. Also, the creation of electron-induced long-chain branching at the crossing pointor interface might influence the swelling resistance of EBT improved blend system. As of the phase morphology of unmodified and electron-induced PA/FKM blends, a broad variety of the dispersed domain size was found due to mixing under dynamic settings. It was discovered that the domain size of unmodified blends was much bigger than that of electron-induced blends. Dn of the scattered domain was 2.5 m in the case of the untreated blend. For the 25 kGy electron-induced blends, this value was reduced to 1.1 m. This microstructural analysis confirmed a wide spread of the scattered domain as well as a considerable reduction in size (by 50%). The decrease in domain size of electron-modified blends was caused by decreased interfacial tension and improved adhesion between the blend constituents. The morphology of electron modified blends corresponds to their physical and mechanical characteristics. Detailed investigations showed that the increase of these features is mostly due to the creation of electron-induced long-chain branching (Banerjee, Gohs et al. 2016).

In addition, the impact of EBT on the nano-mechanical characteristics and crystalline structure of PA/FKM blends with modern peak force quantitative nano-mechanical mapping atomic force microscopy (AFM) was successfully studied. Further, utilizing small and wide-angle X-ray scattering (SAXS and WAXS), the effect of EBT on the long period of PA lamellae and its crystallinity in the blends was also investigated. Using WAXS, it was proven that PA in PA/FKM blends had a different crystalline structure than pure PA. In the blend without EBT, the c-axis of PA was somewhat dropped. This impact was clearly emphasized after EBT of these blends. Moreover, from these comprehensive studies, the AFM-based DMT moduli of the PA matrix phase along with the FKM dispersion phase were observed to rise with increasing dose, but force of adhesion, energy of dissipation, and deformation behaviour of both phases reduced from 12.5–25 kGy as shown in Figure 10.4d–g. The variations in nanomechanical characteristics of PA/FKM blends were connected with the parameter of interaction between the two phases of the mixed blend constituents, the significant duration of the PA lamellae in the PA/FKM blend, the crystallinity, and the crystallite size of the PA/FKM blends formed. Theparameter of interaction of the unmodified blend was great before EBT and declined afterward. However, by using WAXS, it was proven that PA in PA/FKM blend had a different crystalline structure than pure PA. This impact was clearly emphasized after EBT of these blends. Eventually, increased inter-domain contact of phase components and a longer period of PA lamellae, as well as crystallinity, were found to be responsible for better DMT modulus and decreased force of adhesion, energy of dissipation, and distortion following EBT (Banerjee, Janke et al. 2017, Banerjee, Janke et al. 2018).

Few additional TPEs developed by RET technique are also discussed in this section. For example, Chattopadhyay et al. used EBT todevelop a novel TPE from polyethylene (PE) and ethylene–vinyl acetate (EVA) blends with ditrimethylol propane tetraacrylate, at various doses from 20–500 kGy (Chattopadhyay, Chaki et al. 2001). According to their findings, electron-induced cross-linking of PE–EVA blends resulted in stronger interfacial adhesion, as well as increased tensile strength and modulus. Similarly, Rooj et al. used EBT to investigate the compatibilized effect of non-polar polypropylene (PP) and polar epoxidized natural rubber (ENR) along with triallyl cyanurate (Rooj, Thakur et al. 2011). As per their findings, detailed FTIR analysis revealed the formation of phase-coupling and chemical compatibilization. Along with this, DMA studies revealed increased high-temperature modulus, and rheological evaluation revealed that modulus increased in the low-frequency range. Further, DSC analysis revealed that EBT and the addition of the rubber phase into the non-polar PP matrix had a substantial influence depending on the crystallinity of the blend system. With these findings, they drew the conclusion that the EBT of these blends performed the best modification. Based on the same concept of modification, Fray et al. scrutinize the sway of various doses on the chemical structure-of a multiblock thermoplastic elastomer composed of poly (aliphatic/aromatic–ester) (El Fray, Piątek-Hnat et al. 2012). The experimental results revealed that EBT increased the molecular-weight and tensile-strength of the poly (aliphatic/aromatic–ester) significantly. Furthermore, DSC analysis revealed that cross-linking occurs mostly within the hard phase, as evidenced by a small rise in melting temperature (T_m) and a decrease in crystallinity. Further, the mechanical characteristics of these multiblock copolymers were improved by the development of an extra electron-induced cross-linked network. Further, Anagha et al. inspected the sway of EBT on cross-linking of SEBS/TPU and SEBS-g-MA/TPU blends in the dosage range from 0 to 100 kGy (Anagha, Chatterjee et al. 2022). According to their findings, the complex-viscosity of both blends increased significantly, indicating the presence of highly cross-linked networks, which resulted in the development of a three-dimensional cross-linked network. In addition, electron-induced cross-linking demonstrated enhanced mechanical properties in blends of SEBS-g-MA/TPU up to a dosage of 25-kGy. On the other hand, even though cross-linking occurs in the SEBS/TPU system, the lack of component compatibility has hampered the increase in performance characteristics.

10.5 FACTORS AFFECTING REACTIVE ELECTRON TREATMENT FOR THE DEVELOPMENT OF TPEs

Reactive electron treatment of a TPV is a more advanced approach than typical thermal and chemical methods. Electron-induced modifications in thermoplastic elastomeric materials i.e., either in a plastic domain or elastomeric domain or both may result in certain beneficial physicochemical features in the finished product. This treatmentis used in a variety of industrial processes such as polymerization, cross-linking, graft copolymerization, curing of lacquers and coatings, and so on (Banerjee, Gohs et al. 2016). Because of the benefits, such as when compared to other radiation treatments such as gamma or X-rays, EBT has a higher throughput and does not use radioactive isotopes making EBT less risky than a gamma ray. An electron accelerator is the most beneficial equipment for the production of ionizing radiation. The electron energy is stepwise transformed to the material and leads to excitation or ionization of atoms or molecules. Finally, highly reactive polymer radicals are formed (Kashiwagi 2004). The chemical modifications caused by these reactive species are used in a number of applications and are the driving force of the electron beam treatment sector. Based on this, the aforementioned are the main factors that affectthe modification of polymers via EBT.

10.5.1 Significance of absorbed dose

The absorbed dosage is a critical parameter that might affect the characteristics and morphology of high-temperature TPEs during reactive electron treatment. Generally, absorbed dose refers to the quantity of energy absorbed by the polymer materials. Its unit is expressed in gray (Gy) (Sengupta, Banik et al. 2008). Controlling this parameter is important because it controls the cross-linking characteristics of TPEs. Simultaneously, it has been proved that the higher the dose, the greater will be the mechanical properties. For instance, Thakur and coworkers studied the impact of absorbed dose and electron dose/rotations on PP-EPDM blends (Thakur, Gohs et al. 2012). In their studies, three distinct doses such as 50, 75, 100 kGy and three different dosages on each rotations values like 70, 90, and 110 kGy per rotation were chosen to investigate their impact on the production of TPVs. From their studies, absorbed dosage, as well as dose on each rotations or gyrations, showed a significant impact on the tensile strength of the TPVs. When the dosage was raised from 50 to 75 kGy, TPVs treated at 70, and 110 kGy per gyrations exhibited a considerable improvement in tensile strength. Surprisingly, when the dosage was raised to 100 kGy, the TPV's tensile strength jumped vividly to 14MPa, the highest for all TPVs generated. In addition to this, the variance in elongation break-was very close to the variation in mechanical characteristics at various doses-and dosage per rotation-values. As the-dosage isamplified from 50 to 100 kGy during 70 kGy per gyration, the elongation at break values enhanced. The TPVs under treatment with 100 kGy-absorbed dosage at 90 kGy per gyration had the greatest elongation break value of 618%. As a result, they conclude that the TPVs under the treatment of 100 kGy and 90 kGy per gyration had the best combination of mechanical characteristics. So the optimal absorbed dosage meant for the formation of PP-EPDM TPVs appears to be 100 kGy. Similarly, Das et al. reconnoitered the effect of dose on the hydrogenated nitrile rubber (HNBR) and nylon-6 TPVs and found that the electron-induced cross-linking and the mechanical strength of the blend system progressively rises with dosage. In their study, blend system with excessive nylon have the highest mechanical characteristics at 60 kGy, whereas blends with excessive HNBR have the highest mechanical characteristicsat 40 kGy. The lower tensile strength of HNBR rich blends is due to the increase in particle size of the HNBR phases at higher doses and nylon degradation by electron beam at that particular dose. Because of this, the tensile strength of a plastic elastomer blend is proportional to the strength of the plastic component, so the blend with a higher nylon percentage has a higher tensile strength (Das, Ambatkar et al. 2006, Das, Ambatkar et al. 2006). Additionally, with increasing absorbed dose, the elongation at break of the TPVs rises progressively, andthe increased elongation reflects improved stress transmission over the blend system's interface. As a result, during stretching, the potential of fracture formation and propagation at the contact decreases. Furthermore, they conclude that the increase in elongation at break is most likely related to the presence of significant contact at the interface as a result of electron-induced grafting and cross-linking. Along this line, Chatttopady et al. used this concept to study the effect of electron dose on tan(δ) of TPE (Chattopadhyay, Chaki et al. 2001). They used doses of 20, 50, 100, 200, and 500 kGy to modify a 50/50 PE-EVA blend. According to their results, the glass transition temperature (T_g) rapidly increases as the dose increases. As a result, the degree of cross-linking develops, therebyan increasing potential barrier to relaxation. The tanδ_{max} corresponding to T_g first falls, reaches a minimum at 50 kGy, and then abruptly intensifies. There is an initial drop in tanδ_{max} which might be attributed to an increase in cross-linking. Increased mobility of chain branches occurs at higher doses, probably due to the loss of the vinyl acetate group. Additionally, the dynamic storage modulus first increases with a dose to reach its maximum value at a dose of 50 kGy, and subsequently falls as the dose increases. The first rise in storage-modulus is most likely due to cross-linking and improved interfacial adhesion. The steady decrease in storage modulus with increasing doses may be attributed to a change in structural properties of the TPEs produced by EBT. In the same manner, Ratnam et al. designed a 50:50:2 combination of polyvinylchloride (PVC)-epoxidized natural rubber blend (ENR) and carbon nanotubes (CNTs) nanocomposites.

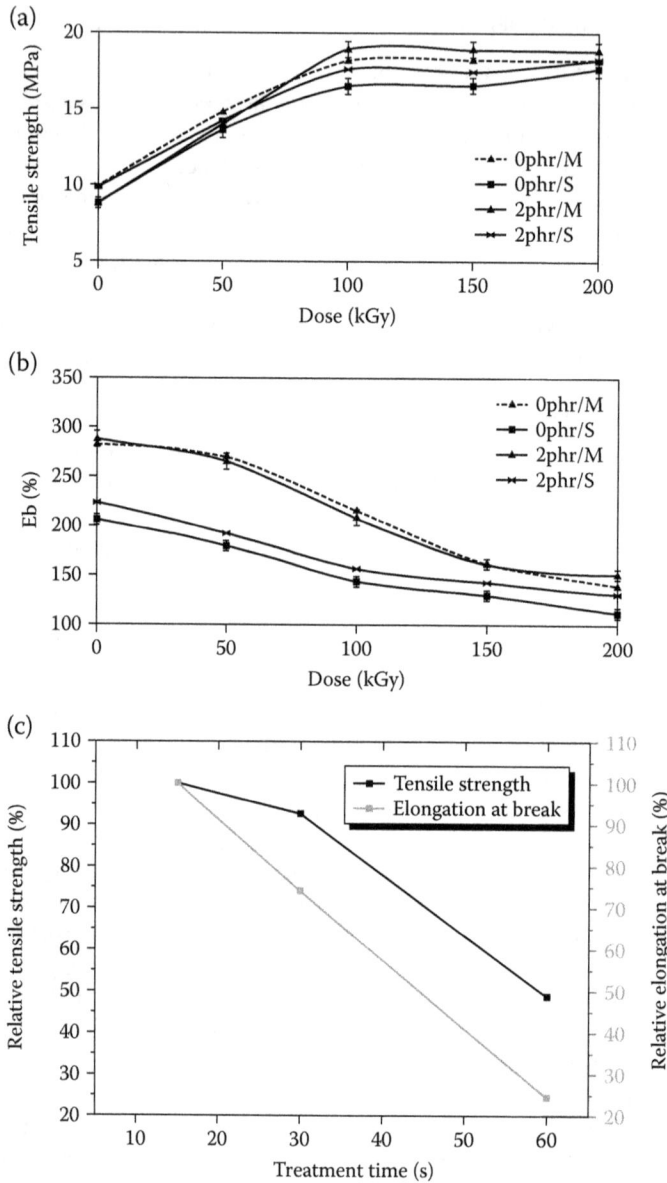

FIGURE 10.5 (a) Tensile strength of TPVs subjected to varying dosages, (b) the elongation at break of TPVs subjected to varying dosages, (c) tensile strength and elongation at break with different time of treatment prior to EBT at 100 kGy.

EBT was applied to the nanocomposites at dosages from a scale of 0 to 200 kGy. Their findings indicate that the influence of tensile characteristics (Figure 10.5a), i.e., by the inclusion of 2 phr of CNT preparatory to EBT reduced the mechanical features of the PVC/ENR blend, thereby suggesting inadequate CNT dispersion in the blend system. However, after increasing the dosages, the melt-blending nanocomposites had higher mechanical values than the immaculate PVC/ENR blend due to the presence of reactive electron cross-linking in the medium (Ratnam, Ramlee et al. 2015). Additionally, drop in percentage elongation with respect to increasing dosage was observed which wasinitiated by the same reactive electron cross-linking in the blend system (Figure 10.5b). It is evident that, as the dosage rises, more cross-links are formed in the blend, which limits structural rearrangement upon material distortion, resulting in a decreased elongation. Further, as seen in Figure 10.5b, the addition of CNT to the blend also resulted in a decrease in elongation, which evinces that there is an improved CNT filler dispersion in the PVC-ENR blend system. So, the higher dosage will diminish CNTs aggregates, which function as stress concentration sites under load and result in failure at low strain levels. Furthermore, better CNT dispersion leads to improved filler matrix interaction, which lowers chain mobility thereby reducing the elongation (Ratnam, Ramlee et al. 2015).

10.5.2 SIGNIFICANCE OF PREMIXING TIME

Premixing time is another important factor which affects TPVs properties by electron treatment. This section describes the effect of the mixing period prior to electron treatment on the final characteristics of TPVs. Generally, there are four processes for the evolution of morphology in TPVs during dynamic vulcanization (Goharpey, Katbab et al. 2001, Banerjee, Janke et al. 2018). The first process is rubber phase melt dispersion in a thermoplastic matrix, which is linked with rubber-phase globule breakage and amalgamation prior to vulcanization. Then the agglomeration of cross-linked rubber particles occurs concurrently with globule breakage during dynamic vulcanization. In addition to this, the interfacial adhesion between the two phases rises as vulcanization progresses and the agglomerates break down as a consequence of an increase in interfacial adhesion and applied shear forces, resulting in the dispersion of rubber particles into the continuous phase. Simultaneously, an effective balance between rubber-phase breakage and coalescence in melt mixing with a thermoplastic matrix is critical to the development of TPVs, which defines their final functional properties (Goharpey, Katbab et al. 2001, Goharpey, Katbab et al. 2003, Wu, Tian et al. 2014, Banerjee and Bhowmick 2016). Therefore, mixing time before dynamic vulcanization is crucial. The role of premixing time before reactive electron treatment was carried out by Thakuret al. on the PP-EPDM blend (Thakur, Gohs et al. 2012). To investigate the influence of premixing time, the mixing time before electron treatment was adjusted to 8, 11, and-14 minutes, while the overall mixing-timeremained fixed at 16 minutes. Along with this, an absorbed-dose of 100 kGy and a dose per-rotation of 90 kGy per rotation was chosen-because these settings gave the optimum combination of mechanical characteristics for TPVs. The mechanical characteristics of the respective TPVs revealed a noticeable influence on the mixing time prior to reactive electron treatment. The change in tensile-strength and elongation at-break for TPVs made with mixing durations before electron treatment of 8, 11, and 14 minutes. Tensile-strength and elongation-at-break values for the TPV were 9MPa and-240% for 8 minutes premixing period. When the premixing time was increased to 11 minutes, the tensile strength and elongation at break values improved to 10 MPa and 320%, respectively. For a premixing period of 14 minutes, the tensile strength and elongation at break values achieved a maximum of 14 MPa and 618%, respectively. This pattern demonstrates the significance of the premixing step in TPV creation prior to reactive electron treatment. When EPDM and PP are melt mixed, the blend shows a co-continuous phase where neither EPDM nor PP is the continuous phase. As for mixing proceeds, an equilibrium between rubber-phase breakage and coalescence is formed, determining the final phase morphology shortly before EBT. Because of applied shear stresses and increased interfacial-adhesion between the dualistic phases thru melt-mixing, the cross-linked rubber-particles break down into minor disseminated particles during EBT. The size of these dispersed rubber particles is also determined by the shape and dispersion of the rubber phase prior to EBT. The mechanical characteristics of TPVs increased as the premixing-time increased from 8 to 14 minutes, demonstrating this dependency. These findings show that the time spent for premixing PP and EPDM before reactive electron treatment has a substantial impact on the morphology, and henceforth the mechanical characteristics, of final TPVs. Similarly, followed by this same design principle of PP-EPDM TPVs, Gohs et al. concluded that electron treatment time, which influences dosage rate and absorbed dose per gyration, is an extra considering parameter determining the stress-strain performance of the TPVs (Gohs, Leuteritz et al. 2010). In their study, they used electronenergy of 1.5 MeV and a dose of 100 kGy. Their experimental resultsindicated that lowering the treatment time from 60–15 seconds enhanced tensile strength-and elongation-break while keeping modulus unchanged. Furthermore, the highest-tensile-strength and maximum-elongation-break were obtained during a 15-second electron treatment time. Similarly, Mondal et al. analyzed the performance of electron treatment time during EBT of dosage 100 kGy on tensile strength and elongation at break for PP/NR TPVs, as shown in Figure 10.5c. From their scrutiny, at constant absorbed dosage and rotor speed, the performance of PP/NR TPVs drops vividly with increasing electron treatment duration. When the treatment period was increased from 16–32-seconds, the values of tensile strength and elongation a break were reduced by 10% and 25%, respectively. A further increase in treatment time to 64 seconds lowered the output of PP/NR TPVs by over almost 50%. In conclusion, they estimated that an ideal electron treatment period of 11 seconds for an ingested dosage at 100 kGy at such a consistent rotation speed of 90 revolutions per minute was required to prevent a curing rate mismatch during the manufacturing of PP-NR-TPVs (Mondal, Gohs et al. 2013).

10.5.3 SIGNIFICANCE OF DIFFERENT BLEND RATIOS OF TPEs

This is yet another crucial aspect influencing the characteristic features and morphological behavior of typical TPVs. In TPVs, before a reactive electron treatment, the rubber phase will be in a dispersed phase at a specific blend ratio, and after a reactive electron treatment, that rubber phase will still be in the dispersed phase but with smaller particle size. In addition, the differences in mechanical characteristics were caused by differences in the morphologies of these two distinct phases of TPVs. Moreover, the shape of immiscible elastomer/thermoplastic blends following dynamic vulcanization is mostly determined by the viscosity and composition ratio of the two blend components, as seen in Figure 10.6. From Figure 10.6, if we consider pristine thermoplastic and elastomer, the rubber/thermoplastic viscosity ratio was approximately 10, but the viscosity ratio grew approximately 100 after thermoplastic and rubber phaseswere melted and mixed for a

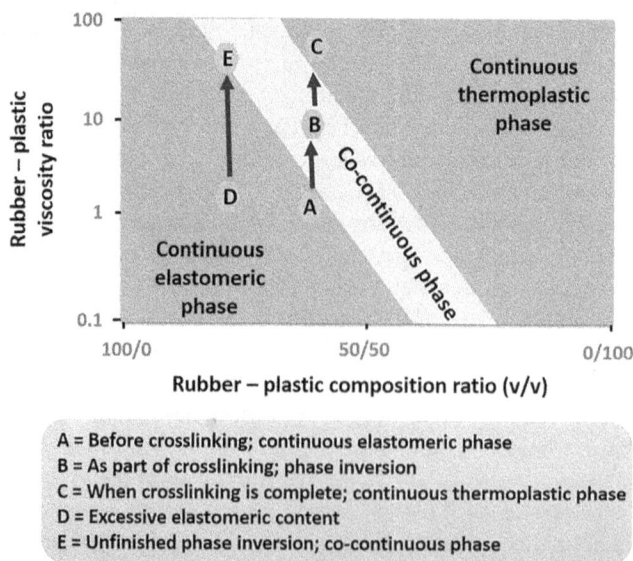

A = Before crosslinking; continuous elastomeric phase
B = As part of crosslinking; phase inversion
C = When crosslinking is complete; continuous thermoplastic phase
D = Excessive elastomeric content
E = Unfinished phase inversion; co-continuous phase

FIGURE 10.6 The effect of composition and viscosity ratio on the morphology of an elastomer/thermoplastic blend during melt processing.

particular time, respectively. This time was chosen based on the electron treatment time as it took to premix the thermoplastic and the rubber before the reactive electron treatment. Because the elastomer concentration was high and the viscosity ratio before electron treatment was around 100 at a 30/70 blend ratio. Here, point D switches to a region of co-continuous phase, and point E remains in the co-continuous phase after electron treatmentdue to incomplete phase inversion. Further, points A and B switch to the continuous thermoplastic phase before and after reactive electron treatment, respectively, at a 50/50 blend ratio, whereas point C remains in the continuous thermoplastic phase (Jordhamo, Manson et al. 1986, Mekhilef and Verhoogt 1996, Shariatpanahi, Nazokdast et al. 2002). Based on this concept, Thakur et al. studied the impact of various PP-EPDM blend ratioson the mechanical, morphological, and impact properties of the TPVs (Thakur, Gohs et al. 2012). In their study, two different blend ratios ofPP-EPDM i.e., 30/70 and 50/50 were chosen. In addition, a dose of 100 kGy, 90 kGy per revolution, and a 14-minute premixing period were chosen for their study. Further, from the variance in tensile strength and elongation break for a PP-EPDM TPVs using reactive electron treatment, it is clearly understood that the influence of the blend ratio in TPVs i.e., when compared to the matching control sample (0 kGy), for example, TPVs made with a 30/70 blend ratio demonstrated a marginal gain in tensile strength of 5.6 MPa. Similarly, when compared to the 50/50 control sample, the tensile strength and elongation at break of the electron treated samples were about 14 MPa and 618%. As a result, they concluded that 50/50 PP/EPDM blend ratio gave the optimum set of mechanical characteristics with reactive electron treatment at the experimental parameter used for EBT. Additionally, from the morphological behavior through SEM studies, they mentioned that in 50/50 blend, EPDM was already dispersed in the continuous PP phase before electron treatment, and EPDM was still scattered in the dispersed phase morphology even after electron treatment, albeit with smaller particle sizes. Furthermore, the distinct deformation processes acting in the two TPVs with differing blend ratios are linked to the discrepancies in tensile strength and elongation at break. Along with this, the improvement in mechanical characteristics when the rubber-particle size is <1 µm is known to be related to crazing in the rubber particles under deformation. Because under this condition, the deformation process in the 30/70 TPV is mostly dependent on the crazing-of the rubber particles (Jain, Nagpal et al. 2000, Bucknall 2007). When the rubber-particle size is <1 µm, however, shear deformation is the primary source of mechanical property improvement. Under this condition, the shear yielding of the PP matrix dominates the deformation process of the 50/50 TPV. Additionally, localized yielding of the PP matrix is due to the thin rubber ligaments that enclose the PP matrix, which dominates the deformation behavior of TPV. It is well known that rubber particles smaller than 1 µm cannot produce crazes.Hence shear yielding is the primary deformation process in such particles. Rubber-particles with a diameter of >1 µm, on the other hand, are able to internally cavitate, resulting in the local release of applied stress during tensile deformation. Another notable aspect of this blend ratio is that the addition of the rubber phase to the thermoplastic matrix improves the impact strength of the TPVs substantially. This behavior is assumed to be caused by the addition of a higher volume of rubber component, which is known for its superior damping characteristics (Thakur, Gohs et al. 2012).

As part of this study, Chattopadhyay et al. conducted a similar investigation using thermoplastic elastomeric blends of PE and EVA, and the influence of the blend ratio on the characteristics was evaluated at 100 kGy and 0 wt% monomer level. According to their findings, because of the increasing crystallinity of the blends, the tensile strength and 100% modulus rise as the PE component of the blend increases. However, when the EVA content is increased, the effective energy dissipation process gets simpler, resulting in an increase in the elongation break. Additionally, increased EVA content increases cross-link density, resulting in a lower ratio of hysteresis loss and permanent set. Also, the temperature-dependent fluctuation of tan d for various blend ratios treated at 100 kGy. Tan δmax (at T_g) rises as the blend's EVA content increases. However, at room temperature and higher, the tan d value increases as the PE percentage of the blend increases. Finally, due to the melting of the PE crystallites, it tends to a maximum at higher temperatures (α-transition). The storage modulus, on the other hand, increases as the blend's PE concentration increases (Chattopadhyay, Chaki et al. 2001).

10.5.4 SIGNIFICANCE OF CROSS-LINKING BY ELECTRON BEAM TREATMENT

Reactive electron treatment is commonly used to cross-link TPEs. The electron beam radiation on the commercially available polymers is illustrated in Figure 10.7a (Mizera, Manas et al. 2016). Typically, this cross-linking requires at least one α-hydrogen, and the absorbed electron energy must be larger than the bond energy of the labile bonds (Sengupta, Banik et al. 2008). The dose mentioned in the above section is the primary factor that influences the cross-linking or cross-linking efficiency of thermoplastic elastomeric materials. Apart from this, the addition of external agents known as a cross-linking booster or cross-link promoters can also stimulate cross-linking of TPE materials. They are mainly divided into two types: indirect cross-link boosters and direct cross-link boosters. Indirect cross-link boosters areexternal agents that will not directly enter the cross-linking reaction but rather enhance the production of radical intermediates such as free radicals, which then lead to the formation of cross-links via reaction conditions. A few examples which come under this category are: alkyl halides, nitrous oxide, sulphur monochloride, ammonia, amines, and water (Okada 1967, Yunshu, Yibei et al. 1998, Lappan, Geißler et al. 2000, Nagasawa, Kaneda et al. 2005). On the other hand, direct cross-link boosters engage the cross-linking process directly and form an interconnected molecular bond in their own way. This category mainly includes alkyl and aryl maleimides such as N-phenylmaleimide and m-phenylene dimaleimide.

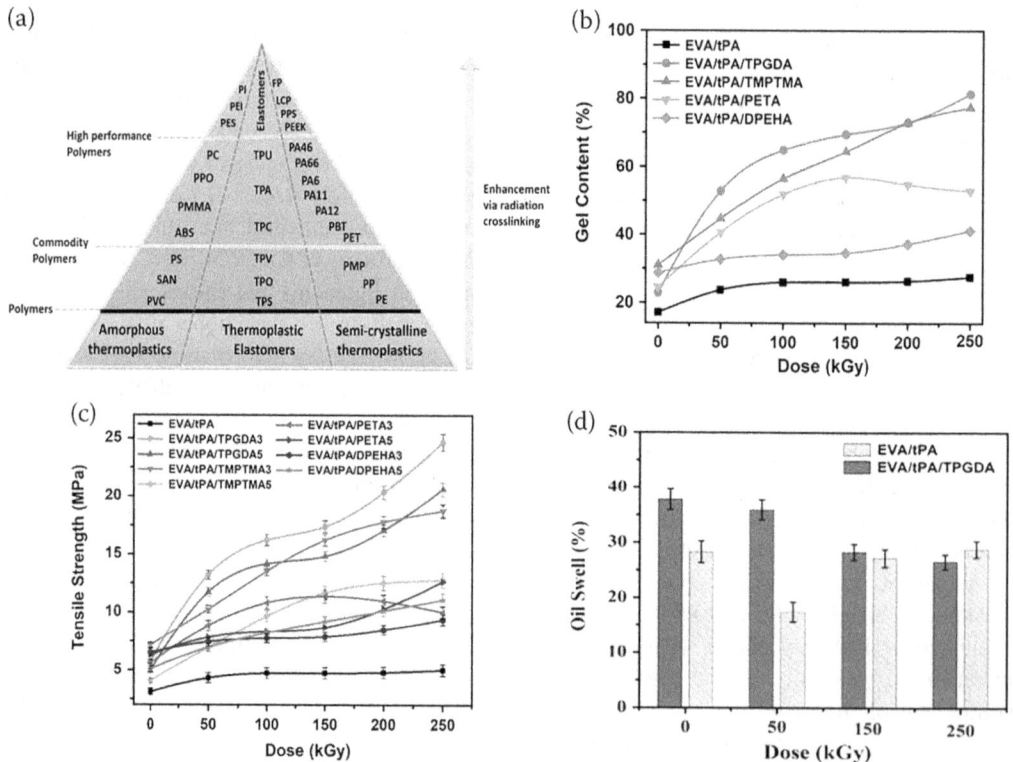

FIGURE 10.7 (a) Commercially available polymers that can be treated with radiation cross-linking. (b) Effect of dose (kGy) on gel content of co-agent modified and unmodified EVA/tPA blends. (c) Effect of dose (kGy) on tensile strength of co-agent modified and unmodified EVA/tPA blends. (d) Effect of dose (kGy) on oil swelling of TPGDA modified and unmodified EVA/tPA blend.

Evena very small quantity of these maleimides will considerably improve the cross-link yield of electron-treated TPVs (Miller, Roberts et al. 1962, Odian, Lamparella et al. 1967). Mohamad et al. investigated this notion in natural rubber (NR)/recycled rubber (RR)/HDPE blend by sodium dodecyl sulphate (SDC) where SDC acts as a cross-link booster. After the preparation of the blends, the TPE blends were compressed into sheets and then electron treated with doses from 50 to 200 kGy (Mohamad, Osman et al. 2021). According to their result, the gel content is increased with increasing dose. Additionally, the incremental patterns indicate that the electron treatment improved the cross-linking. Cross-linking increases TPE stiffness and, as a result, prevents it from dissolving in the solvent. Based on these findings, it is possible to deduce that a larger gel content implies a higher-degree of cross-linking between NR/RR/HDPE, which happens during the mixing stage and is produced by thetreatment. The dose rate used to modify a specific polymer determines the degree of cross-linking in the presence of oxygen. Similarly, Ahmed et al. studied the impacts of EB cross-linking on the mechanical and thermal characteristics of ethylene-vinyl acetate co-polymer (EVA)/ternarycopolyamide (tPA) blends treated with cross-linking co-agents such as tripropylene glycol diacrylate (TPGDA), trimethylolpropane trimethacrylate (TMPTMA), pen-taerythritol tetraacrylate (PETA), and dipenterythritolhexaacrylate (DPEHA) (Ahmed, Wu et al. 2020). In this study, dose ranges from 50 to 250 kGy were used. According to their conclusions, the tensile strength of the blends containing co-agents enhanced effectively when the dosage was raised compared to the neat blend without any co-agents (Figure 10.7b). Further, with 250 kGy of EBT, the gel content of the blend with 3 phr TPGDA drastically increased to 81%, which was significantly more than the 22% of EVA/tPA without co-agent (Figure 10.7b). In addition, EVA/tPA/TPGDA demonstrated the least amount of oil swelling. After treatment at doses up to 250 kGy, the oil swelling of an EVA/tPA blend decreased by 26.5% in IRM 903 oil. EVA/tPA/TPGDA had the lowest oil swelling, which reduced from 28.3% to 17.4% at the lower dosage of 50 kGy, but its oil swelling increased as the dose increased. The rise and decrease in oil swelling are associated with cross-linking, and chain scission that occurs as dosages fluctuate. So, in general, cross-link promoters along with electron dosage are playing an important role in enhancing the characteristic features of a typical TPE material.

10.6   APPLICATIONS

The use of reactive electron treatment in the development of commercial products is a relatively new technological development. Today's electron technology employs electron beams supplied via machines as a processing tool in the industry to create a variety of high-quality and tailored products. Sterilized medical products, heat resistance insulations of cables, heat-shrinkable films and tubing, cross-linked polyethylene foam, radial automobile tyres, ion-exchange membrane for battery separator, computer floppy disc, panel circuit board (PCB), and others are among the products (Yamamoto 1990, Datta, Chaki et al. 1998, Kashiwagi 2004, El-Nemr, Khattab et al. 2011, Ray Chowdhury, Sharma et al. 2016, Loo, Sin et al. 2017). These applications with respect to electron energy are depicted in Figure 10.8. Electron-induced chemical and biological mod-ifications were commercialized since this technology fulfills the principles of green processing and high societal need.

10.6.1   ELECTRON BEAM TREATMENT IN BIOMEDICAL SECTORS

The use of thermoplastic elastomers for biomedical applications has continued to grow, and the horizon is still expanding (Basak 2021). Accelerated electrons are commonly used for the pro-duction of medical devices and sterile medical devices. Surgical dressings, wound care products, intravenous administration kits, dialyzers, endoscopic loops, catheters, artificial muscles, contact lenses,diapers,and implants are among the devices regularly and ultimately used by electron treatment (Sandle 2013, Ataee, Li et al. 2017, Ramskogler, Warchomicka et al. 2017, Nune, Li

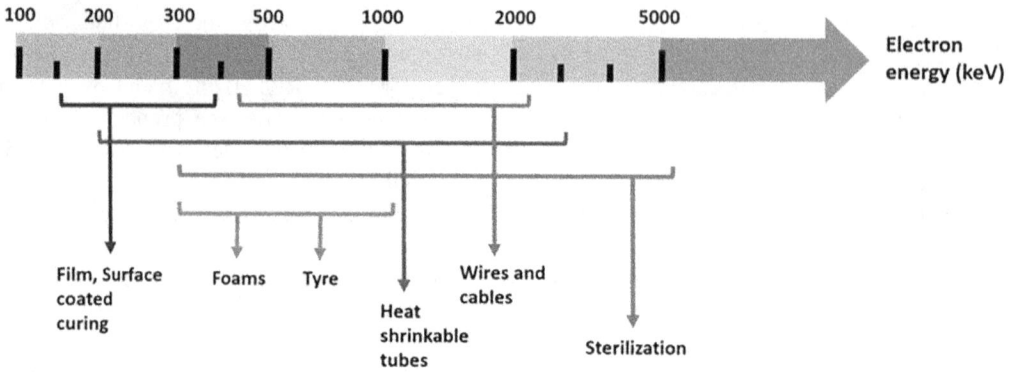

FIGURE 10.8   The relation between electron energy and its application.

et al. 2018, Wei, Anniyaer et al. 2019). Aside from the aforementioned applications, electron beam sterilization of medical products is well developed since other traditional treatments such as autoclaving, heat index, including the use of olefins gases use heat, are delayed processes, and involve the use of a noxious gas, when on exposure is hard to manage. One of the benefits of electron beam sterilization is that the sterilizing procedure is fast, low temperature, and standardized by ISO 11137. For example, surface sterilization such as aseptic filling of pharmaceuticals based on electron beam sterilizing procedure which is fast due to the fact that low energy electrons deposit energy at or near the surface, resulting in an exceptionally rapid rate of dose delivery. The rapid dosage also shields the material being sterilized. Following are the advantages of electron treatment in biomedical sectors.

- Even though the installation of the electron beams sterilizing facility is somewhat expensive, the operational/running cost is low.
- The technique ensures sterilization with a high degree of uniformity.
- Most materials, particularly plastics, are compatible with electron treatment.
- The material is safe to handle after sterilizing and does not require confinement.

The sterilization dose to be used in the sterilization process depends on the amount and species of microorganisms existing on the manufactured goods (bio-burden), the sterilization method lethality, and, in some instances, the environmental condition prior and during electron treatment. The efficacy of the sterilization process is verified by validation of the complete production process, including the microbiologicalstatus of raw materials, the microbiological barrier characteristics of the packaging, the maintenance of the environment in which the product is manufactured, assembled, packaged, and put in storage as well as the sterilization process. Further, the electron-induced changes in product properties are be studied to define the maximum acceptable product-specific dose.

Based on these advantages, Balaji et al. investigated the effect of electron beam treatment on PP/EPDM blends in order to produce TPV-based medical devices to be sterilized by accelerated electrons in the final packaging. The blends were treated in air at ambient temperature using a 3.0 MeV electron beam accelerator at doses ranging from 0 to 100 kGy. According to their findings, the improvement in mechanical characteristics is probably related to enhanced compatibility between PP and EPDM via electron-induced cross-linking. Furthermore, in vivo tests indicate that no sensitivity remained after exposure to dose levels of 25 and 100 kGy. Thus, the PP/EPDM blend modified by E-beam gives us the impression that these blends might be a viable substitute for PVC in the manufacture of medical equipment (Balaji, Ratnam et al. 2018).

10.6.2 ELECTRON BEAM TREATMENT IN WIRES AND CABLE APPLICATIONS

Wire and cable cross-linking through electron treatment is one of the most effective applications of EBT and continues to be an appealing technology for modifying a wide range of novel polymer materials used in other industries (Cleland and Galloway 1812, Yamamoto 1990). Electron beams used for EBT of electrical wire and cable are most widely employed within the electron energy range of 0.5–3 MeV and, sometimes, up to 10 MeV. The quality of EB cross-linking strongly depends on geometrical parameters of cable and wires as well as extraction device and synchronization between the beam current and transportation system. Consequently, multiple-sided treatment systemsare available as well as improve the homogeneity of dose application and the completeeffectiveness of energy absorption.

The throughput is indirectly related to the overall-efficiency of the energy absorption and the required dose and directly related to beam power, which can vary between 20–500 kW depending on the accelerator type. In wire and cable treatment, the absorbed dose typically ranges from 50 to 200 kGy (Sabharwal 2013, Zimek, Przybytniak et al. 2014). Following are the advantages of electron treatment in wire and cable application (Hanisch, Maier et al. 1987, Yamamoto 1990, Sabet, Hassan et al. 2012):

- enhanced heat resistance such as enhanced soldering resistance, enhanced dimensional stability under heat influence, enhanced of hot-set, enhanced flame EB Tardancy, enhanced resistance against wire heating
- good mechanical characteristics such as an increase in modulus and strength enhanced bending stability
- superior resistance against solvents, oil, and water, as well as enhanced swelling resistance
- another great advantage of cross-linking is that it improves long-term heat resistance i.e., the rated temperature of normal polyethylene containing TPV is around 80°C. However, EBT cross-linked polyolefinic TPVs have a higher-rated temperature, such as ~105°C. Moreover, temperatures between 125°C and 150°C Urethane-jacketed car control cable cannot be used at 200°C, although EBT cross-linked cable can be used at 200°C in short intervals. Table 10.1

10.7 SUMMARY AND FUTURE PERSPECTIVES

As the need for polymeric materials and sustainable processing intensifies, numerous physical techniques for chemical modification of polymers were developed to satisfy the demands of different of applications and sustainability. In this sense, electron beam processing is a potent tool for modifying and developing novel high-performance TPEs with improved characteristics over traditional processing methods (Berejka, Cleland et al. 2014, Banerjee, Burbine et al. 2019, Chaudhary, Singh et al. 2020, Utrera-Barrios, Verdejo et al. 2020, Mohammadi, Matinfar et al. 2021, Surmeneva, Grubova et al. 2021). There are various difficulties in developing high-temperature TPEs using the conventional dynamic vulcanization approach. Electron beam treatment addresses these challenges, making it an ideal alternative to the static and dynamic cross-linking of typical TPEs because static cross-linking results in the partial connection of rubber particles by chemical bonding of grafted thermoplastic chains, whereas dynamic cross-linking involves the cross-linking of elastomers during the melt mixing of the molten thermoplastic. The linked structures by dynamic cross-linking conditions will cause to breakdown resulting in considerable improvement in the mechanical characteristics as well as melt processability. Electron beam treatmentis mainly used for cross-linking, embedding, polymerizing, curing, and degradation of polymers (Figure 10.1). The electron treatment, in contrast to conventional reactive processing, utilizes the precise spatial and temporal input of electron energy to vary polymers and compounds (Naskar, Gohs et al. 2009). The main processing parameters of this treatment are electron energy,

TABLE 10.1

Other Applications of Reactive Electron Treatment in TPEs

Polymers	Application Types	Changes in Property	Areas	Ref.
PP/EPDM, PP/EOC, PE/EVA, EVA/TPU, LLDPE/PDMS	Wire and cable insulation	Tolerance to high temperatures, higher resistance to wear and structural rigidity, cold flow minimization, and solvent and corrosive chemical resistance	Automobiles, military equipment, airplane	(Giri, Sureshkumar et al. 2008, Naskar, Gohs et al. 2009, Rajeshbabu, Gohs et al. 2011, Thakur, Gohs et al. 2012, Dutta, Chatterjee et al. 2015, Sabet and Soleimani 2017)
PE/EVA	Heat-shrinkable tube and film	Maintains the tube's original dimensions, increased mechanical properties	Electrical and electronic components	(Chattopadhyay, Chaki et al. 2000, Borhani, Mirjalili et al. 2007)
TPU, polyester TPEs, Polyamide TPEs, PP/EPDM	Hydrogels, catheters, mMedical products such as plastic bags and tubes	Increases in molecular weights and viscosity, higher tensile and melting strengths	medical sectors	(El Fray, Piątek-Hnat et al. 2012, Murray, Kennedy et al. 2013, Balaji, Ratnam et al. 2018, Basak 2021)
PA/FKM, PA/HNBR, PA/ EVA	O-rings and hoses	Resistance to chemicals, oil, improved mechanical properties etc.	Petrochemicals, aerospace, automotive sectors	(Das, Ambatkar et al. 2006, Banerjee and Bhowmick 2013, Banerjee, Gohs et al. 2016, Ahmed, Wu et al. 2020)
PP/EOC, PE/EVA	Foams	Increased mechanical and thermal characteristics	Automotive seat cushioning, foamed gaskets.	(Chattopadhyay, Chaki et al. 2001, Rajeshbabu, Gohs et al. 2011, Entezam, Aghjeh et al. 2017)
NR/recycled rubber/PE, ground tire rubber (GTR)/ Thermoplastic blends	Tyre components, roofing applications, processing aid, binder	Enhanced mechanical properties, superior thermal stability	Automobile sectors	(Sengupta, Banik et al. 2008, Mohamad, Osman et al. 2021, Saeb, Wiśniewska et al. 2022)
EVA/TPU	Footwear sole	Optimized mechanical properties, good thermal properties, improved hardness	Footwear sectors	(Dutta, Chatterjee et al. 2015)

beam current, dose, and dose rate. In reactive electron treatment, the influence of the properties of polymers to be modified (constitution and configuration of the polymers, cross-linking additives, and antioxidants) and the chemical environment during electron treatment (gas atmosphere, water, pH value, and temperature) is taken into account to enable a tailored chemical modification of polymers (Ismail, Galpaya et al. 2010, Banerjee, Bhowmick et al. 2015, Anagha, Chatterjee et al. 2022, Dawant, Seils et al. 2022).

The well-established prominence of reactive electron treatment, as well as current requirements from science and industry, provide a strong foundation for future growth and the development of the instruments and processes that surround it. Electron beam processing is regarded as a cutting-edge technique that may increase manufacturing, processing, and production efficiency. Additionally, it has been generally recognized that it can improve product quality and increase market value.

REFERENCES

Abdou-Sabet, S., R. Puydak, C. J. R. c. Rader and technology (1996). "Dynamically vulcanized thermoplastic elastomers." *Rubber Chemistry and Technology,* **69**(3): 476–494.

Aghjeh, M. R., H. A. Khonakdar, S. H. Jafari, C. Zschech, U. Gohs and G. J. R. a. Heinrich (2016). "Rheological, morphological and mechanical investigations on ethylene octene copolymer toughened polypropylene prepared by continuous electron induced reactive processing." *RSC advances* **6**(29): 24651–24660.

Ahmed, J., J. Wu, S. Mushtaq and Y. J. M. T. C. Zhang (2020). "Effects of electron beam irradiation and multi-functional monomer/co-agents on the mechanical and thermal properties of ethylene-vinyl acetate copolymer/polyamide blends." *Materials Today Communications* **23**: 100840.

Anagha, M., T. Chatterjee, F. Picchioni and K. J. J. o. A. P. S. Naskar (2022). "Exploring the influence of electron beam cross-linking in SEBS/TPU and SEBS-g-MA/TPU thermoplastic elastomer blends." *Journal of Applied Polymer Science* 51721.

Ataee, A., Y. Li, G. Song and C. Wen (2017). *Metal scaffolds processed by electron beam melting for biomedical applications.* Metallic Foam Bone, Elsevier, Woodhead Publishing: 83–110.

Awasthi, P. and S. S. J. A. M. Banerjee (2021). "Fused Deposition Modeling of Thermoplastics Elastomeric Materials: Challenges and Opportunities." *Additive Manufacturing* 102177.

Balaji, A. B., C. T. Ratnam, M. Khalid and R. J. J. o. b. a. Walvekar (2018). "E-beam sterilizable thermoplastics elastomers for healthcare devices: Mechanical, morphology, and in vivo studies." *Journal of Biomaterials Applications* **32**(8): 1049–1062.

Banerjee, S. S., A. K. J. I. Bhowmick and E. C. Research (2015). "Tailored nanostructured thermoplastic elastomers from polypropylene and fluoroelastomer: Morphology and functional properties." *Industrial & Engineering Chemistry Research* **54**(33): 8137–8146.

Banerjee, S. S. and A. K. J. o. M. S. Bhowmick (2016). "An effective strategy to develop nanostructured morphology and enhanced physico-mechanical properties of PP/EPDM thermoplastic elastomers." *Journal of Materials Science* **51**(14): 6722–6734.

Banerjee, S. S. and A. K. J. P. Bhowmick (2013). "Novel nanostructured polyamide 6/fluoroelastomer thermoplastic elastomeric blends: Influence of interaction and morphology on physical properties." *Polymer* **54**(24): 6561–6571.

Banerjee, S. S. and A. K. J. P. Bhowmick (2015). "Dynamic vulcanization of novel nanostructured polyamide 6/fluoroelastomer thermoplastic elastomeric blends with special reference to morphology, physical properties and degree of vulcanization." *Polymer* **57**: 105–116.

Banerjee, S. S., A. K. J. R. C. Bhowmick and Technology (2017). "High-temperature thermoplastic elastomers from rubber–plastic blends: A state-of-the-art review." *Rubber Chemistry and Technology* **90**(1): 1–36.

Banerjee, S. S., S. Burbine, N. Kodihalli Shivaprakash and J. J. P. Mead (2019). "3D-printable PP/SEBS thermoplastic elastomeric blends: Preparation and properties." **11**(2*Polymers*): 347.

Banerjee, S. S., U. Gohs, C. Zschech and G. J. E. P. J. Heinrich (2016). "Design and properties of high-performance polyamide 6/fluoroelastomer blends by electron-induced reactive processing." *European Polymer Journal* **85**: 508–518.

Banerjee, S. S., A. Janke, U. Gohs, A. Fery and G. J. E. P. J. Heinrich (2017). "Some nanomechanical properties and degree of branching of electron beam modified polyamide 6." *European Polymer Journal* **88**: 221–230.

Banerjee, S. S., A. Janke, U. Gohs and G. J. E. P. J. Heinrich (2018). "Electron-induced reactive processing of polyamide 6/polypropylene blends: Morphology and properties." *European Polymer Journal* **98**: 295–301.

Banerjee, S. S., A. Janke, D. Jehnichen, U. Gohs and G. J. P. Heinrich (2018). "Influence of electron-induced reactive processing on structure, morphology and nano-mechanical properties of polyamide 6/fluoroelastomer blends." *Polymer* **142**: 394–402.

Banerjee, S. S., K. D. Kumar, A. K. J. M. M. Bhowmick and Engineering (2015). "Distinct melt viscoelastic properties of novel nanostructured and microstructured thermoplastic elastomeric blends from polyamide 6 and fluoroelastomer." *Macromolecular Materials and Engineering* **300**(3): 283–290.

Banerjee, S. S., K. D. Kumar, A. K. Sikder, A. K. J. M. C. Bhowmick and Physics (2015). "Nanomechanics and origin of rubber elasticity of novel nanostructured thermoplastic elastomeric blends using atomic force microscopy." **216** *Macromolecular Chemistry and Physics*(15): 1666–1674.

Basak, S. J. J. o. M. S., (2021). "Thermoplastic elastomers in biomedical industry–evolution and current trends." *Journal of Macromolecular Science, Part A* 1–15.

Becker, R., J. Bly, M. Cleland, J. J. R. P. Farrell and Chemistry (1979). "Accelerator requirements for electron beam processing." *Radiation Physics and Chemistry* **14**(3-6): 353–375.

Berejka, A., M. Cleland, M. J. R. P. Walo and Chemistry (2014). "The evolution of and challenges for industrial radiation processing—2012." *Radiation Physics and Chemistry* **94**: 141–146.

Bhattacharya, A. B., T. Chatterjee and K. J. M. T. C. Naskar (2020). "Dynamically vulcanized blends of UHM-EPDM and polypropylene: Role of nano-fillers improving thermal and rheological properties." *Materials Today Communications* **25**: 101486.

Bhowmick, A. K. (2008). *Current topics in elastomers research*, CRC press.

Borhani, M., G. Mirjalili, F. Ziaie and M. J. N. Bolorizadeh (2007). "Electrical properties of EVA/LDPE blends irradiated by high energy electron beam." *ukleonika* **52**(2): 77–81.

Bucknall, C. J. J. o. P. S. P. B. P. P. (2007). "Quantitative approaches to particle cavitation, shear yielding, and crazing in rubber-toughened polymers." *Journal of Polymer Science Part B: Polymer Physics* **45**(12): 1399–1409.

Chattopadhyay, S. (2001). *Development and properties of electron beam modified thermoplastic elastomeric polyolefin blends, Journal of adhesion science and technology,* .IIT, Kharagpur

Chattopadhyay, S., T. Chaki and A. K. J. o. a. p. s. Bhowmick (2001). "Development of new thermoplastic elastomers from blends of polyethylene and ethylene–vinyl acetate copolymer by electron-beam technology." *Journal of applied polymer science* **79**(10): 1877–1889.

Chattopadhyay, S., T. Chaki, A. K. J. R. P. Bhowmick and Chemistry (2000). "Heat shrinkability of electron-beam-modified thermoplastic elastomeric films from blends of ethylene-vinylacetate copolymer and polyethylene." *Radiation Physics and Chemistry* **59**(5-6): 501–510.

Chaudhary, N., A. Singh, D. Aswal and A. J. C. A. Sharma (2020). "Electron beam induced tailoring of electrical characteristics of organic semiconductor films." *Chemistry Africa* 1–22.

Chaudhary, N., A. Singh, A. Debnath, S. Acharya and D. Aswal (2015). *Electron beam modified organic materials and their applications*. Solid State Phenomena, Trans Tech Publ.

Cleland, M., L. Parks, S. J. N. I. Cheng, M. i. P. R. S. B. B. I. w. Materials and Atoms (2003). "Applications for radiation processing of materials." *Nuclear Instruments and Methods in Physics Research Section B: Beam Interactions with Materials and Atoms* **208**: 66–73.

Cleland, M. R. (2006). "Industrial applications of electron accelerators."

Cleland, M. R. and R. A. J. I. I. W. P. Galloway (1812). "Electron Beam Crosslinking of Wire and Cable Insulation." *IBA Industrial—White Paper.*

Das, P., S. Ambatkar, K. Sarma, S. Sabharwal and M. J. P. i. Banerji (2006). "Electron-beam processing of nylon 6 and HNBR blends. Part II: development of high strength, heat-and oil-resistant thermoplastic elastomers." *Polymer international* **55**(6): 688–693.

Das, P., S. Ambatkar, K. Sarma, S. Sabharwal and M. J. P. i. Banerji (2006). "Electron beam processing of nylon 6 and hydrogenated nitrile rubber (HNBR) blends: 1. Development of high strength heat-and oil-resistant thermoplastic elastomers." *Polymer international* **55**(1): 118–123.

Datta Sarma, A., H. R. Padmanathan, S. Saha, S. Shankar Banerjee and A. K. J. J. o. A. P. S. Bhowmick (2017). "Design and properties of a series of high-temperature thermoplastic elastomeric blends from polyamides and functionalized rubbers." *Journal of Applied Polymer Science* **134**(39): 45353.

Datta, S. K., T. K. Chaki and A. K. Bhowmick (1998, P. Cheremisinoff Nicholas). Electron beam processing of polymers. *Advanced polymer processing operations*, Cheremisinoff (Ed.). William Andrew Publishing: 157–186.

Dawant, R., R. Seils, S. Ecoffey, R. Schmid, D. J. J. o. V. S. Drouin, N. Technology B, P. Microelectronics: Materials, Measurement, and Phenomena (2022). "Hybrid cross correlation and line-scan alignment strategy for CMOS chips electron-beam lithography processing." *Journal of Vacuum Science & Technology B* **40**(1): 012601.

Drobny, J. G. (2014). *Handbook of thermoplastic elastomers*, Elsevier.

Dutta, J., T. Chatterjee, G. Dhara and K. J. R. A. Naskar (2015). "Exploring the influence of electron beam irradiation on the morphology, physico-mechanical, thermal behaviour and performance properties of EVA and TPU blends." *RSC Advances* **5**(52): 41563–41575.

El Fray, M., M. Piątek-Hnat, J. E. Puskas and E. J. P. J. o. C. T. Foreman-Orlowski (2012). "Influence of e-beam irradiation on the chemical and crystal structure of poly (aliphatic/aromatic-ester) multiblock thermoplastic elastomers." *Polish Journal of Chemical Technology* **14**(2): 70–74.

El-Nemr, K. F., M. M. Khattab, H. J. J. o. a. s. Abdel-Rahman and technology (2011). "Effect of Electron Beam Irradiation on Physico-Mechanical and Chemical Properties of NBR–PVC Loaded with Cement Kiln Dust." *Journal of adhesion science and technology* **25**(9): 1017–1034.

Entezam, M., M. K. R. Aghjeh, M. J. R. P. Ghaffari and Chemistry (2017). "Electron beam irradiation induced compatibilization of immiscible polyethylene/ethylene vinyl acetate (PE/EVA) blends: Mechanical properties and morphology stability." *Radiation Physics and Chemistry,* **131**: 22–27.

Fang, C., Y. Zhang, W. Wang, Z. Wang, F. Jiang, Z. J. I. Wang and E. C. Research (2015). "Fabrication of copolymer-grafted multiwalled carbon nanotube composite thermoplastic elastomers filled with un-modified MWCNTs as additional nanofillers to significantly improve both electrical conductivity and mechanical properties." *Industrial & Engineering Chemistry Research* **54**(50): 12597–12606.

Fazli, A. and D. J. M. Rodrigue (2020). "Waste rubber recycling: A review on the evolution and properties of thermoplastic elastomers." **13**(3): 782.

Frank, N. W. and S. Hirano (1993). M. Penetrante Bernie The history of electron beam processing for environmental pollution control and work performed in the United States. *Non-Thermal Plasma Techniques for Pollution Control*, Springer: 1–26.

Ghosh, P., B. Chattopadhyay and A. K. J. P. Sen (1994). "Thermoplastic elastomers from blends of poly-ethylene and ethylene-propylene-diene rubber: influence of vulcanization technique on phase mor-phology and vulcanizate properties." *Polymer* **35**(18): 3958–3965.

Ghumman, A. S. M., M. M. Nasef, M. R. Shamsuddin, A. J. P. Abbasi and P. Composites (2021). "Evaluation of properties of sulfur-based polymers obtained by inverse vulcanization: Techniques and challenges." *Polymers and Polymer Composites* **29**(8): 1333–1352.

Giri, R., M. Sureshkumar, K. Naskar, Y. Bharadwaj, K. Sarma, S. Sabharwal and G. J. A. i. P. T. J. o. t. P. P. I. Nando (2008). "Electron beam irradiation of LLDPE and PDMS rubber blends: studies on the physicomechanical properties." *Journal of the Polymer Processing Institute* **27**(2): 98–107.

Goharpey, F., A. Katbab and H. J. J. o. a. p. s. Nazockdast (2001). "Mechanism of morphology development in dynamically cured EPDM/PP TPEs. I. Effects of state of cure." *Journal of applied polymer science,* **81**(10): 2531–2544.

Goharpey, F., A. Katbab, H. J. R. c. Nazockdast and technology (2003). "Formation of rubber particle ag-glomerates during morphology development in dynamically cross-linked EPDM/PP thermoplastic elastomers. Part 1: effects of processing and polymer structural parameters." *Rubber chemistry and technology,* **76**(1): 239–252.

Gohs, U., H. Dorschner, G. Heinrich, M. Stephan, U. Wagenknecht, R. Bartel and O. Röder (2006). Requirements on Electron Accelerators for Innovative Applications in Polymer Industry. Russian Accelerator Conference (RuPAC), Novosibirsk, Russia.

Gohs, U., A. Leuteritz, K. Naskar, S. Volke, S. Wiessner and G. Heinrich (2010). Reactive EB processing of polymer compounds. *Macromolecular symposia*, Wiley Online Library.

Hanisch, F., P. Maier, S. Okada, H. J. I. J. o. R. A. Schönbacher, I. P. C. R. Physics and Chemistry (1987). "Effects of radiation types and dose rates on selected cable-insulating materials." *Radiation Physics and Chemistry* **30**(1): 1–9.

Hidayah, I., M. Mariatti, H. Ismail, M. J. P. Kamarol, Rubber and Composites (2015). "Evaluation of PP/ EPDM nanocomposites filled with SiO2, TiO2 and ZnO nanofillers as thermoplastic elastomeric in-sulators." *Plastics, Rubber and Composites* **44**(7): 259–264.

Hochleitner, G., E. Fürsattel, R. Giesa, J. Groll, H. W. Schmidt and P. D. J. M. r. c. Dalton (2018). "Melt electrowriting of thermoplastic elastomers." *Macromolecular rapid communications* **39**(10): 1800055.

Ismail, H., D. Galpaya, Z. J. J. o. V. Ahmad and A. Technology (2010). "Electron-beam irradiation of blends of polypropylene with recycled acrylonitrile-butadiene rubber." *Journal of Vinyl and Additive Technology* **16**(2): 141–146.

Jain, A., A. Nagpal, R. Singhal and N. K. J. J. o. a. p. s. Gupta (2000). "Effect of dynamic cross-linking on impact strength and other mechanical properties of polypropylene/ethylene-propylene-diene rubber blends." *Journal of applied polymer science* **78**(12): 2089–2103.

Jordhamo, G., J. Manson, L. J. P. E. Sperling and Science (1986). "Phase continuity and inversion in polymer blends and simultaneous interpenetrating networks." *Polymer Engineering & Science* **26**(8): 517–524.

Kashiwagi, M. (200414). 10Electron beam processing system (No. JAERI-CONF--2004-007). Kear, K. E. (2003). "Developments in thermoplastic elastomers

Kongmon, E. Kangrang, N. Rhode, M. W. Wichaisirimongkol, P Saisut, J. Rimjeam, S. Thongbai, C. (2016). Studies on electron linear accelerator system for polymer reseach, https://doi.org/10.18429/JACoW-IPAC2016-TUPOY039

Kear, K. E. (2003). "Developments in thermoplastic elastomers."

Kongmon, E., N. Kangrang, M. Rhode, J. Saisut, C. Thongbai, P. Wichaisirimongkol and S. Rimjeam (2016). "Studies on electron linear accelerator system for polymer reseach."

Lappan, U., U. Geißler, K. J. R. P. Lunkwitz and Chemistry (2000). "Changes in the chemical structure of polytetrafluoroethylene induced by electron beam irradiation in the molten state." *Physics and Chemistry* **59**(3): 317–322.

Leo, K., R. Chulan, S. Hashim, A. Baijan, R. Sabri, M. Mohtar, H. Glam, L. Lojius, M. Zahidee and A. Azman (2016). Study on the parameters of the scanning system for the 300 keV electron accelerator. AIP Conference Proceedings, AIP Publishing LLC.

Liao, J., N. Brosse, A. Pizzi, S. Hoppe, X. Xi and X. J. P. Zhou (2019). "Polypropylene blend with poly-phenols through dynamic vulcanization: mechanical, rheological, crystalline, thermal, and UV pro-tective property." *Polymers* **11**(7): 1108.

Loo, K.-H., L. T. Sin, S.-T. Bee, C. Ratnam, J.-Y. Tey, T.-T. Tee, A. J. P.-P. T. Rahmat and Engineering (2017). "Current development of electron beam irradiation of natural rubber–polymer blends." *Polymer-Plastics Technology and Engineering* **56**(17): 1874–1897.

Makhlis, F. A. e. (1972). "Radiation physics and chemistry of polymers."Atomizdat Moscow,

Manaila, E., G. Craciun, D. Ighigeanu, I. B. Lungu, M. Dumitru and M. D. J. P. Stelescu (2021). "Electron Beam Irradiation: A method for degradation of composites based on natural rubber and plasticized starch." *Polymers* **13**(12): 1950.

Mandal, A. K., D. Chakraborty, M. Das and S. K. J. J. o. P. M. Siddhanta (2021). "On the Engineering Properties of TPV derived from Hypalon, PP and a Compatibilizer (PMES-MA) prepared by Dynamic Vulcanization." *Journal of Polymer Materials,* **38**.

Mandal, A. K., D. Chakraborty and S. K. J. J. o. A. P. S. Siddhanta (2014). "Effect of the compatibilizer, on the engineering properties of TPV based on Hypalon® and PP prepared by dynamic vulcanization." *Journal of Applied Polymer Science* **131**(11).

Mekhilef, N. and H. J. P. Verhoogt (1996). "Phase inversion and dual-phase continuity in polymer blends: theoretical predictions and experimental results." *Polymer* **37**(18): 4069–4077.

Miller, S., R. Roberts and R. J. J. o. P. S. Vale (1962). "Use of dimaleimides as accelerators for the radiation-induced vulcanization of hydrocarbon polymers." *Journal of Polymer Science* **58**(166): 737–754.

Mizera, A., M. Manas, D. Manas, P. Stoklasek, M. Bednarik and L. Hylova (2016). Mechanical properties change of thermoplastic elastomer after using of different dosage of irradiation by beta rays. MATEC Web of Conferences, EDP Sciences.

Mohamad, S. F., H. N. B. Osman, M. N. B. Karoji, P. Ibrahim, S. S. O. Al Edrus, L. S. Hua, N. M. F. H. Abd Rahim and C. T. J. M. R. E. Guan (2021). "Insight on the properties of thermoplastic elastomer-based natural rubber and recycled rubber post-treated with electron beam irradiation." *Materials Research Express* **8**(2): 025302.

Mohammadi, X., G. Matinfar, A. M. Khaneghah, A. Singh and A. J. W. M. J. Pratap-Singh (2021). "Emergence of cold plasma and electron beam irradiation as novel technologies to counter mycotoxins in food products." *World Mycotoxin Journal* **14**(1): 75–83.

Mondal, M., U. Gohs, U. Wagenknecht, G. J. M. C. Heinrich and Physics (2013). "Additive free thermo-plastic vulcanizates based on natural rubber." *Materials Chemistry and Physics* **143**(1): 360–366.

Monem, N. A., Z. Ali, H. M. Said, H. Youssef, H. J. P.-P. T. Saleh and Engineering (2005). "Optical properties and morphological structure of electron beam irradiated low-density polyethylene/poly-propylene blends." *Polymer-Plastics Technology and Engineering,* **44**(5): 1025–1047.

Murray, K. A., J. E. Kennedy, B. McEvoy, O. Vrain, D. Ryan, R. Cowman and C. L. J. j. o. t. m. b. o. b. m. Higginbotham (2013). "Effects of gamma ray and electron beam irradiation on the mechanical, thermal, structural and physicochemical properties of poly (ether-block-amide) thermoplastic elastomers." *Journal of the mechanical behavior of biomedical materials,* **17**: 252–268.

Nagasawa, N., A. Kaneda, S. Kanazawa, T. Yagi, H. Mitomo, F. Yoshii, M. J. N. I. Tamada, M. i. P. R. S. B. B. I. w. Materials and Atoms (2005). "Application of poly (lactic acid) modified by radiation cross-linking." *Nuclear Instruments and Methods in Physics Research Section B: Beam Interactions with Materials and Atoms* **236**(1-4): 611–616.

Naskar, K., U. Gohs, U. Wagenknecht and G. J. E. P. L. Heinrich (2009). "PP-EPDM thermoplastic vul-canisates (TPVs) by electron induced reactive processing." *Express Polym. Lett* **3**(11): 677–683.

Nune, K. C., S. Li and R. J. S. C. M. Misra (2018). "Advancements in three-dimensional titanium alloy mesh scaffolds fabricated by electron beam melting for biomedical devices: Mechanical and biological as-pects." *Science China Materials,* **61**(4): 455–474.

Odian, G., D. Lamparella and J. Canamare16, 6 (1967). 3619-3623Radiation effects in polypropylene and ethylene–propylene copolymers. Journal of Polymer Science Part C: Polymer Symposia Wiley Online Library

Okada, Y. (1967). *Irradiation of Hydrocarbon Polymers in Nitrous Oxide Atmosphere*, ACS Publications.

Rajeshbabu, R., U. Gohs, K. Naskar, V. Thakur, U. Wagenknecht, G. J. R. P. Heinrich and Chemistry (2011). "Preparation of polypropylene (PP)/ethylene octene copolymer (EOC) thermoplastic vulcanizates (TPVs) by high energy electron reactive processing." *Radiation Physics and Chemistry* **80**(12): 1398–1405.

Ramskogler, C., F. Warchomicka, S. Mostofi, A. Weinberg, C. J. M. S. Sommitsch and E. C (2017). "Innovative surface modification of Ti6Al4V alloy by electron beam technique for biomedical appli-cation." *Materials Science and Engineering* **78**: 105–113.

Ratnam, C. T., N. A. Ramlee, S. Appadu, S. M. Manshor, H. J. P.-P. T. Ismail and Engineering (2015). "Preparation and electron beam irradiation of PVC/ENR/CNTs nanocomposites." *Polymer-Plastics Technology and Engineering* **54**(2): 184–191.

Ray Chowdhury, S., B. K. Sharma, P. Mahanwar and K. S. J. J. o. A. P. S. Sarma (2016). "Vinyl acetate content and electron beam irradiation directed alteration of structure, morphology, and associated properties of EVA/EPDM blends." *Journal of Applied Polymer Science,* **133**(21).

Ristić, I., I. Krakovsky, T. Janić, S. Cakić, A. Miletić, M. Jotanović, T. J. J. o. T. A. Radusin and Calorimetry (2018). "The influence of the nanofiller on thermal properties of thermoplastic polyurethane elasto-mers." *Journal of Thermal Analysis and Calorimetry* **134**(2): 895–901.

Rooj, S., V. Thakur, U. Gohs, U. Wagenknecht, A. K. Bhowmick and G. J. P. f. A. T. Heinrich (2011). "In situ reactive compatibilization of polypropylene/epoxidized natural rubber blends by electron induced reactive processing: novel in-line mixing technology." *Polymers for Advanced Technologies* **22**(12): 2257–2263.

Sabet, M., A. Hassan, C. T. J. P. d. Ratnam and stability (2012). "Electron beam irradiation of low density polyethylene/ethylene vinyl acetate filled with metal hydroxides for wire and cable applications." *Polymer degradation and stability,* **97**(8): 1432–1437.

Sabet, M. and H. Soleimani (2017). The impact of electron beam irradiation on Low density polyethylene and Ethylene vinyl acetate. IOP Conference Series: Materials Science and Engineering, IOP Publishing.

Sabharwal, S. (2013). Electron beam irradiation applications. Proceedings of the PAC, Citeseer.

Saeb, M. R., P. Wiśniewska, A. Susik, Ł. Zedler, H. Vahabi, X. Colom, J. Cañavate, A. Tercjak and K. J. M. Formela (2022). "GTR/Thermoplastics Blends: How Do Interfacial Interactions Govern Processing and Physico-Mechanical Properties?" *Materials* **15**(3): 841.

Sandle, T. (2013). *Sterility, sterilisation and sterility assurance for pharmaceuticals: technology, validation and current regulations*, Elsevier.

Sbrescia, S., J. Ju, T. Engels, E. Van Ruymbeke and M. J. J. o. P. S. Seitz (2021). "Morphological origins of temperature and rate dependent mechanical properties of model soft thermoplastic elastomers." *Journal of Polymer Science ,* **59**(6): 477–493.

Sengupta, R., I. Banik, P. S. Majumder, V. Vijayabaskar and A. K. BhowmickElectron beam processing of rubber, Bhowmick, A. K. (2008). . *Current topics in elastomers research*, CRC Press: 851–916.

Seymour, R. B. and G. B. J. o. c. e. Kauffman (1992). "Elastomers: III. thermoplastic elastomers." *Journal of chemical education* **69**(12): 967.

Shariatpanahi, H., H. Nazokdast, B. Dabir, K. Sadaghiani and M. J. o. a. p. s. Hemmati (2002). "Relationship between interfacial tension and dispersed-phase particle size in polymer blends. I. PP/ EPDM." *Journal of applied polymer science* **86**(12): 3148–3159.

Spenadel, L. J. R. P. and Chemistry (1979). "Radiation cross-linking of polymer blends." *Radiation Physics and Chemistry* **14**(3-6): 683–697.

Surmeneva, M., I. Grubova, N. Glukhova, D. Khrapov, A. Koptyug, A. Volkova, Y. Ivanov, C. M. Cotrut, A. Vladescu and A. J. M. Teresov (2021). "New Ti–35Nb–7Zr–5Ta alloy manufacturing by electron beam melting for medical application followed by high current pulsed electron beam treatment." *Metals,* **11**(7): 1066.

Thakur, V., U. Gohs, U. Wagenknecht and G. J. P. j. Heinrich (2012). "Electron-induced reactive processing of thermoplastic vulcanizate based on polypropylene and ethylene propylene diene terpolymer rubber." *Polymer journal,* **44**(5): 439–448.

Utrera-Barrios, S., R. Verdejo, M. A. López-Manchado and M. H. J. M. H. Santana (2020). "Evolution of self-healing elastomers, from extrinsic to combined intrinsic mechanisms: a review." *Materials Horizons* **7**(11): 2882–2902.

Wang, W., R. Schlegel, B. T. White, K. Williams, D. Voyloy, C. A. Steren, A. Goodwin, E. B. Coughlin, S. Gido and M. J. M. Beiner (2016). "High temperature thermoplastic elastomers synthesized by living anionic polymerization in hydrocarbon solvent at room temperature." *Macromolecules,* **49**(7): 2646–2655.

Watts, A., N. Kurokawa and M. A. J. B. Hillmyer (2017). "Strong, resilient, and sustainable aliphatic polyester thermoplastic elastomers." *Biomacromolecules* **18**(6): 1845–1854.

Wei, D., A. Anniyaer, Y. Koizumi, K. Aoyagi, M. Nagasako, H. Kato and A. J. A. M. Chiba (2019). "On microstructural homogenization and mechanical properties optimization of biomedical Co-Cr-Mo alloy additively manufactured by using electron beam melting." *Additive Manufacturing,* **28**: 215–227.

Wu, H., M. Tian, L. Zhang, H. Tian, Y. Wu and N. J. S. m. Ning (2014). "New understanding of micro-structure formation of the rubber phase in thermoplastic vulcanizates (TPV)." *Soft matter* **10**(11): 1816–1822.

Yamamoto, S. (1990). *Crosslinking of wire and cables with electron beam*(No. JAERI-M--90-194) Machi, Sueo . (Ed.). Japan.

Yuan, R., S. Fan, D. Wu, X. Wang, J. Yu, L. Chen and F. J. P. C. Li (2018). "Facile synthesis of polyamide 6 (PA6)-based thermoplastic elastomers with a well-defined microphase separation structure by melt polymerization." *Polymer Chemistry* **9**(11): 1327–1336.

Yunshu, X., F. Yibei, F. Yoshii, K. J. R. P. Makuuchi and Chemistry (1998). "Sensitizing effect of poly-functional monomers on radiation cross-linking of polychloroprene." *Radiation Physics and Chemistry* **53**(6): 669–672.

Zanchin, G. and G. J. P. i. P. S. Leone (2021). "Polyolefin thermoplastic elastomers from polymerization catalysis: Advantages, pitfalls and future challenges." *Progress in Polymer Science* **113**: 101342.

Zimek, Z., G. Przybytniak, A. Nowicki, K. Mirkowski, K. J. R. P. Roman and Chemistry (2014). "Optimization of electron beam cross-linking for cables." *Radiation Physics and Chemistry* **94**: 161–165.

Zschech, C., M. Pech, M. T. Mueller, S. Wiessner, U. Wagenknecht, U. J. R. P. Gohs and Chemistry (2020). "Continuous electron-induced reactive processing–A sustainable reactive processing method for polymers." *Radiation Physics and Chemistry* **170**: 108652.

11 Selective laser melting of CoCr alloys in biomedical application: A review

J. S. Saini*

Department of Mechanical Engineering, Thapar Institute of Engineering and Technology, Patiala, Punjab, India

Luke Dowling and Daniel Trimble

Department of Mechanical and Manufacturing Engineering, Trinity College, The University of Dublin, Dublin-2, Ireland

Sachin Singh

Department of Mechanical Engineering, Thapar Institute of Engineering and Technology, Patiala, Punjab, India

*Corresponding author: jsaini@thapar.edu

CONTENTS

11.1 INTRODUCTION

Additive manufacturing (AM) or 3D printing is used to describe the fabrication of functional, end use components in a layer-by-layer manner based upon a 3D computer-aided design (CAD) model (Gibson et al., 2010). AM enables the fabrication of complex geometries which are difficult to manufacture using alternative methods (Bibb et al., 2006; Kok et al., 2018). AM has been recognized as a transformative technology for realizing the on-demand manufacturing across multiple industries ranging from nuclear, chemical, medical, and petrochemical applications (Gu et al., 2012; Averyanova et al., 2012).

    The development in AM technology led to the creation of three typical processes in terms of laser sintering (LS), laser melting (LM) and laser metal deposition (LMD) (Gu et al., 2012). Among them, the selective laser melting (SLM) is the most in-demand approach, as researchers

DOI: 10.1201/9781003321910-11

have demonstrated its great potential of rapid manufacturing of complex, unique, and low-volume metallic parts with equivalent or superior-quality properties in comparison to traditional processes (Thijs et al., 2010).

SLM is now being used in different industries such as aerospace, automotive, and medical, to manufacture complex and non-machinable parts for advanced applications (Vandenbroucke and Kruth, 2007; Gebhardt et al., 2010). The most attractive feature of SLM relevant to the medical applications is the ability to produce fully customized parts, with arbitrary geometric complexity, and without the need for expensive patient-specific tooling (Yager et al., 2015).

11.2   SELECTIVE LASER MELTING PROCESS

The SLM technique can fabricate metal components from CAD data with layer-by-layer addition of metal powder and selectively melting successive layers of metal powder on top of each other, using thermal energy supplied by a focused and accurately controlled laser beam (Osakada and Shiomi, 2006; Yang et al., 2012). SLM is enjoying its ever-growing popularity (Frazier, 2014; Li et al., 2015; Xu et al., 2015; Wang et al., 2018).

Figure 11.1 shows the basic components used in the SLM process and describe the principle of operation of the SLM process. The important parts of SLM machine are: the building platform, roller, feed piston, build piston, the laser system, precision optics, a high-speed scanner, and a computer with process software. During the SLM process, a laser beam irradiates the metal powder along the sliced layer. Once the layer is scanned, the building platform is lowered by the predefined distance given by the user and a new layer of powder is deposited by the roller. Again the laser is moved along the predefined path according to the cross-section. The complete process is performed in cyclic way until the complete component is fabricated. In order to minimize the reaction of the outer atmosphere with the molten metal the complete SLM process is performed in the inert gas atmosphere.

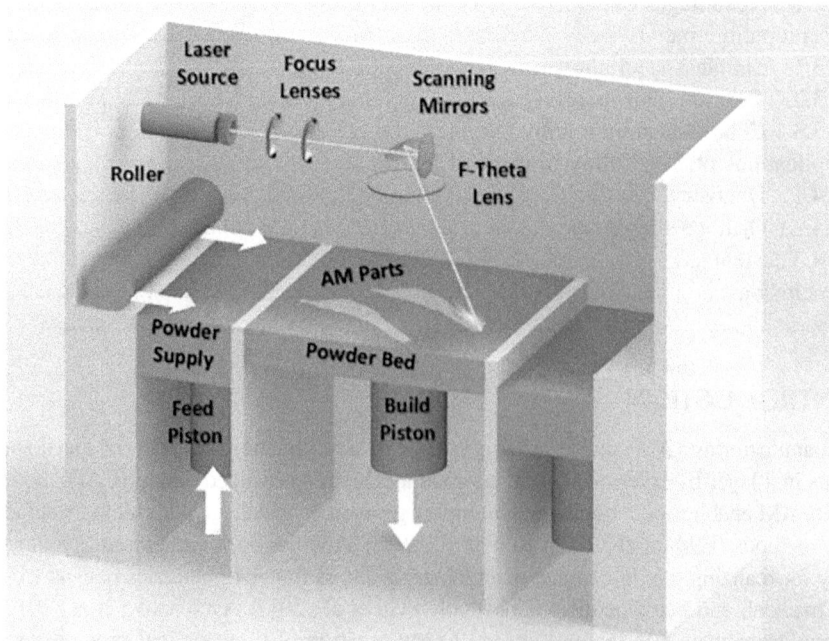

FIGURE 11.1   Schematic diagram of selective laser melting process (Shipley et. al., 2018).

Input SLM process parameters plays a very vital role in determining the output responses viz. porosity, microstructure, mechanical properties, residual stresses of the 3D-printed component. Thomas (2009) stated the existence of more than 130 process parameters affecting the process. These are broadly divided into pre-process, in-process, and post-process parameters.

i. Pre-process parameters: Metal powder characteristics viz. its morphology, chemistry, and microstructure primarily comes under the pre-process parameters. Individual particle comprising the metal powder influences the final part quality as it determines the flowability, thermal conductivity, density, and penetration of the laser beam during the SLM process.

ii. In-process parameters: It is impossible for the researchers to have an in-depth study and control over all the SLM input parameters. There are 13 crucial toward gaining part quality. The main among these 13 factors are the laser scanning speed, laser power, layer thickness and hatching distance (Yadroitsev and Smurov, 2010). Therefore, researchers studied the individual effect of the major in-process parameters (Figure 11.2) as well to optimize the properties of the as-built component. Due to the independency between the various input parameters it was observed that individual parameter study produces erroneous results. Therefore, researchers adopted the volumetric energy density method which is given as:

$$E = \frac{P}{v.\,h.\,t}$$

(11.1)

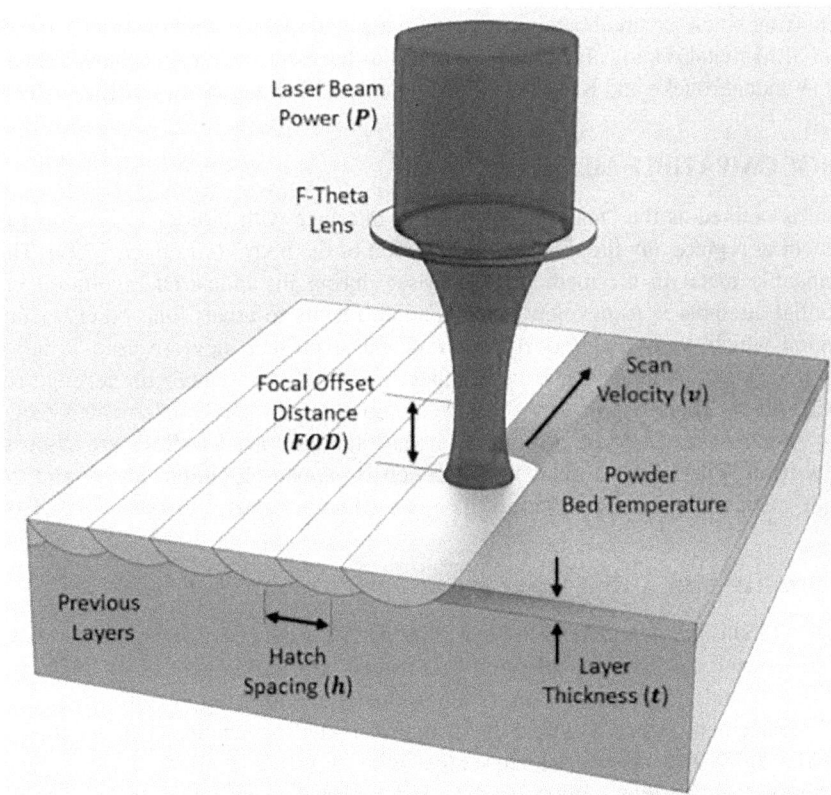

FIGURE 11.2   Major input parameters of the selective laser melting process (Shipley et. al., 2018).

where E is volumetric energy density, P is laser power, v is the speed at which laser is moving over the powder bed (known as scanning speed), h is the distance between the centre lines of two successive laser scans (known as hatch spacing), and t is the layer thickness of the powder for each laser scan.

iii. Post-process parameters: In order to improve the quality of the in-built component in terms of improving density, microstructure or relieving residual stresses various post-process operations such as post-process heat treatment (annealing) or thermo-mechanical operations (hot isostatic pressing (HIP)) are performed. Temperature, residence time, and cooling rate are some of the important process parameters that influence the output responses.

11.2.1 Advantages of selective laser melting process

Compared to traditional casting methods, the SLM technique offers advantages such as low manufacturing cost, high density of products, high-dimensional accuracy, reduced operator errors, accuracy of the products, improved electrochemical characteristics and prevention of casting defects (Vandenbroucke and Kruth, 2007; Yadroitsev et al., 2007; Ucar et al., 2009; Yves-Christian et al., 2010, Xiang et al., 2012; Xin et al., 2012; Sun and Zhang, 2012). It also produces finer grains as compared to the traditional manufacturing processes due to the rapid melting and high cooling rate of the metal powder (Satoh et al., 2013; Simchi and Asgharzadeh, 2004; Kempen et al., 2011). It can even reduce the design life cycle and lower the cost of complex parts.

SLM has the capability of forming nearly 100% dense metallic parts as compared to other AM technologies (Yang et al., 2012). SLM is capable of fabricating complex components, is appropriate for making single or small batches of metal parts, and that too with materials efficiency (Xin, 2012). The SLM metal copings have been reported to have satisfactory mechanical and chemical properties (Vandenbroucke and Kruth, 2007; Wu et al., 2014; Yang et al., 2014; Zeng et al., 2014).

11.3 BIOCOMPATIBLE METALS

Biomaterial is defined as the "material intended to interface with biological systems to evaluate, treat, augment or replace any tissue, organ or function of the body" (Grainger, 1999). The demand of biocompatible metal in the medical field arises during the industrial revolution in the 19th century. Initial attempts is to develop biocompatible metals to repair long bone fractures by internal fixation which were continuously evolved and at present they are used in total joint replacement (hip, knee, shoulder etc.), bone, plate, screw, stents, dental, tissue engineering, and organ regeneration. There are mainly three well known and vastly implemented biocompatible metals viz. stainless steel, cobalt chromium alloys, titanium, and its alloys are discussed in the following section. Thereafter, a detailed discussion on cobalt-chromium alloys and their application in the medical field is presented in the current book chapter.

11.3.1 Stainless steel alloy

Stainless steel (SS) is the first metal alloy that is properly introduced as the biomaterial. Gollwitzer et al. (2003) reduced the bacterial infections associated with stainless steel implants by applying Poly (D, L-lactic acid) coating. Alt et al. (2006) reduced the infection in cementless prostheses in total joint replacement (steel K-wires) by gentamicin–hydroxyapatite (HA) coatings. Authors (Baena et al., 2006) also alloyed stainless steel with copper and niobium to improve its antimicrobial properties. Various authors studies and reported a considerable number of SS316L stainless-steel implants failure that is primarily due to crevice and attacks of pitting corrosion (Sivakumar et al., 1995). As a result research for developing various other biomaterials started.

11.3.2 TITANIUM AND ITS ALLOY

Initially, commercially pure titanium (CP-Ti) is mainly used in medical application. However, due to low mechanical strength the use of CP-Ti is restricted to orthopedic applications and is primarily used in the dental applications (Elias et al., 2013). Various efforts are made to enhance the mechanical properties of the CP-Ti. Researchers performed severe plastic deformation of CP-Ti that leads to its grain refinement that resulted in enhanced fatigue, wear, and corrosion properties (Mora-Sanchez et al., 2016). Later, aluminium (Al) and vanadium (V) are alloyed with Ti to obtain the most vastly used α + β type Ti alloy in the medical field i.e., Ti6Al4V. Researchers developed Ti64 alloy hip implant (Harrysson et al., 2008), dental implants (Figliuzzi et al., 2012), disc implants (de Beer and van der Merwe, 2013), fusion cage for spine disorders (Lin et al., 2007), and facial implants (Salmi et al., 2012). However, it is reported that Al and V ions released from Ti64 alloy can case long-term health problems such as Alzheimer's and neuropathy [2]. Moreover, considerable high young's modulus (110 GPa) and low wear resistance leads to the loosening of implants restricting its life span to 10–15 years. Subsequently to eliminate such shortcomings low modulus β-titanium alloys are developed and currently under investigation. Several studies showed encouraging results for β-titanium alloys that can replace α and α + β type Ti alloys in the medical field. Chlebus et al. (2011) used selective laser melting process (SLM) to investigate Ti6Al7Nb as an implant material. Later, Marcu et al. (2012) developed Ti6Al7Nb-hydroxyapatite composite. Zhang et al. (2011) examined the properties of the Ti-24Nb-4Zr-8Sn samples fabricated by the SLM process as an improvement over Ti64. The lower modulus of Ti-24Nb-4Zr-8Sn compared to that of Ti64 prevented bone resorption, which results in implant loosening.

11.3.3 COBALT-CHROMIUM ALLOY

Recently, SLM-processed cobalt-chromium (CoCr) alloys have gained a lot of interest in the research community due to their greater hardness, mechanical strengths, and biocompatibility (Takaichi et al., 2013; Hazlehurst et al., 2014). CoCr alloys are widely used in biomedical applications as they are the hardest known biocompatible alloy along with good tensile and fatigue properties (Learmonth et al., 2007; Monroy et al., 2013).

Cobalt-chromium-molybdenum (CoCrMo) alloys have been defined as effective metallic biomaterials in the ASTM Standards (2008 Annual Book of ASTM Standards, Section Thirteen, Medical Devices and Services). CoCrMo is the hardest known biocompatible metal alloy with good tensile strength, good fatigue properties, and excellent corrosion resistance, mainly due to its high chromium content that forms a thin oxide layer that protects the underlying material (Ameer et al., 2004; Gaytan et al., 2011; Xiang et al., 2012; Xin et al., 2012; Xin et al., 2013; Monroy et al., 2014; Al Jabbari et al., 2014). Oxide layers on the alloy surfaces are formed spontaneously as a result of the rapid intake of oxygen from the air after polishing (Muñoz and Julián, 2010). The oxide layers are mostly formed with the ionic compounds, such as chromium oxide (Cr_2O_3) and chromium hydroxide ($Cr(OH)_3$) (Muñoz and Julián, 2010; Hanawa et al., 2001; Surviliene et al., 2008). This oxide layer acts like an isolated barrier to electron flow between the surface of the alloy and the electrolyte so that it cannot corrode (Hodgson et al., 2004). The alloy also derives its wear resistance from the volume fraction, size, and distribution of its carbide particles (Monroy et al., 2014).

CoCrMo alloys have been used for removable partial dentures, metal frames, and porcelain-fused-to-metal crowns (Craig et al., 2000; Powers and Sakaguchi, 2006; Wataha, 2002). CrCoMo alloys are also adopted for dental restorations such as customized abutments, crown and bridges in the anterior and posterior region, telescope and conical crowns, and screw-retained restorations (Wataha and Schmalz, 2009).

Originally, the CrCo alloys were adopted for dental applications and lately they are being used for body joints and fracture fixation applications such as hips, knees, and shoulders (Montero-Ocampo et al., 1999; Davis, 2003; Learmonth et al., 2007; Vandenbroucke and Kruth, 2007

Wataha and Schmalz, 2009; Monroy et al., 2013). Compared to other metallic biomaterials such as titanium and stainless steels, the excellent wear resistance properties of CoCr alloys makes them best suited for sliding parts in artificial joints (Gaytan et al., 2011).

However, the fabrication processes for these alloys, such as casting, cutting, and plastic works, are usually difficult because of their high melting point, hardness, and limited ductility (Davis, 2003; Montero-Ocampo et al., 1999). Moreover, the method which is used to manufacture the alloy plays a very important role in the mechanical and metallurgical properties of the resulting component which needs to be accurately controlled for desired quality (Monroy et al., 2014). SLM has been used to manufacture the components from CoCr alloys due to its capability to control the different process parameters accurately. SLM poses a certain number of controlled parameters that make the successful manufacturing of complex parts with the desirable properties possible (Vandenbroucke and Kruth, 2007; Averyanova et al., 2012). The quality of parts formed using SLM depends upon the process parameters selected in the machine (Fu and Guo, 2014; Huang et al., 2015a; Pupo et al., 2015; Hong et al., 2016).

11.4 APPLICATIONS OF CoCr ALLOY IN MEDICAL FIELD

Nowadays, many people are suffering from diseases like osteoarthritis, coronary artery disease, fracture of any bone which can't be repaired. For treatment of these problems replacement surgery is among the best options. As the treatment is related to direct contact with internal body parts so the materials applicable to surgical implants must possess high corrosion and wear resistance, high strength, non-toxicity, and bio-inertness. Cobalt-chromium alloys are one of the most reliable materials because they inhibit good mechanical, biological properties and high corrosion resistance and are also comparatively low in cost with chromium wt.% more than 25wt.%. The high performance, good aesthetics, and high density of SLM-manufactured products make the process suitable for CoCr medical applications (Xiang et al., 2012). These medical applications range from dental, orthopedic implants to stents, to name a few.

11.4.1 Dentistry

Over a last few decades, rapid prototyping techniques have been employed in dentistry with the manufacturing of crowns and bridges (Duret et al., 1996; Willer et al., 1998; Van Der Zel et al., 2001). Bibb et al. (2006) used SLM to the production of patient specific, custom-fitting removable partial denture (RPD) alloy frameworks. It was found that SLM is a viable RM method that can be used in dentisty. SLM-produced RPD frameworks with CoCr alloy were comparable in terms of accuracy, quality of fit and function to the existing methods in dental technology.

The SLM-fabricated CoCr alloy dental crowns had significantly smaller marginal fit discrepancies as compared to traditional cast crowns (Xu et al., 2015; Huang et al., 2015b) but milling specimens had the best internal and marginal fit (Nesse et al., 2015).

Biofunctionality and biocompatibility are important considerations for the selection of dental alloys. Based on the results of the agar diffusion test (ADT), and dye exclusion test (DET) tests, it was concluded that the Co–Cr alloy used in SLM does not exhibit cytotoxic potential (Jevremovic et al., 2011). The corrosion behavior is of high interest to ensure biocompatibility as corrosion is characterised by electrochemical phase boundary reactions which cause the release of metal ions (Geurtsen, 2002; Upadhyay et al., 2006).

Vandenbroucke and Kruth (2007) tested CoCrMo for their corrosion behavior. The corrosion characteristics were examined by static immersion tests according to the DIN EN ISO 10271 standard. Test specimens were stored in a corrosion solution (sodium chloride and lactic acid, each 0.1 mol/l with a pH value of 2.3) for 14 days with the solution being exchanged after 1, 2, 7, and 14 days and analyzed by ICP-OES (inductively coupled plasma-optic emission spectrometry analysis) to determine the different ion emissions as a function of time. It was shown that the SLM manufactured

TABLE 11.1

Metal ion release in artificial saliva solution (A) and Dulbecco's modified Eagle's medium (B) (µg/cm^2). 'dl' indicates 'detection limit'. (Xin et al., 2012)

	Co	Cr	Mo
A			
CAST	0.49 (0.13)	dl	0.12 (0.015)
SLM	0.12 (0.04)	dl	0.015 (0.007)
	$P < 0.05$	–	$P < 0.05$
B			
CAST	1.04 (0.38)	0.03 (0)	0.19 (0.06)
SLM	0.09 (0.006)	0.02 (0.006)	dl
	$P < 0.05$	$P > 0.05$	$P < 0.05$

test specimens showed lower emissions than the cast specimens because the laser molten material is more homogeneous, contains fewer pores, and has a finer microstructure.

Xin et al. (2012) found that the corrosion behavior of SLM CoCrMo specimens was not significantly different in artificial saliva solution with pH of 5 as compared to traditional casting. But the SLM specimens showed significantly increased corrosion resistance at an acidic pH of 2.5 as compared to traditional cast specimens and met the clinical dental requirements. Xin et al. (2012) also examined the release of toxic metal ions from, and the cell response to, CoCrMo alloy fabricated using traditional casting and SLM techniques. Specimens were immerged in artificial saliva solution for 7 days. Table 11.1 shows the quantity of metal ions released in artificial saliva solution over 7 days from traditional-cast and SLM specimens. It can be seen that SLM specimens have lower level of ion release as compared to the cast specimens. There was more release of Co ions as compared to Cr ions (Zhang et al., 2012; Takaichi et al., 2013). The amount of Co from the build was half of that from the as-cast alloy (Takaichi et al., 2013). The release of a higher amount of Co ions was associated with the lower pH of the test solution (Biao et al., 2012).

The quantity of ions released increased significantly with increasing fluoride concentration when immersed in an artificial saliva solution containing fluoride (Yang et al., 2014). The ion release from cast specimen was greater than that of SLM manufactured CoCrMo alloy.

Table 11.2 shows the amount of ion release from the cast specimens and from the SLM specimens when immersed in an artificial saliva solution containing fluoride (Yang et al., 2014)

The corrosion resistance and the amount of metal release in SLM formed CoCrMo alloy also depends upon the scanning strategies used in SLM which helps to build a dense component (Hedberg et al., 2014).

The corrosion resistance of the SLM-fabricated CoCrMo alloy measured before and after ceramic firing in modified Fusayama artificial saliva was adequate for the clinical use (Zeng et al., 2014). The SLM manufactured CoCrMo alloy were also studied for corrosion properties before and after porcelain-fused-to-metal (PFM) firing with better corrosion resistance that cast specimens at pH 2.5 (Xin et al., 2014).

There was a microstructural difference between the SLM and cast specimens that have accounted for different Co ion release characteristics. Figure 11.3 shows that the SLM specimen had homogeneous and compact structure as compared to the cast specimen, which had a heterogeneous structure.

SLM-manufactured CoCrMo alloy had no difference in their electrochemical corrosion properties in the artificial saliva through potentiodynamic curves and EIS, and no significant difference via XPS as compared to the cast specimens (Xin et al., 2013).

TABLE 11.2

Amount of metal released from unit area of specimen surface (μg/cm^2) in fluoride-containing artificial saliva at pH 5 (mean ± standard deviation) (Yang et al., 2014)

Fluoride concentration, %	Cr	Co	Mo
Cast specimens			
0	0.001 ± 0.001	1.066 ± 0.704	0.030 ± 0.025
0.05	0.001 ± 0.001	2.209 ± 0.576	0.074 ± 0.022
0.1	0.013 ± 0.013	2.53 ± 3.589	0.362 ± 0.299
0.2	0.973 ± 0.325	11.222 ± 5.016	0.448 ± 0.143
SLM specimens			
0	0.001 ± 0.001	0.052 ± 0.005	0.005 ± 0.001
0.05	–	0.112 ± 0.04	0.126 ± 0.016
0.1	–	0.125 ± 0.048	0.111 ± 0.004
0.2	0.18 ± 0.326	0.534 ± 0.054	0.106 ± 0.006

(a)
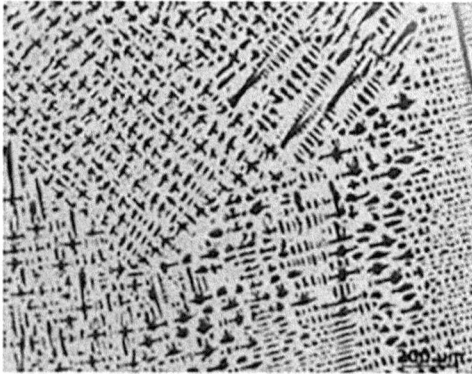
(b)
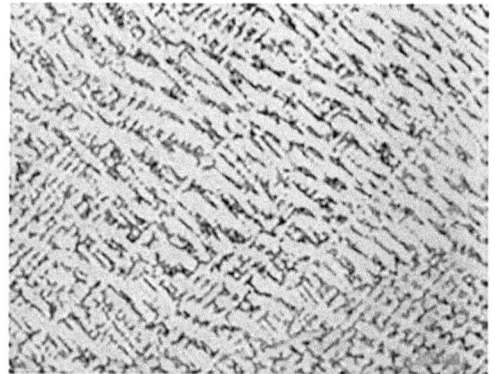
(c)
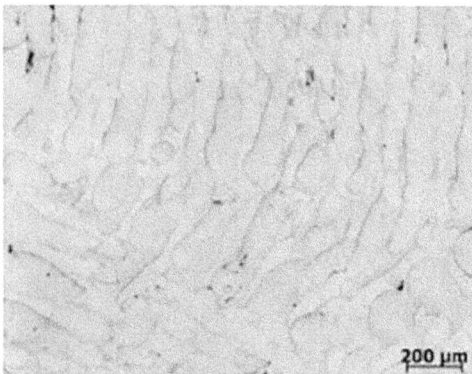
(d)
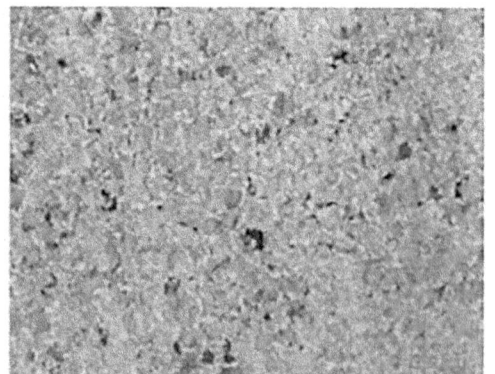

FIGURE 11.3 Metallographic photographs of cast specimen before (**a**) and after (**b**) firing and SLM specimen before (**c**) and after (**d**) firing (magnification 100×) (Xin et al., 2014).

Images as shown in Figure 11.3 were also taken to study the microstructure before and after porcelain fused to metal (PFM) firing (Xin *et al.*, 2014).

The unfired cast specimens, shown in Figure 11.3(a), displayed a typical inhomogeneous dendritic solidification microstructure consisting of dendrites (light areas), interdendritic regions (dark areas), and a third portion manifesting as dark lines in the light areas. After firing, the cast specimens shown in Figure 11.3(b) retained a dendritic solidification microstructure with more pronounced dendrites. Figure 11.3(c) and (d) shows the microstructure of the SLM specimens before and after firing. The SLM specimens before firing exhibited a homogeneous and compact finger-like structure which consisted of Co–Cr solid solution and a $Co_3Mo(W)$ phase. After firing, there was no finger-like structure but the samples remained homogeneous and compact (Xin et al., 2014). Lu et al. (2015) suggested that the specimens formed using the island strategy showed slightly better corrosion resistance than that formed using the line strategy. The results were based on the electrochemical and metal release tests conducted on the specimens built using island and line strategies.

Elemental Cu is an alloying element which can increase the wear and corrosion resistance of the metals. Copper ions are also known for its antimicrobial properties (Ren et al., 2016a). The excellent antibacterial effect of SLM-fabricated CoCrCu is due to the copper ions released from its surface.

Guoqing et al. (2018) determined both the blood compatibility and the release rate of ions of SLM manufactured and cast CoCrMo alloy implants following submersion in various corrosive solutions. It was found that the concentrations of Co and Cr ions released by SLM-fabricated specimen were both lower than that of cast parts under the different corrosion conditions. The release of Cr ions was greater in artificial saliva than in sodium chloride. The higher ion release rate from CoCrMo alloy in NaCl compared to that in artificial saliva was due to the reduction in pH of the NaCl solution over time. Table 11.3 demonstrates that both the absorbance and hemolysis ratio of SLM-manufactured parts of CoCrMo alloy were 8.3 and 41.21% lower than that of cast parts, respectively. The SLM manufactured specimens possessed better blood compatibility than the cast ones. It is due to the reason that the microstructure and density of SLM formed specimens were more uniform than that of the cast parts. It was also found that that the average consumption rates after contact with SLM-manufactured parts were 13.1, 40, 60.6, and 60.8% lower than that of cast parts for leukocytes, erythrocytes, hemoglobin, and platelets, respectively.

Heat treatment has been used to improve the different properties of cast alloys (Brantley and Alapati, 2012). Heat treatment can also be used in SLM-fabricated parts to reduce the internal stresses and to improve the different mechanical properties (Thöne et al., 2012). But heat treatment can increase the corrosion in cast alloys (Craig, 1985; Sorensen et al., 1990; Al-Hity et al., 2007). The comparison of corrosion of the SLM-manufactured heat-treated (HRx) and non-heat-treated (NHRx) CoCr alloys were done by Alifui-Segbaya et al. (2015). Specimens were immersed in an artificial saliva solution suspended by a nylon thread for 42 days at 37°C.

It was shown that although ion release in both HRx and NHRx specimens were within the safe level as recommended by ISO but there was reduced corrosion resistance due to heat treatment in the SLM-fabricated CoCr alloys (Alifui-Segbaya et al., 2015).

The reduced corrosion resistance of the heat treated alloy is due to the change in the microstructure that occurred at high temperature (Thöne et al., 2012).

TABLE 11.3

Absorbance and hemolysis rate of CoCrMo alloy

	Positive control group	Negative control group	SLM parts	Casting parts
Absorbance	0.933	0.067	0.077	0.084
Hemolysis rate	–	–	1.154	1.963

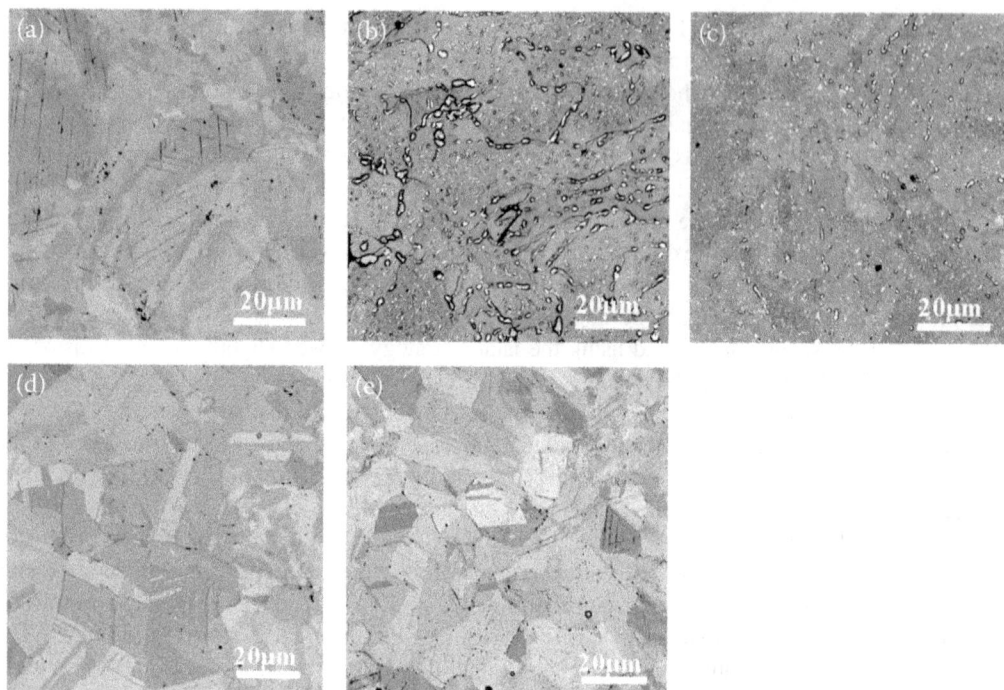

FIGURE 11.4 The backscattered electron images of the CCW alloys (Lu et al., 2016, 2018): (a) as-SLM, (b) 1150FC, (c) 1100WC, (d) 1150WC, (e) 1200WC.

Heat treatment conditions plays a very important role for corrosion resistance of CoCr alloys. The influence of different heat treatment conditions on the corrosion resistance of SLMed CoCrW alloys was investigated using electrochemical tests in the 0.9% NaCl solution with and without fluoride ion concentration (Lu et al., 2018). Figure 11.4 shows the microstructure of the CoCrW alloys for different conditions.

As can be seen from Figure 11.4(b), at 1150°C followed by the furnace cooling, a large amount of continuous blocky precipitates were randomly dispersed inside the grain and at the grain boundaries. At 1100°C following the water cooling, the microstructure shows a distribution of small precipitates within the grain and at the grain boundary, shown in Figure 11.4(c). The microstructure appears to be fine when the temperature increases to 1150°C and 1200°C both followed by water cooling, shown in Figure 11.4(d) and (e). This fine microstructural homogeneity gave the highest resistance to corrosion at 1150°C water cooled as compared to other heat treatment conditions (Lu et al., 2016, 2018).

The *in vitro* corrosion behavior of dental alloys is dependent on several parameters, which include composition, and treatment of alloys, pH, the clinical situations of rest and chewing simulated in the test design and instrumentation (Zhao et al., 2008). The corrosion resistance of CoCr alloys is mainly due to the existence of a passive oxide layer (Ameer and Khamis, 2004; Nascimento et al., 2007; McGinley et al., 2011), which is spontaneously formed when the alloy is exposed to atmospheric oxygen after polishing (Hanawa et al., 2001). These oxide layers protect against corrosion by acting as a barrier to electron flow (resistor) between the alloy and the surrounding electrolyte (Muñoz and Julián, 2010). The composition of the oxide layers on the surface of Co–Cr alloy is mainly Cr_2O_3 and $Cr(OH)_3$ (Hanawa et al., 2001; Hodgson et al., 2004; Kocijan et al., 2004). Although cobalt is the predominant constituent of the alloy, its concentration in the oxide films is only 5% (Qiu et al., 2011). Hence, changes in the concentration of Co and Cr may change the composition of the passive oxide layer on the alloy surface and affect the corrosion resistance. The enhanced corrosion resistance of SLM samples in an acidic environment may be due to the thicker oxide layer observed on SLM samples (Xin et al., 2012).

The corrosion resistance of SLM specimen is more than the cast specimens as the SLM group showed a compact homogeneous cellular microstructure, finer grain size, and have fewer pores than cast group. The cast group microstructure showed a heterogeneous two-phase mixture consisting of a solid solution and a crystalline phase which indicates that the cast specimens had a less dense structure compared to the SLM group (Xin et al., 2012).

The rapid cooling during SLM and the strong temperature gradient induced a fine cellular microstructure with Mo enriched at the cell boundaries which is beneficial for the corrosion resistance of alloys (Kobayashi et al., 2000; Metikoš-Huković et al., 2006; Takaichi et al., 2013).

The metal oxide layer also plays a very important role in metal-porcelain bonding (Biao et al., 2012). The combination of porcelain and alloy can be improved by holding CoCr alloy under vacuum at 980°C for 60 s. This will make Cr ion active which will then gather around the metal-porcelain interface to take part in chemical reaction (Junkun et al., 2011).

The SLM metal–ceramic system exhibited a shear bond strength of 44 MPa for CoCr alloy that exceeded the requirement of ISO 9691:1999(E) (Xiang et al., 2012). It showed a better behavior in porcelain adherence test as compared to traditional cast methods.

The bond strengths of the SLM alloy was found to be 55.78 MPa and that of the cast alloy was 54.17 MPa with the help of three-point bend test. The bond strengths of SLM and cast alloy were almost similar but were well above the minimum acceptable value of 25 MPa as per ISO 9693-1.6 (Wu et al., 2014). Results also showed no statistical differences between the metal-ceramic bond strength of SLM CoCrWMo alloy and that of conventionally cast groups after firing three, five, and seven times (Ren et al., 2016b). As shown in Table 11.4, the SLM group showed significantly more porcelain adherence than that of the cast group which even exceeded the requirements of ISO 9 691:1 999 (E). A more regular arrangement in the SLM specimens is the reason behind a better chemical bond of SLM fabricated specimens with porcelain (Ren et al., 2016b).

The metal-ceramic bond characteristics were also evaluated for CoCr alloys fabricated by casting, computer numerical control milling, and SLM techniques (Wang et al., 2016). Table 11.5 shows that the mean values of the bond strength for the SLM specimen was better than the cast and CNC specimens.

Bond strength is directly related to the thickness of the oxide layer i.e., if the oxide layer is absent, thin, or excessively thick, poor bond strength will result (Serra-Prat et al., 2014; Li et al., 2015). The bond strength of the cast group was lowest due to the thicker oxide layer within the three groups.

TABLE 11.4

Mean ±SD metal-ceramic bond strength results of cast and SLM groups after multiple firings (Ren et al., 2016b)

No. of firings	Group	Number	Mean (MPa)	SD	P
3	Cast	6	31.79	±5.74	.290
	SLM	6	34.73	±2.91	
5	Cast	6	33.03	±3.4	.213
	SLM	6	35.07	±1.62	
7	Cast	6	33.22	±4.35	.778
	SLM	6	33.96	±4.43	

TABLE 11.5

Descriptive statistics for metal-ceramic bond strength (MPa) of three groups (Wang et al., 2016)

Group	No.	Mean ± SD	Minimum	Maximum
CAST	10	37.7 ± 6.5	24.6	44.4
CNC	10	43.3 ± 9.2	24.5	51.3
SLM	10	46.8 ± 5.1	40.7	55.4

But Li et al. (2015) found that the metal-ceramic bond strength of CoCr alloy is independent of the manufacturing methods; alloys produced by milling and SLM behaved better in the porcelain adherence test than the cast specimens. The results were consistent with that given by Akova et al. (2008), Serra-Prat et al. (2014), and Lee et al. (2015).

The effect of thermo-mechanical cycling on shear-bond-strength (SBS) of dental porcelain to CoCr alloys fabricated by casting, milling, and SLM process were determined by Antanasova et al. (2018). It was found that the excessive surface oxide formation, due to exposure to high temperatures during metal processing, does not appear to be an issue for CoCr alloys.

11.4.2 ORTHOPAEDICS

Due to its versatility and durability, Co–Cr alloys has been used as an orthopedic implant material. Lately these alloys have been used for body joints and fracture fixation applications as well (Wataha and Schmalz, 2009).

Malhotra et al. (2019) performed an in-vitro study to evaluate the ability of five orthopedic biomaterials i.e., cobalt-chromium, highly cross-linked polyethylene, stainless steel, trabecular metal and titanium alloy to resist the adherence of *Staphylococcus aureus, Staphylococcus epidermidis, Escherichia coli, Klebsiella pneumoniae*, and *Pseudomonas aeruginosa* bacteria and formation of biofilm onto the surface of materials. Machined disks having diameter of 5 mm and thickness of 5 mm out of all five biomaterials were used as test specimen. SEM was used to observe the different bacterial strains adhered at different levels on the five biomaterials. It was observed that cobalt-chromium has highest resistance to bacterial adherence in comparison to other biomaterials. Whereas highest level of adherence was observed on highly cross-linked polyethylene, followed by titanium, stainless steel, and trabecular metal. However, bacterial adhesion occurred on all material specimens. Among the bacterial strains tested, the ability for high adherence was observed with *S. epidermidis* and *K. pneumoniae* followed by *P. aeruginosa* and *E. coli*, whereas *S. aureus* showed the least adherence.

España et al. (2010) discussed the characteristics of CoCrMo alloy structures, which had the potential to overcome long-standing difficulties in load-bearing implants like total hip prostheses and enhance their in-vivo life.

Murr et al. (2011) employed EBM to produce open cellular mesh and foam components, shown in Figure 11.5, representing total knee replacement implants, made of a Co–Cr–Mo alloy (ASTM F–75) and Ti–6Al–4V for a variety of densities and stiffnesses. The Co–29Cr–6Mo containing 0.22% C formed columnar (directional) Cr23C6 carbides spaced 2 μm in the build direction, while HIP annealed CoCr alloy exhibited an intrinsic stacking fault microstructure, according to a comparison of characteristic microstructures for solid, mesh, and foam Ti–6Al–4V as well as Co–29Cr–6Mo prototypes.

Song et al. (2014) used SLM to built a customized femoral component using CoCrMo alloy that was checked for accuracy using a contrastive detection method. The component was built with a standard deviation of 0.03 mm. Four different stiffness configurations for a CoCrMo alloy femoral stem manufactured using SLM were performed by Hazlehurst et al. (2014). Three

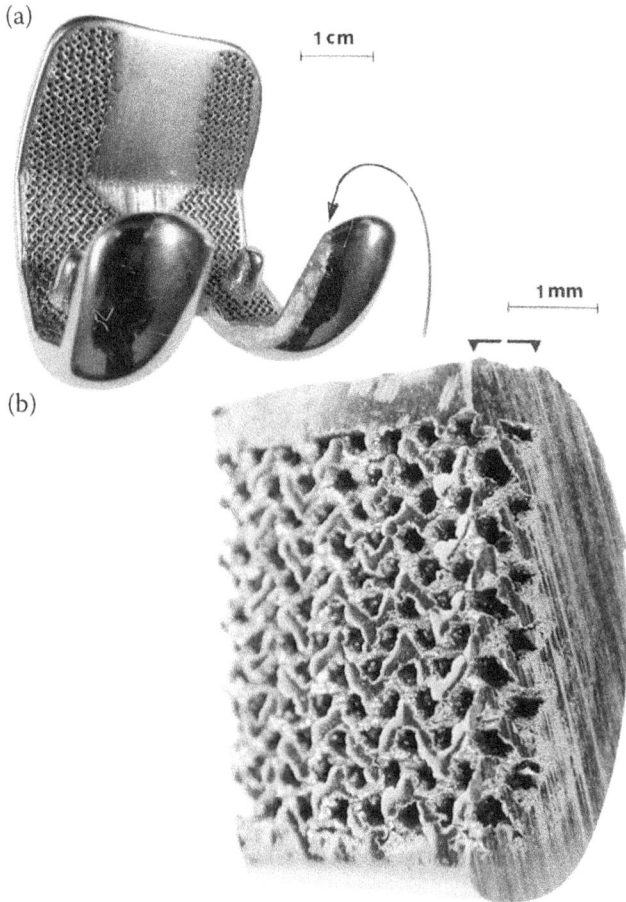

FIGURE 11.5 Co–29Cr–6Mo femoral prototype with mesh structure fabricated by EBM and HIP-annealed using ASTM-F75 standard, and finish machined and partially polished as shown in (a). (b) Shows a magnified view of mesh structure section (Murr et al., 2011).

designs (PC1, PC2, PC3) were based upon a functionally graded approach where the stems had a 1 mm thick fully dense CoCrMo outer skin that encased a porous core comprised of a square pore cellular structure and the fourth design was a fully dense CoCrMo stem (FDS) for comparison. Cantilever bend tests were performed to investigate the flexural properties of each stiffness configuration. Table 11.6 shows the comparison of the stiffness values obtained using experimental and numerical analysis.

The results confirmed that laser-melted functionally graded femoral stems are lighter and more stiff when compared to a traditional fully dense CoCrMo prosthesis.

Three possible ankle prosthesis designs, shown in Figure 11.6, were made by Liverani et al. (2016) using the optimized process parameter of SLM-fabricated CoCrMo alloy. The process parameters were the laser power of 126 W, laser scanning speed of 700 mm/s, and hatch spacing of 0.05 mm based upon the density, tensile strength, and hardness of the produced specimens. A repeatable intra-specimen path of motion was observed in the natural ankle and also after joint replacement, both at the ankle and subtalar joints.

Limmahakhun (2017a) evaluated the possibility of applying graded porous cobalt chromium to cementless femoral stems. Stress shielding in the femoral bone was evaluated using von Mises stresses along the medial and lateral paths using 3D finite element analysis. Three-point bending

TABLE 11.6

Comparison of the stiffness values using experimental and numerical analysis Hazlehurst et al. (2014)

Femoral stem	Mean stiffness from physical test (N/mm)	Stiffness from numerical analysis (N/mm)	% Difference
FDC	869.94	911.14	4.7
PC1	394.45	431.43	9.4
PC2	351.57	419.25	19.2
PC3	309.32	394.27	27.3

(a)   (b)   (c)

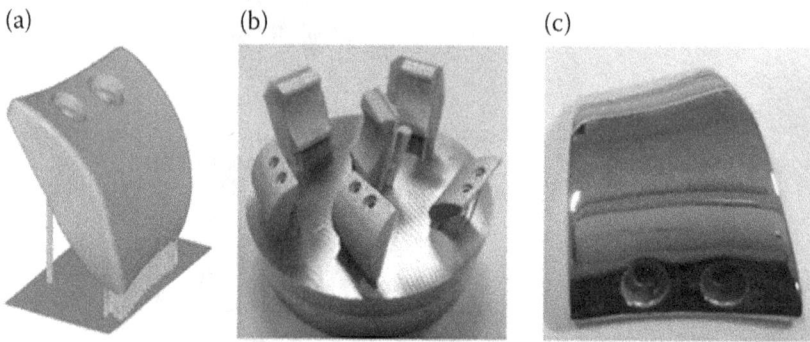

FIGURE 11.6 (a) Example of orientation in AutoFab of the talar component of the STC prosthesis; (b) final arrangement of the STC (skewed truncated cone), TC (truncated cone), and CYL (cylindrical) prosthesis components on the platform; (c) example of the same talar component after the polishing (Liverani et al., 2016).

experiments were performed to compare the mechanical properties of the CoCr graded femoral stem (proximal AGS) to the CoCr solid femoral stem and femur. The flexural stiffness of CoCr solid femoral stems was 17,734 ± 3558 N/mm, which was nearly ten times rigid than the graded femoral stems. Limmahakhun (2017b) used SLM to manufacture the four types of graded cellular CoCr alloy structures based on grading orientations along radial and longitudinal planes. Implants with pore sizes ranging from 5 and 300 μm were created to match the requirements for bone osseointegration. It was found that the stiffness and compressive strength of the CoCr cellular structures were comparable to those of cortical bone tissues and demonstrated a greater energy absorption than bone tissues.

Han et al. (2017) designed and fabricated the CoCrMo alloy porous scaffolds in four different topologies, *i.e.,* cubic close packed (CCP), face-centered cubic (FCC), body-centered cubic (BCC), and spherical hollow cubic (SHC), via SLM process. It was found that the FCC unit cell has the minimum stress concentration due to its inclined bearing struts and horizontal arms. Simulations and experiments both indicated that the compression modulus and strengths of FCC, BCC, SHC, and CCP scaffolds with the same cell size presented in descending order. The compression properties satisfy the load-bearing requirements of cortical and trabecular bones.

Chethan et al. (2019) have analyzed and proposed best possible design and material combination for stem and acetabular cup by computational analysis. Modeling was done using CATIA V-6. Static structural analysis was done using ANSYS R-19 to evaluate von mises stress, deformation in load acting direction and total deformation. Circular, oval, ellipse, and trapezoidal designs each having three individual cross-sections leading to 12 sample sizes were used to arrive at optimum design. As

CoCr or Ti-4Al-6V materials are used commercially for stem implants so these two materials were considered for analysis with each material having 12 samples. For acetabular cup CoC (ceramic on ceramic), CoPE (ceramic on polyethylene), and ultra-high molecular weight polyethylene (UHMWPE) materials were considered. The boundary conditions are applied as per the ASTM F2996-13 and loading conditions are considered as per ISO 7206-4:2010(E).

It was found that CoCr was found to be preferred choice for stem design as it showed superior mechanical properties. Whereas for acetabular cup CoC material exhibited better mechanical properties and is best suited for hip joint implants. Irrespective of material considered for stem the analysis of profile 2 of trapezoidal stem design showed lesser deformation and von-mises stress.

Nagdeve et al. (2020) aimed to reduce time and provide a more consistent surface quality, an effort had been made to optimise the parameters in nano-finishing of Co-Cr-Mo alloy knee joint. With a very low standard deviation of the final Ra value range over the surface of the femoral, the total finishing time was lowered from 64.7 h to 2.0 h. The minimal roughness, which was initially 190.5 nm was reduced to 50 nm. The surface roughness was also reduced by using an abrasive flow machining (AFM) technique in a hip joint manufactured using CoCr alloy by Subramanian et al. (2016). The major input parameters of the AFF process were improved in order to achieve nanometric component finishing.

Iatecola et al. (2021) reported the osseointegration of the additively manufactured Co-Cr-Mo alloy in three conditions: as-built, after plasma immersion ion implantation (PIII), and coated with titanium (Ti) followed by PIII. The metallic samples were developed to have a solid half and a porous half to study bone ingrowth in various surfaces. The results demonstrated that cortical bone development was present in all situations. The titanium-coated sample had the best biomechanical results due to a larger percentage of bone ingrowth, with practically all medullary canals filled with neoformed bone and the implant pores filled and surrounded by bone ingrowth. The metal alloys employed in AM are biocompatible and induce bone neoformation, particularly when the Co-28Cr-6Mo alloy with a Ti-coated surface, nanostructured, and anodized by PIII is utilized, whose technology was demonstrated to improve the osseointegration potential of the implant.

Ahirwar et al. (2021) proposed the plate fixation strategy for the designed biocompatible plates to be used in different fracture models. 3D solid fracture models of a femur bone was created which were fixed by two bone plates of similar material placed side by side over fractured femur using screws made of Ti-6Al-4V material. FEA was used to evaluate interface deformation, stress and strain generated at bone-bioimplant interface. Ti-6Al-4V, SS 316L and Co-Cr alloy were used as the representative metallic biomaterials for the design of the bone plates. The length and average width was considered as 100 mm and 15 mm, respectively. Computer tomography (CT) images were used to extract 3D models of the femur bone. For FEA, a load of 2,800 N was applied during the typical gait on the femoral head which is four times the bodyweight. Whereas load applied on femoral head in standing position was 1/3rd of human bodyweight i.e., 230 N. The bone was considered an anisotropic material.

FEA results show that the bone-bioimplant interface stress and deformation values were reduced in a double plate assembly compared to that of the established single plate assembly to the fractured femur. For consideration of femur bone as naturally anisotropic SS 316L and Ti-6Al-4V demonstrated as the best fit material of choice for design and development of the prosthetic bone plate bioimplants to fix oblique and transverse fracture models, respectively. Considering the bone as an anisotropic material model, the prosthetic bone plate bioimplant is mechanically more stable and safe design to fix the fractured femur models.

11.4.3 STENTS

SLM can be used to produce stent precursors directly from powder which reduces minitube manufacturing and laser microcutting steps in a single process. It also removes the constraints of a tubular precursor and the presence of oxides in laser cut stent surfaces as SLM is being done in the inert atmosphere (Finazzi et al., 2020).

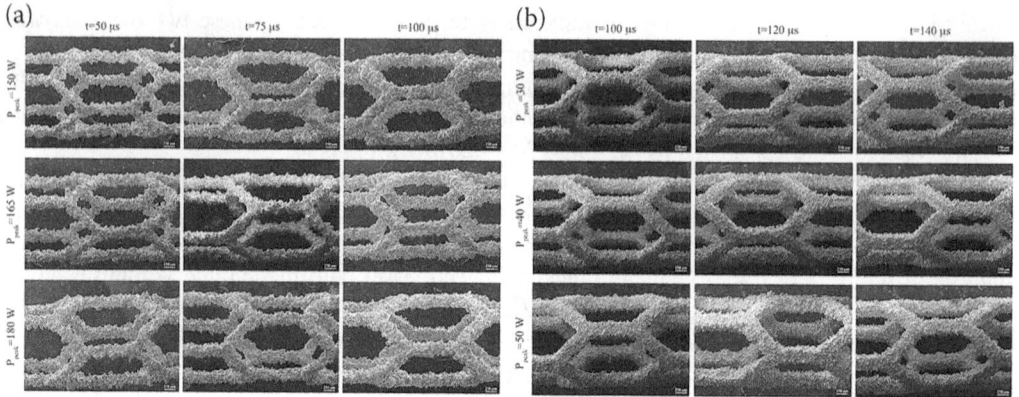

FIGURE 11.7 SEM images of the CoCr prototype stents obtained with (a) hatching and (b) concentric scanning strategy. Build orientation is from left to right with respect to the orientation of the stents in the images (Demir et al., 2017).

Demir et al. (2017) investigated the feasibility of fabricating a CoCr stent precursor using SLM as an alternative to the traditional manufacturing cycle, which is based on creating microtubes and subsequent laser micro-cutting. A simple prototype design for additive manufacturing is proposed, and design rules for manufacturability. The results indicate that SLM can be used to shape precursors in-stent fabrication instead of microtube manufacturing and laser micro-cutting. Electrochemical polishing was used to improve the surface quality of prototype stents with acceptable geometrical precision. Except for a slight rise in the oxide concentration, the chemical makeup remained unchanged. Figure 11.7 shows the SEM images of the stents that were developed.

The properties of cobalt super alloy stents, such as strength, surface finish, and orientation, SLM technology was also explored by Omar et al. (2017) to utilize cobalt chromium materials in the stent manufacturing. Finazzi et al. (2020) looked at making balloon-expandable stents out of a cobalt-chromium alloy utilizing a new mesh and an industrial SLM machine. Using balloon expansion, the SLM-optimized stent mesh was designed, manufactured, finished, and functionally tested. Electropolishing was required to extract sintered particles from the stent surface and improve the surface finish. Figure 11.8 shows the SEM images of the stents. Roughness measurements were taken on stents in both their as-built and electropolished forms, and mechanical qualities were evaluated using tensile testing on dogbone specimens made with the exact specifications and dimensions as the stents. Finally, the expansion behavior of the CoCr stents was assessed utilizing a balloon catheter and monitoring the diameter variation as a function of inflation pressure in both as-built and electropolished circumstances. The results indicate that SLM-made stents may be balloon inflated successfully without losing their characteristic mesh.

Kumar et al. (2021) evaluated the stent performance characteristics of recoil, foreshortening, radial force, and dog-boning and compared the results obtained experimentally with simulation done using finite element analysis in ABAQUS software. For experimental analysis the L605 CoCr alloy tube was used having outer diameter of 1.6 mm and thickness of 110 um. The stent was laser cut on a CoCr tube. A tri-folded balloon of 18 mm length and nominal diameter of 3 mm was used to expand the stent. The stent was crimped onto the balloon.

Whereas for finite element analysis, a finite element model geometry for stent, catheter, tri-folded balloon, crimping cylinder, and compression cylinder were made. Stent with an initial outer diameter of 1.6 mm was crimped to an outer diameter of 1.2 mm. A gradually increasing pressure was applied on the inner surface of the balloon in pseudo time step of 0.5 s to mimic the inflation of an actual stent during deployment, then the pressure was reduced to below zero (-0.1 bar) to mimic the deflation of a stent. Expansion recoil, foreshortening and dog-boning were then calculated from simulation.

(a)

(b)

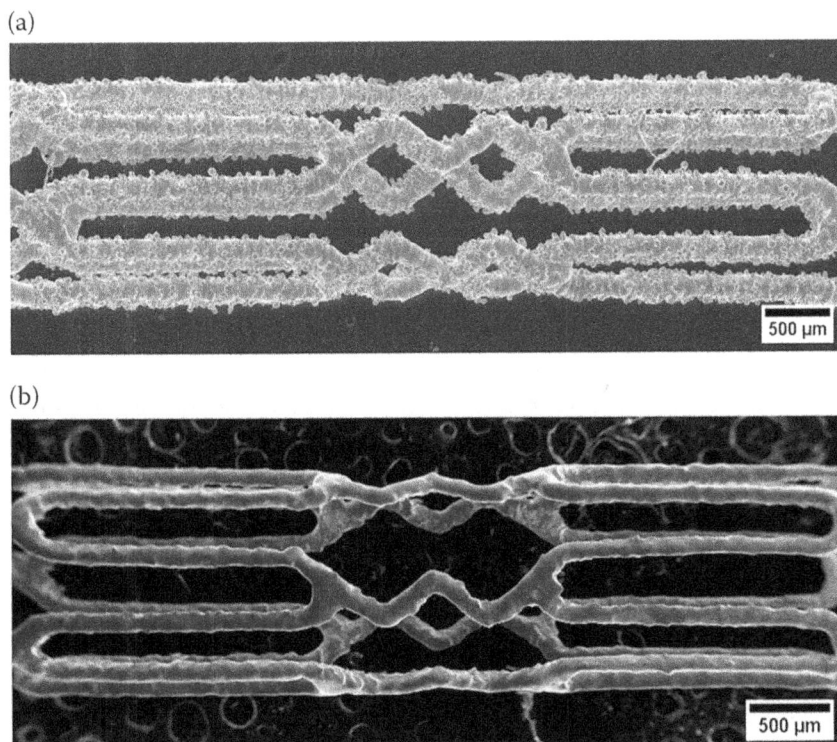

FIGURE 11.8 SEM image of a produced stent (a) in the as-built condition and (b) after electropolishing (P = 50W, t = 90 μm) (Finazzi et al., 2020).

The simulated results were in conformity with experimental results and predicted the same trends for all design variations. It was concluded that radial strength, maximum dog-boning ratio, the corresponding expansion pressure, the pressure required to expand the stent to a predetermined diameter (3.0 mm) increases as the width increases and radial recoil increases as the width decreases. Finite model was also studied by Wang at al. (2021) based on existing cobalt-chromium (Co-Cr) alloy L605 coronary stent to simulate the stent expansion test and crush resistance test. Also, the effects of different parameters on performance and flexibility of stents were analyzed. From the results it was observed that the average stress, axial shrinkage, and the expansion uniformity of stents were improved by increasing the number of circumferential support bodies, while the radial shrinkage and the radial resistance to compression decreased. The radial shrinkage and the radial resistance to compression improved by decreasing the length of support body. Flexibility of stents was not influenced by changing length of support body and number of circumferential support bodies.

11.5 CONCLUSIONS

The present work gives the review of the applications of SLM-manufactured CoCr alloys in the medical field. Based upon the above review, the following conclusions can be drawn:

i. SLM has the capability to fabricate fully customized patient-specific complex parts with nearly 100% density. The parts have finer grains and homogeneous microstructure that results in greater mechanical strength as compared to the traditional manufacturing processes due to the rapid melting and high cooling rate of the metal powder.

ii. SLM-processed CoCr alloys have gained a lot of interest in the medical field due to its good hardness, mechanical strengths, fatigue and biocompatibility due to its high chromium strength.

iii. The good corrosion resistance behavior of SLM-manufactured CoCr alloys due to the existence of a thicker passive oxide layer make it favorable for dental applications. The oxide layers protect against corrosion by acting as a barrier to electron flow between the alloy and the surrounding electrolyte.

iv. High stiffness, compressive strength, and load bearing capacity of the SLM-manufactured CoCr cellular structures makes it suitable for orthopedic and stents applications.

REFERENCES

Ahirwar, H., Gupta, V.K., & Nanda, H.S. (2021). Finite element analysis of fixed bone plates over fractured femur model. *Computer Methods in Biomechanics and Biomedical Engineering*, 24(15), 1742–1751.

Alifui-Segbaya, F., Lewis, J., Eggbeer, D., & Williams, R. J. (2015). In vitro corrosion analyses of heat treated cobalt-chromium alloys manufactured by direct metal laser sintering. *Rapid Prototyping Journal*, 21(1), 111–116.

Akova, T., Ucar, Y., Tukay, A., Balkaya, M.C., & Brantley, W.A. (2008). Comparison of the bond strength of laser-sintered and cast base metal dental alloys to porcelain. *Dental Materials*, 24, 1400–1404 10.1016/j.dental.2008.03.001.

Alt, V., Bitschnau, A., Österling, J., Sewing, A., Meyer, C., Kraus, R., ... & Schnettler, R. (2006). The effects of combined gentamicin–hydroxyapatite coating for cementless joint prostheses on the reduction of infection rates in a rabbit infection prophylaxis model. *Biomaterials*, 27(26), 4627–4634.

Al-Hity, R. R., Kappert, H. F., Viennot, S., Dalard, F., & Grosgogeat, B. (2007). Corrosion resistance measurements of dental alloys, are they correlated? *Dental Materials*, 23, 679–687 10.1016/j.dental.2006.06.008.

Altun, H., & Sen, S. (2004). Studies on the influence of chloride ion concentration and pH on the corrosion and electrochemical behaviour of AZ63 magnesium alloy. *Materials & Design*, 25(7), 637–643.

Antanasova, M., Kocjan, A., Kovač, J., Žužek, B., & Jevnikar, P. (2018). Influence of thermo-mechanical cycling on porcelain bonding to cobalt–chromium and titanium dental alloys fabricated by casting, milling, and selective laser melting. *Journal of Prosthodontic Research*, 62(2), 184–194.

Ameer, M.A., Khamis, E., & Al-Motlaq, M. (2004). Electrochemical behaviour of recasting Ni–Cr and Co–Cr non-precious dental alloys. *Corrosion Science*, 46, 2825–2836 10.1016/j.corsci.2004.03.011.

Al Jabbari, Y.S., Koutsoukis, T., Barmpagadaki, X., & Zinelis, S. (2014). Metallurgical and interfacial characterization of PFM Co–Cr dental alloys fabricated via casting, milling or selective laser melting. *Dental Materials*, 30, e79–e88 10.1016/j.dental.2014.01.008.

Averyanova, M., Cicala, E., Bertrand, Ph., & Grevey, Dominique (2012). Experimental design approach to optimize selective laser melting of martensitic 17-4 PH powder: part I – single laser tracks and first layer. *Rapid Prototyping Journal*, 18, 28–37 10.1108/13552541211193476.

Baena, M. I., Marquez, M. C., Matres, V., Botella, J., & Ventosa, A. (2006). Bactericidal activity of copper and niobium–alloyed austenitic stainless steel. *Current Microbiology*, 53(6), 491–495.

Bibb, R J, Eggbeer, D, Williams, R J, & Woodward, A (2006). Trial fitting of a removable partial denture framework made using computer-aided design and rapid prototyping techniques. *Proceedings of the Institution of Mechanical Engineers, Part H: Journal of Engineering in Medicine*, 220, 793–797 10.1243/09544119jeim62.

Brantley, W. A., & Alapati, S. B. (2012). Heat treatment of dental alloys: A review. *Metallurgy-Advances in Materials and Processes*.

Chethan, K.N., Zuber Mohammad, Shyamasunder Bhat, N., Satish Shenoy, B., & Chandrakant, R. Kini, (2019). Static structural analysis of different stem designs used in total hip arthroplasty using finite element method. *Heliyon*, 5(6), e01767.

Chlebus, E., Kuźnicka, B., Kurzynowski, T., & Dybała, B. (2011). Microstructure and mechanical behaviour of Ti—6Al—7Nb alloy produced by selective laser melting. *Materials Characterization*, 62(5), 488–495.

Craig, R. G. (1985). "Cast and wrought base metal alloys", in *Restorative Dental Materials*, 7th ed., CV Mosby, St. Louis.

Craig, R.G. , Powers, J. M. , Wataha, J. C. (2000). *Dental Materials: Properties and Manipulation*, 7th ed. Mosby, Missouri, 229–231.

Davis, J. R. (2003).*Handbook of Materials for Medical Devices*. 2nd ed. ASM International, OH, 31–37.

de Beer, N., & van der Merwe, A. (2013). Patient-specific intervertebral disc implants using rapid manufacturing technology. *Rapid Prototyping Journal*, 19(2), 126–139.

Duret, F., Preston, J. & Duret, B. (1996). Performance of CAD/CAM crown restorations.*Journal of the California Dental Association*, 24(9), 64–71.

Elias, C. N., Meyers, M. A., Valiev, R. Z., & Monteiro, S. N. (2013). Ultrafine grained titanium for biomedical applications: An overview of performance. *Journal of Materials Research and Technology*, 2(4), 340–350.

España, F. A., Balla, V. K., Bose, S., & Bandyopadhyay, A. (2010). Design and fabrication of CoCrMo alloy based novel structures for load bearing implants using laser engineered net shaping. *Materials Science and Engineering: C*, 30(1), 50–57.

Figliuzzi, M., Mangano, F., & Mangano, C. (2012). A novel root analogue dental implant using CT scan and CAD/CAM: Selective laser melting technology. *International Journal of Oral and Maxillofacial Surgery*, 41(7), 858–862.

Finazzi, V., Demir, A. G., Biffi, C. A., Migliavacca, F., Petrini, L., & Previtali, B. (2020). Design and functional testing of a novel balloon-expandable cardiovascular stent in CoCr alloy produced by selective laser melting. *Journal of Manufacturing Processes*, 55, 161–173.

Frazier, W. E. (2014). Metal additive manufacturing: A review. *Journal of Materials Engineering and Performance*, 23(6), 1917–1928.

Fu, C., & Guo, Y. (2014). Three-dimensional temperature gradient mechanism in selective laser melting of Ti-6Al-4V. *Journal of Manufacturing Science and Engineering*, 136(6), 061004.

Gaytan, S. M., Murr, L. E., Ramirez, D. A., Machado, B. I., Martinez, E., Hernandez, D. H., Martinez, J., Medina, F., & Wicker, R. (2011). A TEM study of cobalt-base alloy prototypes fabricated by EBM. *Materials Sciences and Applications*, 2(5), 355.

Geurtsen, W. (2002). Biocompatibility of dental casting alloys. *Critical Reviews in Oral Biology & Medicine*, 13(1), 71–84.

Gebhardt, A., Schmidt, F.-M., Hötter, J.-S., Sokalla, W., & Sokalla, P. (2010). Additive Manufacturing by selective laser melting the realizer desktop machine and its application for the dental industry. *Physics Procedia*, 5, 543–549 10.1016/j.phpro.2010.08.082.

Gibson, I., Rosen, D. W., & Stucker, B. (2020). *Additive manufacturing technologies*. (Vol. 17). Cham, Switzerland: Springer.

Gibson, I., Rosen , D. W., & Stucker, B. (2010). *Additive manufacturing technologies: rapid prototyping to direct digital manufacturing*. 1st ed., Springer, Heildelberg , Germany.

Gollwitzer, H., Ibrahim, K., Meyer, H., Mittelmeier, W., Busch, R., & Stemberger, A. (2003). Antibacterial poly (D, L-lactic acid) coating of medical implants using a biodegradable drug delivery technology. *Journal of Antimicrobial Chemotherapy*, 51(3), 585–591.

Grainger, D. W. (1999). The Williams dictionary of biomaterials. *Materials Today*, 3(2), 29.

Gu, D. D., Meiners, W., Wissenbach, K., & Poprawe, R. (2012). Laser additive manufacturing of metallic components: Materials, processes and mechanisms. *International Materials Reviews*, 57(3), 133–164.

Guoqing, Z., Yongqiang, Y., Changhui, S., Fan, F., & Zimian, Z. (2018). Study on biocompatibility of CoCrMo alloy parts manufactured by selective laser melting. *Journal of Medical and Biological Engineering*, 38(1), 76–86.

Han, C., Yan, C., Wen, S., Xu, T., Li, S., Liu, J. & Shi, Y. (2017). Effects of the unit cell topology on the compression properties of porous Co-Cr scaffolds fabricated via selective laser melting.*Rapid Prototyping Journal*, 23(1), 16–27.

Hanawa, T., Hiromoto, S., & Asami, K. (2001). Characterization of the surface oxide film of a Co–Cr–Mo alloy after being located in quasi-biological environments using XPS. *Applied Surface Science*, 183(1–2), 68–75.

Harrysson, O. L., Cansizoglu, O., Marcellin-Little, D. J., Cormier, D. R., & West II, H. A. (2008). Direct metal fabrication of titanium implants with tailored materials and mechanical properties using electron beam melting technology. *Materials Science and Engineering: C*, 28(3), 366–373.

Hazlehurst, K. B., Wang, C. J., & Stanford, M. (2014). An investigation into the flexural characteristics of functionally graded cobalt chrome femoral stems manufactured using selective laser melting. *Materials & Design*, 60, 177–183.

Hedberg, Y. S., Qian, B., Shen, Z., Virtanen, S., & Wallinder, I. O. (2014). In vitro biocompatibility of CoCrMo dental alloys fabricated by selective laser melting. *Dental Materials*, 30(5), 525–534.

Hodgson, A., Kurz, S., Virtanen, S., Fervel, V., Olsson, C. O., & Mischler, S. (2004). Passive and transpassive behaviour of CoCrMo in simulated biological solutions. *Electrochimica Acta*, 49(13), 2167–2178.

Hong, M.-H., Min, B., & Kwon, T.-Y. (2016). The Influence of Process Parameters on the Surface Roughness of a 3D-Printed Co–Cr Dental Alloy Produced via Selective Laser Melting. *Applied Sciences*, 6, 401. 10.3390/app6120401.

Huang, Y., Leu, M. C., Mazumder, J., & Donmez, A. (2015a). Additive manufacturing: Current state, future potential, gaps and needs, and recommendations. *Journal of Manufacturing Science and Engineering*, 137(1), 014001.

Huang, Z., Zhang, L., Zhu, J., & Zhang, X. (2015b). Clinical marginal and internal fit of metal ceramic crowns fabricated with a selective laser melting technology. *The Journal of prosthetic dentistry*, 113(6), 623–627.

Iatecola, A., Longhitano, G. A., Antunes, L. H. M., Jardini, A. L., Miguel, E. D. C., Béreš, M., & da Cunha, M. R. (2021). Osseointegration Improvement of Co-Cr-Mo Alloy Produced by Additive Manufacturing. *Pharmaceutics*, 13(5), 724.

Juan-Kun, L., Jian-Tao, Y., Feng, Z., Cui-Cui, Z., Xiao-Shan, W., Yi-Ping, Z., & Bo-Hua, L. (2011). Effect of different heat treatment on the bonding strength of porcelain and Co-Cr alloy. *Shanghai Journal of Stomatology*, 20(6), 567–571.

Jevremovic, D., Kojic, V., Bogdanovic, G., Puskar, T., Eggbeer, D., Thomas, D., & Williams, R. (2011). A selective laser melted Co-Cr alloy used for the rapid manufacture of removable partial denture frameworks: Initial screening of biocompatibility. *Journal of the Serbian Chemical Society*, 76, 43–52 10.2298/jsc100406014j.

Junkun, L., Jiantao, Y., & Feng, Z. (2011). The Effect of Different Thermal Treatment on Bond Strength of Co-Cr Alloy and Ceramic. *Shanghai J Stomatology*, 20, 567–571.

Kempen, K., Yasa, E., Thijs, L., Kruth, J. P., & Van Humbeeck, J. (2011). Microstructure and mechanical properties of Selective Laser Melted 18Ni-300 steel. *Physics Procedia*, 12, 255–263.

Kobayashi, Y., Virtanen, S., & Böhni, H. (2000). Microelectrochemical studies on the influence of Cr and Mo on nucleation events of pitting corrosion. *Journal of the Electrochemical Society*, 147(1), 155.

Kocijan, A., Milošev, I., & Pihlar, B. (2004). Cobalt-based alloys for orthopaedic applications studied by electrochemical and XPS analysis. *Journal of Materials Science: Materials in Medicine*, 15(6), 643–650.

Kok, Y., Tan, X. P., Wang, P., Nai, M., Loh, N. H., Liu, E., & Tor, S. (2018). Anisotropy and heterogeneity of microstructure and mechanical properties in metal additive manufacturing: A critical review. *Materials & Design*, 139, 565–586.

Kumar, A., & Bhatnagar, N. (2021). Finite element simulation and testing of cobalt-chromium stent: A parametric study on radial strength, recoil, foreshortening, and dogboning. *Computer Methods in Biomechanics and Biomedical Engineering*, 24(3), 245–259.

Learmonth, I. D., Young, C., & Rorabeck, C. (2007). The operation of the century: Total hip replacement. *The Lancet*, 370(9597), 1508–1519.

Lee, D. H., Lee, B. J., Kim, S. H., & Lee, K. B. (2015). Shear bond strength of porcelain to a new millable alloy and a conventional castable alloy. *The Journal of Prosthetic Dentistry*, 113(4), 329–335.

Li, X., Wang, X., Saunders, M., Suvorova, A., Zhang, L., Liu, Y., Fang, M. H., Huang, Z. H., & Sercombe, T. B. (2015). A selective laser melting and solution heat treatment refined Al–12Si alloy with a controllable ultrafine eutectic microstructure and 25% tensile ductility. *Acta Materialia*, 95, 74–82.

Limmahakhun, S., Oloyede, A., Chantarapanich, N., Jiamwatthanachai, P., Sitthiseripratip, K., Xiao, Y., & Yan, C. (2017a). Alternative designs of load-sharing cobalt chromium graded femoral stems. *Materials Today Communications*, 12, 1–10.

Limmahakhun, S., Oloyede, A., Sitthiseripratip, K., Xiao, Y., & Yan, C. (2017b). Stiffness and strength tailoring of cobalt chromium graded cellular structures for stress-shielding reduction. *Materials & Design*, 114, 633–641.

Lin, C. Y., Wirtz, T., LaMarca, F., & Hollister, S. J. (2007). Structural and mechanical evaluations of a topology optimized titanium interbody fusion cage fabricated by selective laser melting process. *Journal of Biomedical Materials Research Part A: An Official Journal of The Society for Biomaterials, The Japanese Society for Biomaterials, and The Australian Society for Biomaterials and the Korean Society for Biomaterials*, 83(2), 272–279.

Liverani, E., Fortunato, A., Leardini, A., Belvedere, C., Siegler, S., Ceschini, L., Ascari, A. (2016). Fabrication of Co–Cr–Mo endoprosthetic ankle devices by means of selective laser melting (SLM). *Materials & Design*, 106, 60–68.

Lu, Y., Ren, L., Xu, X., Yang, Y., Wu, S., Luo, J., Yang, M., Liu, L., Zhuang, D., & Yang, K. (2018). Effect of Cu on microstructure, mechanical properties, corrosion resistance and cytotoxicity of CoCrW alloy fabricated by selective laser melting. *Journal of the Mechanical Behavior of Biomedical Materials*, 81, 130–141.

Lu, Y., Wu, S., Gan, Y., Li, J., Zhao, C., Zhuo, D., et al. (2015). Investigation on the microstructure, mechanical property and corrosion behavior of the selective laser melted CoCrW alloy for dental application. *Materials Science and Engineering: C*, 49, 517–525.

Lu, Y., Wu, S., Gan, Y., Zhang, S., Guo, S., Lin, J., & Lin, J. (2016). Microstructure, mechanical property and metal release of As-SLM CoCrW alloy under different solution treatment conditions. *Journal of the Mechanical Behavior of Biomedical Materials*, 55, 179–190.

Malhotra, R., Dhawan, B., Garg, B., Shankar, V., & Nag, T. C. (2019). A comparison of bacterial adhesion and biofilm formation on commonly used orthopaedic metal implant materials: An in vitro study. *Indian journal of Orthopaedics*, 53(1), 148–153.

Marcu, T., Todea, M., Maines, L., Leordean, D., Berce, P., & Popa, C. (2012). Metallurgical and mechanical characterisation of titanium based materials for endosseous applications obtained by selective laser melting. *Powder Metallurgy*, 55(4), 309–314.

Metikoš-Huković, M., Pilić, Z., Babić, R., & Omanović, D. (2006). Influence of alloying elements on the corrosion stability of CoCrMo implant alloy in Hank's solution. *Acta Biomaterialia*, 2(6), 693–700.

McGinley, E. L., Coleman, D. C., Moran, G. P., & Fleming, G. J. (2011). Effects of surface finishing conditions on the biocompatibility of a nickel–chromium dental casting alloy. *Dental Materials*, 27(7), 637–650.

Monroy, K., Delgado, J., & Ciurana, J. (2013). Study of the pore formation on CoCrMo alloys by selective laser melting manufacturing process. *Procedia Engineering*, 63, 361–369.

Montero-Ocampo, C., Lopez, H., & Talavera, M. (1999). Effect of alloy preheating on the mechanical properties of as-cast Co-Cr-Mo-C alloys. *Metallurgical and Materials Transactions A*, 30(3), 611–620.

Monroy, K. P., Delgado, J., Sereno, L., Ciurana, J., & Hendrichs, N. J. (2014). Effects of the selective laser melting manufacturing process on the properties of CoCrMo single tracks. *Metals and Materials International*, 20(5), 873–884.

Mora-Sanchez, H., Sabirov, I., Monclus, M. A., Matykina, E., & Molina-Aldareguia, J. M. (2016). Ultra-fine grained pure Titanium for biomedical applications. *Materials Technology*, 31(13), 756–771.

Muñoz, A. I., & Julián, L. C. (2010). Influence of electrochemical potential on the tribocorrosion behaviour of high carbon CoCrMo biomedical alloy in simulated body fluids by electrochemical impedance spectroscopy. *Electrochimica Acta*, 55(19), 5428–5439.

Murr, L. E., Amato, K. N., Li, S. J., Tian, Y. X., Cheng, X. Y., Gaytan, S. M., & Wicker, R. B. (2011). Microstructure and mechanical properties of open-cellular biomaterials prototypes for total knee replacement implants fabricated by electron beam melting. *Journal of the Mechanical Behavior of Biomedical Materials*, 4(7), 1396–1411.

Nagdeve, L., Jain, V. K., & Ramkumar, J. (2020). Optimization of process parameters in nano-finishing of Co-Cr-Mo alloy knee joint. *Materials and Manufacturing Processes*, 35(9), 985–992.

Nascimento, M. L., Mueller, W.-D., Carvalho, A. C., & Tomás, H. (2007). Electrochemical characterization of cobalt-based alloys using the mini-cell system. *Dental Materials*, 23(3), 369–373.

Nesse, H., Ulstein, D. M. Å., Vaage, M. M., & Øilo, M. (2015). Internal and marginal fit of cobalt-chromium fixed dental prostheses fabricated with 3 different techniques. *The Journal of Prosthetic Dentistry*, 114(5), 686–692.

Omar, M. A., Baharudin, B. H., & Sulaiman, S. (2017, December). Stent manufacturing using cobalt chromium molybdenum (CoCrMo) by selective laser melting technology. In AIP Conference Proceedings, AIP Publishing LLC, 1901(1), 040001.

Osakada, K., & Shiomi, M. (2006). Flexible manufacturing of metallic products by selective laser melting of powder. *International Journal of Machine Tools and Manufacture*, 46(11), 1188–1193.

Powers, J. M. (2006). Craig's restorative dental materials. *Mechanical Properties*, 51–96. Elsevier.

Pupo, Y., Monroy, K. P., & Ciurana, J. (2015). Influence of process parameters on surface quality of CoCrMo produced by selective laser melting. *The International Journal of Advanced Manufacturing Technology*, 80(5), 985–995.

Qiu, J., Yu, W. Q., & Zhang, F. Q. (2011). Effects of the porcelain-fused-to-metal firing process on the surface and corrosion of two Co–Cr dental alloys. *Journal of Materials Science*, 46(5), 1359–1368.

Ren, L., Memarzadeh, K., Zhang, S., Sun, Z., Yang, C., Ren, G., et al. (2016a). A novel coping metal material CoCrCu alloy fabricated by selective laser melting with antimicrobial and antibiofilm properties. *Materials Science and Engineering: C*, 67, 461–467.

Ren, X. W., Zeng, L., Wei, Z. M., Xin, X. Z., & Wei, B. (2016b). Effects of multiple firings on metal-ceramic bond strength of Co-Cr alloy fabricated by selective laser melting. *The Journal of Prosthetic Dentistry*, 115(1), 109–114.

Salmi, M., Tuomi, J., Paloheimo, K. S., Björkstrand, R., Paloheimo, M., Salo, J., & Mäkitie, A. A. (2012). Patient-specific reconstruction with 3D modeling and DMLS additive manufacturing. *Rapid Prototyping Journal*, 18(3), 209–214.

Satoh, G., Yao, Y. L., & Qiu, C. (2013). Strength and microstructure of laser fusion-welded Ti–SS dissimilar material pair. *The International Journal of Advanced Manufacturing Technology*, 66(1), 469–479.

Serra-Prat, J., Cano-Batalla, J., Cabratosa-Termes, J., & Figueras-Àlvarez, O. (2014). Adhesion of dental porcelain to cast, milled, and laser-sintered cobalt-chromium alloys: Shear bond strength and sensitivity to thermocycling. *The Journal of Prosthetic Dentistry*, 112(3), 600–605.

Shipley, H., McDonnell, D., Culleton, M., Coull, R., Lupoi, R., O'Donnell, G., & Trimble, D. (2018). Optimisation of process parameters to address fundamental challenges during selective laser melting of Ti-6Al-4V: A review. *International Journal of Machine Tools and Manufacture*, 128, 1–20.

Simchi, A., & Asgharzadeh, H. (2004). Densification and microstructural evaluation during laser sintering of M2 high speed steel powder. *Materials Science and Technology*, 20(11), 1462–1468.

Sivakumar, M., Kumar Dhanadurai, K. S., Rajeswari, S., & Thulasiraman, V. (1995). Failures in stainless steel orthopaedic implant devices: A survey. *Journal of Materials Science Letters*, 14(5), 351–354.

Sorensen, J. A., Engelman, M. J., Daher, T., & Caputo, A. A. (1990). Altered corrosion resistance from casting to stainless steel posts. *The Journal of Prosthetic Dentistry*, 63(6), 630–637.

Song, C., Yang, Y., Wang, Y., Wang, D., & Yu, J. (2014). Research on rapid manufacturing of CoCrMo alloy femoral component based on selective laser melting. *The International Journal of Advanced Manufacturing Technology*, 75(1), 445–453.

Subramanian, K. T., Balashanmugam, N., & Shashi Kumar, P. V. (2016). Nanometric finishing on biomedical implants by abrasive flow finishing. *Journal of The Institution of Engineers (India): Series C*, 97(1), 55–61.

Sun, J., & Zhang, F. Q. (2012). The application of rapid prototyping in prosthodontics. *Journal of Prosthodontics: Implant, Esthetic and Reconstructive Dentistry*, 21(8), 641–644.

Survilienė, S., Jasulaitienė, V., Češūnienė, A., & Lisowska-Oleksiak, A. (2008). The use of XPS for study of the surface layers of Cr–Co alloy electrodeposited from Cr (III) formate–urea baths. *Solid State Ionics*, 179(1–6), 222–227.

Takaichi, A., Nakamoto, T., Joko, N., Nomura, N., Tsutsumi, Y., Migita, S., Doi,H., Kurosu, S., Chiba, A., & Wakabayashi,N. (2013). Microstructures and mechanical properties of Co–29Cr–6Mo alloy fabricated by selective laser melting process for dental applications. *Journal of the Mechanical Behavior of Biomedical Materials*, 21, 67–76.

Thijs, L., Verhaeghe, F., Craeghs, T., Van Humbeeck, J., & Kruth, J. P. (2010). A study of the microstructural evolution during selective laser melting of Ti–6Al–4V. *Acta Materialia*, 58(9), 3303–3312.

Thomas, D. (2009). *The development of design rules for selective laser melting*. University of Wales.

Thöne, M., Leuders, S., Riemer, A., Tröster, T., & Richard, H. (2012). Influence of heat-treatment of selective laser melting products-eg Ti6Al4V. Paper presented at the 2012 International Solid Freeform Fabrication Symposium.

Ucar, Y., Akova, T., Akyil, M. S., & Brantley, W. A. (2009). Internal fit evaluation of crowns prepared using a new dental crown fabrication technique: Laser-sintered Co-Cr crowns. *The Journal of Prosthetic Dentistry*, 102(4), 253–259.

Upadhyay, D., Panchal, M. A., Dubey, R., & Srivastava, V. (2006). Corrosion of alloys used in dentistry: A review. *Materials Science and Engineering: A*, 432(1–2), 1–11.

Vandenbroucke, B., & Kruth, J. P. (2007). Selective laser melting of biocompatible metals for rapid manufacturing of medical parts. *Rapid Prototyping Journal*, 13(4), 196–203.

Van Der Zel, J. M., Vlaar, S., de Ruiter, W. J., & Davidson, C. (2001). The CICERO system for CAD/CAM fabrication of full-ceramic crowns. *The Journal of Prosthetic Dentistry*, 85(3), 261–267.

Wang, H., Feng, Q., Li, N., & Xu, S. (2016). Evaluation of metal-ceramic bond characteristics of three dental Co-Cr alloys prepared with different fabrication techniques. *The Journal of Prosthetic Dentistry*, 116(6), 916–923.

Wang, H., Wang, X., Qian, H., Lou, D., Song, M., & Zhao, X. (2021). The optimal structural analysis of cobalt-chromium alloy (L-605) coronary stents. *Computer Methods in Biomechanics and Biomedical Engineering*, 24(14), 1566–1577.

Wang, Y. M., Voisin, T., McKeown, J. T., Ye, J., Calta, N. P., Li, Z., Zeng, Z., Zhang, Y., Chen, W., & Roehling, T. (2018). Additively manufactured hierarchical stainless steels with high strength and ductility. *Nature Materials*, 17(1), 63–71.

Wataha, J. C. (2002). Alloys for prosthodontic restorations. *The Journal of Prosthetic Dentistry*, 87(4), 351–363.

Wataha, J. C., & Schmalz, G. (2009). Dental alloys. In *Biocompatibility of Dental Materials*. Springer, pp. 221–254.

Willer, J., Rossbach, A., & Weber, H.-P. (1998). Computer-assisted milling of dental restorations using a new CAD/CAM data acquisition system. *The Journal of Prosthetic Dentistry*, 80(3), 346–353.

Wu, L., Zhu, H., Gai, X., & Wang, Y. (2014). Evaluation of the mechanical properties and porcelain bond strength of cobalt-chromium dental alloy fabricated by selective laser melting. *The Journal of Prosthetic Dentistry*, 111(1), 51–55.

Xiang, N., Xin, X.-Z., Chen, J., & Wei, B. (2012). Metal–ceramic bond strength of Co–Cr alloy fabricated by selective laser melting. *Journal of Dentistry*, 40(6), 453–457.

Xin, X., Chen, J., Xiang, N., Gong, Y., & Wei, B. (2014). Surface characteristics and corrosion properties of selective laser melted Co–Cr dental alloy after porcelain firing. *Dental Materials*, 30(3), 263–270.

Xin, X., Chen, J., Xiang, N., & Wei, B. (2013). Surface properties and corrosion behavior of Co–Cr alloy fabricated with selective laser melting technique. *Cell Biochemistry and Biophysics*, 67(3), 983–990.

Xin, X., Xiang, N., Chen, J., & Wei, B. (2012). In vitro biocompatibility of Co–Cr alloy fabricated by selective laser melting or traditional casting techniques. *Materials Letters*, 88, 101–103.

Xu, W., Brandt, M., Sun, S., Elambasseril, J., Liu, Q., Latham, K., Xia, K., & Qian, M. (2015). Additive manufacturing of strong and ductile Ti–6Al–4V by selective laser melting via in situ martensite decomposition. *Acta Materialia*, 85, 74–84.

Yadroitsev, I., Bertrand, P., & Smurov, I. (2007). Parametric analysis of the selective laser melting process. *Applied Surface Science*, 253(19), 8064–8069.

Yadroitsev, I., & Smurov, I. (2010). Selective laser melting technology: From the single laser melted track stability to 3D parts of complex shape. *Physics Procedia*, 5, 551–560.

Yager, S., Ma, J., Ozcan, H., Kilinc, H., Elwany, A., & Karaman, I. (2015). Mechanical properties and microstructure of removable partial denture clasps manufactured using selective laser melting. *Additive Manufacturing*, 8, 117–123.

Yang, X., Xiang, N., & Wei, B. (2014). Effect of fluoride content on ion release from cast and selective laser melting-processed Co-Cr-Mo alloys. *The Journal of Prosthetic Dentistry*, 112(5), 1212–1216.

Yang, Y., Lu, J. b., Luo, Z. Y., & Wang, D. (2012). Accuracy and density optimization in directly fabricating customized orthodontic production by selective laser melting. *Rapid Prototyping Journal*, 18(6), 482–489.

Yves-Christian, H., Jan, W., Wilhelm, M., Konrad, W., & Reinhart, P. (2010). Net shaped high performance oxide ceramic parts by selective laser melting. *Physics Procedia*, 5, 587–594.

Zeng, L., Xiang, N., & Wei, B. (2014). A comparison of corrosion resistance of cobalt-chromium-molybdenum metal ceramic alloy fabricated with selective laser melting and traditional processing. *The Journal of Prosthetic Dentistry*, 112(5), 1217–1224.

Zhang, B., Huang, Q., Gao, Y., Luo, P., & Zhao, C. (2012). Preliminary study on some properties of Co-Cr dental alloy formed by selective laser melting technique. *Journal of Wuhan University of Technology-Mater. Sci. Ed.*, 27(4), 665–668.

Zhang, L. C., Klemm, D., Eckert, J., Hao, Y. L., & Sercombe, T. B. (2011). Manufacture by selective laser melting and mechanical behavior of a biomedical Ti–24Nb–4Zr–8Sn alloy. *Scripta Materialia*, 65(1), 21–24.

Zhao, M.-C., Liu, M., Song, G.-L., & Atrens, A. (2008). Influence of pH and chloride ion concentration on the corrosion of Mg alloy ZE41. *Corrosion Science*, 50, 3168–3178 10.1016/j.corsci.2008.08.023.

12 Radiation curing of epoxy composites and coatings

Poornima Vijayan P.
Department of Chemistry, Sree Narayana College for Women (affiliated to University of Kerala), Kollam, Kerala

Anu Surendran and Sabu Thomas*
International and Inter University Centre for Nanoscience and Nanotechnology, Mahatma Gandhi University, Kottayam, Kerala, India

*Corresponding author: sabuthomas@mgu.ac.in

CONTENTS

12.1 INTRODUCTION

Radiation induced curing of thermoset is an interesting and stimulating application of radiation processing in the polymer industry (Chmielewski, Haji-Saeid, Ahmed 2005) (Tamada 2018) (Rao 2009). The irradiation of monomers/oligomers initiates curing process to make highly cross-linked thermosetting materials. Radiation induced curing is considered as a green and energy efficient alternative to chemical curing methods. The radiation curing offers several advantages over conventional curing methods such as high rate of polymerization, ambient temperature curing, minimal emission of toxic chemicals, low processing cost, etc. (Dickson and Singh 1988). Radiation curing can be performed using different types of ionising radiations like ultraviolet (UV), gamma (γ), electron beam (EB), and X-rays and heating radiations like microwaves, laser, etc. (Hay and O'Gara 2006). The major radiation curable thermosetting resins investigated are unsaturated polyester resins (Shi and Rånby 1994) (Mahmoud, Tay, Rozman 2011), poly-methyl methacrylate (Lhl, Jacob, Boey 1995) (Usanmaz, Eser, Doğan 2001), polyurethane acrylates (Koshiba, Hwang, Foley 1982), and epoxy resins (Dickson and Singh 1988) (Alessi, Calderaro, Parlato 2005).

Among radiation curable thermosets, epoxies, and their composites were well explored. They are the popular thermosetting materials having great commercial importance. The cured epoxy composites have an inherent property of brittleness due to the high cross-link density (Vijayan, Puglia, Kenny

2013) (Jyotishkumar, Abraham, George 2013). Usually, composite structures with a secondary phase are preferred for the functional applications. Epoxy resins form the matrix embedded with fillers, fibers, and nanoparticles for advanced applications. Curing parameters for epoxy resins are critical, as they are the matrix material for many structural components in aircrafts, marine vessels and boats; adhesive and encapsulating sealant for the microelectronics and electronics industries; and protective coating for metal parts in oil and gas industry, automobiles, and buildings. Temperature curing using toxic curing agents/solvents has not been considered an energy-efficient and eco-friendly approach. Conventional composite processing methods waste huge thermal energy and need expensive autoclaves. As a green alternative, the radiation curing of epoxy-based composites and coatings have much attention among researchers. Moreover, radiation curing offers less cure time, less thermally induced stresses, and high controllability over cure (Kumar, Saini, Bhunia 2020).

The current chapter focuses on the various aspects of radiation curing of epoxies and epoxy-based composites. This chapter begins with a basic discussion on the chemistry and generally accepted mechanism of radiation curing in epoxies. Thereafter, the chapter gives a brief discussion on various radiation sources used for the curing of epoxies and their composites. The radiation curing of fiber-reinforced epoxy composites and epoxy nanocomposites will be discussed by emphasizing the interference of fibers and matrix on the mechanism and kinetics of radiation induced curing. Moreover, a detailed discussion on radiation curing of epoxy-based coatings is provided with prime importance. The final performance of the radiation-cured epoxy composites has been discussed. The advantages of radiation curing on curing processing, cross-link density of the cured product, and their end use properties over thermal curing are discussed.

12.2   THE CHEMISTRY BEHIND RADIATION CURING OF EPOXY

Most of the epoxy resins observed to have radiation reactivity. Earlier studies revealed that the molecular structure of the epoxy monomer or oligomer influenced the mechanism of radiation curing (Al-Sheikhly and McLaughlin 1996). Radiation-induced curing undergoes either by free radical or by cationic mechanism. Generally, free-radical polymerization is inhibited by oxygen as the growing free radical have high reactivity with oxygen. Due to this reason, the radiation induced cure reaction undergoes via. free radical mechanism needs to process in an inert atmosphere (Boey, Rath, Ng 2002). Dinkson and Singh (Dickson and Singh 1988) reviewed that most of the epoxy resins cross-linked upon irradiation by a cationic mechanism. In such cases, curing of epoxy systems were normally initiated by cationic photoinitiators such as diaryliodonium salts, triarylsulfonium salts, phenacylsulfonium salts etc. (Ghosh and Palmese 2005). However, in the presence of moisture even in trace amount, the cationic polymerization inhibited due to the proton transfer from the chain-propagating epoxy cation to water molecule. In acrylic derivatized epoxies, the radiation-induced cross-linking proceeds by free-radical mechanism (using free radical initiators), where the produced free radicals add to the unsaturated sites of the molecule (Gotoda, Miyashita, Mori 1975) (Chattopadhyay, Panda, Raju 2005), which won't inhibit by traces of water in the resin system.

In general, the conventional resin-curing agent system is not suitable for curing using ionising radiations. Even though no heat energy is required to initiate ionizing radiation curing, the exothermic cross-linking reaction causes uncontrolled heating of the system, which spoils the quality of cross-linked material. It requires special care to control the generation of heat energy during radiation curing of epoxy (Alessi, Calderaro, Parlato 2005). High heat energy is produced at high radiation dose rate and low curing time. Hence, careful selection of the irradiation parameters avoids the deterioration of the property of the cured resin due to unexpected temperature hike. Meanwhile, in heating radiation (like microwave radiation) induced curing, the conventional epoxy-curing agent mixture undergoes a cross-linking reaction using the heat energy produced by the molecular rotational of the monomer upon irradiation. It was found that shorter dose rate and prolonged exposure ensured higher degree of cure (Raghavan 2009). Other important variables in radiation curing technology that controls the overall properties of the final cured network structure are the type of radiation source, chemical nature of epoxy resin, the radiation curing mechanisms, kinetics of cure, etc. (McGinniss 2000).

12.3 RADIATION SOURCES USED TO CURE EPOXY

Various epoxy-based structural composites, adhesives, and coatings have been evaluated for their radiation curing efficiency. Commonly used radiation sources for this purpose are UV (Ceccia, Turcato, Maffettone 2008), EB (Lopata, Saunders, Singh 1999), gamma (Hoffman and Skidmore 2009), X-ray (Berejka, Cleland, Galloway 2005), and microwave (Boey, Yap, Chia 1999) radiations. The characteristics of cure reaction by each type of radiation are discussed below.

12.3.1 ULTRAVIOLET (UV) RADIATION

UV radiation has immense industrial applications, especially to cure epoxy thin films, coatings, and adhesives. The UV curing technology in coating industry is rapidly advancing as it is safe to environment (Bajpai, Shukla, Kumar 2002). Western countries are practicing this solvent-free curing approach in coating industry to preserve the environment.

The major highlights of UV curing of epoxy are the very fast cure time and simplicity of curing equipment. The most frequently used radiation sources are mercury arc lamps, incandescent lamps, light emitting diodes, excimer lasers etc. (Endruweit, Johnson, Long 2006). The selection of UV source depends on the chemical nature of the photo initiator and penetration thickness. UV radiation has the less penetration through matter, which limits its applications in the composite industry dealing with large structures.

In general, the curing mechanism involved in UV-irradiated epoxy resins is cationic polymerization. Figure 12.1 shows the cationic polymerization mechanism of cure reaction in UV-irradiated epoxy resin (Malucelli, Bongiovanni, Sangermano 2007) (Sangermano, Roppolo, Chiappone 2018). The initiation reaction involves the photo-excitation of onium salts (photo initiator) followed by the decay of the excited singlet state via. both heterolytic and homolytic cleavages. Cations and arylcations generated during photolysis react with monomers to form a Brönsted acid, which protonates the epoxy monomer (Fouassier, Burr, Crivello 1994). Chain growth reaction takes place by the addition of further epoxy monomer molecules. Apart from conventional onium salts, iodonium or sulfonium salts containing tetrakis (perfluoro-tbutyloxy) aluminate anion (Klikovits, Knaack, Bomze 2017), ferrocenium salt (Zhang, Campolo, Dumur 2016) etc. were discovered with much higher ability to initiate cationic polymerization.

It is important to discuss the heterogeneity issues in radiation cured epoxy matrix. Kowandy et al. (Kowandy, Ranoux, Walo 2018) reported the presence of several relaxation domains in dynamic mechanical analysis (DMA) and dielectric spectroscopic (DS) curves of UV visible or

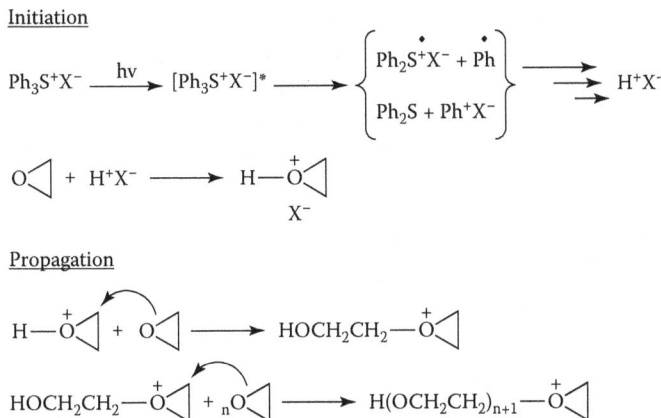

FIGURE 12.1 General scheme of the photopolymerization of epoxy resins (Malucelli, Bongiovanni, Sangermano 2007).

electron beam radiation-cured epoxy matrix due to the presence of heterogeneities. The formation of glassy nanoclusters, as evident from AFM studies, found to be highly influenced by the initiator content, irradiation dose, and post-irradiation heating.

Surface coatings over plastics and their composites are critically important to protect them from severe environmental conditions such as UV, acidic or saline rain, dusty wind, etc. UV-cured epoxy coatings could be better choice for the protection of plastic materials. Barletta et al. (Barletta, Vesco, Puopolo 2015) studied the efficiency of a UV-curable cycloaliphatic epoxy resinreinforced with amino organo-silane modified reduced graphene oxide as protective coating over polycarbonate. The above coating provided high anti-scratch performance, anti-soiling, and chemical barrier to the plastic substrate.

12.3.2 Electron beam (EB)

In EB curing of thermosets, high-energy electrons from an accelerator are used to initiate cure reaction (Sui, Zhang, Chen 2003). An energy transfer from accelerated electrons to the resin resulted in the curing of the monomer or oligomer. In the late 1970, Aerospatiale, a French company started research on EB curing of thermoset composites. Researchers attempted to develop EB cured epoxies and their composites with performance similar or even superior to that of thermally cured one (Lopata, Saunders, Singh 1999). EB radiation has higher depth of penetration than UV radiation. Hence, EB curing offers effective curing of thick epoxy coatings and composites over UV-induced curing. Also, highly pigmented and opaque composites, which are inactive towards UV curing, can be effectively cured using EB radiations. However, it was found that curing degree showed an accelerated decrease along the depth direction (Sui, Zhong, Yang 2009). According to Sui et al. (Sui, Zhong, Yang 2009), a properly optimised irradiation energy, irradiation dose, and initiator concentration could effectively eliminate homogeneity in the cross-linking structure.

EB curing technology offers the development of composite structures containing different types of fibers in complex shapes to use in aerospace and other industries without the need for expensive fabrication tools (Nho, Kang, Park 2004) (Saunders, Lopata, Barnard 2000). The early studies explored the mechanism of curing in EB induced cationic polymerization in the presence of onium salts. Also, epoxy resins can be cured at a low dose of EB irradiation to achieve desired cross-link density (Crivello, Fan, Bi 1992). For thick resin products, Sui et al. (Sui, Zhong, Yang 2009) proposed a layer upon layer irradiation or irradiation from two sides to obtain a homogeneous cross-linking structure.

Different studies evaluated the various factors affecting electron beam curing of the epoxy and its composites and the peculiar microstructure formed in the composite structures. Sui et al. (Sui, Zhang, Chen 2003) reported that the electron-donating solvent could inhibit the EB radiation curing of epoxy systems. They established the formation of hemispheroidal-like cured fields around incident EB (Figures 12.2a and b), the dimension of the cured field increases with increase in radiation dosage. Cured fields consist of many lamellar structures (Figures 12.2c and d) by the spreading of EB induced curing via. layer upon layer around an active center.

12.3.3 Gamma and X-rays

Compared to UV and EB irradiation methods, gamma and X-rays can penetrate deeply into matter. Gamma and X-rays, owing to their high electromagnetic energy, are capable of curing thick (up to 300 mm) composite for the large structures used in aeronautic, transport, and marine industries. UV and EB radiation penetrate and cure a few micrometres and to several centimetres respectively, gamma radiation can be up to 30 cm. A recent study showed that an epoxy-bonded ceramic composites wall diameter of 20 cm was reasonably cured by X-ray (Puchleitner, Riess, Kern 2017). By considering the advantage of high penetration thickness, many industrial EB facilities allow for the conversion of electron radiation to X-rays (Berejka, Cleland, Walo 2014). However,

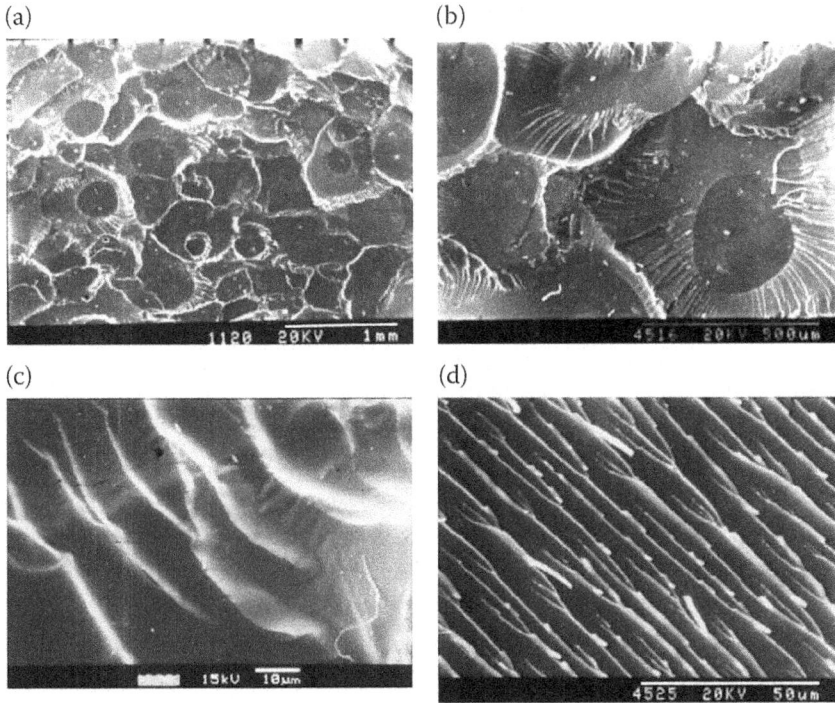

FIGURE 12.2 Scanning electron microscopy (SEM) micrographs of radiation-cured surface of epoxy matrix (a) lamellar structures in the surface of cured resin, (b) the partially enlarged micrograph of (a), (c) the enlarged micrograph for border of lamellar structure, (d) the cross-sectional microstructure parallel with the direction of EB radiation (Sui, Zhang, Chen 2003).

the restriction on using low dose rates of gamma and X-rays resulted in a longer cure time when compared to that of electron beam curing. It was reported that in epoxy/carbon fiber composite, gamma irradiation at room temperature give rise to different cross-link densities in epoxy matrix (heterogeneous matrix), which could be homogenised by post-irradiation thermal treatment to get a composite with an increased glass transition temperature (T_g) (Dispenza, Alessi, Spadaro 2008). The potential safety risk associated with X-rays and gamma rays limits their practical application in many industries (Abliz, Duan, Steuernagel 2013).

12.3.4 MICROWAVE RADIATION

Microwave radiation curing is considered as a cost-effective alternative to conventional thermal curing (Johnston, Pavuluri, Leonard 2015). Different from the above-described ionising radiations, the microwave radiation is unable to drive the cure reaction by ionisation of the resin material due to its extremely low energy. Here, the heat required for the cross-linking of monomers is generated inside the system rather than externally transferred. The processing mechanism in this case is the conversion of microwave radiation energy into heat by the molecular rotation of the monomer. Thus, produced heat is sufficient to activate cross-linking reaction between epoxy and the curing agents as in conventional thermal curing (Le Van and Gourdenne 1987). The advantage of using microwave radiations than conventional heat cure is that uniform cure could be achieved especially in thick composite structures (Marand, Baker, Graybeal 1992). The high depth of penetration of microwave in polymers makes it as an efficient volume-heating mode. An increase in curing rate at lower cure times for the microwave reaction compared to thermal curing (Figure 12.3) was reported by Johnston et al. (Johnston, Pavuluri, Leonard 2015) based on an experiment carried out in a

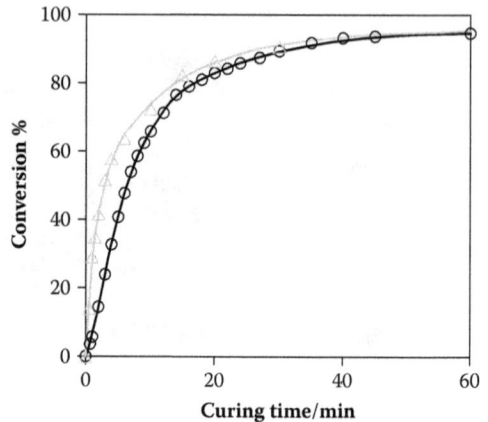

FIGURE 12.3 Time–conversion relationship at a curing temperature of 120°C for thermally cured EO1080 (○) and microwave cured EO1080 (Δ) (Johnston, Pavuluri, Leonard 2015).

commercially available one component epoxy resin filled with silica and carbon additives (EO1080) for improved microwave absorption.

While using epoxies as adhesives, microwave irradiation offers a selective heating which restricts the damage of other material components (Horikoshi, Arai, Serpone 2021). The selective nature of microwave radiations also offers the incorporation of extraneous substrate like carbon nanotube, activated carbon powder etc. with high microwave response into epoxy as additional heat source. (Rangari, Bhuyan, Jeelani 2010).

In advanced attempt, a radiator which facilitated the curing of epoxy resin via. the use of both UV and microwave has been used (Wu, Xie, Xiao 2018). The microwave combined ultraviolet radiator utilized non-uniform slotted coaxial line structure to cure epoxy resin. When compared with curing time of 1.5 h with single UV curing method, the MW combined UV radiator offered a shorter curing time of 25 min for epoxy. Apart from the processing advantages, the combined irradiation helps to develop superior coatings with toughness and adhesiveness when compared to their singly cured counterparts.

12.4 RADIATION CURING OF EPOXY BASED SYSTEMS

12.4.1 FIBER-REINFORCED EPOXY COMPOSITES

Fiber-reinforced epoxy composites have a major role in civil engineering and aerospace applications. The good interfacial adhesion between the fiber and the matrix are essential in fiber reinforced composites for achieving the high transverse properties. The fiber-reinforced epoxy composites possess high mechanical strength, low weight, high durability ,and resistant to corrosion ((Jin, Lee, Park 2013) (Stewart and Douglas 2012) (Grammatikos, Evernden, Mitchels 2016) (Uthaman, Xian, Thomas 2020). The interfacial adhesion between the fibers and epoxy matrix played a critical role in increasing the overall properties of the epoxy-based fiber composites. The radiation curing of epoxy/fiber composites decreased the processing time without the requirement of the complicated fabrication parameters such as high temperature and pressure. The fiber-reinforced epoxy composites are generally thick and hence require high-energy radiation for inducing curing reaction. Hence, scientists suggest a dual-cure process (irradiation followed by thermal curing) to achieve desired cross-link density in epoxy composites to meet the demand (Alessi, Calderaro, Parlato 2005) (Alessi, Parlato, Dispenza 2007).

Crivello et al. (Crivello, Walton, Malik 1997) devised a strategy for EB curing of epoxy/carbon fiber composites by employing diaryliodonium and triarylsulfonium salt as photo initiators. They developed high performance composites by photocuring with higher mechanical, thermal properties, and oxygen plasma resistance. Crivello (Crivello 2002) demonstrated curing in epoxy/

carbon fiber composites by UV, gamma rays, and EB induced cationic polymerizations. He reported the feasibility of using low-dose EB radiation for the fast and effective curing of fiber reinforced epoxy-functional silicone monomers. He used diaryliodonium and triarylsulfonium salts as initiators for inducing cationic polymerization. Alessi et al. (Alessi, Conduruta, Pitarresi 2010) demonstrated the effect of hydrothermal ageing in radiation cured epoxy/polyether sulfone (PES)/ carbon fiber composites. Two compositions of the composites with 10 wt% of PES and 20 wt% of PES were analyzed and found that composites with 20 wt% PES exhibited better resistance to aging. Spardo et al. (Spadaro, Alessi, Dispenza 2014) fabricated epoxy/carbon fiber composites by electron beam irradiation using a pulsed 10 MeV electron beam accelerator. Dynamic mechanical thermal analysis revealed a wide range of relaxation temperature due to the formation of a non-uniform cross-linked network. However, a post-thermal treatment after radiation curing rectified this defect and resulted in a uniform cross-linked network. Fracture toughness and inter laminar shear strength were maximum for composites which were thermally treated after irradiation.

Lee and Boey (Boey and Lee 1991) observed that an optimum curing process achieved at a faster rate in epoxy/glass fiber composites by microwave radiation than by thermal curing. A significantly higher flexural modulus was also observed in those composites obtained by microwave curing. Xu et al. (Xu, Wang, Cai 2016) observed a 39% reduction in cure time and 22% increment in the compression strength for epoxy/carbon fiber composites manufactured by vacuum bagging using microwave and thermal curing. Colangelo et al. (Colangelo, Russo, Cimino 2017) compared microwave radiation curing and thermal curing in epoxy/glass fiber composites. Comparable cross-linking degree was obtained when the composites were subjected to three different curing procedures such as room temperature curing for 7 days (T_{amb}), oven curing at 100°C for 1 h (oven) and microwave curing at 1500 W for 10 min (MW). The observed flexural modulus and strength were maximum for microwave cured composites (Figure 12.4) and was found suitable for civil engineering applications. Pull-out strength of microwave cured epoxy/glass fiber composites especially with respect to the conventional concrete (CC) support was reported as outstanding.

Duan and co-workers (Zhang, Duan, Wang 2020) used microwave irradiation initially to enhance the interfacial interaction between carbon fibers and epoxy resins which was later cured by EB radiation. The electrically driven polymer-carbon fiber interaction was induced by microwave radiation and theoretical simulation predicted the mechanical interlock between the dielectric polymer

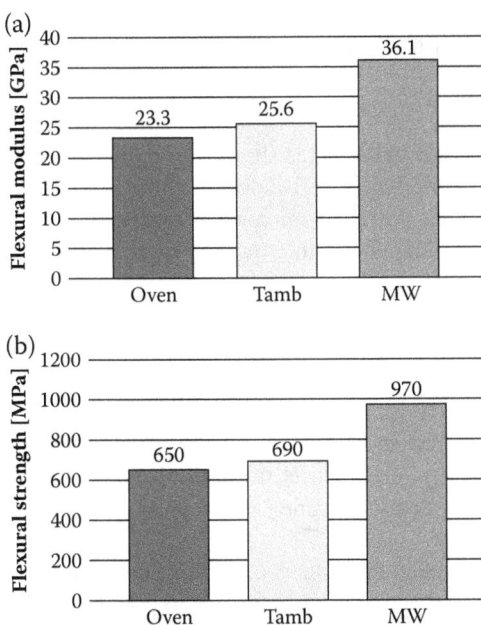

FIGURE 12.4   a) Flexural modulus (Ef) and b) flexural strength (σf) for epoxy/glass fiber composites cured under three different conditions (Colangelo, Russo, Cimino 2017).

(a)

(b)

MW 0 s

MW 30 s (30 s × 1)

(c)

(d)

MW 60 s (30 s × 2)

MW 90 s (30 s × 3)

(e)

MW 180 s (30 s × 6)

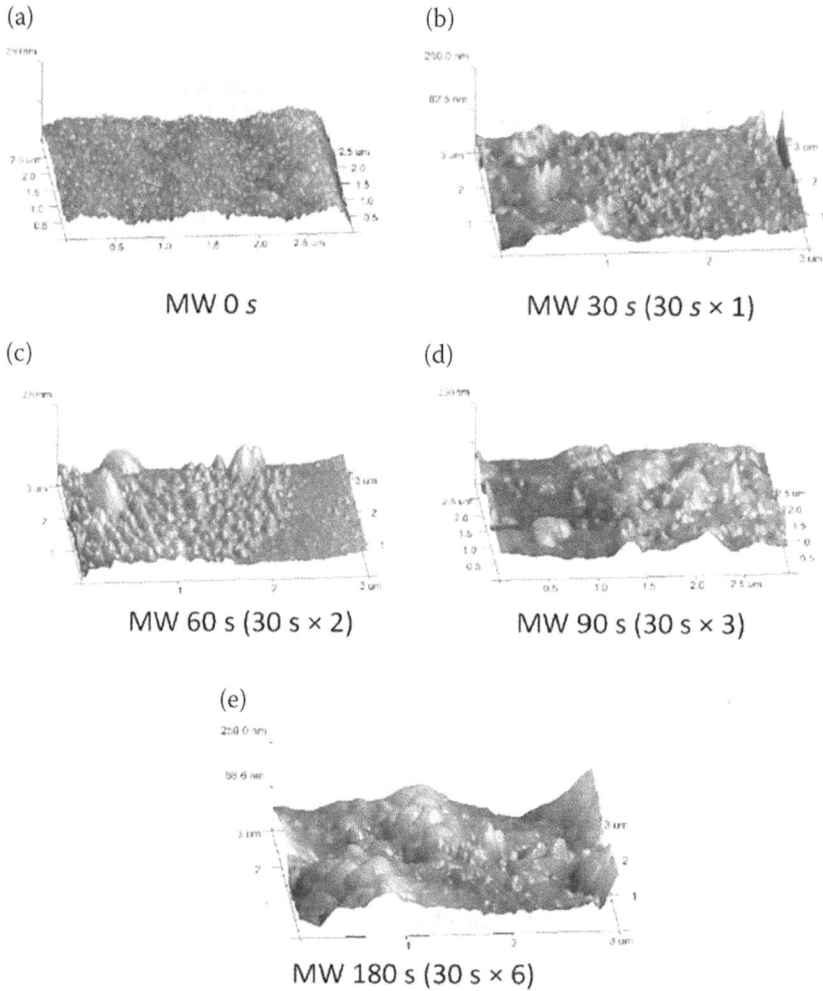

FIGURE 12.5 AFM micrographs of carbon fibers extracted from microwave irradiated prepregs vs irradiation time (Zhang, Duan, Wang 2020).

and the fiber. Microwave radiation enhanced the surface roughness (as shown in Figure 12.5) and energy, which further increased the wettable surface favorable for EB radiation curing. The interfacial shear strength (IFSS) was increased by 31% after 180 s of microwave exposure and the interlaminar shear strength (ILSS) was improved by 22% after 90 s of microwave radiation exposure.

12.4.2 EPOXY NANOCOMPOSITES

Another important class of epoxy hybrid materials are epoxy nanocomposites, which comprise of a nanoparticle or nanofiller embedded within an epoxy matrix. The studies on radiation curing on epoxy nanocomposites are well explored. The main idea is to understand the mechanism and kinetics of curing reaction which was followed from the in-situ Fourier transform infrared (FTIR) spectroscopy, small-angle X-ray scattering (SAXS), dynamic scanning calorimetry (DSC), and rheological analysis.

Sharif et al. (Sharif, Pourabbas, Sangermano 2017) studied the influence of graphene oxide (GO) on the UV curing of epoxy resins. While, GO has been proved to catalyze the photo polymerization but retard the cure conversion. The free radical scavenging by the GO sheets and the opaqueness of

GO to UV radiation retarded the curing reaction at higher concentration of GO. 3 wt% loading of GO facilitated a conductive network and was ideal for electrical properties. But, when a high intensity UV radiation of 150 mW/cm^2 was used, real time FTIR analysis indicated that the nanocomposites with even 3 wt% loading of GO showed complete conversion of epoxy groups.

Chen and Curliss (Chen and Curliss 2003) fabricated intercalated epoxy/organoclay nano-composites with a low dosage EB radiation using diaryliodonium salt as the initiator. Decker and co-workers (Benfarhi, Decker, Keller 2004) synthesized epoxy/clay nanocomposites by UV light intensity of 500 mW cm^2 at ambient temperature in a short span of time. The UV cured epoxy/3 wt % organoclay nanocomposites with high impact resistance and flexibility were found to suitable for coating applications. Uyanik (Uyanık, Erdem, Can 2006) used microwave curing for preparing epoxy/organoclay nanocomposites. α–ω diacrylate poly-(dimethylsiloxane) was incorporated to modify the toughness before curing reaction. It was observed that the nanocomposites containing 5% organoclay and 5% organosloxane were beneficial for the mechanical properties. It was also observed that the microwave curing supported the intra-gallery reaction of the epoxy groups, which facilitated the exfoliation of the clay.

Przybytniak et al. (Przybytniak, Nowicki, Mirkowski 2016) employed gamma rays for radiation curing for the preparation of epoxy/carbon nanotubes (CNT) nanocomposites with iodonium salt as cationic initiator. The prepared composites showed uniform dispersion of nanoparticles in the epoxy matrix. The initiator got adsorbed on the on the CNT surface which cause the polymerization to occur around the vicinity of the nanofiller and in turn contributed the good adherence to the matrix. The mechanical properties and Tg was improved in the gamma radiation cured epoxy/CNT nanocomposites. Chen et al. (Chen, Chen, Hsu 2006) developed a low shrinkage, visible light curable epoxy/55 wt% nanosilica nanocomposites suitable for use as a dental restorative material. 3, 4-epoxycyclohexylmethyl-(3, 4-epoxy) cyclohexane carboxylate and GPS (γ-glycidoxypropyltrimethoxysilane) was used as the resin and surface modifier for silica, respectively. The developed nanocomposites exhibited a low thermal expansion coefficient, high microhardness (62 KHN), high tensile strength (47 MPa) and high degree of conversion for irradiation of 60 s. Shin et al. (Shin, Jeun, Kang 2009) fabricated epoxy/MWCNT nanocomposites by EB curing using triarylsulfoniumhexafluoroantimonate (TASHFA) as initiator. The storage modulus (G') and the thermal properties were increased with increase in MWCNT content and radiation dose. The thermal properties were observed to increase with increasing the amount of initiator, which was evident from the thermo gravimetric analysis (TGA) of the prepared nanocomposites (Figure 12.6). The TGA plots also gave evidence for better thermal stability of epoxy/MWCNT nanocomposites in nitrogen atmosphere than in air (Figure 12.6d).

Researchers also studied epoxy composites cured by radiation followed by thermal curing and found better performance than their thermally or radiation alone cured counterparts. Epoxy based on Bis (4-glycidyloxyphenil) methane (DGEBF) obtained by EB irradiation followed by thermal post-curing (DGEBFirr-pc) showed a best compromise in water diffusion behaviour and development of swelling stresses with respect to both pure irradiation (DGEBFirr) and pure thermal curing (DGEBFt) (Alessi, Toscano, Pitarresi 2017). DGEBFirr-pc has low stresses in desorption and absorption with relatively high T_g and a lower water absorption. Kumar et al. (Kumar, Saini, Bhunia 2020) evaluated the load-bearing performance of the EB radiation cured MWCNT filled carbon/epoxy composite pin joints. It was found that failure loads of pin joints greatly influenced by geometric parameters (W/D and E/D ratios) and the presence of MWCNT.

Santos et al. (dos Santos, Opelt, Lafratta 2011) investigated the mechanical and thermal properties of photocurable epoxy-acrylate-based nanocomposites containing 0.25 wt% and 0.75 wt% MWCNT. The nanocomposites were cured for 12 and 24 h with UV (355 nm) lamps of 800W/m^2 intensity. A nanoscale dispersion MWCNTs was observed in the cured epoxy matrix. Hence, a significant enhancement in T_g and nano- and micro-hardness were obtained for the nanocomposites.

Lee and coworkers (Im, Jeong, In 2010) studied the effect of fluorinated MWCNT on the viscoelastic properties of EB cured epoxy nanocomposites. EB curing was carried out under a dose

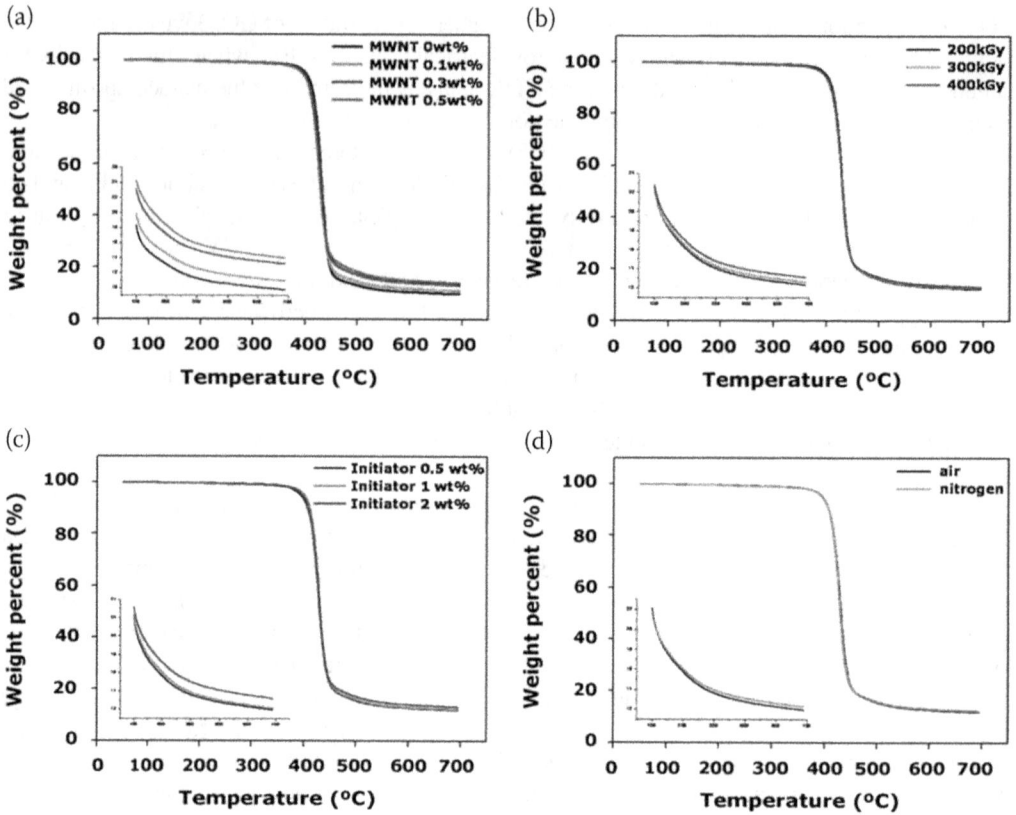

FIGURE 12.6 Thermogravimetric analysis of EB cured epoxy/MWCNT nanocomposites (Shin, Jeun, Kang 2009).

rate of 300 kGy/h for 10 min. Fluorination improved both interfacial adhesion and dispersion of the nanofillers. Bongiovanni et al. (Bongiovanni, Casciola, Gianni 2009) prepared epoxy nanocomposites by cationic polymerization of a mixture of 3,4-epoxycyclohexylmethyl-30,40-epoxycyclo-hexane carboxylate (BCDE) monomer and amino alcohols functionalized zirconim phosphate (ZrP). In general, chain transfer reactions occur in cationic polymerization when hydroxyl groups are present in the form of alcohols, polyols, and moisture (Crivello and Liu 2000). In the curing process of epoxy-ZrP nanocomposite, the presence of –OH groups in the modified ZrP involved in the chain transfer step (Figure 12.7). The attack of an –OH group present in the functionalized ZrP towards the

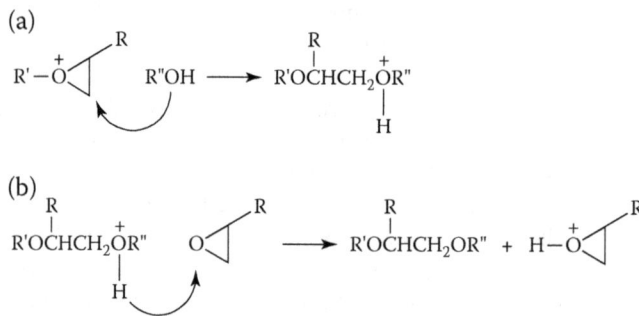

FIGURE 12.7 Mechanism of chain transfer reaction in epoxy-ZrP nanocomposite (Bongiovanni, Casciola, Gianni 2009).

positively charged carbon atom of epoxy growing chain hinders the chain growth and releases a proton which can further accelerate the growth of a new chain resulting in a flexible and shorter network of epoxy chains. The nanofillers accelerated the cure kinetics and increased the thermal stability of the nanocomposites.

Microwave curing of epoxy nanocomposites with graphitic nanofillers interestingly reported with shortest cure time owing to the high microwave absorption of graphitic fillers. Epoxy (EPON 862)/CNT nanocomposites were reported to cure in few minutes instead of several hours with thermal curing where CNTs assist the cure by maximum absorption of microwave radiations (Rangari, Bhuyan, Jeelani 2010).

The understanding of dielectric property of the material would help to explore the mechanism of curing and Pal et al. (Pal, Jha, Akhtar 2017) elucidated the dielectric properties of epoxy/carbon black (CB) nanocomposites before and after the microwave curing as a parameter of the particle size of the filler. Advanced cavity perturbation method (CPM) was employed to measure the dielectric properties during curing. It was observed that the smaller sized filler decreased the curing time, increased the T_g, tensile strength, and flexural modulus. CB of 15 nm particulate size had the highest dielectric loss factor and could efficiently enhance the heating process during microwave processing.

Yagci et al. (Yagci, Sahin, Ozturk 2011) synthesized epoxy/silver nanocomposites by visible light irradiation. The reaction between highly conjugated thiophene derivative 3,5-bis (4-methoxyphenyl) dithieno[3,2-b;2,3-d]thiophene) (P-DDT) and silver salt resulted in the cationic ring opening of epoxy ring and reduction of silver ion. Spherical type silver nanoparticles (25–50 nm) were dispersed in the epoxy matrix. Sangermano and co-workers (Martin-Gallego, Lopez-Manchado, Calza 2015) fabricated epoxy nanocomposites embedded with gold functionalized graphene sheets (Au–GNP). The nanocomposites showed considerable enhancement in the electrical conductivity due to the charge transfer mechanism facilitated by the Au nanoparticles. dell'Erba et al. (dell'Erba, Martínez, Hoppe 2017) elucidated the mechanism involved in the particle formation in epoxy/silver nanocomposites formed by a visible light assisted in situ synthesis. A wide range of experimental techniques were used for the analysis such as in situ time-resolved SAXS, rheology experiments combined with near Infrared (NIR) and UV-vis spectroscopies to follow the formation of nanocomposites. In the elucidated mechanism, the nucleation occurred in less than 5 seconds. Ostwald ripening and dynamic coalescence were the dominant mechanisms governing the particle growth formation. After gelation, diffusion constraints of the epoxy matrix hinder the motion of the particles, and were further characterized by the reduction of Ag^+ ions on the surface of the formed nuclei. Thermal post curing further improved the surface quality of the nanoparticles.

12.4.3 EPOXY-BASED COATINGS

Epoxy-based surface coating provides excellent protection for various substrates including metals, concrete, plastics, etc. owing to their excellent anticorrosive, tribological properties, and solvent resistance. The main incentive for the fabrication of radiation-cured epoxy coatings is the absence of solvent and high temperature/pressure conditions for curing. Coatings are relatively thinner, so radiations without high penetrating power were usually used for curing coatings and thin films. The environmental considerations increased the demand for greener protective coatings. So, solvent less and less energy consuming protective coatings are required to meet the demand.

A lot of research works were focussed on epoxy-based coatings, which possess both adhesive and protective function (Decker 1983) (Morawetz 1981) (Decker and Moussa 1990). A suitable radiation source to cure thin epoxy coatings is UV visible radiation via. cationic polymerization. Czajlik et al. (Czajlik, Hedvig, Ille 1996) observed from the calorimetric studies that cationic polymerization was relatively slow compared with the radical initiated polymerization. Vabrik et al. (Vabrik, Ille, Víg 1996) studied UV curing of an epoxy monomer, penta-erithritol-tetra-glycidyl ether by a cationic photoinitiator bis [4-(diphenylsulfonio)-phenyl]sulphide-bis-hexa-fluorophosphat. The effect of UV radiation curing and irradiation time was investigated. They observed an increase in T_g with the increase in the concentration of photoinitiator. Decker et al. (Decker, Viet, Decker 2001) synthesized an epoxy/acrylate interpenetrating polymer network by UV-induced curing in the presence of both cationic and radical photo initiators. From real-time FTIR studies, they observed that free radical polymerization of acrylate monomers was faster than the cationic polymerization of epoxy monomers. The application was mainly focused on the fast-drying protective coating, composite structures, and optical components at ambient temperature.

Flame-retardant properties are required especially for wood coating and cables. Some of the flame-retardant additives are less compatible with the polymer matrix while some are retarding the UV-induced curing of the epoxides. Camino and co-workers (Randoux, Vanovervelt, Bergen 2002) synthesized flame-retardant monomers and oligomers having halogen-free radical groups attached to the polymer backbone.

Ceccia et al. (Ceccia, Turcato, Maffettone 2008) synthesized epoxy/layered silicate nano-composite coatings with intercalated morphology by UV curing. 1, 2 diol, ethylene glycol was added to enhance the interfacial interactions of the layered silicate with the epoxy matrix. Rheological measurements concluded the positive effect of cure kinetics and effective dispersion of the filler in presence of ethylene glycol. Gianni et al. (Gianni, Amerio, Monticelli 2008) prepared organic si-lylated montmorillonite/epoxy composites for coating application. The silanisation was done at the –OH reactive sites of the montmorillonite, which increased the interlayer distance in the nano-composites and produced a mixed intercalated/exfoliated structure. Moreover, the thermal properties and scratch resistance were enhanced in the nanocomposites. Sangermano et al. (Sangermano, Borlatto, Bytner 2007) developed transparent UV curable coatings with no response to weathering on exposure to sun. The in-situ generation of TiO_2 after photopolymerization did not interfere with the photopolymerization and imparted a screening effect to light monitored by FTIR. Also, TiO_2 enhanced the mechanical properties of the nanocomposites. Bongiovanni et al. (Bongiovanni, Turcato, Gianni 2008) developed transparent UV curable epoxy/organoclay (Cloisite 30B) coatings. They observed the -OH groups in the nanoclay could act as a chain transfer agent in the cationic photopolymerization and increase the kinetics.

Amerio et al. (Amerio, Sangermano, Malucelli 2005) fabricated a nanocomposite hybrid coating by employing a dual curing mechanism consisting of cationic photopolymerization of epoxy groups and condensation of alkoxysilane groups. Real-time FTIR analysis revealed complete conversion of epoxide groups after 2 min of UV irradiation. Decker et al. (Decker, Keller, Zahouily 2005) synthesized a novel epoxy/organoclay nanocomposites by UV curing that is suitable for coating application. The prepared nanocomposites were both flexible and impact resistant containing 3 wt% loading of organoclay. Organoclay increased the surface roughness of the na-nocomposites and also provided resistance to solvent, moisture, and weathering. Sangermano et al. (Sangermano, Malucelli, Amerio 2005) developed an epoxy coating via. UV radiation containing nanosilica (SiO_2) particles. The incorporation of nanosilica increased the T_g, modulus and surface roughness. Real-time FTIR studies revealed the acceleration of kinetics of epoxide conversion with the addition of nanosilica, which was due to the charge transfer mechanism. The –OH group on the surface of SiO_2 nanoparticles interact with the carbocationic growing chain during the photo-polymerization curing reaction.

Işin et al. (Işın, Kayaman-Apohan, Güngör 2009) fabricated UV curable epoxy/silica coatings containing 0 to 6 wt% silica. Silica surface was modified with glycidoxypropyltrimethoxysilane (GPTMS), and curing was performed with a diaryliodonium photoinitiator and thioxanthone photosensitizer. The properties of the coatings such as gloss, impact, and hardness were improved for the nanocomposites. No damage on the coatings was observed after impact tests. The crosscut adhesion tests implied a relatively 100% value for the coatings.

Sangermano and coworkers (Martin-Gallego, Verdejo, Lopez-Manchado 2011) prepared UV-curable epoxy/functionalized graphene coatings. The real-time FTIR conversion studies showed that the UV shielding effect of graphene slowed down the reaction rate in the nanocomposites, as shown in Figure 12.8. Hence, a considerable increase in cure time and radiation intensity is

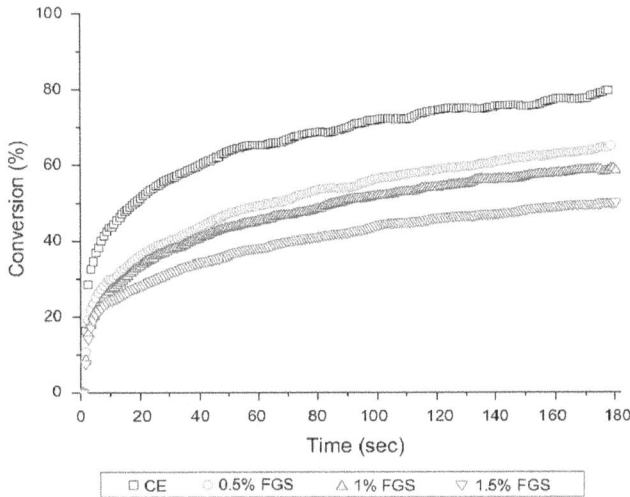

FIGURE 12.8 Real-time FTIR conversion curves as a function of UV irradiation time for the pristine Bis-cycloaliphatic diepoxy (3, 4-epoxycyclohexylmethyl-3',4'-epoxycyclohexyl carboxylate) resin (CE) and for the epoxy resin formulations containing 0.5, 1 and 1.5 wt% of the functionalized graphene sheets (FGS). Light intensity on the surface of the sample 35 mW/cm^2. Film thickness 25 mm (Martin-Gallego, Verdejo, Lopez-Manchado 2011).

required for complete conversion. An increase in stiffness, surface hardness and T_g was observed in the presence of uniformly dispersed nanometre thick graphene platelets in the matrix.

Wang et al. (Wang, Liu, Xue 2013) synthesized epoxy acrylate (EA)/vinyl-polyhedral oligomeric silsesquioxane (POSS) nanocomposite via. UV curing for the coating application. The incorporation of vinyl POSS improved the thermal stability of the nanocomposites. Kumar et al. (Kumar, Misra, Paul 2013) fabricated epoxy/SiO$_2$ nanocomposite coatings by a solvent-free, environmentally friendly EB curing process. The reduced gloss at 60° of epoxy/SiO$_2$ nanocomposite coating indicated the matting effect of the silica, which is advantageous for wood coating or floor finish where glossy finish is not desirable. Silica particles were surface modified with ^{60}Co-gamma radiation induced grafting of glycidyl methacrylate (GMA) and 2-hydroxyethyl methacrylate (HEMA). The modified silica in the coatings improved the T_g, abrasion resistance, hardness, chemical, and steam-resistance properties.

Gaidukovs et al. (Gaidukovs, Medvids, OnufrijevsandGrase 2018) report that the UV curing of branched novolac epoxy resin with the photo initiator bis (4-dodecylphenyl) iodonium hexaflurorantimonate. The optimal photoinitiator content and the irradiation times determined for epoxy novolac were 1.5% and 3 min, respectively. An enhancement in Vickers hardness revealed the formation of densely cross-linked structure. It was noted that irradiation longer than 6 minutes resulted in photodegradation of epoxy novolac.

The attempts to achieve sustainable coating materials via. the exploitation of biobased epoxy resins using UV curable coating technology are of considerable attention due to the elimination of toxic components in coating components over the green profile of UV curing. Yan et al. (Yan, Yang, Zhang 2018) synthesized UV-curable lignin/epoxy acrylate coatings. The in situ synthesis was performed by an initial etherification reaction between lignin and epoxy followed by the esterification with the epoxy-based lignin and acrylic acid. The synthesized coating showed high pencil hardness, thermal stability, chemical resistance, and adhesion properties.

Jena et al. (Jena, Alhassan, Arora 2019) prepared coating film by UV irradiation of a mixture of ZnO decorated MWCNTs and an acrylated epoxy novolac polymer. The formation of the cross-link involves both Michael addition and free radical polymerization reaction. In the presence of photo-initiator (2-hydroxy 2-methyl phenyl propane 1-one), the acrylated epoxy novolac is subjected to free radical reaction during UV irradiation. In addition, the amine group on the 3-aminopropyl triethoxysilane (APTES) modifier of ZnO reacted with acrylic double bond via. Michael addition reaction, which grafted the hybrid filler in the network. The hybrid acrylated epoxy novolac coatings had improved tensile strength, thermal stability, T_g, and storage modulus. The antibacterial properties were also recorded, as the *S. aureus* bacteria showed a higher degree of inhibition compared to *E. coli* bacteria.

Prolongo et al. (Prolongo, Moriche, Jiménez-Suárez 2020) fabricated printable self-heating epoxy coatings doped with carbon nanotubes and graphitic nanoparticles (CNTs or GNPs) via. UV radiation curing. The possible applications are as thermo electric devices with potential de-icing and icing functions. The fabricated coating was demonstrated to be an effective low-power (~10 W) resistive heating, where the commercially available current technologies use ~240–280 W. Aung et al. (Aung, Li, Lim 2020) developed an anticorrosion coating based on Jatropha-oil-based epoxy acrylate (AEJO) resin and nano ZnO via. UV curing. The anticorrosion properties were confirmed in a salt spray test in 5 wt% NaCl solution for 792 h. Also, ZnO improved the barrier properties and increased the contact angle of the coating suitable for surface and interface engineering.

12.5   ADVANTAGES OF RADIATION-CURED EPOXY MATERIALS

Radiation-induced curing of epoxies and their composites could significantly reduce the cure time, environmental concerns due to the volatile emission, internal stress during processing, and overall manufacturing costs. The scientific studies discussed here revealed that properly controlled radiation-cured epoxy composites and coatings showed improved quality and performance. The major limitation of thermal curing of epoxy composites is the internal stress produced because of the difference between the thermal expansion coefficients of the matrix and that of the additives (Dickson and Singh 1988). This internal stress causes volume shrinkage in the composites and adversely affects their performance. At the same time, in radiation curing of epoxy composites, where curing happens at ambient temperature, the action of internal stress is negligible. Hence, the volume shrinkage in epoxy composites associated with the radiation curing is less when compared with that of thermal curing. The high volatile emissions during thermal curing could be effectively controlled with radiation curing (Fengmei, Jianwen, Xiangbao 2002). Only composite structures with limited dimension could be fabricated using thermal curing, as the use of large-scale heating panels or autoclaves is not practical. This could be overcome by radiation curing using cost-effective processing techniques.

It was reported that fiber structure would be distorted during processing of fiber-reinforced epoxy composites at high temperature. Meanwhile, radiation curing of fiber-reinforced epoxy composites could be done by retaining fiber structure (Fengmei, Jianwen, Xiangbao 2002) (Chattopadhyay, Panda, Raju 2005) (Singh 2001) (Kumar, Saini, Bhunia 2020) (Saunders, Lopata, Barnard 2000) (Nishitsuji, Marinucci, Evora 2007). Another important advantage of radiation curing is in the coating industry, where large-scale curing of epoxy coatings can be done economically at ambient temperature to protect huge structures.

12.6   CONCLUSIONS

The manufacturing of epoxy composite is an ever-demanding industry; unfortunately the most energy consuming one, where there is an obvious need for energy efficient and eco-friendly curing technologies. The radiation curing offers great advantages over thermal curing for the development of various types of epoxy composites for versatile applications. Based on the applications, various type of radiation curing technology has been explored for the development of epoxy composites. While gamma and X-rays with high penetration depth are used to cure thick epoxy composite structures, UV radiations are used to cure thin epoxy protective coatings. The low dosage of gamma and X-rays used due to the safety restrictions make the cure process lengthy. However, in most cases, scientists suggest a dual-cure process (irradiation followed by thermal curing) to achieve desired cross-link density of epoxy composites. Microwave radiations also offer an uniform cure even in thick composite structures.

The radiation dose and intensity could effectively modify the curing and thereby reflect on the ultimate properties of the epoxy composites. In fiber-reinforced epoxies, apart from curing, the radiation can enhance the interaction between the fiber and matrix. In epoxy nanocomposites, certain nanofillers alter the radiation curing mechanism or the curing kinetics of epoxy chains, especially functionalized nanofillers. Fast and energy-efficient radiation curing enhances the demand for epoxy composites for protective coatings, microelectronics, and lighter structural components in automotive and aerospace. Beyond the above processing advantages, radiation-cured epoxy composites showed superior performance, too. Reports showed that radiation-cured epoxy composites have superior tensile strength, modulus, interfacial shear strength, interlaminar shear strength, hardness, thermal stability, aging resistance, etc. The radiation-cured epoxy-based coatings have excellent solvent, scratch, moisture, and weathering resistance along with superior adhesion strength. A controlled processing is recommended for radiation curing to avoid degradation of cross-links. The innovative and more economical processing strategies using radiation open up more demanding applications of epoxy composites. The use of simple instruments for radiation curing will be explored for the processing of composite and coating materials. Radiation curing methods need to be further explored for the fabrication of advanced functional epoxy nanocomposites.

REFERENCES

Abliz D, Duan Y, Steuernagel L, Xie L, Li D, Ziegmann G (2013) Curing methods for advanced polymer composites – A Review. *Polymers and Polymer Composites* 21(6):341–348. 10.1177/096739111302100602

Alessi S, Calderaro E, Parlato A, Fuochi P, Lavalle M, Corda U, Dispenza C, Spadaro G (2005) Ionizing radiation induced curing of epoxy resin for advanced composites matrices. *Nuclear Instruments and Methods in Physics Research Section B: Beam Interactions with Materials and Atoms* 236:55–60. 10.1016/j.nimb.2005.03.250

Alessi S, Conduruta D, Pitarresi G, Dispenza C, Spadaro G (2010) Hydrothermal ageing of radiation cured epoxy resin-polyether sulfone blends as matrices for structural composites. *Polymer Degradation and Stability* 95:677–683. 10.1016/j.polymdegradstab.2009.11.038

Alessi S, Parlato A, Dispenza C, De Maria M, Spadaro G (2007) The influence of the processing temperature on gamma curing of epoxy resins for the production of advanced composites. *Radiation Physics and Chemistry* 76:1347–1350. 10.1016/j.radphyschem.2007.02.029

Alessi S, Toscano A, Pitarresi G, Dispenza C, Spadaro G (2017) Water diffusion and swelling stresses in ionizing radiation cured epoxy matrices. *Polymer Degradation and Stability* 144:137–145. 10.1016/j.polymdegradstab.2017.08.009

Al-Sheikhly M, McLaughlin WL (1996) On the mechanisms of radiation-induced curing of epoxy-fiber composites. *Radiation Physics and Chemistry* 48:201–206. 10.1016/0969-806X(95)00415-T

Amerio E, Sangermano M, Malucelli G, Priola A, Voit B (2005) Preparation and characterization of hybrid nanocomposite coatings by photopolymerization and sol–gel process. *Polymer* 46:11241–11246. 10.1016/j.polymer.2005.09.062

Aung MM, Li WJ, Lim HN (2020) Improvement of anticorrosion coating properties in bio-based polymer epoxy acrylate incorporated with nano zinc oxide particles. *Ind Eng Chem Res* 59:1753–1763. 10.1021/acs.iecr.9b05639

Bajpai M, Shukla V, Kumar A (2002) Film performance and UV curing of epoxy acrylate resins. *Progress in Organic Coatings* 44:271–278. 10.1016/S0300-9440(02)00059-0

Barletta M, Vesco S, Puopolo M, Tagliaferri V (2015) High performance composite coatings on plastics: UV-curable cycloaliphatic epoxy resins reinforced by graphene or graphene derivatives. *Surface and Coatings Technology* 272:322–336. 10.1016/j.surfcoat.2015.03.046

Benfarhi S, Decker C, Keller L, Zahouily K (2004) Synthesis of clay nanocomposite materials by light-induced cross-linking polymerization. *European Polymer Journal* 40:493–501. 10.1016/j.eurpolymj.2003.11.009

Berejka AJ, Cleland MR, Galloway RA, Gregoire O (2005) X-ray curing of composite materials. *Nuclear Instruments and Methods in Physics Research Section B: Beam Interactions with Materials and Atoms* 241:847–849. 10.1016/j.nimb.2005.07.188

Berejka AJ, Cleland MR, Walo M (2014) The evolution of and challenges for industrial radiation processing—2012. *Radiation Physics and Chemistry* 94:141–146. 10.1016/j.radphyschem.2013.04.013

Boey F, Rath SK, Ng AK, Abadie MJM (2002) Cationic UV cure kinetics for multifunctional epoxies. *Journal of Applied Polymer Science* 86:518–525. 10.1002/app.11041

Boey FYC, Lee TH (1991) Electromagnetic radiation curing of an epoxy/fiber glass reinforced composite. *International Journal of Radiation Applications and Instrumentation Part C Radiation Physics and Chemistry* 38:419–423. 10.1016/1359-0197(91)90118-L

Boey FYC, Yap BH, Chia L (1999) Microwave curing of epoxy-amine system—Effect of curing agent on the rate enhancement. *Polymer Testing* 18:93–109. 10.1016/S0142-9418(98)00014-2

Bongiovanni R, Casciola M, Di Gianni A, Donnadio A, Malucelli G (2009) Epoxy-nanocomposites containing exfoliated zirconium phosphate: Preparation via cationic photopolymerization and physicochemical characterisation. *European Polymer Journal* 45:2487–2493. 10.1016/j.eurpolymj.2009.06.022

Bongiovanni R, Turcato EA, Di Gianni A, Ronchetti S (2008) Epoxy coatings containing clays and orga-noclays: Effect of the filler and its water content on the UV-curing process. *Progress in Organic Coatings* 62:336–343. 10.1016/j.porgcoat.2008.01.014

Ceccia S, Turcato EA, Maffettone PL, Bongiovanni R (2008) Nanocomposite UV-cured coatings: Organoclay intercalation by an epoxy resin. *Progress in Organic Coatings* 63:110–115. 10.1016/j.porgcoat.2008.04.012

Chattopadhyay DK, Panda SS, Raju KVSN (2005) Thermal and mechanical properties of epoxy acrylate/me-thacrylates UV cured coatings. *Progress in Organic Coatings* 54:10–19. 10.1016/j.porgcoat.2004.12.007

Chen C, Curliss D (2003) Thermally-cured and e-beam-cured epoxy layered-Silicate nanocomposites. *Polymer Bulletin* 49:473–480. 10.1007/s00289-003-0128-1

Chen M-H, Chen C-R, Hsu S-H, Sun S-P, Su W-F (2006) Low shrinkage light curable nanocomposite for dental restorative material. *Dental Materials* 22:138–145. 10.1016/j.dental.2005.02.012

Chmielewski AG, Haji-Saeid M, Ahmed S (2005) Progress in radiation processing of polymers. *Nuclear Instruments and Methods in Physics Research Section B: Beam Interactions with Materials and Atoms* 236:44–54. 10.1016/j.nimb.2005.03.247

Colangelo F, Russo P, Cimino F, Cioffi R, Farina I, Fraternali F, Feo L (2017) Epoxy/glass fibers composites for civil applications: Comparison between thermal and microwave cross-linking routes. *Composites Part B: Engineering* 126:100–107. 10.1016/j.compositesb.2017.06.003

Crivello JV (2002) Advanced curing technologies using photo- and electron beam induced cationic poly-merization. *Radiation Physics and Chemistry* 63:21–27. 10.1016/S0969-806X(01)00476-5

Crivello JV, Fan M, Bi D (1992) The electron beam-induced cationic polymerization of epoxy resins. *Journal of Applied Polymer Science* 44:9–16. 10.1002/app.1992.070440102

Crivello JV, Liu S (2000) Photoinitiated cationic polymerization of epoxy alcohol monomers. *Journal of Polymer Science Part A: Polymer Chemistry* 38:389–401. 10.1002/(SICI)1099-0518(20000201)38:3<389::AID-POLA1>3.0.CO;2-G

Crivello JV, Walton TC, Malik R (1997) Fabrication of epoxy matrix composites by electron beam induced cationic polymerization. *Chem Mater* 9:1273–1284. 10.1021/cm9700312

Czajlik I, Hedvig P, Ille A, Dobó J (1996) Calorimetric study of cationic photopolymerization. *Radiation Physics and Chemistry* 47:453–455. 10.1016/0969-806X(95)00140-S

Decker C (1983) Ultra-fast polymerization of epoxy-acrylate resins by pulsed laser irradiation. *Journal of Polymer Science: Polymer Chemistry Edition* 21:2451–2461. 10.1002/pol.1983.170210827

Decker C, Keller L, Zahouily K, Benfarhi S (2005) Synthesis of nanocomposite polymers by UV-radiation curing. *Polymer* 46:6640–6648. 10.1016/j.polymer.2005.05.018

Decker C, Moussa K (1990) Kinetic study of the cationic photopolymerization of epoxy monomers. *Journal of Polymer Science Part A: Polymer Chemistry* 28:3429–3443. 10.1002/pola.1990.080281220

Decker C, Nguyen Thi Viet T, Decker D, Weber-Koehl E (2001) UV-radiation curing of acrylate/epoxide systems. *Polymer* 42:5531–5541. 10.1016/S0032-3861(01)00065-9

dell'Erba IE, Martínez FD, Hoppe CE, Eliçabe GE, Ceolín M, Zucchi IA, Schroeder WF (2017) Mechanism of particle formation in silver/epoxy nanocomposites obtained through a visible-light-assisted in situ synthesis. *Langmuir* 33:10248–10258. 10.1021/acs.langmuir.7b01936

Di Gianni A, Amerio E, Monticelli O, Bongiovanni R (2008) Preparation of polymer/clay mineral nano-composites via dispersion of silylated montmorillonite in a UV curable epoxy matrix. *Applied Clay Science* 42:116–124. 10.1016/j.clay.2007.12.011

Dickson LW, Singh A (1988) Radiation curing of epoxies. *International Journal of Radiation Applications and Instrumentation Part C Radiation Physics and Chemistry* 31:587–593. 10.1016/1359-0197(88) 90231-7

Dispenza C, Alessi S, Spadaro G (2008) Carbon fiber composites cured by γ-radiation-induced polymerization of an epoxy resin matrix. *Advances in Polymer Technology* 27:163–171. 10.1002/adv.20127

dos Santos MN, Opelt CV, Lafratta FH, Lepienski CM, Pezzin SH, Coelho LAF (2011) Thermal and mechanical properties of a nanocomposite of a photocurable epoxy-acrylate resin and multi-walled carbon nanotubes. *Materials Science and Engineering: A* 528:4318–4324. 10.1016/j.msea.2011.02.036

Endruweit A, Johnson MS, Long AC (2006) Curing of composite components by ultraviolet radiation: A review. *Polymer Composites* 27:119–128. 10.1002/pc.20166

Fengmei L, Jianwen B, Xiangbao C, Huaying B, Huiliang W (2002) Factors influencing EB curing of epoxy matrix. *Radiation Physics and Chemistry* 63:557–561. 10.1016/S0969-806X(01)00620-X

Fouassier JP, Burr D, Crivello JV (1994) Photochemistry and photopolymerization activity of diaryliodonium salts. *Journal of Macromolecular Science, Part A* 31:677–701. 10.1080/10601329409349748

Gaidukovs S, Medvids A, Onufrijevs P, Grase L (2018) UV-light-induced curing of branched epoxy novolac resin for coatings. *Express Polym Lett* 12:918–929. 10.3144/expresspolymlett.2018.78

Ghosh NN, Palmese GR (2005) Electron-beam curing of epoxy resins: Effect of alcohols on cationic polymerization. *Bull Mater Sci* 28:603–607. 10.1007/BF02706350

Gotoda M, Miyashita Y, Mori K (1975) Japan, Radiation curing of epoxy-acrylate prepolymers and their mixtures with vinyl monomers. IV. Effects of addition of pigments to epoxy-acrylate/vinyl monomer mixtures upon the electron beam curing of their coatings on steel.

Grammatikos SA, Evernden M, Mitchels J, Zafari B, Mottram JT, Papanicolaou GC (2016) On the response to hygrothermal aging of pultruded FRPs used in the civil engineering sector. *Materials & Design* 96:283–295. 10.1016/j.matdes.2016.02.026

Hay JN, O'Gara P (2006) Recent developments in thermoset curing methods. *Proceedings of the Institution of Mechanical Engineers, Part G: Journal of Aerospace Engineering* 220(3):187–195 10.1243/095441 00JAERO35

Hoffman EN, Skidmore TE (2009) Radiation effects on epoxy/carbon-fiber composite. *Journal of Nuclear Materials* 392:371–378. 10.1016/j.jnucmat.2009.03.027

Horikoshi S, Arai Y, Serpone N (2021) In search of the driving factor for the microwave curing of epoxy adhesives and for the protection of the base substrate against thermal damage. *Molecules* 26:2240. 10.3390/molecules26082240

Im JS, Jeong E, In SJ, Lee Y-S (2010) The impact of fluorinated MWCNT additives on the enhanced dynamic mechanical properties of e-beam-cured epoxy. *Composites Science and Technology* 70:763–768. 10.1016/j.compscitech.2010.01.007

Işın D, Kayaman-Apohan N, Güngör A (2009) Preparation and characterization of UV-curable epoxy/silica nanocomposite coatings. *Progress in Organic Coatings* 65:477–483. 10.1016/j.porgcoat.2009.04.007

Jena KK, Alhassan SM, Arora N (2019) Facile and rapid synthesis of efficient epoxy-novolac acrylate / MWCNTs-APTES-ZnO hybrid coating films by UV irradiation: Thermo-mechanical, shape stability, swelling, hydrophobicity and antibacterial properties. *Polymer* 179:121621. 10.1016/j.polymer.2019.121621

Jin F-L, Lee S-Y, Park S-J (2013) Polymer matrices for carbon fiber-reinforced polymer composites. *Carbon Letters* 14:76–88. 10.5714/CL.2013.14.2.076

Johnston K, Pavuluri SK, Leonard MT, Desmulliez MPY, Arrighi V (2015) Microwave and thermal curing of an epoxy resin for microelectronic applications. *Thermochimica Acta* 616:100–109. 10.1016/j.tca.2015.08.010

Jyotishkumar P, Abraham E, George SM, Elias E, Pionteck J, Moldenaers P, Thomas S (2013) Preparation and properties of MWCNTs/poly (acrylonitrile- styrene-butadiene)/epoxy hybrid composites. *Journal of Applied Polymer Science* 127:3093–3103. 10.1002/app.37677

Klikovits N, Knaack P, Bomze D, Krossing I, Liska R (2017) Novel photoacid generators for cationic photopolymerization. *Polym Chem* 8:4414–4421. 10.1039/C7PY00855D

Koshiba M, Hwang KKS, Foley SK, Yarusso DJ, Cooper SL (1982) Properties of ultra-violet curable polyurethane acrylates. *J Mater Sci* 17:1447–1458. 10.1007/BF00752259

Kowandy C, Ranoux G, Walo M, Vissouvanadin B, Teyssedre G, Laurent C, Berquand A, Molinari M, Coqueret X (2018) Microstructure aspects of radiation-cured networks: Cationically polymerized aromatic epoxy resins. *Radiation Physics and Chemistry* 143:20–26. 10.1016/j.radphyschem.2017.09.006

Kumar M, Saini JS, Bhunia H, Chowdhury SR (2020) The effect of radiation curing on mechanical joints prepared from carbon nanotubes added carbon/epoxy laminates. *Polymer Composites* 41:4260–4276. 10.1002/pc.25709

Kumar V, Misra N, Paul J, Bhardwaj YK, Goel NK, Francis S, Sarma KSS, Varshney L (2013) Organic/inorganic nanocomposite coating of bisphenol A diglycidyl ether diacrylate containing silica nanoparticles via electron beam curing process. *Progress in Organic Coatings* 76:1119–1126. 10.1016/j.porgcoat.2013.03.010

Le Van Q, Gourdenne A (1987) Microwave curing of epoxy resins with diaminodiphenylmethane—I. General features. *European Polymer Journal* 23:777–780. 10.1016/0014-3057(87)90121-2

Lhl C, Jacob J, Boey F (1995) Radiation curing of poly-methyl-methacrylate using a variable power microwave source. *Journal of Materials Processing Technology* 48:445–449. 10.1016/0924-0136(94)01681-P

Lopata VJ, Saunders CB, Singh A, Janke CJ, Wrenn GE, Havens SJ (1999) Electron-beam-curable epoxy resins for the manufacture of high-performance composites. *Radiation Physics and Chemistry* 56:405–415. 10.1016/S0969-806X(99)00330-8

Mahmoud AH, Tay GS, Rozman HD (2011) A preliminary study on ultraviolet radiation-cured unsaturated polyester resin based on palm oil. *Polymer-Plastics Technology and Engineering* 50:573–580. 10.1080/03602559.2010.543740

Malucelli G, Bongiovanni R, Sangermano M, Ronchetti S, Priola A (2007) Preparation and characterization of UV-cured epoxy nanocomposites based on o-montmorillonite modified with maleinized liquid polybutadienes. *Polymer* 48:7000–7007. 10.1016/j.polymer.2007.10.008

Marand E, Baker KR, Graybeal JD (1992) Comparison of reaction mechanisms of epoxy resins undergoing thermal and microwave cure from in situ measurements of microwave dielectric properties and infrared spectroscopy. *Macromolecules* 25:2243–2252. 10.1021/ma00034a028

Martin-Gallego M, Lopez-Manchado MA, Calza P, Roppolo I, Sangermano M (2015) Gold-functionalized graphene as conductive filler in UV-curable epoxy resin. *J Mater Sci* 50:605–610. 10.1007/s10853-014-8619-z

Martin-Gallego M, Verdejo R, Lopez-Manchado MA, Sangermano M (2011) Epoxy-graphene UV-cured nanocomposites. *Polymer* 52:4664–4669. 10.1016/j.polymer.2011.08.039

McGinniss VD (2000) Radiation curing. In: *Kirk-OthmerEncyclopedia of Chemical Technology. American Cancer Society*, John Wiley & Sons, 1-24.

Morawetz H (1981) *Developments in polymer photochemistry*, Norman S. Allen, Ed., Applied Science Publishers, London, 1980, 223 pp. Price: $42.50. Journal of Polymer Science: Polymer Letters Edition 19:375–376. 10.1002/pol.1981.130190709

Nho YC, Kang PH, Park JS (2004) The characteristics of epoxy resin cured by γ-ray and E-beam. *Radiation Physics and Chemistry* 71:243–246. 10.1016/j.radphyschem.2004.03.047

Nishitsuji DA, Marinucci G, Evora MC, de Andrade e Silva LG (2007) Study of electron beam curing process using epoxy resin system. *Nuclear Instruments and Methods in Physics Research Section B: Beam Interactions with Materials and Atoms* 265:135–138. 10.1016/j.nimb.2007.08.039

Pal R, Jha AK, Akhtar MJ, Kar KK, Kumar R, Nayak D (2017) Enhanced microwave processing of epoxy nanocomposites using carbon black powders. *Advanced Powder Technology* 28:1281–1290. 10.1016/j.apt.2017.02.016

Prolongo SG, Moriche R, Jiménez-Suárez A, Delgado A, Ureña A (2020) Printable self-heating coatings based on the use of carbon nanoreinforcements. *Polymer Composites* 41:271–278. 10.1002/pc.25367

Przybytniak G, Nowicki A, Mirkowski K, Stobiński L (2016) Gamma-rays initiated cationic polymerization of epoxy resins and their carbon nanotubes composites. *Radiation Physics and Chemistry* 121:16–22. 10.1016/j.radphyschem.2015.11.037

Puchleitner R, Riess G, Kern W (2017) X-ray induced cationic curing of epoxy-bonded composites. *European Polymer Journal* 91:31–45. 10.1016/j.eurpolymj.2017.03.036

Raghavan J (2009) Evolution of cure, mechanical properties, and residual stress during electron beam curing of a polymer composite. *Composites Part A: Applied Science and Manufacturing* 40:300–308. 10.1016/j.compositesa.2008.12.010

Randoux T, Vanovervelt J-C, Van den Bergen H, Camino G (2002) Halogen-free flame retardant radiation curable coatings. *Progress in Organic Coatings* 45:281–289. 10.1016/S0300-9440(02)00051-6

Rangari VK, Bhuyan MS, Jeelani S (2010) Microwave processing and characterization of EPON 862/CNT nanocomposites. *Materials Science and Engineering: B* 168:117–121. 10.1016/j.mseb.2010.01.013

Rao V (2009) Radiation processing of polymers. In: Thomas S, Weimin Y (eds) *Advances in Polymer Processing*. Woodhead Publishing, pp. , Abington Hall, Granta Park, Great Abington, Cambridge CB21 6AH, UK, 402–437.

Sangermano M, Borlatto E, D'HérinBytner FD, Priola A, Rizza G (2007) Photostabilization of cationic UV-cured coatings in the presence of nanoTiO2. *Progress in Organic Coatings* 59:122–125. 10.1016/j.porgcoat.2007.01.020

Sangermano M, Malucelli G, Amerio E, Priola A, Billi E, Rizza G (2005) Photopolymerization of epoxy coatings containing silica nanoparticles. *Progress in Organic Coatings* 54:134–138. 10.1016/j.porgcoat.2005.05.004

Sangermano M, Roppolo I, Chiappone A (2018) New horizons in cationic photopolymerization. *Polymers* 10:136. 10.3390/polym10020136

Saunders C, Lopata V, Barnard J, Stepanik T (2000) Electron beam curing—Taking good ideas to the manufacturing floor. *Radiation Physics and Chemistry* 57:441–445. 10.1016/S0969-806X(99)00411-9

Sharif M, Pourabbas B, Sangermano M, Moghadam FS, Mohammadi M, Roppolo I, Fazli A (2017) The effect of graphene oxide on UV curing kinetics and properties of SU8 nanocomposites. *Polymer International* 66:405–417. 10.1002/pi.5271

Shi W, Rånby B (1994) UV curing of composites based on modified unsaturated polyester. *Journal of Applied Polymer Science* 51:1129–1139. 10.1002/app.1994.070510619

Shin J-W, Jeun J-P, Kang P-H (2009) Fabrication and characterization of the mechanical properties of multi-walled carbon nanotube-reinforced epoxy resins by e-beam irradiation. *Journal of Industrial and Engineering Chemistry* 15:555–560. 10.1016/j.jiec.2009.01.012

Singh A (2001) Radiation processing of carbon fiber-reinforced advanced composites. *Nuclear Instruments and Methods in Physics Research Section B: Beam Interactions with Materials and Atoms* 185:50–54. 10.1016/S0168-583X(01)00753-4

Spadaro G, Alessi S, Dispenza C, Sabatino MA, Pitarresi G, Tumino D, Przbytniak G (2014) Radiation curing of carbon fiber composites. *Radiation Physics and Chemistry* 94:14–17. 10.1016/j.radphyschem.2013.05.052

Stewart A, Douglas EP (2012) Accelerated testing of epoxy-FRP composites for civil infrastructure applications: Property changes and mechanisms of degradation. *Polymer Reviews* 52:115–141. 10.1080/15583724.2012.668152

Sui G, Zhang Z-G, Chen C-Q, Zhong W-H (2003) Analyses on curing process of electron beam radiation in epoxy resins. *Materials Chemistry and Physics* 78:349–357. 10.1016/S0254-0584(01)00553-3

Sui G, Zhong W-H, Yang X-P (2009) The revival of electron beam irradiation curing of epoxy resin—Materials characterization and supportive cure studies. *Polymers for Advanced Technologies* 20:811–817. 10.1002/pat.1292

Tamada M (2018) Radiation Processing of Polymers and Its Applications. In: Kudo H (ed) *Radiation Applications*. Springer, Singapore, pp. 63–80.

Usanmaz A, Eser Ö, Doğan A (2001) Thermal and dynamic mechanical properties of γ-ray-cured poly (methyl methacrylate) used as a dental-base material. *Journal of Applied Polymer Science* 81:1291–1296. 10.1002/app.1552

Uthaman A, Xian G, Thomas S, Wang Y, Zheng Q, Liu X (2020) Durability of an epoxy resin and its carbon fiber-reinforced polymer composite upon immersion in water, acidic, and alkaline solutions. *Polymers* 12:614. 10.3390/polym12030614

Uyanık N, Erdem AR, Can MF, Çelik MS (2006) Epoxy nanocomposites curing by microwaves. *Polymer Engineering & Science* 46:1104–1110. 10.1002/pen.20574

Vabrik R, Ille A, Víg A, Czajlik I, Rusznák I (1996) Thermomechanical investigations of UV cured epoxy coatings. *Radiation Physics and Chemistry* 47:457–459. 10.1016/0969-806X(95)00141-J

Vijayan PP, Puglia D, Kenny JM, Thomas S (2013) Effect of organically modified nanoclay on the miscibility, rheology, morphology and properties of epoxy/carboxyl-terminated (butadiene-co-acrylonitrile) blend. *Soft Matter* 9:2899–2911. 10.1039/C2SM27386A

Wang Y, Liu F, Xue X (2013) Synthesis and characterization of UV-cured epoxy acrylate/POSS nanocomposites. *Progress in Organic Coatings* 76:863–869. 10.1016/j.porgcoat.2013.02.007

Wu Y, Xie T, Xiao W, Zhang WC, Ma WQ, Zhu HC, Yang Y, Huang KM (2018) A novel microwave combined ultraviolet radiator based on slotted coaxial line for epoxy resin curing. *AIP Advances* 8:115021. 10.1063/1.5048530

Xu X, Wang X, Cai Q, Wang X, Wei R, Du S (2016) Improvement of the compressive strength of carbon fiber/epoxy composites via microwave curing. *Journal of Materials Science & Technology* 32:226–232. 10.1016/j.jmst.2015.10.006

Yagci Y, Sahin O, Ozturk T, Marchi S, Grassini S, Sangermano M (2011) Synthesis of silver/epoxy nanocomposites by visible light sensitization using highly conjugated thiophene derivatives. *Reactive and Functional Polymers* 71:857–862. 10.1016/j.reactfunctpolym.2011.05.012

Yan R, Yang D, Zhang N, Zhao Q, Liu B, Xiang W, Sun Z, Xu R, Zhang M, Hu W (2018) Performance of UV curable lignin based epoxy acrylate coatings. *Progress in Organic Coatings* 116:83–89. 10.1016/j.porgcoat.2017.11.011

Zhang J, Campolo D, Dumur F, Xiao P, Fouassier JP, Gigmes D, Lalevée J (2016) Visible-light-sensitive photoredox catalysis by iron complexes: Applications in cationic and radical polymerization reactions. *Journal of Polymer Science Part A: Polymer Chemistry* 54:2247–2253. 10.1002/pola.28098

Zhang J, Duan Y, Wang B, Zhang X (2020) Interfacial enhancement for carbon fiber reinforced electron beam cured polymer composite by microwave irradiation. *Polymer* 192:122327. 10.1016/j.polymer.2020.122327

13 Microwave-assisted activated carbon: A promising class of materials for a wide range of applications

M. N. M. Ansari*

Department of Mechanical Engineering, Universiti Tenaga + National, Kajang, Selangor, Malaysia

M. A. Sayem

Institute of Power Engineering, Universiti Tenaga Nasional, Kajang, Selangor, Malaysia

*Corresponding author: ansari@uniten.edu.my

CONTENTS

DOI: 10.1201/9781003321910-13

13.1   INTRODUCTION

Activated carbon (AC) is said to have originated in Ancient Egypt in 1500 B.C., when the Egyptians used its adsorbent properties for water purification and medical uses. Karl Wilhelm [Bubanale et al., 2017], a Swedish scientist, has published on the adsorption of gases on charcoal in recent years. In recent decades, the AC has been utilized as a decolorizing agent in the industry. During World War I, however, the potential of AC was realized when it was utilized in gas masks to protect against hazardous gases [Schanz and Parry, 1962]. Egyptians utilized activated charcoal as early as 3750 B.C. in the production of bronze when it was first recorded. It was also used by the Egyptians about 1500 B.C. for digestive disorders, odor absorption, and writing on papyrus. Antibacterial qualities of activated charcoal were discovered about 400 B.C. by the Ancient Hindus and Phoenicians, who started using it to filter drinking water. The habit of storing water in burned barrels was well-known during lengthy sea voyages. Activated charcoal was pioneered by Hippocrates and Pliny around 50 AD, who used it to cure a variety of conditions, including epilepsy, chlorosis, and vertigo. Charcoal reemerged in medicinal treatments in the 1700s and 1800s after being suppressed throughout the Dark Ages, both for its absorbent and disinfecting characteristics of fluids and gases. This historical period saw the common usage of charcoal poultices and charcoal powders to treat fetid stomach ulcers, acidity in the stomach, and even the occasional nosebleed, if sub-sulfate of iron had failed to ease these symptoms. Charcoal was also being offered as lozenges, biscuits, and tooth powders by the turn of the century [Sanchez et al., 2020]. Figure 13.1 shows the history of activated carbon from 3750 B.C. until the current time.

Activated carbon's market size seems to be increasing, according to a 2018 report, the worldwide activated carbon market was valued at around $4.72 billion. During the projection period, it is predicted to grow at a CAGR of 17.5%. An important development driver is predicted to be an increase in the need for water treatment and sewage treatment applications in the future. The surface of activated carbon commonly referred to as activated charcoal, is covered with many microscopic holes [Wong et al., 2018]. Adsorption capacity rises as the number of holes increases since the surface area is bigger when there are more pores. It is utilized extensively in water purification and sewage treatment facilities, among other applications. Manufacturing is likely to continue to develop as a result of recent technological breakthroughs. Raw materials such as wood, peat, coconut shell charcoal, wood chips, petroleum pitch, phenolic resin, and viscose rayon are used to produce various versions of the product. When a lot of raw material is used, it creates a lot of carbon dioxide gas. Raw material quality is a major concern for manufacturers since it has a direct influence on the final product's performance. Raw material prices have risen significantly because of the widening supply-demand disparity. Particulate and dissolved contaminants may be removed by activated carbon, which is widely employed in water treatment facilities [Wong et al., 2018]. It also removes chlorine from water, as well as some organic contaminants. It is used in greenhouses and industrial plants to eliminate hazardous gases, smells, and dangerous dust particles from the air. The product may also be used in the food and beverage, pharmaceutical, and automobile sectors. Governments throughout the world are enforcing tough air and water pollution

THE HISTORY OF ACTIVATED CARBON

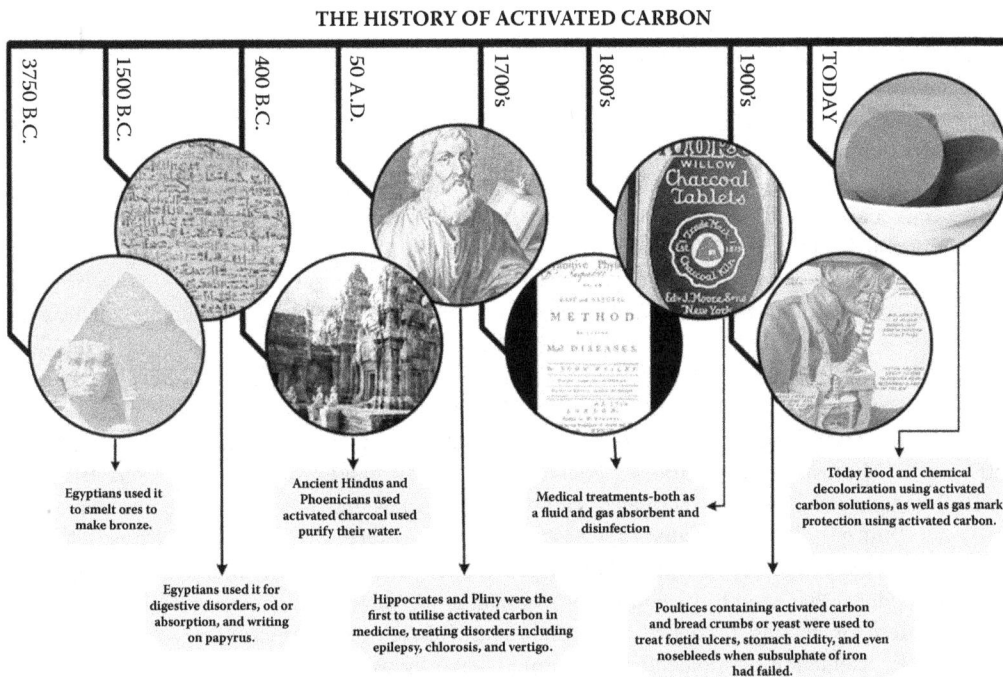

3750 B.C. Egyptians used it to smelt ores to make bronze.

1500 B.C. Egyptians used it for digestive disorders, od or absorption, and writing on papyrus.

400 B.C. Ancient Hindus and Phoenicians used activated charcoal used purify their water.

50 A.D. Hippocrates and Pliny were the first to utilise activated carbon in medicine, treating disorders including epilepsy, chlorosis, and vertigo.

1700's

1800's Medical treatments-both as a fluid and gas absorbent and disinfection

1900's Poultices containing activated carbon and bread crumbs or yeast were used to treat foetid ulcers, stomach acidity, and even nosebleeds when subsulphate of iron had failed.

TODAY Today Food and chemical decolorization using activated carbon solutions, as well as gas mark protection using activated carbon.

FIGURE 13.1   The evolution of activated carbon [Bubanale *et al.*, 2017; Schanz and Parry, 2002; Schanz *et al.*, 2020].

management and waste treatment regulations because of the growing levels of air and water pollution. Activated carbon has been suggested by the EPA and other government agencies as the best material for eliminating mercury and other chemical contaminants [Mohanty *et al.*, 2008].

Activated carbon is a common name for carbon materials that consist of charcoal. It is also known as activated charcoal or activated coal. Coconut shells, lignite coal, bituminous coal, and other carbonaceous materials are used to make activated carbon (AC). It's a type of carbon material with tiny, low-volume holes that enhance the amount of surface area accessible for chemical reactions or adsorption [Wickramaratne and Jaroniec, 2013]. The complex pore structure aids in a variety of AC applications and also serves as a good medium for harboring functional groups [Treeweranuwat *et al.*, 2020]. With increasing specific surface area and pore structure, AC's adsorption capabilities and reaction activities improve [Benjamin *et al.*, 2018]. The majority of the inner surface area is made up of micropores and mesopores. Macropores are commonly thought of as carbon particle freeways. The size and distribution of pores affect the distribution of functional groups on the surface of AC [Hu *et al.*, 2001]. The groups and pieces that make up the pores' interface are considered functional. AC's fine structure enhanced the surface area of pores (>1,000 m^2/g), resulting in the presence of strong adsorptive characteristics. Carbon can be found in three different forms: powder, granular, and pellet. Nonetheless, granular, and powdered AC is the most often utilized [Dabrowski *et al.*, 2005; Jjagwe *et al.*, 2021; Mohammed *et al.*, 2014]. Many additional types of AC have attracted the interest of researchers. Fiber, which is mostly made from petroleum pitch and isotropic coal, felts, and clothing are among them. Because of its large surface area, the AC is effective in removing numerous pollutants from both potable water and wastewater [Boopathyand Karthikeyan, 2013]. Low-cost materials with high carbon content and low inorganic content have been reported to be used as raw materials for AC synthesis [Bae *et al.*, 2014; Ioannidou *et al.*, 2007]. Several studies have shown a wide range of basic materials that may be utilized to make AC. Agro-industrial by-products, for example, are frequently utilized because of their renewability, high mechanical strength, low cost, availability, and low ash content. A range of organic and inorganic contaminants, polar and nonpolar chemicals, and aqueous and gaseous environments have been successfully removed using activated carbon (AC) as an adsorbent and it has also been employed in energy storage areas [Wickramaratne and Jaroniec, 2013]. AC has a large porous surface area, tunable pore structure and surface chemistry, good thermostability at high temperatures in inert or reduction atmospheres, low acid-base reactivity, and receives much attention owing to its superior and efficient ability in air pollution control solvent recovery, food processing, wastewater treatment (e.g., dyes, heavy metals, detergents, herbicides, pesticides, and polyaromatic hydrocarbons). In general, AC is made by pyrolysis and activation from woody biomass, agricultural wastes, and coal, such as sawdust, coconut shells, fruit stones and peels, peat, and bituminous coal [Boopathy and Karthikeyan, 2013]. Figure 13.2 shows the process of preparing activated carbons using coconut shells in two different forms using two individual machines.

Microwave radiation has recently been utilized to create activated carbon as a unique method. For microwave-assisted activation of agricultural wastes, activating chemicals such as KOH, H_3PO_4, and $ZnCl_2$ were utilized [Hu et al., 2001]. To achieve an SSA of 700–2,500 m^2g^{-1}, microwave powers of 600–700 W are used to create activated carbon. By treating rice husk and cotton stalk waste with potassium hydroxide, we created activated carbon with a surface area specific gravity (SSA) of 750 m^2g^{-1}. Microwave activation of coal tar pitch char yielded activated carbon. Based on KOH, biomass ratios of 5–8 activated carbon SSA values have been reported to vary from 2,500–4,100 m^2g^{-1}. The SSA value of 1,335 m^2g^{-1} was achieved by the microwave activation of NaOH on pomelo skin. Microwaves and furnaces have various heat transfer methods, which produce different products that need to be compared [Namazi et al., 2016].

The pattern of heating produced by a traditional furnace and a microwave oven differs significantly. Reorientation of molecules under microwave fields generates heat in microwaves. In a traditional furnace, heat is transported through the walls and generates a temperature gradient. Due to direct heating of chemicals, the removal of temperature gradients in a single material, and selective heating, microwaves have been found to enhance a wide range of heating processes. Selective heating is another benefit of using microwaves in chemical processes. The study on conventional chemical activation in a furnace has been considerable, but the research on microwave-assisted chemical activation has been very restricted. There is a possibility that microwaves might have a significant impact on the activation process [Błażewicz and Świątkowski, 1999].

Microwave radiation is poorly absorbed by certain biomass materials. To ensure that these biomass materials are heated evenly, microwave absorbers must be added. When using microwaves to

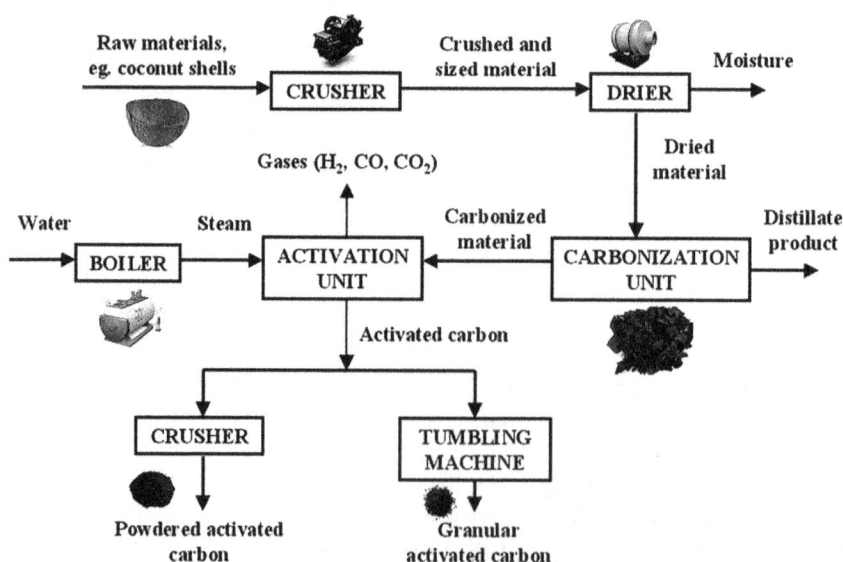

FIGURE 13.2 Shows the activation process [Lin et al., 2015; Hu et al., 2001].

activate chemical reactions, activating agents are anticipated to be the primary microwave radiation absorbers in the early stages. Faster activation temperature attainment might be achieved as a result of this phenomenon. In a previous study, this team looked at the feasibility of making activated carbon only from biomass [Laine *et al.*, 1989]. NaOH was added to dried pulp mill sludge and heated in a microwave oven. The combined pyrolysis and activation yielded activated carbon with a surface area of 300–660 m^2g^{-1} [Laine *et al.*, 1989]. Activated carbon characteristics were improved and microwave effects on chemical activation reactions were investigated in this study. Biochar chemical activation and improved activated carbon properties were the main goals of this research, which used microwave radiation. KOH and NaOH activation differ in microwave ovens compared to furnaces, according to this study [Shukla *et al.*, 2019]. The setup microwave to activate carbon is quite different from a conventional microwave to heat or prepare food, as illustrated in Figure 13.3, since it requires extra components.

In recent years, a variety of methods have been used to remove harmful components such as anions, heavy metals, chemical compounds, and dyes from water sources [Jjagwe *et al.*, 2021]. Adsorption is one of the most practical and straightforward ways of water purification. When it comes to procedures like industrial effluent purification [Mohammed *et al.*, 2014], groundwater treatment [Boopathy and Karthikeyan, 2013], and the removal of volatile organic compounds from the air and mercury vapors from a gas mixture [Bae *et al.*, 2014], activated carbon (AC) is often utilized as a possible adsorbent. Its high micropore volume (V_{mic}), large specific surface area (S_{BET}), good pore size distribution, thermal stability, fast adsorption capabilities, and low acid/base reactivity all contributed to AC's significant adsorption in both gas and liquid phases. The high cost of AC production is a major obstacle for commercial producers, and current research has focused on employing low-cost raw materials with high carbon content and low quantities of inorganic chemicals to make low-cost AC. Low-cost precursors for the manufacturing of AC include rice husks, coconut husks, and oil palm fibers from agricultural waste [Hoseinzadeh *et al.*, 2013].

AC's quality, characteristics, and price are all influenced by the preparation technique as well as the raw ingredients. Conventional heating is one of the most often used methods for preparing an air conditioner. Convection, conduction, and radiation methods are used to transmit thermal energy to particles in a carbon bed in a typical process. First, the sample's outer layer is heated, followed by its inside. Each particle has a temperature gradient from the outside to the inside [Shukla *et al.*, 2020]. The heat gradient and the high expense of AC preparation may be solved by using a microwave radiation approach. Using a microwave irradiation approach for the manufacture of

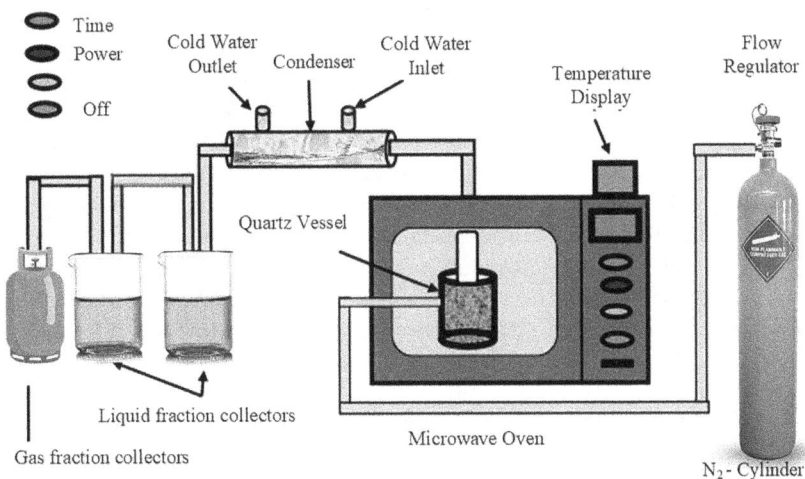

FIGURE 13.3 Shows the system for microwave activation process [Laine *et al.*, 1989; Shukla *et al.*, 2019; Lidström *et al.*, 2001].

reasonably uniform, affordable AC particles with large surface area and considerable adsorption capacity have yielded numerous promising findings in recent years. As there is no external heat source, microwaves may quickly heat a large volume of material because they interact directly with the particles inside the compressed compact material [Namane *et al.*, 2005]. Higher sintering temperatures, faster processing times, and, thus, greater energy savings may be achieved with the application of microwave radiation. The benefits and drawbacks of microwave and traditional heat transfer techniques were discussed in the present study.

There is a temperature gradient that extends from the surface to the inside of each particle in traditional heating. Thermal gradients may be avoided by using a slower rate of heating and retaining constant temperature. The traditional heating method's longer preparation time and higher energy usage are both exacerbated by the sluggish heating rate at intermediate temperatures [Rodriguez-Reinoso *et al.*, 2001; Lafi, 2001]. Some light components may linger within the samples and pyrolyzer as a result of this heat gradient, which makes it more difficult to effectively remove gaseous products from the environment [Elsheikh *et al.*, 2003]. The microporous network may be obstructed by the carbon deposits, resulting in low V_{tot} and BET surface area values [Ahmedna *et al.*, 2000]. Distortion and inhomogeneous microstructure in the produced AC are also caused by the heat gradient [Lafi, 2001]. Contrarily, the standard thermal procedure might take several hours or even up to a week to obtain the necessary degree of activation. As the temperature rises slowly, the procedure becomes more expensive [Dabrowski *et al.*, 2005].

Using a microwave heating technique solves the problem of the conventional quick firing of the thermal heating method [Ioannidou and Zabaniotou, 2007]. As electromagnetic energy is converted into heat transfer inside dielectric materials in the microwave technique, microwave irradiation directly interacts with particles within the compressed compact material. It is not necessary to use an external heater to quickly heat the sample using microwave irradiation [Atwater and Wheeler, 2004]. Since microwave synthesis is a non-contact method of transferring heat to the product through electromagnetic waves, it minimizes the impacts of differential synthesis while yet allowing for substantial quantities of heat to be delivered to the interior of the material [Legrouri *et al.*, 2005; Khalili *et al.*, 2000]. However, microwave radiation is both internal and volumetric, allowing the microwave-induced reaction to continue more rapidly and efficiently at a lower bulk temperature, resulting in shorter processing times and less energy consumption [Rozada *et al.*, 2005]. Carbon compounds are among the best microwave absorbers available. Because of this property, microwave heating may change carbonaceous materials into new carbonaceous materials with altered characteristics [Chuayjumnong *et al.*, 2020]. Because the sample's inside is hotter than its surface, the microwave radiation method's thermal gradient diminishes progressively from the sample's center to the sample's surface [Gupta and Deep, 2017; Lozano-Castelló *et al.*, 2002; Bonvin *et al.*, 2016]. For example, the light components are readily released because of the temperature disparity [Manivannan, 1999; Bagheri *et al.*, 2015]. As an alternative to more traditional heating methods, microwave radiation offers the following advantages: energy transfer instead of heat transfer, selective heating, improved efficiency, immediate startup and shutdown, smaller steps, a lower activation temperature, improved safety, simplicity, smaller equipment size, and less automation. Because of mineral impurities in the carbon particles, microwave radiation may cause hot spots (the total temperature of the samples is greater than the hot spots) [Sidheswaran and Destaillats, 2012]. Heterogeneous reactions occur when the sample and the inert gases involved in the reaction are at different temperatures [Sidheswaran and Destaillats, 2012]. Because the sample temperature can't be precisely measured with an infrared pyrometer, only the sample's surface temperature can be measured. There may be tens of degrees or hundreds of degrees difference in the sample's internal and volumetric temperatures owing to microwave heating. As a result, while preparing AC by microwave irradiation, temperature could not be a variable. Finally, this study must be expanded, microwave procedures must be improved, and the creation of AC particles using microwaves must be scaled up [Hoseinzadeh *et al.*, 2013]. It is also important to note that the activation procedure affects the prepared AC's pore shape and adsorption

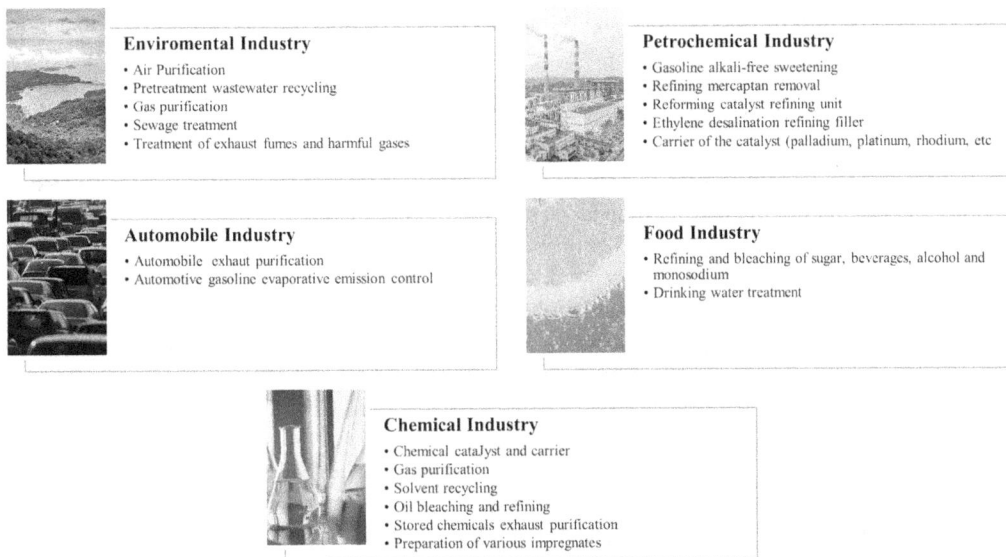

FIGURE 13.4 Shows the application of activated carbon in different industry [Hoseinzadeh *et al.*, 2013; Laine *et al.*, 1989; Bagheri *et al.*, 2015].

capability. To make ACs, a variety of activation methods, including chemical and physical, are used. It's possible to create char with the desired porosity by a process called physical activation, which entails heating carbonaceous material to high temperatures under an inert gas such as nitrogen to partially gasify it. It is then activated by means of oxidizing gases such as steam, CO_2, air, or combinations of these gases to create AC [Laine *et al.*, 1989].

For its many functions and qualities, activated carbon has been used in a variety of industries. There are several fields of use for gifted activated carbon, which are shown in Figure 13.4, including the environment, petrochemicals, automobiles, food, and chemical industries, to name a few examples.

Many studies have categorized the production of AC into four primary processes: pyrolysis, physical and chemical activation, and carbonization and steam/thermal activation. The physical activation process comprises a carbonization and activation stage, with steam and carbon dioxide (CO_2) being the most often employed reagents and having a major impact on the AC's porosity. Chemical activation produces AC in a single step using chemical reagents such as potassium hydroxide, phosphoric acid, and zinc chloride, which may be employed at room temperature. Impurities such as zinc (Zn) and phosphorus (P) can be detected in AC products, depending on the chemical reagents employed, which might raise the operating cost owing to the additional chemical needed [Bae *et al.*, 2014].

The process of physical activation is two-step. Carbonization of carbonaceous materials takes place first, followed by high-temperature activation of the resultant char in the presence of CO_2, steam, air, or three mixes acting as oxidizing gases [Bae *et al.*, 2014]. Multiple two stages occur concurrently in the chemical activation process, with chemical activating chemicals mixing with the precursor as oxidants and dehydrates. Even though environmental concerns may limit the use of chemical agents for activation, performing activation and carbonization concurrently during the chemical activation process at lower temperatures results in a higher porosity structure of AC [Ioannidou and Zabaniotou, 2007]. Carbonization at 500–600 degrees Celsius is followed by activation with steam at 800–1,100 degrees C, which is the underlying concept in all of these processes. These furnaces include rotary (fired directly or indirectly), vertical multi-hearth furnaces, and fluidized bed reactors, as well as retort systems with single- or double-throated retorts for carbonization. Every firm has a preferred method of production. A hopper on top of the retort receives the raw material, which falls into a central duct and is transported to the activation zone by gravity [Ioannidou and Zabaniotou, 2007].

A temperature rise of 800–1,000 degrees C occurs as the raw material slowly goes down the retort. Later in the same kiln, activation, which increases the interior surface area and widens the pore structure, might take place as a second step. High-temperature steam is used to activate the carbonized product. An internal chemical process occurs between the carbon and steam that removes carbon from the walls of the pores, hence increasing their diameter. A tiny portion of the overall retort surface area is used for activation, and here is where the steam is generated and released into the environment. Firebricks on the furnace's bottom are heated by the exothermic nature of this reaction, which allows for self-sustaining operation. For "in situ" heating, the product gas combustion alternates the steam injection site every 15 minutes or such. The length of time the product spends in the activation zone affects its level of activation (or quality) [Lin et al., 2015].

13.2 WHY ACTIVATED CARBON?

Activated carbon is a carbon-bearing substance with a porous structure and a high internal surface area that is non-hazardous. AC has a chemical structure similar to that of graphite, with a random amorphous structure that is extremely porous throughout a wide range of pore sizes, from apparent holes and gaps to those of molecular dimensions. Chemically activated carbon produced from coconut shells, orange peels, and banana peels act as an excellent adsorbent for removing contaminants from water which are very cheap process. More contaminants are removed due to the smaller size and greater concentration of activated carbon. These are low-cost organic materials that can be used to produce low-cost activated carbon for wastewater treatment that will be an alternative to the costly approach and using organic waste to make low-cost activated carbon might reduce the amount of organic waste in landfills [Shukla et al., 2020].

Drinking water purification, ground and municipal water treatment, power plant and landfill gas emissions, and precious metal recovery are just a few of the applications for activated carbon in both industrial and domestic uses. VOC removal and odor control are two features of air purification solutions. Activated carbon is the world's most effective adsorbent, with applications ranging from common water pitcher filters to gold recovery in carbon in pulp/carbon in leach systems [Atwater and Wheeler, 2004].

In water treatment application before reverse osmosis, it is used to remove free chlorine and chloramines to avoid membrane damage from oxidation. To form chlorides, activated carbon interacts fast with free chlorine in the water. Because of this, a tiny amount of carbon may have a big impact. The elimination of chloramines is catalyzed by almost five times the amount of carbon. To remove organic molecules from purified water, high purity activated carbon (HPAC) is used. The feedwater, the system, or the ion exchange resins might all contribute to these contaminants. It is a useful supplement to UV oxidation in keeping TOC levels low. Vent filters, which exploit activated carbon's attraction for organics, may safeguard purified water reservoirs [Rao et al., 2021].

There are several uses for activated carbon, from municipal drinking water and food and beverage processing to odor elimination and industrial pollution prevention. Carbonaceous resources including coconuts, nutshells, coal, peat, and wood are used to make activated carbon. The use of activated carbon generally increases the expense of the treatment. Its cost disadvantage has sparked research on using less expensive raw materials to produce activated carbon [Namane et al., 2005]. As a result, in addition to hardwood and bituminous coal, a wide range of agricultural by-products and wastes have been explored as cellulosic precursors for the synthesis of activated carbon. The following are examples of precursors: coconut shell and wood [Laine et al., 1989], olive stones [Rodriguez-Reinoso et al., 2001; Lafi 2001; Elsheikh et al., 2003], sugarcane bagasse [Ahmedna et al., 2000], pecan shells [Shawabkeh, 1998], palm seed [Legrouri et al., 2005]. Furthermore, some carbonaceous wastes, such as paper mill sludge (Khalili et al., 2000), old newspapers [Khalili et al., 2000), and discarded tires, have gotten increasing attention [Rozada et al., 2005].

13.3   TYPES OF ACTIVATED CARBON

Activated carbons are complex products that are tough to categorize based on their behavior, surface properties, and other fundamental features. For common reasons, however, some broad classification is formed based on their size, preparation techniques, and industrial uses. Activated carbon is generally existing in three forms or shapes including powder, granular, and extruded, whereas every single form is obtainable in many sizes. Based upon the application and requirements, a definite form and size are obtained Figure 13.5.

13.3.1   GRANULAR-ACTIVATED CARBON

Granular-activated carbon (GAC) is a composite material composed of a variety of graphite platelets joined by nongraphic carbon bonds. Due to GAC's adsorptive ability, it is perfect for the removal of a wide range of pollutants from water, air, liquids, and gases. Additionally, GAC is an ecologically friendly substance that can be revived by heat oxidation and reused for the same application numerous times [Menya *et al.*, 2018]. Granular-activated carbon is produced from high-carbon raw organic sources including coconut shells, coal, peat, and wood. Heat is utilized to activate the surface area of the carbon in the absence of oxygen, which is why these filters are frequently referred to as "charcoal" filters. Certain compounds that are dissolved in water flowing through a filter containing GAC are removed by the activated carbon trapping (adsorbing) the chemical in the GAC. The kind and concentration of pollutants, as well as average water usage, must be understood to calculate the proper system size and components. All treatment systems need to be installed correctly and maintained regularly. The GAC's capacity to bind and remove toxins eventually wears out, and it must be replaced. The frequency with which the GAC should be replaced is determined by pollutant levels and water usage. While some filters can survive for years if contaminant levels and/or water

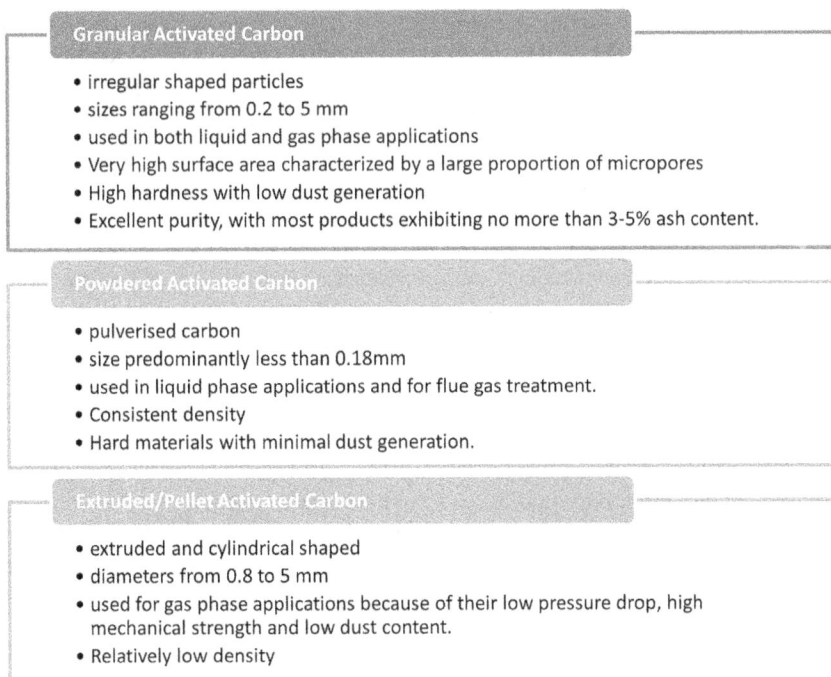

Granular Activated Carbon

- irregular shaped particles
- sizes ranging from 0.2 to 5 mm
- used in both liquid and gas phase applications
- Very high surface area characterized by a large proportion of micropores
- High hardness with low dust generation
- Excellent purity, with most products exhibiting no more than 3-5% ash content.

Powdered Activated Carbon

- pulverised carbon
- size predominantly less than 0.18mm
- used in liquid phase applications and for flue gas treatment.
- Consistent density
- Hard materials with minimal dust generation.

Extruded/Pellet Activated Carbon

- extruded and cylindrical shaped
- diameters from 0.8 to 5 mm
- used for gas phase applications because of their low pressure drop, high mechanical strength and low dust content.
- Relatively low density

FIGURE 13.5   Shows the properties of (a) granular activated carbon, (b) powdered activated carbon, (c) extruded activated carbon [Sidheswaran and Destaillats, 2012; Rao *et al.*, 2021; Bahiraei *et al.,* 2021; Menya *et al.*, 2018].

usage are minimal, greater levels or usage may necessitate more regular filter replacement [Chuayjumnong *et al.*, 2020]. As water passes through a filter containing granular-activated carbon, heat is used to activate the surface area of the carbon, eliminating specific compounds contained in the water. Because of its porous properties, GAC adsorbs the chemical. The adsorption takes place on the activated carbon's interior surface. Liquids or gases flow through the porous structure of activated carbon during adsorption, diffusing the chemicals to be removed to the adsorbent's surface, where they are held due to attraction forces [Gupta, 2016].

There are micro, meso, and macro holes in each carbon source's GAC. There are a lot of micro-pores, but they're tiny—about the size of a normal molecule (5–1,000 Angstroms) [Menya *et al.*, 2018]. Small organic and disinfection byproducts may be removed from water using coconut shell carbon because of its high proportion of micro-pores. De-colorization and removal of bigger organics are better done with wood carbons because they have a higher number of macro-pores. Because of their intermediate pore structure, coal bases are an excellent alternative for the organic reduction in general. The enormous surface area of GAC explains its ability to remove organic matter [LeChevallier *et al.*, 1984]. The surface area of a gramme of GAC might approach 1,000 m^2. Almost 100 football fields may be found on the surface of a pound of carbon. It is not conceivable nor required to use the whole surface area of GAC [Menya *et al.*, 2018]. An organic that is one mole thick, filmed across the whole surface, would only need around 6.25 mL of liquid per pound, or roughly 1/4 teaspoon of liquid. As a result, the film thickens, and adsorption weight rises as the organic dissolves. GAC may be loaded with 0.1 pounds or 45 mL of volumetric GAC per pound of GAC if the system is flowing to saturation (contaminant levels at the outflow and input are equal) [LeChevallier *et al.*, 1984].

The GAC pore structure and surface area available are critical to adsorption site accessibility. The degree of activation of the carbon base is a factor in this. Carbohydrate activation is quantified by its CTC number (CTC). Values greater than 50 are regarded as ideal for water purification [Sidheswaran and Destaillats, 2012]. For drinkable water treatment, the iodine number (Io N) should be in the range of 900 to 1,050, which is the carbon's relative surface area. Choosing a GAC product based on its abrasion number is an essential decision. To determine how well a material can endure being physically tumbled, the abrasion resistance is expressed as an absolute number (AN) (such as during backwashing). The abrasion resistance of shell carbons is the greatest, coming in at 90 AN. Nearly 70 AN is the typical coal basis. It's possible that ratings lower than 70 won't stand up well to repeated backwashing [Menya *et al.*, 2018].

Granular-activated carbon has a larger particle size than powdered activated carbon and, as a result, has a smaller exterior surface area. The adsorbate's diffusion is thus a crucial component. Because gaseous substances disperse quickly, these carbons are ideal for the adsorption of gases and vapors. Granulated carbons are utilized in flow systems and fast mix basins for air filtration and water treatment, as well as general deodorization and component separation [Gupta and Deep, 2017].

Granular-activated carbon is made up of irregularly shaped particles that have been milled and sieved. The sizes of these items range from 0.2 mm to 5 mm. They are tougher and last longer than powdered activated carbons, are easier to handle, filter huge quantities of gas or liquids with uniform quality, and may be reactivated and reused several times [Sidheswaran and Destaillats, 2012]. GACs are employed in both liquid and gas phase applications, as well as in stationary and mobile systems. In liquid phase uses, granular-activated carbon is packed in columns and towers through which liquids flow. GAC is used where there is a single product to be refined or produced continuously in large quantities [LeChevallier *et al.*, 1984]. In gas-phase applications, GACs have the advantage of having sufficient flow with an acceptable pressure drop through the carbon bed. In addition, granular-activated carbons are nearly always regenerated and reused, the period between reactivation varies significantly but is on average 18 months. Loss of material during reactivation ranges from 5% to 15% [Lafi, 2001].

GAC can assist in the removal of the following contaminants: disinfection by-products (DBPs) linked with chlorine and other disinfectants, algal toxins such as microcystin-LR, cylindrospermopsin, and anatoxin-A, endocrine disrupting substances, pharmaceuticals, and personal care items. Compounds that impart flavor and odor, organic components are derived from decaying plants and other naturally occurring stuff that act as precursors of DBPs. The following is a review of the fundamentals of activated carbon, with an emphasis on its application as a filter and adsorbent in potable water treatment [Menya *et al.*, 2018].

GAC has also been found to be a good physical filtering medium in water treatment facilities, with the additional advantage of protecting water quality by adsorbing taste and odor compounds and chemical pollutants. Several practical considerations must be made when retrofitting an existing multimedia filter with GAC or constructing a new GAC filter, including hydraulic requirements, filter on-stream time, and backwash water availability. The GAC's qualities, such as adsorption capacity, abrasion resistance, and density, must also be addressed. Furthermore, the cost of switching the filter to GAC must be calculated [Sidheswaran and Destaillats, 2012]. Backwashing needs, especially when retrofitting a multimedia filter, are another issue that influences how deep the GAC bed should be. The filter should be designed to allow for maximum expansion while yet leaving several inches of space between the enlarged bed and the bottom of the backwash trough. Any GAC that has extended to the level between the troughs will most likely be washed out of the filter when the water velocity rises between the troughs. The mechanics of placing GAC in a retrofit or new filter are quite similar to those of inserting sand or anthracite [LeChevallier *et al.*, 1984]. The filter, as well as the initial gravel/ sand or underdrain/sand support, must be disinfected before the GAC may be installed. This is because disinfectants that are suited for GAC, such as chlorine, will react quickly and leave no disinfectant residue. Another difference is that the GAC must be immersed for 24 hours before the final backwash before the new filter may be used. This soaking period enables air trapped inside the GAC pore structure to escape and the water to properly saturate the interior surface of the GAC [Menya *et al.*, 2018].

GAC filters are susceptible to mechanical operational upsets in the same way as multimedia filters are. Consequently, during production, backwash, or starting online, the filters should not be exposed to abrupt fluctuations in water velocity. Backwash water should be allowed to run freely in the filter for at least 30 seconds, depending on the filter type. When the device is turned back on, it is advised that the filter be refilled for at least 10 minutes [Rao *et al.*, 2021]. Operators should account for seasonal fluctuations in water density while backwashing and monitor a backwash event regularly. Core samples should be obtained once a year to keep the filter in good condition. The procedure's goal is to capture an accurate sample from the filter's core. After that, the GAC may be checked for residual activity using an iodine number assay. The GAC should be reactivated or replaced in the near future after the iodine value is between 450 and 550, according to historical statistics [Menya *et al.*, 2018].

13.3.2 POWDER-ACTIVATED CARBON

Powder-activated carbon (PAC) is a high-performance absorbent aimed to remove hazardous pollutants from the air, gas, and liquid phases. PAC solutions have long been used in the potable water sector to efficiently remove taste and odor chemicals and produce clean, safe drinking water. In the potable water sector, PAC products are widely recognized control technology in the potable water industry. They employ activated carbon technology and have a tiny particle size, so they may be filtered at the treatment plant's head and injected directly into the water to remove contaminants [Bahiraei and Behin, 2021]. Some of the benefits of using PAC products include:

- Straightforward, cost-effective, and reliable selection method
- Cut operating costs
- Negligible waste sludge
- Preservation of resources
- Improved water quality

The PAC material is a finer version of the activated carbon. PAC is composed comprised of crushed or ground carbon particles, with 95–100% of them passing through a mesh sieve. Particles that pass through an 80-mesh sieve (0.177 mm) and are smaller are classified as PAC by the American Society for Testing and Materials (ASTM). Because of the substantial head loss that would occur, it is uncommon to utilize PAC in a specialized vessel. Instead, PAC is commonly supplied to various process units such as raw water intakes, quick mix basins, clarifiers, and gravity filters directly [Bahiraei and Behin, 2021].

Although grainier and finer grades are available, PACs typically have a particle size range ranging from 5 to 150 microns. Powder-activated carbons have the advantages of cheaper processing costs and more operational flexibility. As process conditions change, the dosage of powder-activated carbon can simply be raised or reduced. Powder-activated carbons are mostly utilized for adsorption in the liquid phase. They are introduced to the liquid to be treated, mixed with it, and then removed by sedimentation and filtering following adsorption. In batch processes, powder-activated carbons are commonly utilized since the amount injected may be easily changed and the powder can be quickly withdrawn [Bahiraei and Behin, 2021].

Powder-activated carbon is prepared from a range of materials including wood, lignite, and coal. The most important factor of the cost of using PAC is the cost of the PAC itself, which is very cheap. The effectiveness of elimination of taste and odor-causing compounds by PAC differs, varying on the type and concentration of the mixture causing the problem. One of the key benefits of PAC is its low investment cost. Bagheri *et al.* has presented in his article that cost estimations of the accumulation of PAC at numerous amounts to various plant sizes [Bagheri *et al.*, 2015].

Due to its convenience of use and cost-effectiveness, powder-activated carbon has been employed as a treatment option for a long time. PAC may be used dry or as a wet slurry, which is a carbon and water combination. Powder-activated carbon has the problem of being unable to be reactivated after usage, as well as being difficult to dig out of water treatment reservoirs. The most significant financial benefits, "However, the PAC feed system's cheap construction cost, along with the flexibility to administer PAC seasonally or for periodic difficulties, may make it a cost-effective option on an annual basis." 1 PAC may be dry or wet kept. Wet is carried and kept in a huge slurry, whereas dry may be stored in compact 55-pound bags. If maintained in a slurry condition for an extended amount of time, PAC may solidify. To avoid this, mixing must be done regularly. When PAC is anticipated to be used more often, it is kept in a wet slurry, and when it is expected to be used less frequently, it is stored dry. Dry storage is less costly and allows for more precise dosages. When transporting dry PAC into silos, the facility must account for weight gain owing to the adsorption of moisture from the air [Najm *et al.*, 1991].

The essential design parameters for a PAC feed system are as follows:

- Design flow rate has an impact on dosing needs, storage vessel size, and number, and feed equipment quantity and size.
- Influence of target and background DOC concentrations on the PAC dosage required.
- Treatment goals include identifying target pollutants, determining the target effluent concentration, and determining the dosage and size of equipment.
- Coal, coconut shell, and wood are examples of PACs.
- Design dosage and duration based on pollutants, jar tests, and seasonal/periodic vs. continuous testing.
- Contact time and mixing – quick mix (sometimes referred to as "PAC will stay suspended through the rapid mixer"), flocculation basins, and sedimentation.

The influent concentration, residence duration, external mass transfer, adsorbate characteristics, and other pollutants other than the target molecule all impact the powder-activated carbon. "Increasing the PAC dosage will lower the reactor effluent concentration as more adsorbates can be removed from the solution," says the author. However, the percent reduction in effluent concentration is not proportionate to a corresponding increase in PAC dosage. A dosage of 5 mg/L of PAC may produce a 40% reduction in influent goemin concentration in high TOC water (TOC = 10 mg/L), but it requires around 17 mg/L of PAC to accomplish 90% elimination [Najm et al., 1991].

Chlorine, various disinfectants, and unknown pollutants have been discovered to occupy adsorption sites, lowering the PAC adsorption capacity for target chemicals. This might result in a high concentration of target substances in the effluent. Increasing the contact time improves the efficiency of the removal process [Okajima et al., 2005]. A least 15 minutes of contact time is advised, but longer may be required and helpful to attain the PAC's entire adsorption capacity. Before installing a new feed system, jar tests are usually performed to see how effective the powder-activated carbon will be. Powder-activated carbon is a tried-and-true, low-cost therapy [Najm et al., 1991].

13.3.3 EXTRUDED/PELLETIZED ACTIVATED CARBON

Extruded activated carbon is a type of activated carbon that comes in cylindrical pellets with a diameter ranging from 1 mm to 5 mm. The extrusion process, together with the raw material used, confirms that the end product is durable and appropriate for heavy-duty applications. The extruded pellet shape results in a reduced system pressure drop, which is significant in gas-phase applications. Solvent recovery, gas purification, and automobile pollution control are markets where extruded carbon's high-volume activity, low-pressure drop, and high stock resistance allow it to endure the whole life of the vehicle [Mcdougall, 1991].

Due to the porous nature of activated carbons, they are often in the form of a fluffy powder with a low density. Powdered adsorbents typically complicate scaling-up batch and continuous adsorption operations, as the fine particles require a large volume of adsorbent, a long settling time, hydraulic pressure build-up, and flow resistance inside the close vessel, as well as difficult adsorbent recovery and management. These issues might result in astronomical maintenance and operational costs. Thus, converting activated carbon powder into various forms (cylindrical, spherical, ring, etc.) is critical in a wide variety of industrial applications. An activated carbon pellet may be made by extruding fine activated carbon powder with a binder to stabilize the carbon particles in their compressed condition [Mcdougall, 1991].

While torrefied sawdust and sawdust composite pellets have been extensively explored for use in alternative fuel energy and electrochemical batteries, the use of PAIS as an adsorption binder is still in its infancy. To our knowledge, no investigation has been undertaken to determine the chemical properties and dye adsorption capability of the pellets generated [Tang et al., 2020]. Pellets formed may have distinct physicochemical features, which may alter the adsorptive qualities and process.

Pellets were created with the help of a binder and an epoxy resin that served as an adhesive. This illustration depicts the hand-made pellets that were adhered together using epoxy resin. When activated at high temperatures of 550°C, the pellets, on the other hand, were unable to maintain their original form. It causes the pellet to lose its stability and, as a result of the breakdown, it also burns away. The pellets made from sugarcane juice, on the other hand, can maintain their shape and structure even after being activated. Through the use of zinc chloride impregnation, the pellet-activated carbon is chemically activated. Because the pellets are made from concentrated sugarcane juice, they can behave as an adhesive [Hassan et al., 2020]. The use of concentrated sugarcane juice as a replacement for epoxy resin in the manufacture of pellet-activated carbon is recommended if epoxy resin fails. Pelletizer guns are used to produce pellets that are homogeneous in size and shape and have almost identical dimensions. The pellets are toughened by drying them in the microwave at a temperature of around 105°C for several minutes. Later, these hard pellets are activated at 550 degrees Celsius in a tube furnace in a nitrogen medium. They are then used as fuel [50]. It is important to note that the pellets retain their form and structure after activation. As a result, the sugarcane juice acts as an effective binder for the palletization process. This demonstrated that sugarcane juice may be utilized as a binder for pellet creation and that the pellets created can subsequently be activated at high temperatures without causing structural damage to the pellets. These pellets are manufactured, rinsed, dried, and stored in large numbers using the same processes as are used for other kinds of source materials [Jüntgen, 1986].

13.4   ACTIVATED CARBON THROUGH MICROWAVE METHOD

Microwave heating may speed up various operations and change the characteristics of treated materials since it heats the entire sample directly and uniformly [Jüntgen, 1986]. The microwave method is one of the most promising ways of improving and speeding up chemical processes. Due to excellent heat transfer profiles, the reactions may be performed more quickly than with other conventional techniques. This is quickly becoming one of the most effective technologies in the pyrolysis process; it decreases residence time and speeds up chemical processes, saving energy [Legrouri et al., 2005]. Microwave radiation has recently been employed in a unique procedure to create activated carbon. Chemical activators such as KOH, K_2CO_3, H_3PO_4, and $ZnCl_2$ were utilized to activate agricultural wastes using microwaves. The primary difference between a conventional furnace and the activation of carbon by microwaves is in their heating patterns. Microwaves create heat by reorienting molecules in the presence of microwave radiation. In the conventional furnace, however, heat is conveyed via the furnace walls, creating a temperature gradient across the sample. Microwaves are said to enhance a wide variety of heating processes by directly heating chemicals, removing temperature differences within a single material, and selective heating. Selective heating is another benefit of microwaves for chemical processes. While conventional chemical activation using a furnace has received substantial investigation, microwave-assisted activation has received very little attention. Microwaves may have a significant influence on the mechanism of the activation response. Certain types of biomasses are poor absorbers of microwave radiation. Microwave absorbers are required to heat these biomass sources evenly. In chemical activation using microwaves, it is predicted that the activating agents will be the primary absorbers of microwave radiation during the first phase of the reaction. This process is believed to aid in rapidly attaining the activation temperature (between 500 and 800°C) [Jüntgen, 1986].

There is a temperature gradient from the surface to the inside of each particle in the traditional way of heating. A slower rate of heating with isothermal holding is employed to prevent this thermal gradient at high synthesis temperatures within the material. In the traditional heating technique, this sluggish heating rate at intermediate temperatures lengthens the preparation procedure, resulting in higher energy consumption [Kubota et al., 2009; Oghbaei and Mirzaee, 2010]. Because the heat gradient prevents gaseous products from being effectively removed from the environment [Li et al., 2008], certain light components may linger within the samples and pyrolizer, resulting in carbon deposition. The carbon deposited on the surface may impede the microporous network, resulting in poor Vtot and BET surface area values [Li et al., 2009]. In the prepared AC, the temperature gradient also causes deformation and inhomogeneous microstructure [Li et al., 2009]. The traditional thermal procedure, on the other hand, might take several hours or even a week to obtain the appropriate degree of activation. The cost of the method is increased by the sluggish thermal process [Li et al., 2009; Deng et al., 2010]. Fast firing is another drawback of the thermal heating approach, which may be overcome by adopting a microwave heating method [Li et al., 2009].

Microwave irradiation interacts directly with the particles within the pressed compact material, converting electromagnetic energy into heat transfer inside the dielectric materials in the microwave process. Microwave irradiation from an external heat source is not passed into the sample, allowing for rapid volumetric heating [Xie et al., 1999; Thakur et al., 2007]. Because microwave synthesis is a non-contact technique in which heat is transferred to the product via electromagnetic waves, large amounts of heat can be transferred to the interior of the material, minimizing the effects of differential synthesis [Jones et al., 2002; Kazi, 1999], it overcomes the problems of conventional fast firing. The microwave radiation method, on the other hand, is both internal and volumetric, because the large thermal gradient from the interior of the sample to the cool surface allows the microwave-induced reaction to proceed more quickly and effectively at a lower bulk temperature, saving time and energy [Deng et al., 2010]. Carbon materials, among the various kinds of materials available, are excellent microwave absorbents. This property enables microwave heating to modify carbonaceous materials, resulting in new carbonaceous compounds with altered characteristics [Menéndez et al., 2010]. Due to greater temperatures on the inside than at the surface of the sample, the thermal gradient in the microwave radiation technique diminishes progressively from the core to the surface of the sample [Ji et al., 2007; Foo and Hameed, 2011; Chih-Ju, 1998]. The light components are readily released to produce additional pores as a result of the temperature differential [Yang et al., 2004; Ania et al., 2005]. Other benefits of microwave radiation over traditional heating techniques include energy transfer rather than heat transfer, selective heating, increased efficiency, rapid starting and shutdown, simpler stages, lower activation temperature, enhanced safety, simplicity, reduced equipment size, and less automation [László and Szucs, 2001; Menezes et al., 2007; Foo and Hameed, 2011].

Microwave radiation may cause hot patches within carbon particles (due to mineral impurities) where the temperature is substantially greater than the sample's average temperature [Menéndez et al., 1999]. The sample and the inert gases involved in the reaction are frequently subjected to heterogeneous reactions as a result of the temperature differential [Menéndez et al., 2010]. Furthermore, measuring the sample temperature precisely is almost difficult, and an infrared pyrometer can only detect the sample's surface temperature. Due to the internal and volumetric nature of microwave heating, the sample's interior temperature maybe tens or hundreds of degrees greater than the sample's surface temperature. As a result, in the fabrication of AC employing the microwave irradiation technique [Guo and Lua, 2000], temperature could not be a variable condition. Finally, much more focused labor and investigation are required to extend this research, increase microwave technology performance, and scale up microwave manufacturing of AC particles [Yagmur et al., 2013].

Since microwave heating is faster, consumes less energy, and is a faster process than traditional heating, it is a more energy-efficient method of heating than conventional heating [Xin-Hui et al., 2011; Alslaibi et al., 2013]. Thermal energy is given throughout this process as a result of the conversion of microwave energy into thermal energy. When compared to traditional heating, microwave heating has many advantages. Thermal procedures need a certain amount of time to obtain the necessary degree of activation [Alslaibi et al., 2013; Hernandez and Villanueva, 1941]. Furthermore, surface heating cannot maintain a consistent temperature throughout the sample. Microwave-assisted heating may be used to overcome these drawbacks by acting as a quick and efficient alternate heating method [Gustafsson et al., 2017; Hui, 2015]. In microwave heating, energy is delivered directly to the particles through dipole rotation and ionic conduction occurring inside the particles. The activation period of activated carbon under microwave heating has a significant impact on the surface area and porosity of the carbon. He and colleagues investigated the manufacture of activated carbon by microwave-assisted KOH activation of 'petroleum coke' across a range of activation times, which they found to be effective [Li et al., 2021]. The surface area of freshly generated activated carbon is large, and it contains micropores. With increasing activation time, the surface area and porosity increase as a consequence of the findings. By activating the carbon with KOH and K_2CO_3, Deng et al. were able to manufacture activated carbon from 'cotton stalk' using microwave heating methods. The activation of KOH with microwave assistance results in a gasification process in which KOH is converted to metallic potassium. K_2CO_3 is decomposed into K_2O, CO_2, CO, and K_2O_2 [Deng et al., 2010].

When potassium is heated beyond its boiling point, the potassium diffuses into the carbon matrix, resulting in the continued development of porous structure in the carbon matrix. The findings demonstrate that activation with KOH results in the production of a significant amount of micropores and mesopores, while activation with K_2CO_3 results in the formation of a high number of mesopores [Gustafsson et al., 2017]. In addition, similar research was carried out and improved parameters for the manufacture of activated carbon from mangosteen peel by microwave-assisted K_2CO_3 activation were carried out by Foo and colleagues [Li et al., 2021].

Pores are widening and porosity is being promoted as a result of the potassium molecule produced during K_2CO_3 activation. Excessive K_2CO_3/char ratio, on the other hand, leads to the clogging of pores and the reduction of surface area. In addition to this, high radiation power, such as >800 W, has a negative influence on the environment, causing the burning of carbon and the degradation of pore space. Another important aspect is the amount of time that microwave radiation is exposed to the carbon 4 Characteristics of Activated Carbon 139, which promote the formation of side reactions between carbon and potassium by-products [Lin *et al.*, 2015].

13.4.1 PREPARATION OF ACTIVATED CARBON

Activation is the most essential stage in the synthesis of activated carbon. This is the procedure for converting a specified raw material or carbonized raw material into a finely crystalline form of carbon with the largest number of randomly dispersed holes of varied shapes and sizes. The enlarged interior surface area is due to these pores. It is desirable to produce the highest accessible surface area that is compatible with economic feasibility and product use in the manufacturing of activated carbon. Carbonization and activation are the two major processes that create activated carbon [Pré *et al.*, 2013] Figure 13.6.

Carbonization means the formation of carbon. Through carbonization, the raw material is thermally decomposed in an inert condition, at temperatures below 800°C. Elements such as oxygen, hydrogen, nitrogen, and sulfur are removed from the source material during gasification, leaving biochar as a solid result. To properly establish the pore structure, the carbonized material must now be activated. This is accomplished by oxidizing the carbonized material in the presence of air, carbon dioxide, or steam at temperatures ranging from 800 to 900 degrees Celsius [Zhou *et al.*, 2018]. Physical activation or chemical activation can be used to make activated carbon, depending on the raw material. In any scenario, the charcoal or biochar produced by gasification, pyrolysis, or hydrothermal gasification is treated further to increase the surface area and activity of the biochar, resulting in activated carbon. The chemical process (basically a single-stage process) and the physical process (basically a multi-stage process) are the two most common ways of producing activated carbon. Chemical activation, as predicted, utilizes chemicals to activate, whereas physical activation uses gases (typically carbon dioxide and air), vapors (steam), or combinations of gases and vapors. Any bioresource may be utilized as raw material, including coconut shells, peach and apricot stones, wood, sawdust, cellulose, walnut shells, almond shells, tree leaves, bark, bamboo, rice husks, maize cobs, sugar, bones, and so on. The biochar is rinsed, dried, and then impregnated with a reagent after carbonization. The impregnated biochar is

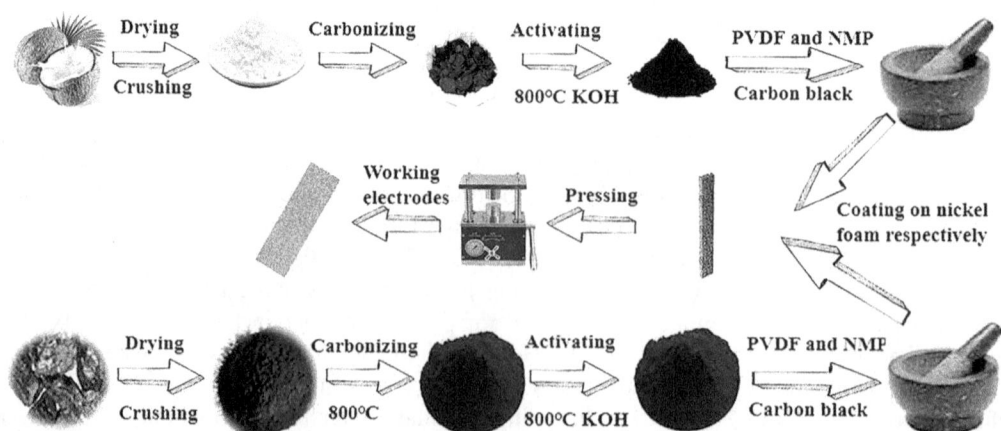

FIGURE 13.6 Shows the process for preparing electrodes using AC [Pré *et al.*, 2013; Ding *et al.*, 2020].

activated via a physical or chemical process under controlled environments. The activated carbon is ready for use once it has been washed and dried [Lamine *et al.*, 2014].

Activated carbon has been synthesized using a variety of methods. In most cases, solid carbon blocks are removed from the precursor by thermal heating, a process known as carbonization or pyrolysis. By thermal breakdown, non-carbon species are removed from biomasses during the carbonization process, resulting in an enrichment of carbon content in carbonaceous materials. The porosity of the char stays low at first, but this is enhanced in the subsequent stages by using the activating procedure. As a result, selecting the right carbonization settings is critical for improving the end product's qualities. The second option is to recover char using wet biomass conversion, commonly known as hydrothermal carbonization. It works similarly to how coal is created naturally. Carbonization aids in the formation of morphological characteristics in activated carbon. A comparative analysis of 'Auricularia'-derived activated carbon from direct carbonization with $ZnCl_2$ activation and hydrothermal alkali carbonization, which results in the creation of extremely porous layered morphology, is carried out in this respect. Özhan et al. used two-step carbonization and $ZnCl_2$ activation of auricular to create ultra-small carbon nanosheets-like structures. $ZnCl_2$ may operate as a spacer to prevent nearby cell walls from agglomerating during activation and carbonization, meaning that it can act as a spacer to prevent adjacent cell walls from agglomerating during activation and carbonization [Özhan *et al.*, 2014]. He et al., on the other hand, used hydrothermal synthesis to create porous layered stacked activated carbon from auricular. The biomolecules are hydrolyzed during hydrothermal treatment in an alkali environment, resulting in an increase in porosity and the creation of a three-dimensional interconnected porous framework [He *et al.*, 2010]. The second phase in the preparation of activated carbon is to build a porous structure and increase the surface area of biochar using chemical and physical methods after it has been extracted using the carbonization procedure Figure 13.7.

Physical and chemical activations are used to generate activated carbon that is tailored to the customer's specifications in terms of pore size distribution and porosity enhancement. A two-stage activation procedure is used, with the first phase consisting of chemical activation, which is typically performed using $ZnCl_2$ and H_3PO_4, followed by physical activation with carbon dioxide [Moreno-Castilla *et al.*, 1998; Pelekani and Snoeyink, 1999]. With just a little quantity of chemical activating agent, narrow micro-porosity may be generated without changing the bulk density of the material. Physical activation is made possible by the presence of gases, which allows for the proper development of the pore structure that was produced during chemical activation [Özhan *et al.*, 2014].

In self-activation, gases are released during the pyrolysis of biomass that has been utilized to activate the process itself. When using this strategy, you may produce activated carbon in an environmentally benign manner that does not need the use of foreign chemicals [Bogdal and Prociak, 2008]. A one-step procedure without the use of chemicals has been developed, in which carbonization and activation have been integrated with a single phase. Because of the savings inactivating agent costs, this approach offers the additional benefit of being environmentally friendly while still being inexpensive in cost of manufacturing [Kar, 2020].

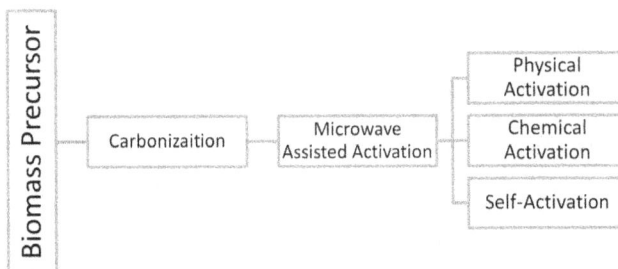

FIGURE 13.7 Shows the method for activating carbon.

13.4.2 MORPHOLOGY OF ACTIVATED CARBON

Morphological analysis of activated carbon is a method for identifying, structuring, and investigating the entire set of likely connections included in a given multidimensional problem complex [Nayebi and Ruhe, 2015]. The surface morphology of activated carbon formulated from the specific raw material is obtained using a scanning electron microscope (SEM). The method of preparation of the activated carbon samples will impact their surface characteristics, hence the necessary to accurately evaluate the effects of pore size on adsorption of the sample [Hidayu et al., 2013]. A scanning electron microscope is used to determine the pore size and morphology of the surface area of AC. Activated carbon morphology is directly dependent on the method of preparation.

The surface morphology of the AC samples was studied using SEM depicts an AC sample with well-developed and ordered hexagonal pores that matched to micropore diameters. Microscope XT measuring software was used to obtain this picture. Physical activation with nitrogen gas produced activated carbon with extensive and well-defined micro porosity. The use of phosphoric acid in the chemical activation procedure resulted in well-organized smooth surfaces on the crystalline carbon's surface, which were characterized by macro porosity. The larger pore diameters of sample sallow for easier adsorption of large molecules [Hidayu et al., 2013].

13.4.3 BET ANALYSIS

The Brunauer-Emmett-Teller (BET) surface area, micropore volume, and micropore surface area of the activated carbons are determined by the application of the BET. The BET surface area is determined in the relative pressure range from 0.01 to 0.15 [Demiral and Demiral, 2008]. The specific surface areas of the carbons were determined from the N_2 adsorption/desorption isotherms by BET equation. Micropore volumes were analyzed using the t-plot method. The BET principle takes to explain the physical adsorption of gas molecules on a solid surface and serves as the foundation for a critical analytical approach for calculating the specific surface area of materials. Physical adsorption, or physisorption, is a term used to describe the phenomenon. In a BET surface area analysis, nitrogen is used because of its availability in high purity and its strong interaction with most solids. Due to the weak contact between gaseous and solid phases, the surface is chilled with liquid N_2 to produce measurable quantities of adsorption. The sample section is then gradually filled with known amounts of nitrogen gas. Partially vacuum circumstances are used to achieve relative pressures lower than atmospheric pressure. There is no further adsorption when the saturation pressure is reached, regardless of how high the pressure is raised. Pressure transducers with high precision and accuracy detect pressure changes caused by the adsorption process [Li et al., 2009].

For the study of physical characteristics, the nitrogen adsorption-desorption isotherm is widely utilized using BET. The nitrogen adsorption isotherms on activated carbon are produced with a microwave power of 400 W, a 5-minute radiation duration, a 72-hour impregnation time, and a 50% $ZnCl_2$ concentration. The isotherm of PCAC is midway between type I of the referred International Union of Pure and Applied Chemistry (IUPAC) classification, which is associated with a combination of microporous, and type II of the mentioned IUPAC classification, which is associated with a combination of microporous [Demiral and Demiral, 2008].

By providing a material characterization service for the samples, the BET analysis machine enables users to get hands-on expertise in particle characterization. The BET surface analyzer will be used to examine the samples. The primarily volumetric approach is enhanced by the use of nitrogen gas. The study is carried out using the most precise multipoint BET area measurement available [Reucroft et al., 1971]. Before testing, degassing samples is usually required to acquire the most accurate findings; nonetheless, the equipment can handle an extremely broad variety of environmental conditions. Furthermore, these may be combined to meet the specific needs of the customer, with degassing temperatures up to 300°C under vacuum and the use of flowing nitrogen gas being possible. In addition to surface area measurements, this analyzer is capable of analyzing micro and mesoporous particles, which may provide useful information regarding the volume and size distribution of the pores in the material under investigation [Reucroft et al., 1971].

The BET area plot, as well as tabular data and the computed surface area of the sample, are often included in the results section. Aside from that, there is also the matter of interpretation of outcomes. Surface area, also known as specific surface area, is a significant component determining the operation and behavior of many materials. The surface area is measured in square inches. A common use of this technique is for concerns such as determining the stability and efficiency of most materials, including catalysts and medicines, among others. Additionally, the measurement of the surface area is an important parameter in the quality control of many materials [Biniak et al., 1997].

13.4.4 Characterization of the Activated Carbons

The performance of activated carbon is condition-dependent. The initial phase of the synthesis technique was the pyrolysis of the biomass, which resulted in the creation of char. Volatile materials and moisture have been eliminated during the carbonization or pyrolysis process. Activated carbon may be made from this char in three distinct ways: physical activation, chemical activation, or physicochemical activation (combination of physical and chemical activation). Physical activation utilizes gaseous activators such as steam and carbon dioxide, while chemical activation utilizes chemical agents such as metal oxides, alkaline metals, and acids. The activation process results in the creation of a porous structure with a high surface area and volume of pores [Sari and Tuzen, 2009]. Thus, activated carbon can be prepared in two steps: the precursor is pyrolyzed in a controlled or inert atmosphere at a temperature between 400 and 900 degrees Celsius to produce biochar, followed by activation using gases or chemical compounds at a temperature between 450 and 900 degrees Celsius [Reza et al., 2020].

The prime objective of an activation process is to enhance the pore volume, diameter, and structure of the activated carbon's pores and porosity. As previously indicated, activating chemicals produce microporous structures during the first step of an activation process. Later in the activation process, the existing holes are broadened by the creation of big pores as a consequence of the wall between the tiny pores being burned away. This results in the formation of micropores in the activated carbon, followed by a reduction in the volume of micropores. Thus, it is worth noting here that the amount to which carbon material is burned, or in other words, the extent to which it is activated, is a critical criterion for producing high-quality activated carbon [Nieto-Delgado and Rangel-Mendez, 2011].

The characteristics of activated carbon are represented by bulk density, ash and moisture contents, pH, conductivity, surface area, micropore and mesopore volumes, and pore size distribution. Ahmedna et al. described a technique for determining bulk density [Ahmedna et al., 2000]. The amount of ash in the sample was measured using a conventional technique [Reza et al., 2020]. The oven-drying technique was used to determine the moisture content [Adekola and Adegoke, 2005]. The pH value was measured using a pH tester by Egwaikhide et al. procedure [Egwaikhide et al., 2007]. The approach of Ahmedna et al. was used to measure electrical conductivity using an EC tester [Ahmedna et al., 2000]. The surface area was calculated using the BET equation and the adsorption-desorption isotherm of N_2 at 77 K [Brunauer et al., 1938]. The Dubinin–Radushkevich equation was used to calculate the micropore volume (Gregg and Sing, 1982). The BJH desorption branch was used to calculate the mesopore volume, and the density functional theory was used to establish the pore size distribution [Lastoskie et al., 1993]. These computations were carried out with the help of the NOVAWin2 data analysis software [Lowell et al., 2004]. Scanning electron microscopy (SEM) was used to evaluate the morphology of activated and raw materials.

The characteristics of activated carbon rely on the physical and chemical properties of the precursor and also on the activation method used [Demiral et al., 2008]. Activated carbon can be prepared from any carbonaceous solid precursor that may be either natural or synthetic. The availability, affordability, and purity of a precursor all have a role in its selection. Agricultural wastes are regarded as a highly significant precursor due to environmental concerns since they are inexpensive, renewable, safe, readily available in huge numbers, and easily traceable sources; they also have a high carbon and low ash content [Kalderis et al., 2008; Mestre et al., 2009].

Physical properties of activated carbon, such as surface area and bulk density, as well as chemical properties such as pH, ash content, and conductivity, can influence its usage and determine its appropriateness for certain applications. High surface area, high bulk density, neutral pH, low ash, and low conductivity are all desired properties for the adsorption of organic molecules like methylene blue (MB) and phenol (Ph). Activated carbon has been widely utilized as an adsorbent in catalysis and separation processes due to its high adsorptive porosity [Ghaedi et al., 2012; Tamai et al., 2009].

The chemical characteristics of activated carbons are basically defined by a certain degree of surface chemical heterogeneity, which is associated to the existence of heteroatoms, i.e., atoms present in the carbon structure that are not carbon, such as oxygen, nitrogen, hydrogen, sulfur, and phosphorus [Shafeeyan et al., 2010]. These components are either obtained from the nature of the starting material or added during the activation process [Snoeyink and Weber, 1967; El-Sayed and Bandosz, 2004]. The acidic or basic character of the activated carbon surface is determined by surface functional groups (which are generated from these heteroatoms) and the delocalized electrons of the carbon structure [Lászlóand Szucs, 2001].

13.4.5 FACTORS DECIDING THE PROPERTIES OF ACTIVATED CARBON

It is critical to maintain appropriate management over the following parameters during the preparation stage to produce high performance activated carbon [Kar, 2020]:

1. Raw materials: Numerous organic compounds containing a significant amount of carbon are employed as precursors in the production of activated carbon. The manufacturing of porous carbon structured activated carbon is highly reliant on many factors, including a high carbon content, a low inorganic or ash content, a high density, a suitable volatile content, stability and minimal degradation during storage, and an affordable price.
2. Surface area and pore size distribution: The temperature and the activating agent used to produce activated carbon are the primary factors that define its physicochemical features. Extreme heat accelerates the breakdown of carbonaceous species and crystallinity. Extreme crystallinity, on the other hand, results in a reduction in surface area.
3. Graphitization extent: High-temperature annealing enhances graphitization by causing the graphite lattice to self-heal, disintegrating the carbon network around the heteroatom. While graphitization results in increased conductivity within the carbon matrix, it also results in the depletion of heteroatoms, lowering the carrier concentration.
4. Heteroatom content: The presence of heteroatoms promotes ion adsorption at the carbon matrix's surface. Over doping heteroatoms, on the other hand, results in a steric effect, reducing the conductivity within the carbon network.
5. Surface heteroatom concentration: The presence of surface-active species enhances the carbon matrix's wettability. A significant number of heteroatoms is required to provide the carbon matrix with the requisite wettability.
6. Type of functional groups: Heteroatom functional groups either boost the reactivity or conductivity of the carbon matrix. For instance, pyridinic-N serves as a Lewis base, but graphitic-N increases conductivity.
7. Activation time: Activation time has a significant influence on carbonization and hence on the activated carbon's ultimate characteristics. Increases in activation time increase the BET surface area while decreasing the percentage yield of activated carbon. This might be explained by the volatilization of organic compounds during the carbonization process.
8. Activation temperature: Along with activation time, activation temperature has a significant effect on the BET surface area and yield of carbonized products. The BET surface

area is shown to increase with an increase in activation temperature. This is because a rise in temperature results in the formation of new pores and the widening of existing holes as a result of the release of volatile substances. Once again, increasing the activation temperature reduces the yield of activated carbon owing to the emission of a large number of volatile materials.

13.5   PROPERTIES OF ACTIVATED CARBON

The properties of activated carbon are divided into types: physical properties and chemical properties. The physical properties include particle density, bulk density, pore-volume, ball-pan hardness, particle size distribution, ash content, moisture content, and carbon content. The chemical properties include surface area, iodine number, carbon tetrachloride number, and benzene number.

Activated carbon's most essential properties are its high specific surface area and variable pore size distribution. Until far, a variety of biomass-derived precursors have been employed to make porous activated carbon with a large surface area. The BET nitrogen adsorption-desorption isotherm is used to compute specific surface area. Activation, on the other hand, causes a porous structure to emerge in carbon material, increasing surface area. In general, activated carbon's nitrogen sorption isotherm has a Type-I profile, indicating a microporous structure, although some activated carbon has a Type-IV profile, indicating a mesoporous structure. A mixture of Type-I and Type-IV isotherms was sometimes observed by 140 P. Elsheikh et al., showing the presence of both mesopores and micropores [Elsheikh et al., 2003]. Reucroft et al. produced activated carbon from 'cotton stalks,' which showed a Type-I/IV isotherm, suggesting that the bulk of adsorption occurs at pressures less than 0.1. This demonstrates the prevalence of micropores in activated carbon as prepared [Reucroft et al., 1971]. In 'willow catkins' derived activated carbon, Elsheikh et al. produced a mix of Type-I and Type-IV properties. The gas uptake isotherm reveals a flat plateau at high pressure, suggesting that microporous structure dominates with a tiny proportion of mesoporous structure [Elsheikh et al., 2003]. Although Reucroft et al. used N_2 and CO_2 sorption tests to investigate the textural qualities of as-prepared activated carbon generated from sugarcane bagasse,' $ZnCl_2$ activation displays a Type-I isotherm for both gases, indicating microporous capabilities. The research demonstrates that when the activation temperature rises, the specific surface area decreases owing to $ZnCl_2$ volatilization above the boiling point. The pore shrinking in carbon structure has been observed following heat treatment in the absence of $ZnCl_2$ particles [Reucroft et al., 1971].

Gopiraman et al. used NaOH activation to create a Type-I isotherm of activated carbon formed from "corn residues". When the concentration of NaOH is increased to 1:2, the surface area of activated carbon rises. A higher concentration of NaOH causes pore deformation owing to severe oxidation [Gopiraman et al., 2017]. For 'Enteromorpha' derived carbon samples, a mixture of Type-I/IV isotherms. Adsorption at low pressure (P/Po 0.01) and knee broadening at medium pressure (0.01 P/Po 0.4) indicate a network of hierarchical micropores and mesopores. Because KOH activation results in an interaction between KOH and C, the creation of CO_2 accelerates the breakdown of K_2CO_3 at higher temperatures, such as >700°C. Carbon gasification produces porous structures as a consequence of this [Wong et al., 2018]. Table 13.1 contains a list of textural features. According to the findings, BET specific surface area declines with KOH concentration at first, from 2,073 m^2g^{-1} for an equal impregnation ratio to 1,532 m^2g^{-1} for a 2:1 KOH: carbonized ratio, before increasing to 1,879 m^2g^{-1}. The quantity of pore rises with an increase in KOH concentration, according to the pore size distribution [Legrouri et al., 2005].

Factors that influence the development of porosity are (i) activating agent, (ii) activation temperature and duration, (iii) gas flow rate, (iv) type of gases employed during pyrolysis and activation, (v) mixing process (solution and mechanical mixing), and (vi) heating rate [Shafeeyan et al., 2010]. Ji et al. investigated the effect of temperature on the generation of activated carbon from a 'sunflower seed shell' using impregnation-activation and carbonization-activation procedures. The surface area and pore volume of the BET test are found to improve as the temperature rises. This is because the

TABLE 13.1

Shows the Properties of Activated Carbon with Two Different Forms (Wu et al., 2013)

Properties	Quality requirements	
	Particle	**Powder**
Volatile matter 950°C, %	Max. 15	Max. 25
Moisture content, %	Max. 4,5	Max. 15
Ash content, %	Max. 2,5	Max. 10
Parts that are not carbonized	0	0
Iodine Number, mg/g	Min. 750	Min. 750
Pure activation carbon, %	Min. 80	Min. 65
Benzene adsorption capacity, %	Min. 25	–
Methylene blue Number, mg/g	Min. 60	Min. 120
Bulk specific gravity, g/mL	0,45–0,55	0,3–0,35
Escaped mesh 325, %	–	Min. 90
Distance, %	90	–
Violence, %	80	–

activating process occurs quickly at higher temperatures. Furthermore, at the same temperature, a high concentration of KOH results in a large BET surface area, demonstrating the impact of activator quantity on activated carbon features (Ji et al., 2007). The carbonization-activation step, of these two synthesis methods, produces the most pore structure, increasing surface area (Márquez et al., 2021). The BET-specific surface area and diverse pore architectures of activated carbon produced from various biomasses are shown in Table 13.1. The quantity of activating agents is proportional to the surface area and pore volume [Schanz and Parry, 2002].

Hidayu et al. obtained a huge BET surface area >2,673 m^2g^{-1} using different 'pollens'-derived activated samples [Hidayu *et al.*, 2013]. Activated carbon generated from Lotus pollen has a surface area of 3,037 m^2g^{-1}, which is much greater than commercial activated carbon, RP20 (1,677 m^2/g). This might be because of the porous structure, which encourages the interaction of KOH with carbon. Table 4.1 lists the specifics of the different attributes. Lotus-HA has a large pore volume of 2.27 cm3/g, with 82% of volume corresponding to mesopores, which is much greater than commercial activated carbon, which has a pore volume of 0.64 cm^3/g (Deryło-Marczewska et al., 2004). Some of the properties are discussed below.

13.5.1 Pore structure

Activated carbon is a rough type of graphite with a random or amorphous structure that is extremely porous in a wide range of pore sizes, from visible cracks and gaps to molecular cracks and gaps. Adsorption occurs when molecules of a liquid or gas are trapped by either an exterior or interior surface of a solid. The effect is analogous to iron filings being kept in place by a magnet. Activated carbon has a large internal surface area of up to 1,500 m^2g^{-1} and is thus an excellent adsorption material. Activated carbon may be made from a wide range of basic materials that possess a high carbon content. The conversion of the raw material into the final adsorbent may be split into two types of processes: chemical and thermal processes, both of which involve the use of high temperatures. The pore structure of activated carbon varies, and this is primarily due to the source material and manufacturing process. Adsorption is enabled by the pore structure in conjunction with attraction forces. The internal surface area of activated carbons is typically higher than $400 \text{ m}^2\text{g}^{-1}$ as determined by the nitrogen BET technique, and the volume of the pores is generally greater than 0.2 mL/g (Budi et al., 2015).

Porous carbon's uses are determined by two key properties: specific surface area and pore structure [Rodrı˘guez-Reinoso and Molina-Sabio, 1998; Figueiredo et al., 1999]. Micropores (less than 2 nm), mesopores (2–50 nm), and macrospores (more than 50 nm) are the three types of pores recognized by the International Union of Pure and Applied Chemistry (IUPAC) [Guo and Lua, 2000; He et al., 2010]. Although a microporous AC is preferred for adsorption, mesopores are useful for the adsorption of big molecules or when a quicker adsorption rate is needed [Huang et al., 2011; Liu et al., 2010]. The pore structure might be controlled by many circumstances and pathways, such as the raw material, activation duration and temperature, and kinds of activation agents [Biniak et al., 1997].

According to the IUPAC (International Union of Pure and Applied Chemistry), three groups of pores are distinguished, according to the pore size:

Macropores: (> 50 nm diameter)
Mesopores: (2–50 nm diameter)
Micropores: (< 2 nm diameter)

Micropores make up a significant portion of the interior surface area. Macro- and mesopores are important for kinetics because they act as highways into the carbon particle [Oshida et al., 1995].

The physical feature of the pore volume is also used to determine the effectiveness of activated carbon. It is the term used to describe the space contained inside a particle of activated carbon powder. The greater the pore capacity of activated carbon, the greater is the efficacy of the material. In contrast, if the size of the pores is incompatible with the size of the molecules to be absorbed, certain pore volumes will not be used [László and Szucs, 2001].

Pore radius is the diameter of the pores, which is generally measured in angstroms and varies depending on the kind of carbon used since the pore radius of activated carbon powder will be different from the pore radius of a grain [László and Szucs, 2001].

13.5.2 HARDNESS/ABRASION

The external integrity against wear along the exterior and fracture of tiny points of activated carbon is measured by the hardness number (DSTM 20). It's measured as a percentage of granule loss on a certain sieve after shaking them under specific circumstances. The abrasion number (AWWA B604) is a measurement of granular activated carbon's structural strength. It is a measure of a particle's capacity to withstand shear pressures generated by particles rubbing against each other or against another surface such as a column wall or supporting screen (Laine et al., 1989; Li et al., 2009; Boopathy and Karthikeyan, 2013; Budi et al., 2015) Table 13.2.

It is a measure of a particle's capacity to withstand shear pressures generated by particles rubbing against each other or against another surface such as a column wall or supporting screen. It is calculated as a percentage decrease in mean particle diameter by shaking granules with steel balls in a container under certain circumstances. Selection is also influenced by hardness/abrasion. Many applications will necessitate activated carbon with excellent particle strength and abrasion resistance (the breakdown of material into fines). Coconut shell-activated carbon has the highest toughness of all activated carbons [Otowa and Nojima, 1997].

13.5.3 ADSORPTIVE PROPERTIES

Adsorption is the process of a solid surface concentrating fluid molecules by physical forces whereas absorption is a process whereby fluid molecules are taken up by a liquid and distributed

TABLE 13.2
Shows the Parameter of Coconut with Its Value (Greenbank et al., 1995 ; Otowa and Nojima, 1997; Shafeeyan and Daud, 2022)

Parameter	Value
Iodine number	Min. 900 mg.g^{-1}
Specific mass	Min. 0.45–0.55 ± 0.05 g.cm^{-3}
Hardness	Min. 95%
Abrasion	Min. 85%
Ash	Max. 10%
Grain size	12 × 40 mesh (0.42–1.40 mm)
Humidity when packing	Max. 3%

throughout that liquid. The London dispersion force, a kind of van der Waals force originating from intermolecular attraction, is the elemental force that causes physical adsorption on activated carbon. Carbon and the adsorbate are thus chemically unchanged in the event of adsorption. Chemisorption, on the other hand, is a process in which molecules chemically react with the carbon's surface and are bound together by chemical bonds that are considerably stronger than London dispersion forces. The absorptive qualities of activated carbon include various features such as adsorptive capacity, adsorption rate, and overall efficiency of activated carbon. These characteristics can be determined by a variety of parameters, including the iodine number, surface area, and carbon tetrachloride activity, depending on the application (liquid or gas) [Otowa and Nojima, 1997].

13.5.4 APPARENT DENSITY

The apparent density of activated carbon, also known as bulk density or volumetric density, is defined as the mass of numerous activated carbon particles divided by the total volume they occupy. Particle volume, inter-particle void volume, and internal pore volume all contribute to the overall volume. Although apparent density has no bearing on adsorption per unit weight, it does have an impact on adsorption per unit volume [Otowa and Nojima, 1997] Table 13.3.

13.5.5 MOISTURE

The moisture content of activated carbon (ASTM D2867) is determined by weighing it after it has been heated to 150°C and dried to a consistent weight usually after 3 hours. During shipping and storage, the moisture content of packed activated carbon will generally rise. The physical moisture content of activated carbon should be between 3% and 6% under ideal conditions [El-Sayed and Bandosz, 2004].

13.5.6 ASH CONTENT

Total ash is a weight-based measurement of the mineral oxide concentration of activated carbon. At 800°C, the mineral components are converted to their respective oxides, which are then measured. The proportion of silica and aluminum in the ash depends on the foundation raw material used to manufacture the product. Coconut shell-based activated carbon has a W/W of 2-3%, wood-based activated carbon has a W/W of 5%, and coal-based activated carbon has a W/W of 8-15%. Total ash analysis can also be used to determine the quality of wasted carbon in groundwater sanitation or drinking water applications. The presence of sand or calcium, aluminum, manganese,

TABLE 13.3
Shows the Surface Properties of GAC (Zaini and Kamaruddin, 2013; Haimour and Emeish, 2006, Mopoung *et al.* 2015; Vanderborght and Van, 1977)

Item name	Granulated Activated Carbon
Base	Coconut shell
Apparent density (Kg/m^3)	480–490
Bulk density (Kg/m^3)	770
BET surface area (m^2/g)	1,000
Particle porosity	0.5
Bed porosity	0.4
Ash content (%)	5 (max)

or iron deposition on the activated carbon might cause a high ash concentration. The inactive, amorphous, inorganic, and useless component of activated carbon is measured by its ash content. The amount of ash in the final product should be as minimal as feasible. As the ash level lowers, the quality of the activated carbon improves [Moreno-Castilla *et al.*, 1998].

13.5.7 pH VALUE

The pH of activated carbon is a measurement of how acidic or basic it is. Activated carbon made from coconut shells is usually indicated for a pH range of 9 to 11. The pH of activated carbon is affected by the activation technique and the reagent employed. In the liquid phase, pH has a considerable impact on adsorption processes. Surface chemistry and surface charge are both influenced by pH value. For example, if the surface is positively charged at low pH, the performance for adsorbing cations will be poor due to electrostatic repulsion. When activated carbon is introduced to liquid, the pH value is frequently tested to estimate prospective changes [Otowa and Nojima, 1997].

13.5.8 PARTICLE SIZE

The adsorption kinetics, flow properties, and filterability of activated carbon are all affected by particle size. The greater the access to the surface area and the faster the rate of adsorption kinetics, the finer the particle size of activated carbon. This must be balanced against pressure drop in vapor phase systems, which affects energy costs. Particle size distribution should be carefully considered since it can bring substantial operational benefits [Budi *et al.*, 2015].

The size range of granular activated carbon (GAC) is often stated in terms of the sieve diameters, either in millimeters or U.S. meshes, that retain the majority of the GAC. It is determined by shaking a granular activated carbon sample through a prescribed set of sieves. The effective size of granular activated carbon is defined as the diameter, given in millimeters, of granules that are 10% by weight smaller. It is a parameter that indicates the pressure drop and filtration efficiency of backwashed and segregated carbon beds. A carbon with a smaller effective size will have a greater pressure drop and filter smaller particles, resulting in greater backwash frequency as compared to carbons with a larger effective size. The mean particle diameter of granular activated carbon is defined as the size, in millimeters, of granules that are 50% by weight smaller [Greenbank *et al.*, 2022]. Typically, the diameter of the pores in the dies of the extruder defines the size of extruded activated carbons or pellets. The size range of powdered or crushed activated carbon is typically defined by the quantity of powdered carbon, given in percent weight, that passes through a 0.075 mm sieve opening throughout a wet screen analysis. Usually, at least 90% of the material passes through 0.075 mm (Wu and Guo, 2013).

13.5.9 Iodine number

The usable surface area in m^2/g of raw carbon is indicated by the iodine number (or "iodine value") (ASTM D4607). Even though the Iodine number has become associated with activated carbon's activity and is commonly employed as quality control (QC) metric in the manufacture and re-activation of activated carbon, it does not always give a measure of the carbon's capacity to adsorb other species [Haimour and Emeish, 2006].

One of the methods for determining the adsorption capacity of activated carbons is to calculate the iodine number. The adsorption of iodine from a solution is used to determine the micropore concentration of activated carbon. The normal range is 500–1,200 mg/g, which corresponds to a carbon surface area of 900–1,100 m^2 per gram. Only the surface area of the few activated carbons affects the iodine number. Iodine adsorption capacity is not related to the adsorption capacity of other molecules, even though the iodine number is proportionate to the surface area of any activated carbon [Mopoung et al., 2015].

13.5.10 Pore, porosity, pore volume, pore diameter

A pore is a tiny hole in the surface of the AC that leads to the interior in a tortuous way. Also refers to the small hole or space that permits liquid to pass through. The void fraction, also known as porosity, is a measurement of the empty spaces in AC. It is a ratio of the volume of voids to the entire volume, ranging from 0 to 1, or a percentage between 0% and 100%. ACs are porous materials that play an important role in a variety of processes. Surface area and pore volume are two physical characteristics that characterize AC. Such physical characteristics are critical in the development of these materials since they will have a direct impact on the material's performance in its application. The technique most often employed for pore volume measurement also employs nitrogen adsorption isotherm data. The total pore volume is derived from the nitrogen adsorption isotherms using the Dubinin-Radushkevich equation, and the micropore volume 13 is obtained using the nitrogen adsorption isotherms. The AC unit is mL/g. The pore diameters of activated charcoal may be classified as micropores (width 2 nm), mesopores (width = 2–50 nm), or macropores (width > 50 nm) depending on the technique of manufacture; the differences in the size of their width openings reflect the pore distance [Pelekaniand Snoeyink, 1999].

13.5.11 Surface area

The Brunauer Emmett-Teller (BET) technique, which uses nitrogen adsorption at various pressures at liquid nitrogen temperature, is commonly used to determine the surface areas of activated carbons. The m^2/g unit of the AC area is used (Otowa et al., 1997). The surface area of activated carbon powder is the most important characteristic. On the whole, more surface area indicates greater efficacy of the carbon form. According to the manufacturer, the surface area of activated carbon may range between 500 and 1,500 m^2/g, and one teaspoon of activated carbon has the surface area of a soccer field. Higher temperatures burn holes in carbonized raw materials, creating surface area, during the process of activation, which is where the surface area is formed. This results in a large number of holes and pores in the carbon matrix, which is then utilized to construct a porous system using phosphoric acid [Moreno-Castilla et al., 1998] Table 13.4.

13.6 ISSUES AND CHALLENGES IN MICROWAVE-ASSISTED ACTIVATED CARBON

There are significant issues concerning the microwave principles that have been figured out. As of its prolonged processing time, energy-intensive, and therefore high costs, the usage of microwaves to assist the preparation of activated carbon has been presented as an alternative to conventional

TABLE 13.4

Shows the Pore Structural Parameter of High Surface Area Activated Carbon (HSAAC) Versus Char [Vanderborght and Grieken, 1977; Zaini and Kamaruddin, 2013; Nüchter et al., 2004; Xia et al., 2015]

Properties	AC	Char
BET surface are (m^2/g)	2,768	61
Micropore surface area (m^2/g)	2,230	0
External surface area (m^2/g)	538	61
Total Pore volume (mL/g)	1.717	0.109
Micropore volume (mL/g)	1.149	0
Mesopore volume (mL/g)	0.568	0.109
Average pore diameter (nm)	2.482	7.15

heating. Usually, microwaves can be selective heating, rapid and high-efficiency heat transfer, short treatment time, and compactness of equipment [Zaini and Kamaruddin, 2013].

13.6.1 TEMPERATURE MEASUREMENT SYSTEM

Most early studies of microwave-assisted activated carbon preparation used modified domestic microwave ovens with pulsed power supplies, which resulted in imprecise temperature control, inhomogeneous microwave fields (multimode), and safety precautions, necessitating the use of off-line temperature bulk measurement. In comparison to the equivalent conventionally heated tests, this has resulted in poor levels of response control and low-quality results. In the end, it has called into question many of the early research results [Zaini and Kamaruddin, 2013].

The bulk temperature monitoring of the carbonaceous precursors has been repeatedly identified as a key issue in the production of microwave-assisted activated carbon. Standard thermocouples have long been prohibited in microwave fields due to increased localized heating around the thermocouple tip, which causes misleading readings. Fiber-optic temperature sensors can be used to measure temperature in situ, but they must be handled with care because they are delicate, costly, and only work at temperatures below 300°C [Nüchter et al., 2004]. Meanwhile, as infrared only detects the temperature at the surface of a microwave heating system, it is not characteristic. The temperature difference between the bulk and the reactor surface is caused by the reactant's and reactor's dielectric characteristics, as well as the heating processes [Bogdal and Prociak, 2008].

The lack of accurate temperature monitoring is thought to be the major source of misunderstanding of the actual microwave activity in microwave-assisted systems. As a result, exact temperature measurement in microwave-assisted experiments is a major scientific issue, as the electric field produced can cause significant inaccuracies or damage to the temperature sensors [Zaini and Kamaruddin, 2013].

13.6.2 DIELECTRIC PROPERTY DATA

The dielectric property data of each material and combination used in microwave-assisted activation is critical. In microwave heating, a dielectric property of a material is typically expressed as dielectric loss, ε'' which is a measure of heat dissipation from the substance. Because a low-loss carbon ($\varepsilon'' < 0.01$) is inefficient at dispersing microwave energy as heat, a strong electric field is needed to ensure a rapid temperature rise. The microwave energy received at the outer layer of a high loss carbon ($\varepsilon'' > 5$) might result in a considerable temperature differential, leaving the interior section inadequately heated [Xia et al., 2015].

Furthermore, small-scale equipment provides little or no indication of how this technique will scale up with the chamber utilized; minimal issues will arise due to microwave dissipation, for example, carbonaceous precursors, temperature, density gradients, or reactor design. To estimate the heating rate, selectivity, heat distribution, and reactor design, a thorough understanding of the dielectric characteristics of reactants, including their sensitivity to frequency and temperature, as well as how these parameters vary during the activation, is required [Xia et al., 2015]. Before microwave-assisted research and scaling-up, dielectric characteristics of precursors, chemical agents, reactant combinations, and products (initial and final) used in the production of activated carbon should be measured [Zaini and Kamaruddin, 2013].

13.6.3 Temperature control and thermal runaway

Thermal runaway, in which the reactant temperature cannot be maintained and routinely deteriorates into arcing [Meredith, 1998], is the major problem in climbing up the microwave heating process. In microwave heating, the heat absorbed is proportional to the dielectric loss, ε''. Thermal runaway can occur if there is a rapid increase in ε'' with rising temperature [Meredith, 1998]. The dielectric loss of most carbon materials has been shown to increase with increasing temperature. Since the value of ''of the dielectric varies not only with frequency, but also with temperature, moisture content, physical state (solid or liquid), and composition [Zaini and Kamaruddin, 2013], thermal runaway is very likely to occur in microwave-assisted activation. This becomes even more critical if the reactant is heterogeneous (i.e., impregnated with chemical agents).

The majority of microwave-assisted studies have been carried out on a limited scale and in batches, using either conventional home microwave ovens or "off-the-shelf equipment" with a sealed chamber as the reactor. Both of these systems do not allow for accurate temperature control. Due to its size and construction, this sort of equipment can provide little or no feedback on response isotherms or the influence of microwave penetration depth [Zaini and Kamaruddin, 2013].

Other than the empirical observation that the activation endpoint is reached in a significantly shorter period than in the conventional, thermal activation case, there has been little true development in understanding the effect that microwave energy brings to the microwave-assisted activated carbon preparation.

13.7 APPLICATION OF ACTIVATED CARBON

Activated carbon is a very flexible material that may be utilized in a wide range of applications due to its excellent adsorption characteristics. AC's large surface area of particles, as well as its adsorptive capabilities, make it an important component in a variety of industries. AC is used in industries such as petroleum, fertilizer plants, nuclear power, pharmaceuticals, cosmetics, textiles, automotive manufacture, and vacuum manufacturing. AC has been discovered to be a porous substance that is highly efficient in adsorbing solutes from aqueous solutions. This was attributed to the presence of a significant specific surface area. It has also been widely utilized for energy storage material, solvent recovery, gas separation, dye removal from industrial effluent, and as a catalyst in the biodiesel manufacturing process [Bagheri et al., 2015].

Activated carbon has a wide range of applications. Food processing, petroleum medicines, and the motor sector all utilize it. This is owing to the large surface area and large pore diameters of the absorptive substance. Recent research has encouraged the use of activated carbon for energy storage systems in batteries, supercapacitors, fuel cells, and solar cells, in addition to the aforementioned use. Microwave absorbers, CO_2 adsorbents, H_2 storage, and wastewater treatment are all applications [El-Sayed and Bandosz, 2004].

Activated carbon's qualities may be adjusted to suit the application, ranging from high porosity to high conductivity. Furthermore, the extraction of activated carbon from biomass results in the doping of heteroatoms mostly with nitrogen and oxygen functions. These qualities improve the carbon matrix's wettability and conductivity. The presence of heteroatoms in activated carbon allows it to be used as electrode material in energy storage systems as well as a catalyst. In addition to the above-mentioned characteristics, choosing the right biomass for a certain application gives the necessary template for achieving the desired shape [Bagheri et al., 2015].

13.7.1 Energy storage

Because of its very large surface area and excellent purity, which may be directly connected to performance in this application, activated carbon is the ideal electrode material in a supercapacitor, asymmetric batteries, and a variety of advanced batteries. Biomass from sustainable sources has been utilized to make lightweight activated carbons with a high specific surface area for a variety of applications, including components in energy storage devices like batteries and supercapacitors. While direct conversion from biomass has received a lot of attention, AC with a large surface area can be generated more efficiently as a co-product from a biorefinery [Sevilla and Mokaya, 2014]. Porous carbons have numerous beneficial properties concerning their use in energy storage applications that require confined space such as in electrode materials for supercapacitors and as solid-state hydrogen stores [Sevilla and Mokaya, 2014].

Due to the availability of carbon-based materials, their chemical and thermal stability, processability, and the ability to tailor their textural and structural features to particular applications, carbon-based materials have gained substantial attention in many energy-related fields. Activated carbons, in particular, are notable for their enormous surface area and pore volume, and inexpensive cost. Since they have a long history as adsorbents for the removal of pollutants from gases and liquids, they have a long history. Despite recent advances in controlling carbon porosity through templating procedures and the development of novel materials, such as graphene or carbon nanotubes, activated carbons remain a primary choice for commercial supercapacitors due to their availability, cost, and simpler production methods. Medium- to long-term stability is anticipated, but more control over the textural qualities (especially pore size) is needed to optimize the energy and power densities to meet future energy requirements. This is the case. Precursors like biomass, which is inexpensive, widely accessible, and renewable, are anticipated to play a crucial role in reducing costs [Sevilla and Mokaya, 2014].

Supercapacitors that employ activated carbons promise to increase the amount of energy stored while maintaining the same degree of cycle stability and power capabilities as EDLCs. The full spectrum of performance between traditional capacitors and electrochemical batteries will be covered by these innovative configurations. In addition, since aqueous electrolytes may be used, these batteries and EDLCs are more cost-effective and ecologically benign. Materials science, particularly the manufacturing of porous carbon, as well as electrochemistry, is still in its infancy and will benefit from additional advances in these technologies. Developing porous materials that can store enough hydrogen to meet onboard application requirements (namely storage temperature and pressure) is still a difficulty. Activated carbons are one of the most investigated materials in this field of study. However, activated carbons' ability to store hydrogen at ambient temperatures is currently significantly below onboard objectives. Textural qualities (surface area and pore size) may increase hydrogen storage at room temperature, although breakthroughs may not be feasible only via textural modifications. For hydrogen storage at ambient temperature, the doping of activated carbons with metal nanoparticles that might cause spillover is a viable technique for improving hydrogen interaction strength with the sorbent. To fully grasp the overflow in hydrogen storage methods, a further basic study is required [Sevilla and Mokaya, 2014] Figure 13.8.

13.7.2 Metals recovery

Activated carbon is a useful tool in the improvement of precious metals such as gold and silver. High-quality activated carbon is crucial to the gold mining process. Using activated carbon of

FIGURE 13.8 Shows the precursor, activated carbon and its application in energy storage field [Xia *et al.*, 2015; Meredith, 1998; Sevilla and Mokaya, 2014].

optimum hardness reduces fines and recovers more gold. Even to generate a few grams of gold, thousands of tons of ore should be mined, crushed, roasted, ground, treated, pulped, and leached. Activated carbon performs a crucial role in securing the gold from its ore. In any gold mine, any gold that is not adsorbed by activated carbon will be lost. To accomplish the maximum possible gold yield, the choice of a suitable activated carbon product is vital. With the focus on quality, minimal fines formation, reliability, and services it can make a difference when it comes to increasing gold recovery in mines and in the gold circuits. A key use of activated carbon in mining is in gold recovery, where granular activated carbon is utilized for adsorption of the gold cyanide complex in carbon-in-pulp (CIP) and carbon-in-leach (CIL) systems, or in carbon-in-column (CIC) systems after a heap leach operation carbon [Han *et al.*, 2000].

13.7.3 Food and beverage

Activated carbon is generally utilized in the food and beverage industry to achieve many goals. This involves decaffeination, and removal of undesirable elements such as odor, taste, color, and more. Food and Beverage manufacturers require very high purity in the water and many times the water supply from local sources needs additional treatment. Activated carbons acquire a diverse range of pore properties that are ideal for the adsorption of numerous chemical compounds. Adsorption is the initial process by which activated carbon works and the primary purpose is commonly used to reduce undesirable taste, odor, and color and to improve the safety of beverage products [Cardona *et al.*, 2012].

13.7.4 Medicinal

Activated carbon can be utilized to cure a variety of ailments and poisonings. Activated carbon is used to treat poisonings and overdoses following oral ingestion. Tablets or capsules of activated carbon are treated in many countries as an over-the-counter drug to treat diarrhea, indigestion, and flatulence. Activated carbon is a strong tool for emergency cleansing of the gastrointestinal tract, perhaps the most effective remedy known these days. It can be applied in cases of poisoning from virtually any toxic substance. AC reduces the absorption of poisonous substances by up to 60%. It adsorbs and improves and eliminates toxins, heavy metals, chemicals, pharmaceutical drugs, morphine, and pesticides from the body. Activated carbons can be also used to whiten teeth [Wu *et al.*, 2013].

13.7.5 Air emission purification

Controlling air emissions, whether organic or inorganic, is becoming an increasingly difficult task in today's industrialized world. It can include components that emit a strong odor or that exceed the manufacturer's operating permit's volatile organic compound (VOC) emission restriction. These emissions can come from a waste storage building, a waste treatment center, a water treatment plant, venting storage tanks or reactors, manufacturing chemical intermediates and end products like polyester, and manufacturing furniture, windows and doors, paint, and chemical/petrochemical products, to name a few examples. The number of components available is limitless, as are the applications. Activated carbon-based systems can mitigate the negative impacts of air pollution [Brady et al., 1996].

13.7.6 Biogas purification

Biogasesneed to be processed before they may be converted into electric power, renewable heat, or biomethane. This eliminates difficulties in the gas engine while also enabling gas quality of more than 99% CH_4 to be attained, making the gas appropriate for feeding into the natural gas grid. H_2S and siloxanes from landfill gas, for example, or H_2S and terpenes from fermentation gas, are examples of components that must be removed. Mobile activated carbon filters are utilized to meet all of these problems [Abatzoglou and Boivin, 2009].

13.7.7 Remediation

Wherever there is or has been industrial activity, soil and/or groundwater may be polluted. It frequently involves historical pollution that predates any consideration of enforced environmental regulation. These polluted sites are treated in stages, allowing them to be reused or repurposed. This frequently necessitates the use of advanced remedial approaches. Mobile activated carbon filters are well-suited for this, and they are commonly used in the cleanup process. Activated carbon filters have been created to clean up historical pollution as well as to clean up disasters on the spot, such as an overturned fuel oil tanker, a leaky petroleum pipeline, or purifying fire water after a blaze on an industrial site [Janssen and Beckingham, 2013].

13.7.8 Chemicals (purification with mobile activated carbon filters)

Plants turn raw materials into finished goods. During the production process, certain reactions might occur, resulting in unwanted contamination of the finished product. Certain goods, on the other hand, may fail to fulfill the manufacturing process's criteria. During the production process, malfunctions can occur, causing a batch of items to fail to satisfy specifications. Furthermore, having a particular level of purity, which may be obtained with activated carbon, increases the market value of some items [Yue et al., 2001].

13.7.9 Wastewater (purification with activated carbon)

When we wish to make something, we frequently need water. During the manufacturing process, this water will get polluted, resulting in the effluent that cannot be easily released back into the environment. This wastewater will then need to be treated before it can be released into a sewer or surface water, or even re-used. On the other hand, it's possible that a factory's water supply has to be treated before it can be utilized in the manufacturing process. Activated carbon is frequently the best technique for treating water at the source, in between phases of production, and at the end of the pipe for example, after upstream biological or physicochemical treatment. The activated carbon technique can even cleanse polluted firewater. AC technology provides a feasible answer to all of these problems (Sumon et al., 2020).

13.8 SUMMARY

This chapter demonstrates a detailed review of microwave-assisted activated carbon. Microwave heating has several pros over conventional heating, such as uniform and internal heating, fast and selective heating, simplicity of control, straightforward set-up, insensitive to particle size and shape, and fewer pre-treatment conditions of the feedstocks. MW technology is a favorable technique for the activation of ACs owing to its unique characteristics. Evaluation of characteristics of ACs prepared from microwave, properties, issues, and challenges, and application of activated carbon has been discussed in this chapter.

MW heating is a stable technology. However, MW-assisted preparation of ACs is still far from development although some experimental and establishment plants are operating in various countries. Small-scale reactors deliver little knowledge about how this technology would scale up and how large-scale MW equipment operate. Dielectric properties measurement on precursors, chemical reagents, reactant mixtures, and products (final and intermediate) engaged in manufacturing ACs is vital for MW management and scale-up. A permanent and dependable AC preparation system under MW irradiation justifies further analyses and energy consumption for producing ACs under MW irradiation should be further examined. Activated carbon is utilized in a wide variety of industrial and domestic applications, including purifying drinking water, ground and municipal water treatment, reducing power station and landfill gas emissions, storing energy, and recovering precious metals. Air purification treatments include the removal of volatile organic compounds (VOCs) and the management of odors. Considered the most effective adsorbent on the planet, activated carbon is used in a variety of applications, from ordinary water pitcher filters to gold recovery in carbon in pulp/carbon in leach systems. Research and development, increased demand for high-performance activated carbons, and the increasing number of strategic partnerships between industry players are all expected to drive the market's expansion over time. As a result, the number of point-of-use carbon devices in people's homes is on the increase as people become more concerned about the quality of their drinking water. Because of its important function in the purification, treatment, and decolorization of liquids, activated carbon is expected to be in very high demand in the pharmaceutical sector as well. There are few published documents and data in these fields, and more research and process development are still necessary.

REFERENCES

Abatzoglou, N., & Boivin, S. (2009). A review of biogas purification processes. *Biofuels, Bioproducts and Biorefining*, 3(1), 42–71. 10.1002/BBB.117

Adekola, F., & Adegoke, H. (2005). Adsorption of blue-dye on activated carbons produced from rice husk, coconut shell and coconut coirpith. *Ife Journal of Science*, 7(1). 151–157 10.4314/ijs.v7i1.32169

Ahmedna, M., Marshall, W. E., & Rao, R. M. (2000). Production of granular activated carbons from select agricultural by-products and evaluation of their physical, chemical and adsorption properties. *Biosource Technology*, 71, 113–123.

Ahmedna, M., Marshall, W. E., Rao, R. M., & Clarke, S. J. (1997). Use offiltration and buffers inraw sugar color measurement. *Journal of the Science of Food and Agriculture* 75(1), 109–116. https://dialnet.unirioja.es/servlet/articulo?codigo=439821

Alslaibi, T. M., Abustan, I., Ahmad, M. A., & Foul, A. A. (2013). A review: Production of activated carbon from agricultural byproducts via conventional and microwave heating. *Journal of Chemical Technology and Biotechnology*, 88(7), 1183–1190. 10.1002/JCTB.4028

Ania, C. O., Parra, J. B., Menéndez, J. A., & Pis, J. J. (2005). Effect of microwave and conventional regeneration on the microporous and mesoporous network and on the adsorptive capacity of activated carbons. *Microporous and Mesoporous Materials*, 85, 7–15.

Atwater, J. E., & Wheeler, R. R. (2004). Microwave permittivity and dielectric relaxation of a high surface area activated carbon. *Applied Physics A: Materials Science and Processing*, 79(1), 125–129. 10.1007/S00339-003-2329-8

Bae, W., Kim, J., & Chung, J. (2014). Production of granular activated carbon from food-processing wastes (walnut shells and jujube seeds) and its adsorptive properties. *J. Air Waste Manage*, 64(8), 879–886.

Bagheri, S., Muhd Julkapli, N., & Bee Abd Hamid, S. (2015). Functionalized activated carbon derived from biomass for photocatalysis applications perspective. *Int. J. Photoenergy*, 2015. DOI: 10.1155/2015/218743.

Bahiraei, A., & Behin, J. (2021). Effect of citric acid and sodium chloride on characteristics of sunflower seed shell-derived activated carbon. *Chemical Engineering and Technology*, 44(9), 1604–1617. 10.1002/CEAT.202100117.

Benjamin, S. L., Mark, K. E., Zaretzky, P., Ferguson, N. N., Dichiara, A., & Reginald, R. E. (2018). Construction of a carbon nanomaterial- based nanocomposite aerogel for the removal of organic compounds from water. *ACS Appl. Nano. Mater.*, 2018(1), 4127–4134.

Biniak, S., Szymański, G., Siedlewski, J., & Światkoski, A. (1997). The characterization of activated carbons with oxygen and nitrogen surface groups. *Carbon*, 35(12), 1799–1810. 10.1016/S0008-6223(97)00096-1

Błażewicz, S., Świątkowski, A., & Trznadel, B. J. (1999). The influence of heat treatment on activated carbon structure and porosity. *Carbon*, 37(4), 693–700. 10.1016/S0008-6223(98)00246-2

Bogdał, D., & Prociak, A. (2008). Microwave-enhanced polymer chemistry and technology. *Microwave-Enhanced Polymer Chemistry and Technology*, 1–275. 10.1002/9780470390276

Bonvin, F., Jost, L., Randin, L., Bonvin, E., & Kohn, T. (2016). Super-fine powdered activated carbon (SPAC) for efficient removal of micropollutants from wastewater treatment plant effluent. *Water Research*, 90, 90–99. 10.1016/J.WATRES.2015.12.001

Boopathy, R., & Karthikeyan, S. (2013). Adsorption of ammonium ion by coconut shell-activated carbon from aqueous solution: kinetic, isotherm, and thermodynamic studies. *Env. Sci Pollut Res*, 20, 533–542.

Brady, T. A., Rostam-Abadi, M., & Rood, M. J. (1996). Applications for activated carbons from waste tires: natural gas storage and air pollution control. *Gas Separation & Purification*, 10(2), 97–102. 10.1016/0950-4214(96)00007-2

Brunauer, S., Emmett, P. H., & Teller, E. (1938). Adsorption of gases in multi molecular layers. *Journal of the American Chemical Society*, 60, 309–319.

Bubanale, S., & Shivashankar, M. (2017). History, method of production, structure and applications of activated carbon. *International Journal of Engineering Research And*, V6(06). 10.17577/IJERTV6IS060277

Budi, E., Nasbey, H., Yuniarti, B. D. P., Nurmayatri, Y., Fahdiana, J., & Budi, A. S. (2015). Pore structure of the activated coconut shell charcoal carbon. *AIP Conference Proceedings*, 1617(1), 130. 10.1063/1.4897121

Cardona, E. D., Del Pilar Noriega, M., & Sierra, J. D. (2012). Oxygen scavengers impregnated in porous activated carbon matrix for food and beverage packaging applications. *Journal of Plastic Film and Sheeting*, 28(1), 63–78. 10.1177/8756087911427730

Chih-Ju, J. G. (1998). Application of activated carbon in a microwave radiation field to treat trichloroethylene. *Carbon*, 36, 1643–1648.

Chuayjumnong, S., Karrila, S., Jumrat, S., & Advances, Y. P.-R. (2020). Activated carbon and palm oil fuel ash as microwave absorbers for microwave-assisted pyrolysis of oil palm shell waste. *Pubs.Rsc.Org*. https://pubs.rsc.org/en/content/articlehtml/2020/ra/d0ra04966b.

Dabrowski, A., Podkościelny, P., Hubicki, Z., & Barczak, M. (2005). Adsorption of phenolic compounds by activated carbon: A Critical Review. *Chemosphere*, 58(8), 1049–1070.

Demiral, H., Demiral, I., Tumsek, F., & Karabacakoglu, B. (2008). Pore structure of activatedcarbon prepared from hazelnut bagasse by chemical activation. *Surface and Inter-face Analysis*, 40, 616–619.

Demiral, H., & Demiral, I. (2008). Surface properties of activated carbon prepared from wastes. *Surface and Interface Analysis*, 40(3–4), 612–615. 10.1002/SIA.2716

Deng, H., Li, G., Yang, H., & Tang, J. (2010). Preparation of activated carbons from cotton stalk by microwave assisted KOH and K$_2$CO$_3$ activation. *Chemical Engineering Journal*, 163, 373–381.

Deng, H., Zhang, G., Xu, X., Tao, G., & Dai, J. (2010). Optimization of preparation of activated carbon from cotton stalk by microwave assisted phosphoric acid-chemical activation. *Journal of Hazardous Materials*, 182, 217–224.

Deryło-Marczewska, A., Goworek, J., Świątkowski, A., & Buczek, B. (2004). Influence of differences in porous structure within granules of activated carbon on adsorption of aromatics from aqueous solutions. *Carbon*, 42(2), 301–306. 10.1016/J.CARBON.2003.10.031

Ding, Y., Wang, T., Dong, D., & Zhang, Y. (2020). Using biochar and coal as the electrode material for supercapacitor applications. *Front. Energy Res.*, 7, 159. doi: 10.3389/FENRG.2019.00159/BIBTEX.

Egwaikhide, P. A., Akporhonor, E. E., & Okieimen, F. E. (2007). Utilization of coconut fibre carbon in the removal of soluble petroleum fraction polluted water. *International Journal of Physical Sciences*, 2(2), 47–49. 10.5897/IJPS.9000615

El-Sayed, Y., & Bandosz, T. J. (2004). Adsorption of valeric acid from aqueous solution onto activated carbons: role of surface basic sites. *Journal of Colloid and Interface Science*, 273(1), 64–72. 10.1016/ J.JCIS.2003.10.006

Elsheikh, A., Newman, A., Al-Daffaee, H., Phull, S., & Crosswell, N. (2003). Characterization of activated carbon prepared from a single cultivar of Jordanian olive stones by chemical and physicochemical techniques. *J. Anal. Appl. Pyrolysis*, 30, 1–16.

Figueiredo, J. L., Pereira, M. F. R., Freitas, M. M. A., & Órfão, J. J. M. (1999). Modification of the surface chemistry of activated carbons, *Carbon*, 37, 1379–1389.

Foo, K. Y., & Hameed, B. H. (2011). Preparation of activated carbon from date stones by microwave-induced chemical activation: Application for methylene blue adsorption. *Chemical Engineering Journal*, 170, 338–341.

Foo, K. Y., Hameed, B. H. (2011). Microwave-assisted preparation of oil palm fiber activated carbon for methylene blue adsorption. *Chemical Engineering Journal*, 166, 792–795.

Ghaedi, M., Biyareh, M. N., Kokhdan, S. N., Shamsaldini, S., Sahraei, R., Daneshfar, A., & Shahriyar, S. (2012). Comparison of the efficiency of palladium and silver nanoparticlesloaded on activated carbon and zinc oxide nanorods loaded on activated carbonas new adsorbents for removal of Congo red from aqueous solution: kinetic andisotherm study. *Materials Science and Engineering*, 32, 725–734.

Ghaedi, M., Hekmati Jah, A., Khodadoust, S., Sahraei, R., Daneshfar, A., Mihandoost, A., & Purkait, M. K. (2012). Cadmium telluride nanoparticles loaded on activated carbon as adsorbent for removal of sunset yellow. *Spectrochimica Acta Part A: Molecular and Biomolecular Spectroscopy*, 90, 22–27. 10.1016/ J.SAA.2011.12.064

Gopiraman, M., Deng, D., Kim, B. S., Chung, I. M., & Kim, I. S. (2017). Three-dimensional cheese-like carbon nanoarchitecture with tremendous surface area and pore construction derived from corn as superior electrode materials for supercapacitors. *Applied Surface Science*, 409, 52–59. 10.1016/J.APSUSC. 2017.02.209

Greenbank, M., Spotts, Steve. (1995). Effects of starting material on activated carbon characteristics and performance. *Industrial Water Treatment*. http://citeseerx.ist.psu.edu/viewdoc/download?doi=10.1. 1.592.7831&rep=rep1&type=pdf

Gregg, S. J., & Sing, K. S. W. Adsorption, surface area and porosity. 2. Auflage, Academic Press, London 1982. 303 Seiten, Preis: $ 49.50. *Berichte Der Bunsengesellschaft Für Physikalische Chemie*, 86(10), 957– 957. 10.1002/BBPC.19820861019

Guo, J., & Lua, A. C. (2000). Preparation of activated carbons from oil-palm-stone chars by microwave-induced carbon dioxide activation. *Carbon*, 38, 1985–1993.

Gupta, S., & Deep, G. (2017). Agricultural waste based-coco peat and coconut shell activated carbon microwave absorber. *2016 IEEE MTTS International Microwave and RF Conference*, IMaRC 2016 – Proceedings. 10.1109/IMARC.2016.7939621

Gustafsson, Å., Hale, S., Cornelissen, G., Sjöholm, E., & Gunnarsson, J. S. (2017). Activated carbon from kraft lignin: A sorbent for in situ remediation of contaminated sediments. *Environ. Technol. Innov.*, 7, 160–168, April 2017. doi: 10.1016/J.ETI.2016.11.001.

Haimour, N. M., & Emeish, S. (2006). Utilization of date stones for production of activated carbon using phosphoric acid. *Waste Management*, 26(6), 651–660. 10.1016/J.WASMAN.2005.08.004

Han, I., Schlautman, M. A., & Batchelor, B. (2000). Removal of hexavalent chromium from groundwater by granular activated carbon. *Water Environment Research*, 72(1), 29–39. 10.2175/106143000X137086

Hassan, M. F., Sabri, M. A., Fazal, H., Hafeez, A., Shezad, N., & Hussain, M. (2020). Recent trends in activated carbon fibers production from various precursors and applications—A comparative review. *J. Anal. Appl. Pyrolysis*, 145, 104715. DOI: 10.1016/J.JAAP.2019.104715.

He X., Geng Y., Qiu J., Zheng M., Long S., & Zhang X. (2010). Effect of activation time on the properties of activated carbons prepared by microwave-assisted activation for electric double-layer capacitors. *Carbon*, 48, 1662–1669.

Hernandez, N. M., & Villanueva, E. P. (1941). Production, purification and utilization of biogas as fuel for internal combustion engine. *AIP Conf. Proc.*, 985, 1166, 1941. doi: 10.1063/1.5028067.

Hidayu, A. R., Mohamad, N. F., Matali, S., & Sharifah, A. S. A. K. (2013). Characterization of activated carbon prepared from oil palm empty fruit bunch using BET and FT-IR techniques. *Procedia Engineering*, 68, 379–384. 10.1016/J.PROENG.2013.12.195

Hoseinzadeh Hess, R., Wan Daud, W. M. A., Sahu, J. N., & Arami-Niya, A. (2013). The effects of a microwave heating method on the production of activated carbon from agricultural waste: A review. *J. Anal. Appl. Pyrolysis*, 100, 1–11, Mar. DOI: 10.1016/J.JAAP.2012.12.019.

Hu, Z., & Srinivasan, M. P. (2001). Mesoporous high-surface-area activated carbon. *Microporous Mesoporous Mater*, 43, 267–275.

Huang L., Sun Y. Wang W., Yue Q., & Yang T. (2011). Comparative study on characterization of activated carbons prepared by microwave and conventional heating methods and application in removal of oxytetracycline (OTC). *Chemical Engineering Journal*, 171, 1446–1453.

Hui, T.M. Z.-C. (2015). Potassium hydroxide activation of activated carbon: A commentary. *Koreascience.or.Kr*, 16(4), 275–280. 10.5714/CL.2015.16.4.275.

Ioannidou, O., & Zabaniotou, A. (2007). Agricultural residues as precursors for activated carbon production: A review. *Renew. Sustain. Energy Rev.* 2007, 11(9), 1966–2005.

Janssen, E. M. L., & Beckingham, B. A. (2013). Biological responses to activated carbon amendments in sediment remediation. *Environmental Science and Technology*, 47(14), 7595–7607. 10.1021/ES4 01142E

Ji, Y., Li, T., Zhu, L., Wang, X., & Lin, Q. (2007). Preparation of activated carbons by microwave heating KOH activation. *Applied Surface Science*, 254, 506–512.

Jjagwe, J., Olupot, P. W., Menya, E., & Kalibbala, H. M. (2021). Synthesis and application of granular activated carbon from biomass waste materials for water treatment: A review. *Journal of Bioresources and Bioproducts*, 6(4), 292–322. 10.1016/J.JOBAB.2021.03.003

Jones, D. A., Lelyveld, T. P., Mavrofidis, S. D., Kingman, S. W., & Miles, N. J. (2002). Microwave heating applications in environmental engineering: a review. *Resources, Conservation and Recycling*, 34, 75–90.

Jüntgen, H. (1986). Activated carbon as catalyst support: A review of new research results. *Fuel*, 65(10), 1436–1446. 10.1016/0016-2361(86)90120-1

Kalderis, D., Bethanis, S., Paraskeva, P., Diamadopoulos, E. (2008). Production of activatedcarbon from bagasse and rice husk by a single stage chemical activation methodat low retention times. *Bioresource Technology*, 99((15) (2008)) 6809–6816.

Kar, K. K. (2020). Handbook of nanocomposite supercapacitor materials I. *Springer Series in Materials Science*, 300, 378. http://link.springer.com/10.1007/978-3-030-43009-2

Kazi, H. E. (1999). Microwave energy for mineral treatment processes: A brief review. *International Journal of Mineral Processing*, 57, 1–24.

Khalili, N. R., Campbell, M., Sandi, G., & Gola, J. (2000). Production of micro and mesoporous activated carbon from paper mill sludge, I: Effect of zinc chloride activation. *Carbon*, 38, 1905–1915.

Kubota, M., Hata, A., & Matsuda, H. (2009). Preparation of activated carbon from phenolic resin by KOH chemical activation under microwave heating. *Carbon*, 47(2009), 2805–2811.

Lafi, W. (2001). Production of activated carbon from acrons and olive seed biomass. *Biomass and Bioenergy*, 20, 57–62.

Laine, J., Calafat, A., & Labady, M. (1989). Preparation and characterization of activated carbons from coconut shell impregnated with phosphoric acid. *Carbon*, 27, 191–195.

Lamine, S. M., Ridha, C., Mahfoud, H. M., Mouad, C., Lotfi, B., & Al-Dujaili, A. H. (2014). Chemical activation of an activated carbon prepared from coffee residue. *Energy Procedia*, 50, 393–400. doi: 10.1016/J.EGYPRO.2014.06.047.

Lastoskie C., Gubbins K. E., & Quirke N. (1993). Pore size distribution analysis ofmicropororous carbons: A density functional theory approach. *The Journal of Physical Chemistry*, 97, 4786–4796.

László, K., & Szucs, A. (2001). Surface characterization of polyethylene terephthalate (PET) based activated carbon and the effect of pH on its adsorption capacity from aqueous phenol and 2,3,4-trichlorophenol solutions. *Carbon*, 39(13), 1945–1953. doi: 10.1016/S0008-6223(01)00005-7.

LeChevallier, M. W., Hassenauer, T. S., Camper, A. K., & McFeters, G. A. (1984). Disinfection of bacteria attached to granular activated carbon. *Applied and Environmental Microbiology*, 48(5), 918–923. 10.1128/AEM.48.5.918-923.1984.

Legrouri, K., Ezzine, M., Ichocho, S., Hannache, H., Denoyel, R., Pailler, R., & Naslain R. (2005). Production of activated carbon from a new precursor: Molasses. *J. Phys. IV France*, 123, 101–104.

Li, D., Yoshida, S., Siritanaratkul, B., Garcia-Esparza, A. T., Sokaras, D., Ogasawara, H., & Takanabe, K. (2021). Transient potassium peroxide species in highly selective oxidative coupling of methane over an unmolten K_2WO_4/SiO_2 catalyst revealed by in situ characterization. *ACS Catalysis*, 11(22), 14237–14248. 10.1021/ACSCATAL.1C04206.

Li, D., Zhang, Y., Quan, X., & Zhao, Y. (2009). Microwave thermal remediation of crude oil contaminated soil enhanced by carbon fiber. *Journal of Environmental Sciences*, 21, 1290–1295.

Li, W., Peng, J., Zhang, L., Yang, K., Xia, H., Zhang, S., & Guo, S. Hui. (2009). Preparation of activated carbon from coconut shell chars in pilot-scale microwave heating equipment at 60 kW. *Waste Management*, 29(2), 756–760. 10.1016/J.WASMAN.2008.03.004

Li, W., Zhang, L.-B., Peng, J.-H., Li, N., & Zhu, X.-Y. (2008). Preparation of high surface area activated carbons from tobacco stems with K_2CO_3 activation using microwave radiation. *Industrial Crops and Products*, 27(2008), 341–347.

Lidström, P., Tierney, J., Wathey, B., & Westman, J. (2001). Microwave assisted organic synthesis—a review. *Tetrahedron*, 57(45), 9225–9283. doi: 10.1016/S0040-4020(01)00906-1.

Lin, P., Zhang, Y., Zhang, X., Chen, C., Xie, Y., & Suffet, I. H. (2015). The influence of chlorinated aromatics' structure on their adsorption characteristics on activated carbon to tackle chemical spills in drinking water source. *Frontiers of Environmental Science and Engineering*, 9(1), 138–146. 10.1007/S11783-014-0725-2.

Liu, Q.-S., Zheng, T., Wang, P., & Guo, L. (2010). Preparation and characterization of activated carbon from bamboo by microwave-induced phosphoric acid activation. *Industrial Crops and Products*, 31, 233–238.

Lowell, S., Shields, J. E., Thomas, M. A., & Thommes, M. (2004). *Characterization of poroussolids and powders: Surface area, pore size and density*. Netherland: Kluwer Academic Publishers.

Lozano-Castelló, D., Cazorla-Amorós, D., & Linares-Solano, A. (2002). Powdered activated carbons and activated carbon fibers for methane storage: A comparative study. *Energy and Fuels*, 16(5), 1321–1328. 10.1021/EF020084S.

Manivannan, A., Chirila, M., Giles, N. C., & Seehra, M. S. (1999). Microstructure, dangling bonds and impurities in activated carbons. *Carbon*, 37(11), 1741–1747. 10.1016/S0008-6223(99)00052-4

Márquez, P., Benítez, A., Hidalgo-Carrillo, J., Urbano, F. J., Caballero, Siles, J. A., & Martín, M. A. (2021). Simple and eco-friendly thermal regeneration of granular activated carbon from the odour control system of a full-scale WWTP: Study of the process in oxidizing atmosphere. *Separation and Purification Technology*, 255, 117782. 10.1016/J.SEPPUR.2020.117782

Mcdougall, G. J. (1991). The physical nature and manufacture of activated carbon. *J. S. Afr. Inst. Min. Met.*, 91(4), 109–120.

Menéndez, J. A., Arenillas, A., Fidalgo, B., Fernández, Y., Zubizarreta, L., Calvo, E. G., & Bermúdez, J. M. (2010). Microwave heating processes involving carbon materials. *Fuel Processing Technology*, 91, 1–8.

Menéndez, J. A., Menéndez, E. M., Iglesias, M. J., García, A., & Pis, J. J. (1999). Modification of the surface chemistry of active carbons by means of microwave-induced treatments. *Carbon*, 37, 1115–1121.

Menezes, R. R., Souto, P. M., & Kiminami, R. H. G. A. (2007). Microwave hybrid fast sintering of porcelain bodies. *Journal of Materials Processing Technology*, 190, 223–229.

Menya, E., Olupot, P. W., Storz, H., Lubwama, M., & Kiros, Y. (2018). Production and performance of activated carbon from rice husks for removal of natural organic matter from water: A review. *Chemical Engineering Research and Design*, 129, 271–296. 10.1016/J.CHERD.2017.11.008

Meredith, R. (1998). *Engineers' Handbook of Industrial Microwave Heating*. 10.1049/PBPO025E

Mestre J., Pires J., Nogueira M. F., & Ania C. O. (2009). Waste derived activated carbons for re-moval of ibuprofen from solution: Role of surface chemistry and pore structure. *Bioresource Technology*, 100(5), 1720–1726.

Mohammed, M. A., Shitu, A., Tadda, M. A., & Ngabura, M. (2014). Utilization of various agricultural waste materials in the treatment of industrial wastewater containing heavy metals: A review. *Int. Res. J. Environ. Sci.* 2014, 3(3), 62–71.

Mohanty, K., Das, D., & Biswas, M. N. (2008). Treatment of phenolic wastewater in a novel multi-stage external loop airlift reactor using activated carbon. *Separation and Purification Technology*, 58(3), 311–319. 10.1016/J.SEPPUR.2007.05.005

Mopoung, S., Moonsri, P., Palas, W., & Khumpai, S. (2015). Characterization and properties of activated carbon prepared from tamarind seeds by KOH activation for Fe(III) adsorption from aqueous solution. *Scientific World Journal*, 2015. 10.1155/2015/415961

Moreno-Castilla, C., Carrasco-Marín, F., Maldonado-Hódar, F. J., & Rivera-Utrilla, J. (1998). Effects of non-oxidant and oxidant acid treatments on the surface properties of an activated carbon with very low ash content. *Carbon*, 36(1–2), 145–151. doi: 10.1016/S0008-6223(97)00171-1.

Najm, I. N., Snoeyink, V. L., Lykins, B. W., & Adams, J. Q. (1991). Using powdered activated carbon: A critical review. *Journal - American Water Works Association*, 83(1), 65–76. 10.1002/J.1551-8833.1991.TB07087.X.

Namane, A., Mekarzia, A., Benrachedi, K., Belhaneche-Bensemra, N., & Hellal, A. (2005). Determination of the adsorption capacity of activated carbon made from coffee grounds by chemical activation with ZnCl2 and H3PO4. *Journal of Hazardous Materials*, 119(1–3), 189–194. 10.1016/J.JHAZMAT.2004.12.006

Namazi, A. B., Allen, D. G., & Jia, C. Q. (2016). Benefits of microwave heating method in the production of activated carbon. *Can. J. Chem. Eng.*, 94(7), 1262–1268, Jul. 2016. DOI: 10.1002/CJCE.22521.

Nayebi, M., & Ruhe, G. (2015). Analytical product release planning. *The Art and Science of Analyzing Software Data*, 555–589. 10.1016/B978-0-12-411519-4.00019-7

Nieto-Delgado, C., & Rangel-Mendez, J. R. (2011). Production of activated carbon from organic by-products from the alcoholic beverage industry: Surface area and hardness optimization by using the response surface methodology. *Industrial Crops and Products*, 34(3), 1528–1537. 10.1016/J.INDCROP.2011.05.014

Nüchter, M., Ondruschka, B., Bonrath, W., & Gum, A. (2004). Microwave assisted synthesis – A critical technology overview. *Green Chemistry*, 6(3), 128–141. 10.1039/B310502D

Oghbaei, M., & Mirzaee, O. (2010). Microwave versus conventional sintering: A review of fundamentals, advantages and applications. *Journal of Alloys and Compounds*, 494(2010), 175–189.

Okajima, K., Ohta, K., & Sudoh, M. (2005). Capacitance behavior of activated carbon fibers with oxygen-plasma treatment. *Electrochimica Acta*, 50(11), 2227–2231. 10.1016/J.ELECTACTA.2004.10.005

Oshida, K., Kogiso, K., Matsubayashi, K., Takeuchi, K., Kobayashi, S., Endo, M., Dresselhaus, M. S., & Dresselhaus, G. (1995). Analysis of pore structure of activated carbon fibers using high resolution transmission electron microscopy and image processing. *Journal of Materials Research*, 10(10), 2507–2517. 10.1557/JMR.1995.2507

Otowa, T., Nojima, Y., & Miyazaki, T. (1997). Development of KOH activated high surface area carbon and its application to drinking water purification. *Carbon*, 35(9), 1315–1319. 10.1016/S0008-6223(97)00076-6

Özhan, A., Şahin, Ö., Küçük, M. M., & Saka, C. (2014). Preparation and characterization of activated carbon from pine cone by microwave-induced ZnCl$_2$ activation and its effects on the adsorption of methylene blue. *Cellulose*, 21(4), 2457–2467. doi: 10.1007/S10570-014-0299-Y.

Pelekani, C., & Snoeyink, V. L. (1999). Competitive adsorption in natural water: Role of activated carbon pore size. *Water Res.*, 33(5), 1209–1219. Apr. 1999, doi: 10.1016/S0043-1354(98)00329-7.

Pré, P., Huchet, G., Jeulin, D., Rouzaud, J., & Carbon, M. S. (2013). A new approach to characterize the nanostructure of activated carbons from mathematical morphology applied to high-resolution transmission electron. *Carbon*. 52, 239–258. Retrieved April 8, 2022, from

Rao, A., Kumar, A., Dhodapkar, R., & Pal, S. (2021). Adsorption of five emerging contaminants on activated carbon from aqueous medium:Kkinetic characteristics and computational modeling for plausible mechanism. *Environmental Science and Pollution Research*, 28(17), 21347–21358. 10.1007/S11356-020-12014-1.

Reucroft, P. J., Simpson, W. H., & Jonas, L. A. (1971). Sorption properties of activated carbon. *Journal of Physical Chemistry*, 75(23), 3526–3531. 10.1021/J100692A007/ASSET/J100692A007.FP.PNG_V03

Reza, M. S., Yun, C. S., Afroze, S., Radenahmad, N., Bakar, M. S. A., Saidur, R., Taweekun, J., & Azad, A. K. (2020). Preparation of activated carbon from biomass and its' applications in water and gas purification, a review. *Arab Journal of Basic and Applied Sciences*, 208–238. 10.1080/25765299.2020.1766799

Rodrı́guez-Reinoso, F., & Molina-Sabio, M. (1998). Textural and chemical characterization of microporous carbons. *Advances in Colloid and Interface Science*, 76–77, 271–294.

Rodriguez-Reinoso, F., Molina-Sobia, M., & Gonzalez, G. C. (2001). Preparation of activated carbon–sepiolite pellets. *Carbon*, 39, 771–785.

Rozada, R., Otero, M., Moran, A., & Garcia, A. I. (2005). Activated carbon from sewage sludge and discarded tyers: Production and optimization. *J. Hazardous Materials*, 124, 181–191.

Sanchez, N., Fayne, R., & Burroway, B. (2020). Charcoal: An ancient material with a new face. *Clinics in Dermatology*, 38(2), 262–264. 10.1016/J.CLINDERMATOL.2019.07.025

Sari, A., & Tuzen, M. (2009). Kinetic and equilibrium studies of biosorption of Pb(II) and Cd(II) from aqueous solution by macrofungus (Amanita rubescens) biomass. *Journal of Hazardous Materials*, 164(2–3), 1004–1011. 10.1016/J.JHAZMAT.2008.09.002

Schanz, J. J., & Parry, R. H. (1962). The activated carbon industry. *Industrial and Engineering Chemistry*, 54(12), 24–28. 10.1021/IE50636A005/ASSET/IE50636A005.FP.PNG_V03

Sevilla, M., & Mokaya, R. (2014). Energy storage applications of activated carbons: Supercapacitors and hydrogen storage. *Energy & Environmental Science*, 7(4), 1250–1280. 10.1039/C3EE43525C

Shafeeyan, M. S., Daud, W. M. A. W., Houshmand, A., & Shamiri, A. (2010). A review on surface modification of activated carbon for carbon dioxide adsorption. *Journal of Analytical and Applied Pyrolysis*, 89(2), 143–151. 10.1016/J.JAAP.2010.07.006

Shawabkeh, Reyad Awwad, Synthesis of novel activated carbon from pecan shells and application to the adsorption of methylene blue, copper, and strontium from aqueous solutions, 1998. Retrieved September 14, 2022, from https://www.proquest.com/openview/3ff0d21b518627b3e568d00a0c5356f0/1?pq-origsite=gscholar&cbl=18750&diss=y.

Shukla, N., Sahoo, D., & Remya, N. (2019). Biochar from microwave pyrolysis of rice husk for tertiary wastewater treatment and soil nourishment. *J. Clean. Prod.*, 235, 1073–1079. doi: 10.1016/J.JCLEPRO.2019.07.042.

Shukla, S. K., Al Mushaiqri, N. R. S., Al Subhi, H. M., Yoo, K., & Al Sadeq, H. (2020). Low-cost activated carbon production from organic waste and its utilization for wastewater treatment. *Appl. Water Sci.* 2020 102, 10(2), 1–9. doi: 10.1007/S13201-020-1145-Z.

Sidheswaran, M. A., Destaillats, H., Sullivan, D. P., Cohn, S., & Fisk, W. J. (2012). Energy efficient indoor VOC air cleaning with activated carbon fiber (ACF) filters. *Building and Environment*, 47(1), 357–367. 10.1016/J.BUILDENV.2011.07.002

Sinyoung, S., Chaiwat, W., & Kunchariyakun, K. (2021). Preparation of activated carbon from bagasse by microwave-assisted phosphoric acid activation. *Walailak Journal of Science and Technology (WJST)*, 18(16), Article 22796 (14 pages). 10.48048/wjst.2021.22796

Snoeyink, V. L., & Weber, W. J. (1967). The surface chemistry of active carbon; A discussion of structure and surface functional groups. *Environmental Science and Technology*, 1(3), 228–234. 10.1021/ES60003A003

Tamai, H., Nobuaki, U., & Yasuda, H. (2009). Preparation of Pd supported mesoporous activated carbons and their catalytic activity. *Materials Chemistry and Physics*, 114(1), 10–13. 10.1016/J.MATCHEMPHYS.2008.09.049

Tang, S. H., & Ahmad Zaini, M. A. (2020). Development of activated carbon pellets using a facile low-cost binder for effective malachite green dye removal. *J. Clean. Prod.*, 253. doi: 10.1016/J.JCLEPRO.2020.119970.

Thakur, S. K., Kong, T. S., & Gupta, M. (2007). Microwave synthesis and characterization of metastable (Al/Ti) and hybrid (Al/Ti+SiC) composites. *Materials Science and Engineering: A*, 452-453, 61–69. 10.1016/J.MSEA.2006.10.156

Treeweranuwat, P., Boonyoung, P., Chareonpanich, M., & Nueangnoraj, K. (2020). Role of nitrogen on the porosity, surface, and electrochemical characteristics of activated carbon. *ACS Omega*, 5, 1911–1918.

Vanderborght, B. M., & Van Grieken, R. E. (1977). Enrichment of trace metals in water by adsorption on activated carbon. *Analytical Chemistry*, 49(2), 311–316. 10.1021/AC50010A032/ASSET/AC50010A032.FP.PNG_V03

Wickramaratne, N. P., & Jaroniec, M. (2013). Activated carbon spheres for CO_2 adsorption. *ACS Appl. Mater. Interfaces*, 5, 1849–1855.

Wong, S., Ngadi, N., Inuwa, I. M., & Hassan, O. (2018). Recent advances in applications of activated carbon from biowaste for wastewater treatment: A short review. *Journal of Cleaner Production*, 175, 361–375. 10.1016/J.JCLEPRO.2017.12.059

Wu, M., Guo, Q., & Fu, G. (2013). Preparation and characteristics of medicinal activated carbon powders by CO2 activation of peanut shells. *Powder Technology*, 247, 188–196. 10.1016/J.POWTEC.2013.07.013

Xia, H., Peng, J., & Zhang, L. (2015). Preparation of high surface area activated carbon from Eupatorium adenophorum using K2CO3 activation by microwave heating. *Green Processing and Synthesis*, 4(4), 299–305. 10.1515/gps-2015-0025

Xie, Z., Yang, J., Huang, X., & Huang, Y. (1999). Microwave processing and properties of ceramics with different dielectric loss. *Journal of the European Ceramic Society*, 19, 381–387.

Xin-hui, D., Srinivasakannan, C., Jin-hui, P., Li-bo, Z., & Zheng-yong, Z. (2011). Comparison of activated carbon prepared from Jatropha hull by conventional heating and microwave heating. *Biomass and Bioenergy*, 35(9), 3920–3926. 10.1016/J.BIOMBIOE.2011.06.010

Yagmur, E., Tunc, M. S., Banford, A., & Aktas, Z. (2013). Preparation of activated carbon from auto-hydrolysed mixed southern hardwood. *Journal of Analytical and Applied Pyrolysis*, 104, 470–478. 10.1016/J.JAAP.2013.05.025

Yang, J., Shen, Z., & Hao, Z. (2004). Preparation of highly microporous and mesoporous carbon from the mesophase pitch and its carbon foams with KOH. *Carbon*, 42, 1872–1875.

Yue, Z., Mangun, C., Economy, J., Kemme, P., Cropek, D., & Maloney, S. (2001). Removal of chemical contaminants from water to below USEPA MCL using fiber glass supported activated carbon filters. *Environmental Science and Technology*, 35(13), 2844–2848. 10.1021/ES001858R

Zaini, M. A. A., & Kamaruddin, M. J. (2013). Critical issues in microwave-assisted activated carbon preparation. *Journal of Analytical and Applied Pyrolysis*, 101, 238–241. 10.1016/J.JAAP.2013.02.003

Zhou, J., Luo, A., & Zhao, Y. (2018). Preparation and characterization of activated carbon from waste tea by physical activation using steam. *Journal of the Air & Waste Management Association*, 68(12), 1269–1277. doi: 10.1080/10962247.2018.1460282.

14 Synthesis of inorganic nanoparticles by using ionizing radiation, their characterization, and applications

Jayashree Biswal*, H. J. Pant, and V. K. Sharma
Isotope and Radiation Application Division, Bhabha Atomic Research Centre, Trombay, Mumbai, India

*Corresponding author: jbiswal@barc.gov.in

CONTENTS

DOI: 10.1201/9781003321910-14

14.1 INTRODUCTION

Materials having size in the range 1 to 100 nm (at least in any one dimension) consisting of clusters of a few atoms or molecules are broadly classified as nanoparticles (NPs). They are often termed as nanoclusters or nanocolloids. By the virtue of their tiny size, they exhibit many intriguing properties that are significantly different from those of the individual atoms and their bulk counterparts. As the size of material is reduced to nano-domain the quantization of electronic states due to spatial confinement become prominent and the surface-to-volume ration increases tremendously [Schmid 1994, Cao 2004]. The unique properties of nanoparticles primarily originate from these two effects. Due to quantization of electronic states the nanoparticle possesses unique optical, electrical and magnetic properties, whereas, high surface-to-volume ratio modifies the thermal, mechanical and chemical properties [Abedini 2013]. Higher surface-to-volume ratio of nanoparticles means number of surface atoms are more than that present in the volume of the nanoparticles. This contributes to higher surface energy of nanoparticles as compared to the bulk counterpart. The physico-chemical and electronic properties of engineered nanoparticles typically lie between isolated small molecules and bulk materials. Nanoparticles have a significantly lower melting point, lower phase transition temperature and smaller lattice constant compared to the bulk because of large fraction of surface atoms [Cao 2004]. The probability of defects decreases; hence the mechanical strength increases in nanoparticles. Optical properties of nanoparticles can be very different from bulk, due to the discrete energy levels. The surface plasmon resonance phenomena in metallic nanoparticles are due to collective oscillation of conduction band electrons in response to the electric field of an incident light. The optical properties of silver and gold nanoparticles is very much useful for signal enhancement in surface enhanced raman scattering (SERS) to several orders of magnitude (10^8–10^{12}) [Le Ru EC, 2007]. These nanoparticle substrates are used in SERS for detection various biomolecules. Similarly, optical absorption peak of semiconductor nanoparticle shifts to shorter wavelength due to an increased band gap. In certain nanoparticles the ferromagnetism in the bulk material can disappear and its corresponding nanoparticle can possess paramagnetism due to huge surface energy [Cao 2004]. By virtue of their unique properties, the nanoparticles have wide range of applications in many frontier areas, such as catalyst, sensor, biomedical, magnetic data storage, optical materials, fuel cell, solar cells, and electronic devices [Campelo 2009, Datsyuk 2008, Du 2015, Kawasaki 2013, Kuila 2012, Marquardt 2011, Okitsu 2000, Cuenya 2010, Wildgoose 2006, Xu 2008, Nguyen 2014].

Depending on the structural confinement, the nanoparticles can be zero, one, and two dimensional. The zero-dimensional nanoparticles are those materials, which are confined in all the three dimensions. The zero-dimensional nanoparticles are known as nanosphere and quantum dot; one-dimensional nanoparticles are nanorods and nanowires; two-dimensional nanoparticles are nanoplates and thin film. The nanoparticles can be classified into inorganic, organic, carbon based and hybrid depending upon their chemical composition (Figure 14.1). Inorganic nanoparticles are metal and metaloxide nanoparticle, semiconductor nanoparticles, etc., whereas organic nanoparticles include polymer nanogels, dendrimers, micelles, liposomes, ferritin, etc.

Fullerene carbon nanotubes and carbon nanofiber are main carbon based nanoparticles. Hybrid nanoparticles are generally constructed from two or more different elements. The hybrid nanoparticles can be synthesized in core-shell or alloy-type structures. Fabrication of nanoparticles can be accomplished either by top-down or bottom-up approach. The top-down approach deals with starting from bulk material and reducing down to nanoscale, whereas the bottom-up approach involves building nanoparticles by assembling atom by atom or molecule by molecule. In bottom-up method, nanoparticles are formed from relatively simpler substances, the final nanoparticles

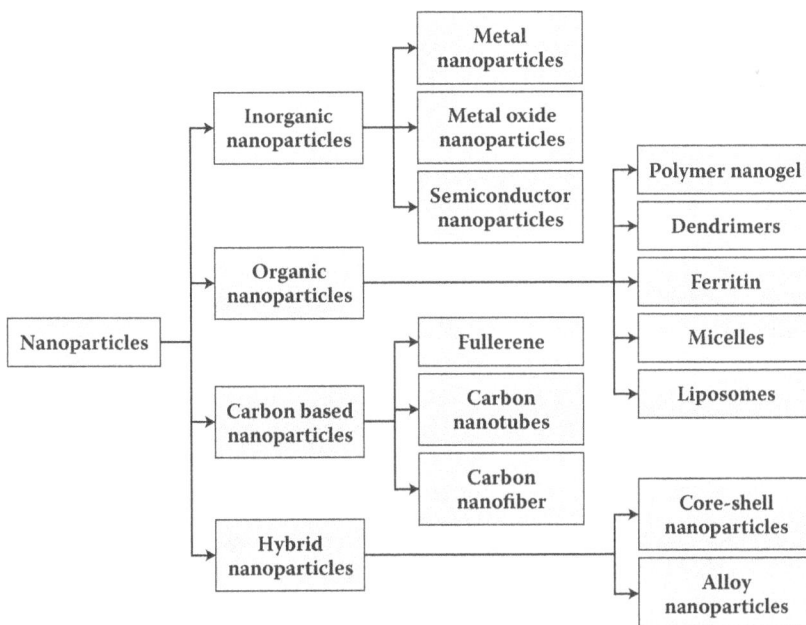

FIGURE 14.1 Different types of nanoparticles.

have less defects, more homogeneous chemical composition and better short and long range ordering. A schematic representation of various synthesis methods of nanoparticles is shown in
Figure 14.2. The focus of this chapter will be on radiolytic synthesis of inorganic nanoparticles,
their characterization. and various applications. Detail discussion related to polymeric and other
nanoparticles and other synthesis methods can be found elsewhere [Dispenza 2015, Baig, 2021,
Tyagi 2011]. Radiation-induced synthesis of inorganic nanoparticles offers a number of advantages, such as (i) it is a room temperature process; (ii) the process is simple and clean; (2)
controlled reduction of metal ions can be carried out without using excess reducing agent or
producing undesired oxidation products of the reductant; (3) the method provides metal nanoparticles in fully reduced, highly pure state; (4) tunable dose rate; and (5) the rate of reaction is
controlled by selectively choosing suitable transient species.

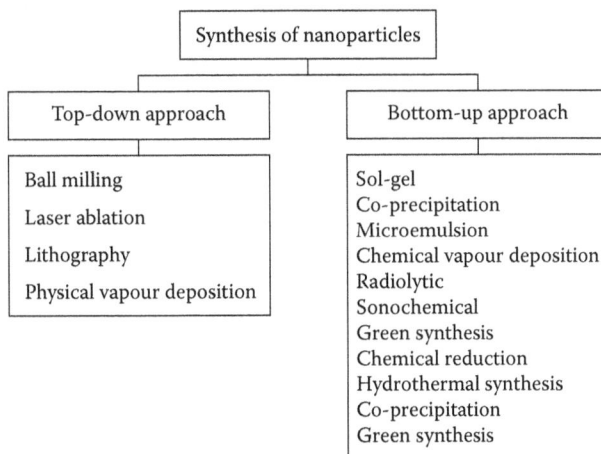

FIGURE 14.2 Different methods for synthesis of nanoparticles.

14.2　NUCLEATION AND GROWTH

The chemical process for synthesis of nanoparticles certainly involves the process of precipitation of a solid phase from the solution. A thorough understanding of these processes helps in controlling and manipulating the parameters to obtain nanoparticles with desired size and shape. The very step of formation of nanoparticles required a supersaturated solution of atoms or molecules (building blocks). The next step is the precipitation process, which consists of a nucleation step followed by particle growth step. The supersaturation stage has high Gibb's free energy. The reduction of overall Gibb's free energy is the driving force for nucleation. The nucleation and growth involve competition between two types of free energy. As a consequence of nucleation, there is a decrease in free energy associated with lower chemical potential of the solid phase (volume energy) and increase in free energy due to positive solid-liquid interfacial tension (surface energy). The volume energy is proportional to volume of particle and the surface energy is proportional to surface area of particle. The overall free energy change associated with the nucleation process is the sum of these two terms. For a spherical particle or cluster having radius R and interfacial tension γ, the overall change in free energy is given in Equation 14.1.

$$\Delta G = -\frac{4\pi R^3}{3v_m}\Delta\mu + 4\pi R^2\gamma \tag{14.1}$$

where $\Delta\mu$ is the difference in chemical potential of the precipitated solid phase and liquid phase, v_m is the molar volume of the precipitated species.

Figure 14.3 shows the change in free energy (ΔG) due to formation of a nanoparticle as a function of the particle radius (R). As it is shown in the figure ΔG has a positive maximum at a critical radius R^*. The critical radius can be obtained by setting $d\Delta G/dR = 0$ and is given in Equation 14.2.

$$R^* = \frac{2\gamma v_m}{\Delta\mu} \tag{14.2}$$

Cluster having size greater that critical radius is thermodynamically stable and will grow to form nanoparticle, whereas, the size below the critical radius being unstable will dissolve. The critical radius of cluster can be controlled by change in degree of supersaturation or by change in solid-liquid interfacial tension. Presence of additives, such as polymer, surfactant or other functional molecules can change the interfacial tension, hence can influence the critical radius of the cluster. The cluster formed in the initial nucleation stage then grows to nanoparticles by the addition of atoms or molecules or small clusters. The final size and size distribution of nanoparticles depend on the fine tuning of the synthesis parameters.

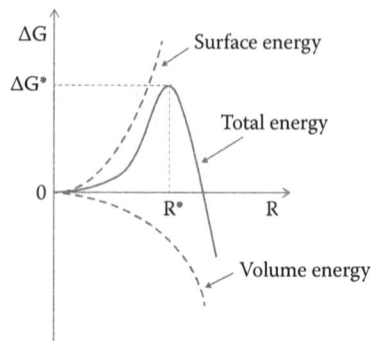

FIGURE 14.3 Schematic representation of change in free energy (ΔG) due to formation of a nanoparticle as a function of the particle radius (R).

14.3 OPTICAL PROPERTIES

The size of the nanoparticles has a pronounced effect on their optical properties. The brilliant color observed in colloidal solution of gold and silver nanoparticles are caused by surface plasmon absorption in the visible frequencies, unlike that for bulk metals where the absorption is in the UV region. The extinction coefficients of noble metal nanoparticles are quite high as compared to traditional dye molecules. For instance, gold nanoparticles have extinction coefficient of the order 10^{11} L mol^{-1} cm^{-1} that enables them to be used as excellent FRET (fluorescence resonance energy transfer) quenchers for study of biomolecule interaction and conformational changes [Haes 2004]. Similarly, the optical behavior of semiconductor nanoparticles is very different from their bulk counterpart due to quantum size effect. The origin of size dependent optical properties of nanoparticles can be classified into two types, such as, intrinsic and extrinsic size effects [Kreibig 1995]. The intrinsic size effect or quantum size effect is pertinent for nanoparticles having a size less than ~10 nm, whereas, above this size the extrinsic effect is valid. The intrinsic size effect is explained by using quantum theory; on the other hand the extrinsic size effect is explained by classical theory. The intrinsic size range, the size of a nanoparticle is smaller than the de-Broglie wavelength of electrons; the electrons and holes are spatially confined and will behave as a particle in a box [Cao 2004, Schmid 1994, Schmid 2004]. As a result, the energy levels will be discrete and the system is referred to as quantum confinement. This phenomenon is typically observed in semiconductor nanoparticles; hence, these are also called quantum dots. A schematic illustration of quantum confinement of semiconductor nanoparticles is represented in Figure 14.4.

As it can be seen from the figure, the electronic energy levels of quantum dots are discrete and the spacing is smaller than the corresponding atoms or molecules. On the other hand, the energy level in bulk solid is a continuum. The band gap of semiconductor nanoparticles is higher than the bulk and it increases with decrease in size of the quantum dot. The quantum size effect can also be observed in metal nanoparticles in a condition the size must be well below 2 nm [Cao 2004]. In case metal quantum dots, the phenomena is known as size induced metal to insulator transition (SIMIT). The spacing between the successive energy levels (δ) depend on the Fermi energy (E_f) of the bulk material and on the number of valence electrons in the nanoparticle (N) and is given in Equation 14.3. This spacing is known as a Kubo gap.

$$\delta = \frac{4E_f}{3N} \tag{14.3}$$

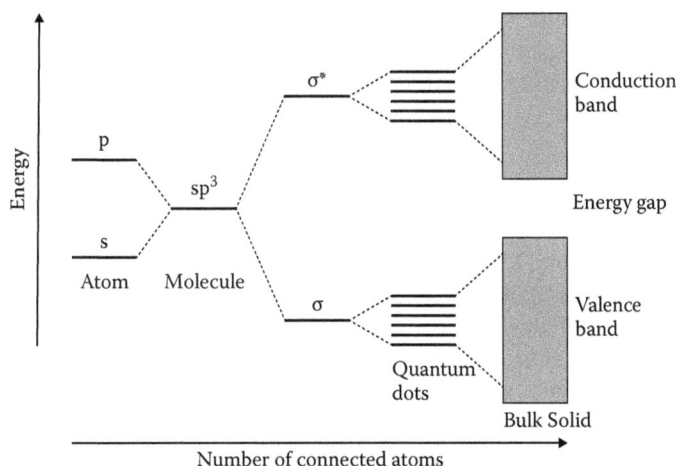

FIGURE 14.4 Schematic illustration of quantum confinement of semiconductor nanoparticles.

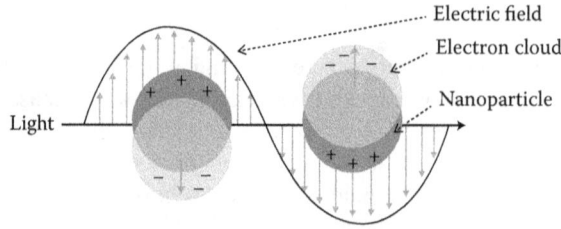

FIGURE 14.5 Schematic illustration of surface plasmon resonance of metal nanoparticles.

The band gap of semiconductor nanoparticles can be estimated using Equation 14.4.

$$E_{nano} = E_{buk} + \frac{h^2\pi^2}{2mR^2} - \frac{1.8e^2}{\varepsilon R} \tag{14.4}$$

where E_{bulk} and E_{nano} are the band gap of bulk and nanomaterial, respectively; e is the electronic charge; m is the effective mass of the exciton; and R and ε are the radius and dielectric constant of the nanoparticle, respectively. Using Equation 14.4, the size of semiconductor nanoparticles can be calculated from the observed absorption spectra [Hassan 2006].

The optical properties of metal nanoparticles are mainly due to the extrinsic size effect [Kreibig 1995] or surface plasmon resonance (SPR). The surface plasmon resonance is the collective oscillation of conduction band electrons in response to the electric field of an incident light. A schematic representation of the SPR of metal nanoparticles is shown in Figure 14.5. The electric field of an incoming light induces a polarization of the conduction band electrons with respect to the heavier ionic core of the metal nanoparticle. Subsequently, a net charge difference is created at the surface of the nanoparticles, which in turn acts as a restoring force. So that a dipolar oscillation is created with certain frequency. The energy of the SPR depends on both the conduction band electron density and the dielectric property of the surrounding medium. When the frequency of the electromagnetic field becomes equal to the plasmon frequency, resonant absorption of the light occurs. The wavelength and width of the absorption peak varies with the size, shape of the metal nanoparticles, and nature of the surrounding medium. The absorption peak of noble metal nanoparticles, such as, silver, gold, and copper in the visible region and the absorption peak of transition metal nanoparticles in the UV region are due the SPR.

In 1908, Gustav Mie developed a theory to explain the SPR by solving Maxwell's equation for interaction of electromagnetic light wave with small metal sphere. Assuming the electronic structure and dielectric constant of the nanoparticles same as that of the bulk, the expression for extinction coefficient (σ) using a simplified Mie theory is given in Equation 14.5.

$$\sigma = \frac{9V\varepsilon_m^{3/2}\varepsilon_2}{\lambda[(\varepsilon_1 + 2\varepsilon_m)^2 + \varepsilon_2^2]} \tag{14.5}$$

where V is the particle volume, λ is wavelength of light, ε_1 and ε_2 are real and imaginary part of the dielectric constant of the metal, respectively; and ε_m is the dielectric constant of the medium. The resonance condition is fulfilled, when $\varepsilon_1 = 2\varepsilon_m$, ε_2 provided is negligible. Mie theory was extended by Gans to explain the SPR in anisotropic metal nanoparticles [Burda 2005]. The absorption spectrum of nanorods split into two bands, due to transverse and longitudinal SPR. The transverse band is due to oscillation of electrons perpendicular to the major axis of the nanorod and occur at lower wavelength, whereas, the longitudinal band is due to oscillation of electrons along the major axis of the nanorod and occur at higher wavelength.

14.4   SUPERPARAMAGNETISM

Generally, any ferromagnetic or ferrimagnetic material undergoes a transition to a paramagnetic state above its Curie temperature. However, when the size of the ferromagnetic or ferrimagnetic materials is reduced down to a critical value, their magnetic behavior changes to paramagnetic even if they are below their Curie temperature. This phenomenon is known as superparamagnetism; it is strongly size dependent and exists only in nanoparticles. The superparamagnism for iron oxide (γ Fe_2O_3 or maghemite) is observed at size less than 20 nm and at 3 nm for pure iron. In these nanoparticles, the magnetization can randomly change the direction under the influence of temperature. The time elapsed in the change in the direction is called the Neel relaxation time. An external magnetic field is able to magnetize the nanoparticles, similar to a paramagnetic material. However, their magnetic susceptibility is much larger than that of a normal paramagnetic material. The superparamagnetism of nanoparticles is a useful property, based on which these nanoparticles have many promising applications in targeted drug delivery, magnetic resonance imaging, magnetic hyperthermia and thermoablation, bioseparation, and magnetic sensors and actuators.

14.5   SYNTHESIS OF METAL NANOPARTICLES BY USING IONIZING RADIATION

The wet chemical synthesis of metal nanoparticles comprises reduction of a precursor metal ion to form neutral metal atoms, creation of nucleation by formation of small clusters, and subsequent growth of these clusters to nanoparticles. In chemical process the reduction is accomplished using a suitable reducing agent. Whereas in the radiolytic method, the reduction is carried out by using transient reducing species produced in-situ by radiolysis of a solvent. Ionising radiation, such as gamma, X-ray, and electron beam cause ionization and excitation of water molecules when they pass through an aqueous medium. Ionising radiation deposits energy in the medium in 10^{-15} seconds, resulting in formation of "spurs" containing positively charged ions, electrons and excited species. In a time period of 10^{-6} seconds, different primary species, such as, e_{aq}^-, H·, OH, H_2, H_2O_2, $H_3O^+$are formed through interactions of the excited and ionized species among each other. The reduction potential values of different radiolytic species are given in Table 14.1. As it can be seen from Table 14.1, e_{aq}^- and H are strong reducing agents, whereas ·OH is a strong oxidizing agent.

When an aqueous solution of metal ion is irradiated with gamma radiation, the metal ions are reduced by e_{aq}^- and H· to generate metal atoms [Belloni 1998a, Belloni 1998b, Braunstein 1999, Belloni 2001, Belloni 2006, Belloni 2008]. The ·OH present in the medium are able to oxidize back the metal atoms into metal ions [101]. To prevent this oxidation, the solution is added with a scavenger of ·OH such assecondary alcohols or formate anions [102]. The secondary radicals formed, α-methyl-hydroxyethyl[$(CH_3)_2$·C(OH)] or formyl [$CO_2^{·-}$], respectively, are also capable of reducing the metal ions. The reduction of metal ions to formation of small metal clusters are

TABLE 14.1
Reduction Potential of Different Radicals Produced Upon Radiolysis of Water

Radicals	Reduction Potential (V vs NHE)
e_{aq}^-	−2.87
H·	−2.31
·OH	1.9

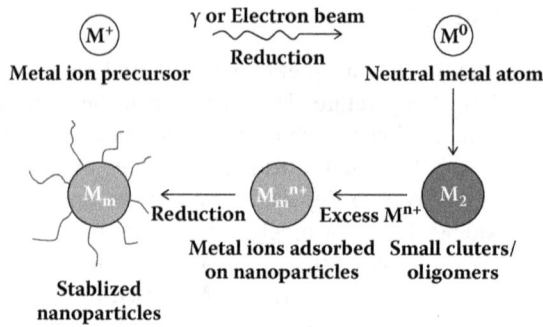

FIGURE 14.6 Different stages of synthesis of metal nanoparticles.

given in Equation 14.6–Equation 14.11. The reactions are often diffusion controlled. The metal in a zero valent state coalesce to form metal nanoparticles in the presence of a suitable stabilizer [Kang 2019]. The process of formation of metal nanoparticles starting from a metal ion precursor is depicted in Figure 14.6. The isolated atoms formed M^0 coalesce into clusters.

These clusters adsorb excess metal ions present in the medium and finally reduced to metal nanoparticles. They are stabilized by ligands, polymers or supports. The mechanism of stabilization of the nanoparticles using suitable stabilizers are discussed in the subsequent section.

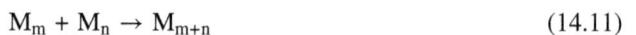

$$M^+ + e_{aq}^- \rightarrow M^0 \tag{14.6}$$

$$M^+ + H^\cdot \rightarrow M^0 \tag{14.7}$$

$$M^+ + (CH_3)_2\,{}^\cdot C(OH) \rightarrow M^0 + CH_3CO\ CH_3 + H^+ \tag{14.8}$$

$$M^+ + CO_2^{\cdot-} \rightarrow M^0 + CO_2 \tag{14.9}$$

$$M^0 + M^0 \rightarrow M_2 \tag{14.10}$$

$$M_m + M_n \rightarrow M_{m+n} \tag{14.11}$$

Multivalent metal ions are reduced to zerovalent state by multistep reduction process, which can be observed by pulse radiolysis studies [Belloni 2006, Belloni 2008, Buxton 1995, Ershov 1990]. Synthesis of various metal nanoparticles, such as, Ag, Au, Cu, Hg, Pt, Pd, Ni, etc. have been reported by different researchers by using ionizing radiation [Chmielewska 2012, Belloni 1998b, Marignier 1985, Biswal 2011, Biswal 2010a, Biswal 2010b, Biswal 2010c, Biswal 2009]. The dissolved oxygen in a solvent can greatly affect the reaction pathway and nanoparticle formation. Oxygen reacts with primary radiolytic products e_{aq}^- and $H^\cdot$ to transform them to hydroperoxyl radical ($HO_2^\cdot$) and superoxide radical anion ($O_2^{\cdot-}$), respectively. Hence, the reduction of metal ions is not possible in presence of oxygen. Hence, the reaction medium is always deaerated either by saturating the solution with nitrogen or argon.

The use of radiation for synthesis of nanoparticles offers many advantages compared to other chemical reduction methods. These include room temperature process, absence of addition of any chemical reducing agent, and tunability of irradiation condition to control the shape, size, and size distribution. Selectively controlling dose, dose rate, and nature of radiolytic species, the desirable size and shape of the nanoparticles can be synthesized. A variety of nanoparticles can be synthesized by using ionizing radiation; hence, it is a versatile technique for the synthesis of nanoparticles.

The spherical shape being thermodynamically most stable, the formation of spherical nano-particles is more probable in any wet chemical synthesis method including radiolytic method. Anisotropic/non-spherical nanoparticles are also very much useful in sensing of different bio-molecules by virtue of their longitudinal SPR band. In chemical reduction method, the anisotropic metal nanoparticles have been synthesized by using step wise growth in presence of long chain branched surfactants. The anisotropic nanoparticles can also be prepared using ionizing radiation using similar surfactant molecules as structure directing agent and the reducing radical can be chosen in such a way that the rate of reduction will be slower to allow the directional growth of nanoparticles [Biswal 2011, Biswal 2010b, Biswal 2010c]. The synthesis of gold nanorods is accomplished in two steps. In the first step a nano cluster/seed solution is prepared at a low temperature by adding drop by drop solution of sodium borohydride into the aqueous metal ion solution. In the second step, a few drops of the seed solution are added to the metal ion precursor solution containing stabilizer (structure directing agent) and is irradiated with gamma radiation at a low dose rate to generate gold nanorods.

In some situations when a monomer (an organic molecule) is mixed with the metal ions and irradiated the transient radicals can be utilized for simultaneous reduction of metal ion as well as polymerization of the monomers. Subsequently, the nanoparticles are stabilized by the polymer produced in-situ. On the other hand, the nanoparticles can be synthesized in a porous solid support [Rojas 2012] by directly irradiating the heterogenous system containing the metal ion solution with the solid support.

The dose rate obtained in EB irradiation is higher than gamma irradiation. The dose rate can play a crucial role in manipulating the size, shape and structure of metal nanoparticles. At higher dose rate, the rate of reduction is higher than the rate of growth. As a result, the localized con-centration of zerovalent metal will be more, and they will coalesce and form small clusters. However, at a low dose rate, the rate of reduction is less; hence, local concentration of zerovalent metal will be less and they will have enough time to grow to bigger nanoparticles. The dose rate dependence of the nanoparticle size is depicted in Figure 14.7. Hence, it can be concluded that smaller nanoparticlesare synthesized when the irradiation is carried out using electron beam and gamma irradiation will result in the formation of bigger nanoparticles [Belloni 2006].

Bimetallic nanoparticles are useful in catalyst applications. The core-shell or alloy types of bi-metallic nanoparticles can be synthesized by controlling the dose rate during the synthesis process. Generally, core-shell nanoparticles are generated at low dose rate, whereas irradiation at high dose rate will result in generation of alloy-type nanoparticles [Remita 2003, Treguer 1998]. At a high dose rate, both metal ions will be reduced simultaneously to generate alloy nanoparticles, whereas at a low dose rate the more noble metal ion will be reduced first and form the core followed by reduction of less noble metal ion to form the shell.

The strength of a reducing agent can control the ultimate shape of nanoparticles in some of the cases [Belloni 2006]. For instance, ascorbic acid being a weak reducing agent can only reduce the metal ion (Au^{III}), which are adsorbed on certain facets of the gold cluster/seed generated by radiolysis [Biswal 2011]. The radiolytic route is not only limited to synthesis of metal nanoparticles, but the ionizing radiation is also reported to be used for synthesis of semiconductor nanoparticle like CdSe, NiS, PbS, ZnS, ZnO [Biswal 2010a, Hu 2007], and metaloxide nanoparticles, such as iron oxide, cobalt oxide, nickel oxide, etc. [Jurkin 2016, Alrehaily 2013, Yang 2021].

14.6   STABILIZATION OF NANOPARTICLES

The nanoparticles are thermodynamically unstable due to high surface energy and they always have a tendency to coalesce. If the coalescence process is not stopped at certain stage, the na-noparticle will coagulation and they will no longer be in the nano size range. Hence it is essential

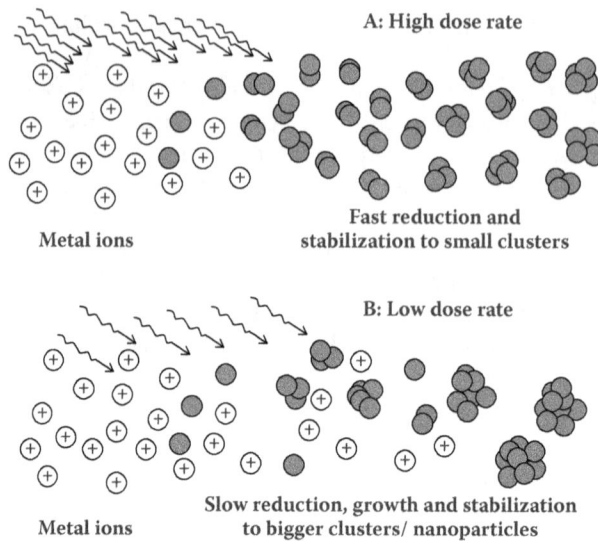

FIGURE 14.7 Schematic representation of effect of dose rate on size of nanoparticles.

to stabilize the nanoparticles for their further utilization in numerous applications. This can be achieved by either electrostatic or steric stabilization [Schmid 2004, Schmid 1994, Cao 2004] by using a capping agent such as polymer [Shan 2007, Khanna 2004], surfactant [Khanna 2004], solid support [Jana 2005], or ligand [Yazid 2010] having suitable functional groups. Electrostatic stabilization is achieved by an electrical double layer arising from the attraction of negatively charged ions to the metal nanoparticles (Figure 14.8) explained by DLVO theory [Yonezawa 1999, Derjaguin 1993]. Electrostatically stabilized NPs are described to have at least one electrical double layer due to a surface charging. The resulting coulomb repulsion forces between the particles decay exponentially with particle-to-particle distance, as shown in Figure 14.8. The resulting maximum of the curve represents the aggregation barrier and determines the colloidal stability. The barrier represents an activation energy for aggregation that the particles have to overcome for coagulation. If the electrostatic repulsion is sufficiently high, it prevents the particles from

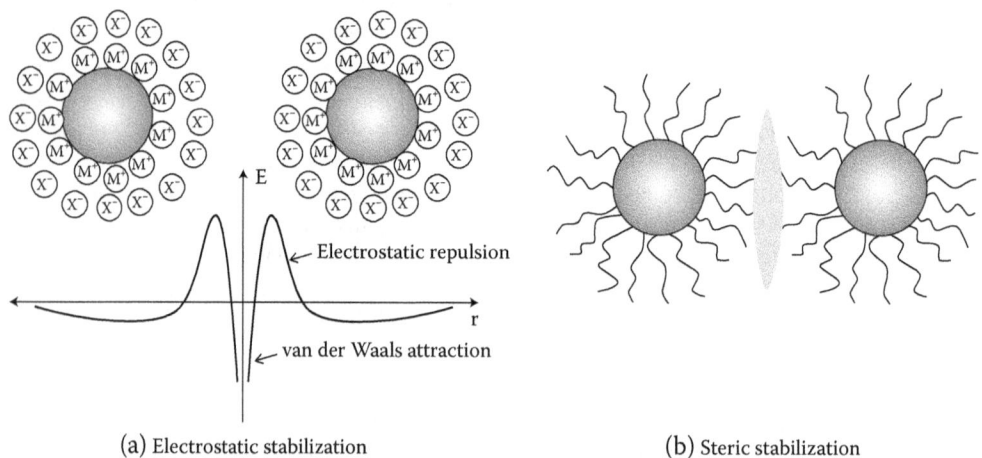

FIGURE 14.8 (a) Electrostatic stabilization and (b) steric stabilization of nanoparticles.

coagulation. The electrostatic stabilization of nanoparticles is controlled by parameters, such as type and concentration of ion, value of the surface potential, and the particle size [Polte 2015].

Steric stabilization is a process in which colloidal particles are prevented from aggregating by adsorption of large molecules at the particle surface, such as polymers or surfactants, providing a protective layer. Polymers having functional group, $-NH_2$, $-COOH$, $-CONH_2$, and $-OH$ have high affinity towards the metal nanoparticles. A variety of polymers are used as stabilizer for metal nanoparticles, such as, poly(vinyl alcohol) (PVA), poly(vinylpyrrolidone)(PVP), sodium poly (vinyl sulphate) (PVS), poly(acrylamide)(PAM), poly(N-methylacrylamide) (PNMAM), poly (ethyleneimine)(PEI), poly(aniline) (PANI), and sodium dodecyl sulphate (SDS) [Belloni 2001, Georgakilas 2007, Jovanovic 2012, Kadlubowski 2014, Biswal 2010a]. In addition to this, some natural polymers having complex structure, such as guar gum are also used to stabilize metal nanoparticles [Biswal 2009]. The prevention of coagulation of these large molecules can be explained via simple mechanisms. The density of the adsorbed molecules in the interparticle space would increase tremendously, if the interparticle distance would become smaller and smaller. This would cause a decrease in entropy, thus an increase of the Gibbs free energy which is thermodynamically not favorable. Due to the increased density, osmotic repulsive forces would also increase. Furthermore, a higher solubility of the stabilizing molecule counteracts agglomeration [Pomogailo 2005].

Long-chain polymers are made to adsorb on the nanoparticle surface to provide steric stabilization, as shown in Figure 14.8. The extent of steric stabilization depends on the polymer concentration, temperature, average chain length, and the solubility of the polymer [Polte 2015].

In some applications, such as catalysts, the use of stabilizers is not desirable. In these cases, supported nanoparticles are prepared. Mesoporous silica, zeolites, titanium dioxide, aluminaand other materials with controlled pore structures are widely used as the support materials for stabilization of nanoparticles [Chettibi 2013, Keghouche 2005, Ramnani 2007]. The metal ions diffuse in the pores of the support and get adsorbed on the surface. The ionizing radiation penetrates through the support the metal ions are reduced to metal atoms by radiolytic radicals. The metal atoms then coalesce and grow to nanoparticles and the functional groups present on the surface of the support anchor the metal nanoparticle.

14.7 CHARACTERIZATION

The characterization of nanoparticles is essential for evaluating the suitability of their specific applications. This requires characterizing equipment with high sensitivity, accuracy and atomic level resolution. The size, size distribution, shape, polydispersity, chemical composition, crystalline or amorphous nature, surface area, and surface charge of the nanoparticles are determined by using UV-visible spectrophotometer, scanning electron microscope (SEM), transmission electron microscope (TEM), dynamic light scattering (DLS), X-ray diffraction (XRD) analyzer, gas adsorption instruments, and X-ray photoelectron spectrometer (XPS). These characterization techniques are discussed in the following section.

14.7.1 ABSORBANCE SPECTROSCOPY

UV-visible absorption spectroscopy technique is used for the characterization of metallic and semiconductor nanoparticles [Creighton, 1991, Biswal 2010a]. Particularly noble metal nanoparticles possess high extinction coefficient and their surface plasmon property is size and shape dependent. A qualitative information about the size and shape of the nanoparticle is obtained using this technique. The technique offers easy sample preparation, rapid analysis, and reliable results. Development of sensors is based on the shift in surface plasmon band or change in intensity of the band of metal nanoparticles.

The absorbance is obtained by employing Beer's Law. The absorbance value (A) depends on nanoparticle concentration (c), path length (l) of measuring cell and extinction coefficient of na-nopaticles (ϵ) (Equation 14.12).

$$A = \epsilon c l \tag{14.12}$$

Factors such as size, shape, extent of aggregation, nature of stabilizer, temperature, presence of any adsorbate on the surface of nanoparticles, as well as nature of the surrounding medium affect the peak maxima in the absorption spectrum [Kreibig 1995]. The FWHM increases with increasing size of the nanoparticle and there is a red-shift as the particle size increases [Link, 2000]. In case of gold nanoparticles, the color of its aqueous solution changes from ruby red to purple and finally blue with increase in size.

14.7.2 Dynamic light scattering

Dynamic light scattering (DLS) is mainly used for measurement of hydrodynamic diameter and aggregation stateof colloidal nanoparticles. The technique is based on the Brownian motion of the particles. The time dependent scattered light contains information about the diffusion speed, thus information about the size distribution. Highly concentrated as well as highly dilute samples can be measured using the technique. Most of the DLS analyzer is equipped to measure polydispersity index (PDI), zeta potential, molecular weight, and concentration. The zeta potential is used to measure the surface charge and stability of nanoparticle solution. Higher the zeta potential more stable the nanoparticles are in the solution. PDI is a dimensionless quantity that measures the broadness of the size distribution. The value of PDI less than 0.1 represents a homogeneous particle size distribution, whereas a PDI greater than 0.3 suggests heterogeneous size distribution.

According to the Stokes-Einstein relationship, the diffusion coefficients (D) of the nanoparticles are inversely proportional to their hydrodynamic diameter (d), as given in Equation 14.13.

$$D = \frac{KT}{3\pi\eta d} \tag{14.13}$$

where K, T, and η are Boltzmann constant, temperature, and viscosity, respectively. These quantities are used to calculated the particle size using Equation 14.13.

14.7.3 X-Ray diffraction

X-ray diffraction (XRD) is a non-destructive technique used for the crystal analysis and phase iden-tification of nanoparticles. XRD is important to characterize bimetallic nanoparticles to confirm the core-shell or alloy structure. In this technique the intensity of the reflected X-ray is analyzed to obtain the inter-atomic spacing (*d* value in Angstrom units) by using Bragg's equation (Equation 14.14).

$$n\lambda = 2d\sin\theta \tag{14.14}$$

where, *n* is an integer, λ and θ *are* the wavelengths of X-rays used, and the angle of incidence of the X-ray beam, respectively.

Nanoparticles have broad peaks in XRD spectra, unlike their bulk counterpart. The broadening provides information about crystallite size by using the Debye-Scherrer equation (Equation 14.15).

$$d = \frac{\kappa\lambda}{\beta\cos\theta} \tag{14.15}$$

where K is the shape factor (0.9), λ is the x-ray wavelength (1.5 A^0 for Cu k_α), β is FWHM of X-ray spectrum peak in radians, and θ is the Bragg's angle.

14.7.4 FOURIER TRANSFORM-INFRARED SPECTROSCOPY

Nanoparticles are anchored by capping agents on their surface by the virtue of the interaction of the functional group of the capping agent with metal surface. Also, the surface chemistry of nanoparticles is modified by selecting suitable functional groups for specific applications. Hence, it is imperative to understand the surface chemistry of nanoparticles. Fourier transform infrared spectroscopy (FTIR) spectroscopy is used for obtaining interaction of the functional groups that are present in a system by measuring the vibrational frequencies of the chemical bonds involved [Baudot, 2010]. The infrared energy is absorbed by the functional groups causes stretching and bending of chemical bonds. Functional groups attached to the metal nanoparticle surface show different FTIR patterns than those of free groups; hence, FTIR provides information related to the surface chemistry of the nanoparticles.

14.7.5 ATOMIC FORCE MICROSCOPY

Atomic force microscopy (AFM) is a technique to characterize the surface morphology and size of nanoparticles by measuring the interacting force between the surface of the sample and an ultra-sharp probe. The principle is based on the deflections of a metal cantilever induced by van der Waals force between the tip of cantilever and the surface of the sample. It can be used to obtain high resolution image of the surface, from which, shape, size, and size distribution of nanoparticles can be known. Low density material is difficult to characterize using electron microscope, however these materials can be efficiently characterized using the AFM technique [Mourdikoudis, 2018]. This technique can also be used for both conducting and nonconducting materials and give a 3D image of the sample. A resolution of about 1 nm can be achieved in this method.

14.7.6 SCANNING ELECTRON MICROSCOPY (SEM) AND ENERGY-DISPERSIVE X-RAY (EDX) ANALYSIS

The optical microscope has limitations of observing particles having size greater than 1 μm due to diffraction of light by material below this dimension. To observe material in a nanometer scale an incident beam having short wavelength beam is required. Hence, to obtain a higher resolution image shorter wavelength electron beam is used and the technique is generally known as electron microscopy. Scanning electron microscopy and transmission electron microscope are two different types of the as electron microscopy. SEM analysis is one of the most widely used techniques for characterization of nanoparticles and it provides information about the topography and morphology of the nanoparticles by detection of secondary electrons or the backscattered electrons. The magnification and resolution of SEM can go up to 1,00,000 and ~3 nm, respectively. Most of the SEM instruments are equipped with EDX modality, which can analyze the chemical composition of the sample by detecting the characteriztic X-ray emitted from the sample.

14.7.7 TRANSMISSION ELECTRON MICROSCOPE (TEM)

A transmission electron microscope (TEM) is the most widely used technique to characterize metal nanoparticles. A TEM instrument equipped with other modalities can provide information regarding topography, morphology, dispersity, composition, and crystallography of the sample. High-resolution TEM can go up to a magnification of ~10,00,000 with a resolution of 0.1 nm. A comparison of different characterization techniques is given in Table 14.2.

TABLE 14.2
Comparison of Different Techniques for Characterization of Nanoparticles

Technique	Detection Limit	Analysis based on	Environment	Material Sensitivity	Parameters Measured
TEM	0.1 nm	Transmission of electrons	Vacuum	Increases with atomic number	Size and shape
SEM	1.5 nm	Scattering of electrons	Vacuum	Increases with atomic number	Size and shape
AFM	1 nm	Physical interaction with sample	Vacuum/ air/ liquid	Equal for all material	Size and shape
DLS	3 nm	Light scattering fluctuation due to diffusion	Liquid	Depends on refractive index	Hydrodynamic size

14.7.8 Small angle X-ray scattering

Small angle X-ray scattering (SAXS) is used for sizes and shapes, particle interactions, and internal structures of nanoparticles. The SAXS has the advantage over other techniques of being able to analyze a wide variety of sample types, including gel, liquid, solids, aerosols, colloidal suspensions, powders, and thin films. This enables the measurements in the native state of samples, hence information about real-life condition of sample can be obtained using this technique. Also, SAXS often requires very little sample preparation time, in contrast to the elaborate sample preparation procedures required for using the electron microscopy methods. The sample is often positioned at a long distance from the detector, hence overall intensity of the scattered beam received by the detector is less. Hence, the use of a high brilliant X-ray source from a synchrotron beam is beneficial to get a higher signal and also it reduces the measurement time, thereby increasing the number of measurements [Agbabiaka, 2013].

14.7.9 X-ray photoelectron spectroscopy

X-Ray photoelectron spectroscopy (XPS) is based on the photoelectric effect. The electronic structure, elemental composition, and oxidation states of elements in a nanoparticle can be determined using XPS. Also, core-shell structure, internal heterostructure, ligand exchange reaction, and surface functionalized nanoparticles can be determined by using this technique. The elemental identification is possible by knowing the electron binding energies of the elements and small shifts, in the binding energies in XPS from those of pure elements can provide information about the chemical states of the elements of interest [Mourdikoudis, 2018].

14.7.10 Superconducting quantum interference device magnetometry

Superconducting quantum interference device magnetometry (SQUID) is a tool for characterizing the magnetic nanoparticles. Different parameters, such as magnetization saturation, magnetization remanence, blocking temperature, and the magnetic response of individual molecules can be measured by SQUID [Mourdikoudis, 2018].

14.7.11 Mossbauer spectroscopy

Mossbauer spectroscopy is also used for characterization of magnetic nanoparticles, mostly iron-based nanoparticles. It is based on the recoil-free resonance absorption or emission of γ-photons in solid lattice. In this technique, a solid sample is exposed to a beam of gamma radiation, and a detector measures the intensity of the beam transmitted through the sample. Mossbauer spectroscopy is used for determination of the oxidation state, symmetry, surface spins, magnetic ordering of Fe atoms, magnetic anisotropy energy, thermal unblocking, and distinguishing between different iron oxides [Mourdikoudis, 2018].

14.8 APPLICATIONS

The nanoparticles have a wide range of applications in a number of fields owing to their superior chemical, physical, and mechanical properties and of their decent formability. Inorganic nanoparticles are being used in different biomedical applications, such as, antimicrobial applications, biosensing, imaging, drug delivery, and cancer therapy. Nanoparticles have wide applications in the automobile industry. TiO_2 nanoparticles are used in paints to provide smooth, non-scratchable coating, as well as for designing self-cleaning glass in automobile industries. Ni-Ti nanoparticle-based shape-memory alloys are used to design powerful electric motors. In addition to this, they have been also used in the fields of optics [So 1997, Lal 2007], catalyst, superconductors [Roumie

2014, Basma 2016], magnetism [Sharrouf 2015, Sharrouf 2016], electronics [Stuart, 1998; Akella, 1997], telecommunications [Ricard, 1985, Elvira, 2013], catalyst, superconductors [Roumie, 2014; Basma, 2016], chemical catalysis [Henry 2000] or biological marking [Burchez 1998], sensor, bioremediation of diverse contaminants [Kanel 2006], water treatment, and production of clean energy. Nanoparticles have unique magnetic properties and sizes comparable to the largest biological molecules, such as enzymes, receptors, or antibodies, that enable their use in diagnostic and therapy [Tri 2019]. Some of the selected applications are discussed in the following section.

14.8.1 Biomedical applications

14.8.1.1 Antimicrobial applications

Silver nanoparticles are known to possess a broad spectrum of antibacterial, antifungal, and antiviral properties. Nanosized silver have the ability to penetrate bacterial cell walls. Subsequently, they release silver ion in the intracellular region, thereby rupturing the cell membranes and causing cell death. Some metal oxides also show antimicrobial activities, such as TiO_2, Bi_2O_3, ZnO, FeO, MnO_2, CuO, Ag_2O, Al_2O_3, and so on, and play significant roles in various medical applications [Gao 2009]. Similarly, metal oxide/semiconductor nanoparticles, such as ZnO and TiO_2, can destroy the pathogenic bacteria byreactive oxygen species (ROS) mechanism under UV light radiation. Core-shell hybrid nanoparticles consisting of both semiconductor oxide nanoparticles and noble metal nanoparticles (Au, Ag, Pd, Pt) are also being used for their antimicrobial action under visible light illumination [Gao 2019, Yaqoob 2020].

14.8.1.2 Imaging

Iron oxide, gadolinium, and gold nanoparticles are used as contrast agents to increase the sensitivity of magnetic resonance imaging (MRI) [Blasiak 2013]. For MRI applications, the nanoparticles are desired to be have high magnetization values, strong plasmon resonance, small size, and narrow particle size distribution. Gold nanoparticles are widely used for optical imaging applications based on their absorption, fluorescence and Raman scattering [Wu 2019]. Semiconductor nanoparticles are fluorescent in nature and they can be used for biological labeling. When surfaces of these particles are functionalized using some specific antibodies, they can be targeted towards specific receptors in biological cells.

14.8.1.3 Drug delivery and cancer therapy

Nanoparticles sizes comparable to the largest biological molecules, such as enzymes, receptors, or antibodies, that enable diagnostic, therapy as well as combined therapy and diagnostic. Both inorganic and organic nanoparticles can be employed successfully for drug delivery. Inorganic magnetic nanoparticles for drug delivery are the most sought-after area of research. These nanoparticles are either functionalized with anticancer drugs or encapsulated with biocompatible polymers that contain the drug to be delivered. Nanoparticles being very small are easy to inject and target towards specific location in the body. Some of the important examples of drug delivery system are in cancer therapy, insulin delivery, and treating Alzheimer's as well as Parkinson's disease.

Phototherapy deals with the heating effects under laser irradiation due to the enhanced absorption induced by localized surface plasmon resonance of nanoparticles. Core-shell silica-gold nanoparticles can be employed to target the cancer cells. The scattered light from the nanoparticles enables detecting the location where the particles are attached. Both imaging as well as treatment is possible using a phototherapy technique. The heat generated by the core-shell particles raises the local temperature, which is sufficient to kill the cancer cell.

Magneto-thermal effect or hyperthermia therapy is another technique for cancer therapy, in which magnetic nanoparticles are utilized to increase the local temperature and destroy cancer cells. Fe_2O_3 superparamagnetic nanoparticles when subjected to an AC magnetic field increase the local temperature around them by 41°C–48°C than the normal body temperature. This can destroy the cancer cells.

14.8.2 ELECTRONICS AND SOLAR CELL

Nanocrystalline ZnSe, ZnS, CdS, etc. zinc selenide, zinc sulfide, cadmium sulfide, and lead telluride are candidates for improving the resolution of personal computers and high-definition television screen by reducing the size of the pixel or phosphors. Nanoparticles are used in computer chips to create devices with compact size and faster performance.

TiO_2 nanoparticles are used as photoanode material in dye-sensitized solar cells to increase the efficiency of solar cell [Fan 2017]. Indium tin oxide nanoparticles are used in high-end electronics and photovoltaic solar cells due to their unique conductivity, transparency, and ease of use. The quantum dot solar cell containing organized nanoparticles of CdS, CdSe, PbS, or PbSe are being studied to replace the dye-sensitized solar cells to increase the efficiency of the solar cell.

14.8.3 CATALYST

Due to a high surface-to-volume ratio, nanoparticle catalysts have high surface activity. Metal oxides are widely used in photoelectrochemical and photocatalytic systems for fuel synthesis and environmental remediation. Supported nanoparticles are useful in sensing and catalyst applications. TiO_2, ZnO, ZnS, SnO_2, Fe_2O_3, CdS, MoS_2, and CdS nanoparticles are widely used for photodegradation of various environmental pollutants. These nanoparticles are synthesized by irradiating a heterogeneous system of solid porous support and metal ion precursor solution. The metal ions are reduced by transient radiolytic radicals in sides the pores of the solid and the nanoparticles generated are impregnated in the porous structure of the solid [Nguyen, 2021].

Supported noble metal catalyst, such as Pt and Pd are used in passive autocatalytic recombiners devices for mitigation of hydrogen generated in nuclear reactors during postulated accidental conditions [Belapurkar 2008, Zhengfeng 2017]. Pd, Pt, and Rh nanoparticles are used as catalysts to remove toxic substances from the exhaust of the vehicles.

14.8.4 HYDROGEN GENERATION AND STORAGE

Hydrogen has the potential to be used as an energy source because of high energy content per unit weight and is a clean fuel. To manage the storage of hydrogen, advanced technology and materials are being developed. Novel nanomaterials, such as carbon nanotubes; metal hydride; aerogels doped with TiO_2, Al_2O_3, MgO, or Fe_3O_4; metal organic frameworks; boron nitride nanotubes; and silicon carbide nanotubes are potential candidate for hydrogen storage by physisorption/chemisorption process due to their high accessible surface area, large free pore volume, strong interactions [Cheng 2001, Free 2021, Froudakis 2011, Kulkarni 2015].

14.8.5 SENSOR

Gas sensors are applied for environmental monitoring, semiconductor, health care, and automobiles. Owing more surface active sites, large surface-to-volume ratios, and high specific surface areas with very high surface reactivity, metal oxide nanoparticles are used in gas sensing applications [Kulkarni 2015, Sun 2012, Chavali 2019]. These nanoparticles sensors have excellent selectivity, fast sensing response, low cost, ease of handling, used for a wide range of target gases, longer lifetimes, and can be recovered and reused.

Noble metal nanoparticles have plasmon resonance in the visible region with high extinction coefficient. This property of the noble metal nanoparticles has been exploited for a wide variety chemical sensors and biosensors [Sharma 2019, Unser 2015]. Gold and silver nanoparticle based surface enhanced raman spectroscopy (SERS) has extremely high sensitivity (single molecule sensitivity) and are considered to be an excellent tool for sensing target DNA molecules, heavy metal ions, organic molecules, and detection of explosives in water and soil.

14.8.6 ANTIWEAR ADDITIVE IN AUTOMOBILE OIL LUBRICANTS

Nanoparticle are used as lubricant additives to the base oil for improvement of lubricating performances of the oil. Relatively small amounts of these additives can provide significant improvement in antioxidation capability and tribological properties (friction, wear), which is of great significance for energy conservation, emission reduction, and environmental protection. CuO, ZrO_2, and ZnO nanoparticles; graphene; and hybrid nanoparticles have proved to provide wear resistance to automobile components [Biswal 2020, Batteza 2008, Zhang 2014]. The mechanism of the antiwear action of the nanoparticles involves three effects: mending effect, rolling effect, and film effect.

14.9 CONCLUSIONS

In this chapter, a detailed overview on synthesis, properties, characterization, and some selected applications of inorganic nanoparticles have been discussed. The focus was on preparation of inorganic nanoparticles by using ionizing radiation, such as gamma and electron beam. Radiolytic route for synthesis of nanoparticles is a clean and room temperature method, so that the synthesized nanoparticles will be free from any chemical residue. The irradiation condition can be fine tuned to obtain tailor-made nanoparticles with respect to size, size distribution, shape, structure, and composition for specific applications. Using radiolytic route bimetallic nanoparticles, such as core-shell and alloy-types, can also be synthesized with controlled size. The advantages of using a radiolytic method have been discussed and the mechanism of generation of nanoparticles from the corresponding precursor salt has been explained. Various thermodynamic parameters responsible for the generation of nucleation center and subsequent growth of nanoparticles have been discussed. Nanoparticles by virtue of their unique properties have been explored for their use in various applications. Some important applications of nanoparticles, such biomedical, sensors, catalysts, and electronics have been highlighted.

REFERENCES

Abedini, A., Daud, A.R., Hamid, M.A.A., Othman, N.K. and Saion, E. 2013. "A review on radiation-induced nucleation and growth of colloidal metallic nanoparticles" *Nanoscale Res. Lett.* **8**: 474.

Agbabiaka, A., Wiltfong, M. and Park, C. 2013. "Small angle X-Ray scattering technique for the particle size distribution of nonporous nanoparticles" *J. Nanoparticles.* 640436.

Akella, A., Honda, T., Liu, A.Y. and Hesselink, L. 1997. "Two photon holographic recording in alumino-silicate glass containing silver particles" Optics Letters **22**: 967.

Alrehaily, L.M., Joseph, J.M., Biesinger, M.C., Guzonas, D. A. and Wren, J.C. 2013. "Gamma-radiolysis-assisted cobalt oxide nanoparticle formation" *Phys. Chem. Chem. Phys.* **15**: 1014.

Baig, N., Kammakakam, I. and Falath, W. 2021. "Nanomaterials: A review of synthesis methods, properties, recent progress, and challenges" *Mater. Adv.* **2**: 1821.

Basma, H., Awad, R., Roumie, M., Isber, S., Marhaba, S. and Abou, A.A. 2016. "Study of the irreversibility line of $GdBa_2Cu_3O_7$-δ added with nanosized ferrite $CoFe_2O_4$" *J. Supercond. Nov. Magn.* **29**: 179.

Batteza, A.H., Gonzaleza, R., Viescaa, J.L., Fernandez, J.E., Diaz, J.M., Machado, A., Chou, R. and Riba, J. 2008. "CuO, ZrO_2 and ZnO nanoparticles as antiwear additive in oil lubricants" *Wear* **265**: 422.

Baudot, C., Tan, C.M. and Kong, J.C. 2010. "FTIR spectroscopy as a tool for nano-material characterization" *Infrared. Phys. Technol.* **53**: 434.

Belapurkar, A. D., Varma, S., Shirole, A. and Sharma, J. 2008. "Cordierite supported platinum catalyst for hydrogen-oxygen recombination for use in nuclear reactors under LOCA" *Power Plant Chemistry* **10**: 461.

Belloni, J. 1998a. "Contribution of radiation chemistry to the study of metal clusters" *Radiat Res.* **150**: 9.

Belloni, J. 2006. "Nucleation, growth and properties of nanoclusters studied by radiation chemistry: Application to catalysis" *Catal. Today.* **113**: 141.

Belloni, J. and Mostafavi, M. 2001. "Radiation chemistry of nanocolloids and clusters" in Jonah, C. D., and Rao, B.S.M. (Eds.) *Radiation Chemistry: Present Status and Future Trends*, 411–452, Elsevier, Amsterdam.

Belloni, J., Mostafavi, M., Remita, H., Marignier, J.L. and Delcourt, M.O. 1998b. "Radiation-induced synthesis of mono- and multi-metallic nanocolloids and clusters" *New J. Chem.* **22**: 1239.

Belloni, J. and Remita, H. 2008. "Metal clusters and nanomaterials" in Spotheim-Maurizot, M., Mostafavi, M., Douki, T. and Belloni, J. (Eds.) *Radiation chemistry: From basics to applications in material and life science*, 97–116, France: EDP Science.

Biswal, J., Pant, H.J., Thakre, G.D., Sharma, S.C. and Gupta, A.K. 2020. "Evaluation of anti-wear properties of automobile lubricant with different additives using thin layer activation technique" *J. Radioanal. Nucl. Chem.* **325**: 795.

Biswal, J., Ramnani, S.P., Shirolikar, S. and Sabharwal, S. 2009. "Synthesis of guar-gum-stabilized nanosized silver clusters with gamma radiation" *J. Appl. Polym. Sci.* **114**: 2348.

Biswal, J., Ramnani, S.P., Shirolikar, S. and Sabharwal, S. 2010a. "Seedless synthesis of gold nanorods employing isopropyl radical generated using gamma radiolysis technique" *Int. J. Nanotechnol.* **7**: 907.

Biswal, J., Ramnani, S.P., Shirolikar, S. and Sabharwal, S. 2011. "Synthesis of rectangular plate like gold nanoparticles by insitu generation of seeds by combining both radiation and chemical methods" *Radiat. Phys. Chem.* **80**: 44.

Biswal, J., Ramnani, S.P., Tewari, R., Dey, G.K. and Sabharwal, S. 2010b. "Short aspect ratio gold nanorods prepared using gamma radiation in the presence of Cetyltrimethyl ammonium bromide (CTAB) as a directing agent" *Radiat. Phys. Chem.* **79**: 441.

Biswal, J., Singh, S., Rath, M.C., Ramnani, S.P. and Sabharwal, S. 2010c. "Synthesis of CdSe quantum dots in PVA matrix by radiolytic methods" *Int. J. Nanotechnol.* **7**: 1013.

Blasiak, B., van Veggel, F.C.J.M. and Tomanek, B. 2013. "Applications of nanoparticles for MRI cancer diagnosis and therapy" *J. Nanomater.* **4**: 12.

Braunstein, P., Oro, L.A. and Raithby, P.R. 1999. "Radiation induced metal clusters. Nucleation mechanisms and chemistry" in Belloni, J. and Mostafavi, M. (Eds.) *Metal Clusters in Chemistry*, 1213, Wiley, New York.

Burchez, M., Moronne, M., Gin, P., Weiss, S. and Alivisatos, A.P. 1998. "Semi-conductor nanocrystals as fluorescent biological labels" *Science* **281**: 2013.

Burda, C., Chen, X., Narayanan, R. and El-Sayed, M.A. 2005. "Chemistry and properties of nanocrystals of different shapes" *Chem. Rev.* **105**(4): 1025.

Buxton, G.V., Mulazzani, Q.G. and Ross, A.B. 1995. "Critical review of rate constants for reactions of transients from metal-ions and metal-complexes in aqueous solution" *J. Phys. Chem. Ref. Data* **24**: 1055.

Campelo, M., Luna, D., Luque, R.L., Marinas, J.M. and Romero, A.A. 2009. "Sustainable preparation of supported metal nanoparticlesand their applications in catalysis" *Chem. Sus. Chem.* **2**(1): 18.

Cao, G. 2004. "Nanostructures and nanomaterials: Synthesis, properties and applications" World scientific publishing Co. Pte. Ltd.

Chavali, M.S. and Nikolova, M.P. 2019. "Metal oxide nanoparticles and their applications in nanotechnology" *SN Appl. Sci.* **1**(6): 1.

Cheng, H.M., Yang, Q. H. and Liu, C. 2001. "Hydrogen storage in carbon nanotubes" *Carbon* **39**: 1447.

Chettibi, S., Keghouche, N., Benguedouar, Y., Bettahar, M.M. and Belloni, J. 2013. "Structural and catalytic characterization of radiation-induced Ni/TiO$_2$ nanoparticles" *Catal. Lett.* **143**: 1166.

Chmielewska, D. and Sartowska, B., 2012. "Radiation synthesis of silver nanostructures in cotton matrix" *Radiat. Phys. Chem.* **81**: 1244.

Creighton, J.A., Eadon, D.G., Ultraviolet-Visible absorption spectra of the colloidal metallic elements, 1991, 87, J. Chem. Soc, 3881.

Datsyuk, V., Kalyva, M., Papagelis, K., Parthenios, J., Tasis, D., Siokou, A., Kallitsis, I. and Galiotis, C. 2008. "Chemical oxidation of multiwalled carbon nanotubes" *Carbon* **46**: 833.

Derjaguin, B. and Landau, L. 1993. "Theory of the stability of strongly charged lyophobic sols and of the adhesion of strongly charged particles in solutions of electrolytes" *Prog. Surf. Sci.* **43**: 30.

Dispenza, C., Grimaldi, N., Sabatino, M. A., Soroka, I.L. and Jonsson, M. 2015. "Radiation-engineered functional nanoparticles in aqueous systems" *J. Nanosci. Nanotechnol.* **15**: 3445.

Du, C., Ao, Q., Cao, N., Yang, L., Luo, W. and Cheng, G. 2015. "Facile synthesis of monodisperse ruthenium nanoparticles supported on graphene for hydrogen generation from hydrolysis of ammonia borane" *Int. J. Hydrog. Energy.* **40**: 6180.

Elvira, D., Braive, R., Beaudoin, G., Sagnes, I., Hugonin, J.P., Abram, I., Philip, I.R., Lalanne, P. and Beveratos, A. 2013. "Broadband enhancement and inhibition of single quantum dot emission in plasmonic nano-cavities operating at telecommunications wavelengths" *Appl. Phys. Lett.* **103**: 061113.

Ershov, B.G. and Sukhov, N.L. 1990. "A pulse radiolysis study of the process of the colloidal metal formation in aqueous solutions" *Radiat. Phys. Chem.* **36**: 93.

Fan, Y., Ho, C.Y. and Chang, Y.J. 2017. "Enhancement of dye-sensitized solar cells efficiency using mixed-phase TiO_2 nanoparticles as photoanode" *Scanning* **1**.

Free, Z., Hernandez, M., Mashal, M. and Mondal, K. 2021. "A review on advanced manufacturing for hydrogen storage applications" *Energies* **14**: 8513.

Froudakis, G.E. 2011. "Hydrogen storage in nanotubes & nanostructures" *Materials Today* **14**: 324.

Gao, J., Gu, H. and Xu, B. 2009. "Multifunctional magnetic nanoparticles: Design, synthesis, and biomedical applications" *Acc. Chem. Res.* **42**: 1097.

Georgakilas, V., Gournis, D., Tzitzios, V., Pasquato, L., Guldi, D.M. and Prato, M. 2007. "Decorating carbon nanotubes with metal or semiconductor nanoparticles" *J. Mater. Chem.* **17**: 2679.

Haes, A.J., Stuart, D.A., Nie, S. and Van Duyne, R.P. 2004. "Using solution-phase nanoparticles, surface-confined nanoparticle arrays and single nanoparticles as biological sensing platforms" *J. Fluoresc.* **14**: 355.

Hassan, P.A. 2006. "Chemistry and technology of nanomaterials" *ISRAPS Bulletin* **18**: 3–8, Mumbai.

Henry, C.R. 2000. "Catalytic activity of supported nanometer-sized metal clusters" *Appl. Surf. Sci.* **164**: 252.

Hu, Y. and Chen, J.F. 2007. "Synthesis and characterization of semiconductor nanomaterials and micro-materials via gamma-irradiation route" *J. Clust. Sci.* **18**: 371.

Jana, N.R. 2005. Gram-scale synthesis of soluble, "Near-Monodisperse Gold Nanorods and Other Anisotropic Nanoparticles" *Small* **1**: 875.

Jovanovic, Z., Radosavljevic, A., Siljegovic, M., Bibic, N., Miskovic-Stankovic, V. and Kacarevic-Popovic, Z. 2012. "Structural and optical characteriztics of silver/poly (Nvinyl- 2-pyrrolidone) nanosystems synthesized by γ-irradiation" *Radiat. Phys. Chem.* **81**: 1720.

Jurkin, T., Gotic, M., Stefanic, G. and Pucic, I. 2016. "Gamma-irradiation synthesis of iron oxide nano-particles in the presence of PEO, PVP or CTAB" *Radiat. Phys. Chem.* **24**: 75.

Kadlubowski, S. 2014. "Radiation-induced synthesis of nanogels based on poly(N-vinyl-2-pyrrolidone) – A review" *Radiat. Phys. Chem.* **102**: 29.

Kanel, S.R., Greneche, J.M. and Choi, H. 2006. "Arsenic (V) removal from groundwater using nano scale zerovalent iron as a colloidal reactive barrier material" *Environ. Sci. Technol.* **40**: 2045.

Kang, H., Buchman, J.T., Rodriguez, R.S., Ring, H.L., He, J., Bantz, K.C. and Haynes, C.L. 2019. "Stabilization of silver and gold nanoparticles: Preservation and improvement of plasmonic function-alities" *Chem. Rev.* **119**: 664.

Kawasaki, H. 2013. "Surfactant-free solution-based synthesis of metallic nanoparticles toward efficient use of the nanoparticles surfaces and their application in catalysis and chemo-/biosensing" *Nanotechnol. Rev.* **2**: 5.

Keghouche, N., Chettibi, S., Latreche, F., Bettahar, M.M., Belloni, J. and Marignier, J.L. 2005. "Radiation-induced synthesis of α-Al_2O_3 supported nickel clusters: Characterization and catalytic properties" *Radiat. Phys. Chem.* **74**: 185.

Khanna, P.K., Gokhale, R. and Subbarao, V.V.V.S. 2004. "Poly(vinyl pyrolidone) coated silver nano powder via displacement reaction" *J. Mater. Sci.* **39**: 3773.

Kreibig Cuenya, R. 2010. "Synthesis and catalytic properties of metal nanoparticles: Size, shape, support, composition, and oxidation state effects" *Thin Solid Films* **518**: 3127.

Kreibig, U. and Vollmer, M. 1995. "Optical properties of metal clusters" in Paul, M. (Ed.) *Springer Series in Materials Science*. Vol. **25**, 532, Springer-Verlag, Berlin.

Kuila, T., Bose, S., Mishra, A.K., Khanra, P., Kim, N.H. and Lee, J.H. 2012. "Chemical functionalization of graphene and its applications" *Prog. Mater. Sci.* **57**: 1061.

Kulkarni, S.K. 2015. "Nanotechnology: Principles and Practices" 3rd Ed. Springer, Capital Publishing Company.

Lal, U., Link, S. and Halas, N.J. 2007. "Nano-optics from sensing to waveguiding" *Nature Photonics* **1**: 641.

Le Ru, E.C., Blackie, E., Meyer, M. and Etchegoin, P.G. 2007. "Surface enhanced Raman scattering en-hancement factors: A comprehensive study" *J. Phys. Chem. C.* **111**: 13794.

Link, S. and El-Sayed, M. A. 2000. "Shape and size dependence of radiative, non-radiative and photothermal properties of gold nanocrystals" *Int. Rev. Phys. Chem.* **19**: 409.

Marignier, J.L., Belloni, J., Delcourt, M.O. and Chevalier, J.P. 1985. "Microaggregates of non-noble metals and bimetallic alloys prepared by radiation-induced reduction" *Nature* **317**: 344.

Marquardt, D., Vollmer, C., Thomann, R., Steurer, P., Mulhaupt, R., Redel, E. and Janiak, C. 2011. "The use of microwave irradiation for the easy synthesis of graphene-supported transition metal nanoparticles in ionic liquids" *Carbon* **49**: 1326.

Mourdikoudis, S., Pallares, R.M. and Thanh, N.T.K. 2018. "Characterization techniques for nanoparticles: Comparison and complementarity upon studying nanoparticle properties" *Nanoscale* **10**: 12871.

Nguyen, K.T. and Zhao, Y. 2014. "Integrated graphene/nanoparticle hybrids for biological and electronic applications" *Nanoscale* **6**: 6245.

Nguyen Vu, D.K. and Vo Nguyen, D.K. 2021. "Gamma irradiation-assisted synthesis of silver nanoparticle embedded graphene oxide TiO_2 nanotube nanocomposite for organic dye photodegradation" *J. Nanomat.* **6679637**: 1.

Okitsu, K., Yue, A., Tanabe, S. and Matsumoto, H. 2000. "Sonochemical preparation and catalytic behavior of highly dispersed palladium nanoparticles on alumina" *Chem. Mater.* **12**: 3006.

Polte, J. 2015. "Fundamental growth principles of colloidal metal nanoparticles – A new perspective" *Cryst. Eng. Comm.* **17**: 6809.

Pomogailo, A.D. and Kestelman, V.N. 2005. "Metallopolymer Nanocomposites" vol. 81, Springer-Verlag Berlin Heidelberg, Berlin/Heidelberg.

Ramnani, S.P., Biswal, J. and Sabharwal, S. 2007. "Synthesis of silver nanoparticles supported on silica aerogel using gamma radiolysis" *Radiat. Phys. Chem.* **76**: 1290.

Remita, H., Etcheberry, A. and Belloni, J. 2003. "Dose rate effect on bimetallic gold-palladium cluster structure" *J. Phys. Chem. B* **107**: 31.

Ricard, D., Roussignol, P. and Flytzanis, C. 1985. "Surface-mediated enhancement of optical phase conjugation in metal colloids" *Opt. Lett.* **10**: 511.

Rojas, J.V. and Castano, C.H. 2012. "Production of palladium nanoparticles supported on multiwalled carbon nanotubes by gamma irradiation" *Radiat. Phys. Chem.* **81**: 16.

Roumie, M., Marhaba, S., Awad, R., Kork, M., Hassan, I. and Mawassi, R. 2014. "Effect of Fe_2O_3 nano-oxide addition on the superconducting properties of the (Bi,Pb)-2223 phase" *J. Supercond. Nov. Magn.* **2**: 143.

Schmid, G. 1994. "Clusters and Colloids: From Theory to Application" Weinheim, VCH.

Schmid, G. 2004. "Nanoparticles: From Theory to Application" Wiley-VCH Verlag GmbH & Co. KGaA: Roumie 22014, Basma Weinheim, Germany.

Schmid, G., Maihack, V., Lantermann, F. and Peschel, S. 1996. "Ligand-stabilized metal clusters and colloids: Properties and applications" *J. Chem. Soc., Dalton Trans.* **5**: 589–595.

Shan, J. and Tenhu, H. 2007. "Recent advances in polymer protected gold nanoparticles: synthesis, properties and applications" *Chem. Commun. (Camb).* **44**: 4580.

Sharma, G., Kumar, A., Sharma, S., Naushad, M., Dwivedi, R.P., ALOthman, Z.A. and Tessema Mola, G. 2019. "Novel development of nanoparticles to bimetallic nanoparticles and their composites: A review" *J. King Saud Univ. Sci.* **31**: 257.

Sharrouf, M., Awad, R., Marhaba, S. and Bakeer, D.E. 2016. "Structural, optical and room temperature magnetic study of Mn-doped ZnO nanoparticles" *Nano.* **11**: 1650042.

Sharrouf, M., Awad, R., Roumie, M. and Marhaba, S. 2015. "Structural, optical and room temperature magnetic study of Mn_2O_3 nanoparticles" *Mater. Sci. Appl.* **5**: 850.

So, D.W.C. and Seshadri, S.R. 1997. "Metal-island-film polarizer" *J. Opt. Soc. Am. B: Opt. Phys.* **14**: 2831.

Stuart, H.R. and Hall, D.G. 1998. "Island size effects in nanoparticle-enhanced photodetector" *Appl. Phys. Lett.* **73**: 3815.

Sun, Y.F., Liu, S.B., Meng, F.L., Liu, J.Y., Jin, Z., Kong, L.T. and Liu, J.H. 2012. "Metal oxide nanostructures and their gas sensing properties: A review" *Sensors (Basel)* **12**: 2610.

Treguer, M., de Cointet, C., Remita, H., Khatouri, J., Mostafavi, M., Amblard, J., Belloni, J. and de Keyzer, R. 1998. "Dose rate effects on radiolytic synthesis of gold-silver bimetallic clusters in solution" *J. Phys. Chem. B* **102**: 4310.

Tri, P.N., Nguyen, T.A., Nguyen, T.H. and Carriere, P., 2019. "Antibacterial Behavior of Hybrid Nanoparticles" in Mohapatra, S., Nguyen, T.A. and Nguyen-Tri, P.P. (Eds.) *Micro and Nano Technologies, Noble Metal-Metal Oxide Hybrid Nanoparticle*, 141–155, Woodhead Publishing.

Tyagi, A.K. 2011. "Chemistry and applications of nanomaterials" *SMC Bulletin*, **2**(No. 1), Mumbai.

Unser, S., Bruzas, I., He, J. and Sagle, L. 2015. "Localized surface plasmon resonance biosensing: Current challenges and approaches" *Sensors* **15**: 15684.

Wildgoose, G.G., Banks, C.E. and Compton, R.G. 2006. "Metal nanoparticles and related materials supported on carbon nanotubes: Methods and applications" *Small* **2**: 182.

Wu, Y., Ali, M.R.K., Chen, K.C., Fang, N. and El-Sayed, M.A. 2019. "Gold nanoparticles in biological optical imaging" *Nano Today* **24**: 120.

Xu, H., Zeng, L., Xing, S., Shi, G., Xian, Y. and Jin, L. 2008. "Microwave-radiated synthesis of gold nanoparticles/carbon nanotubes composites and its application to voltammetric detection of trace mercury (II)" *Electrochem. Commun.* **10**: 1839.

Yang, Y., Johansson, M., Wiorek, A., Tarakina, N.V., Mathieu, R., Jonsson, M. and Soroka, I.L. 2021. "Gamma-radiation induced synthesis of freestanding nickel nanoparticles" *Dalton Trans* **50**: 376.

Yaqoob, A., Ahmad, H., Parveen, T., Ahmad, A., Oves, M., Ismail, I.M.I., Qari, H.A., Umar, K. and Ibrahim, M.N.M. 2020. "Recent advances in metal decorated nanomaterials and their various biological applications: A review" *Front. Chem.* **8**: 1.

Yazid, H., Adnan, R., Hamid, S.A. and Farrukh, M.A. 2010. "Synthesis and characterization of gold nanoparticles supported on zinc oxide via the deposition-precipitation method" *Turk. J. Chem.* **34**: 639.

Yonezawa, T. and Kunitake, T. 1999. "Practical preparation of anionic mercapto ligand-stabilized gold nanoparticles and their immobilization" *Colloid. Surface A* **149**: 193.

Zhang, Z.J., Simionesie, D. and Schaschke, C. 2014. "Graphite and Hybrid Nanomaterials as Lubricant Additives" *Lubricants* **2**: 44.

Zhengfeng, S., Hongzhi, Z., Zhi, Z., Zhenghua, Z., Guohua, M., Xinchun, L., Rong, L., Tao, T., Jun, F. and Bo, G. 2017. "Supported Pd nano-clusters for the hydrogen mitigation application in severe accidents" *Nucl. Eng. Des.* **316**: 93.

Index

For Product Safety Concerns and Information please contact our EU
representative GPSR@taylorandfrancis.com
Taylor & Francis Verlag GmbH, Kaufingerstraße 24, 80331 München, Germany